how things work

how things work

THE PHYSICS OF EVERYDAY LIFE

SECOND EDITION

Louis A. Bloomfield

THE UNIVERSITY OF VIRGINIA

WILEY

John Wiley & Sons, Inc.

NEW YORK CHICHESTER WEINHEIM BRISBANE SINGAPORE TORONTO

ACQUISITIONS EDITOR	Stuart Johnson
MARKETING MANAGER	Bob Smith
PRODUCTION EDITOR	Barbara Russiello
SENIOR DESIGNER	Harry Nolan
INTERIOR DESIGN	Fearn Cutler DeVicq De Cumptich
COVER DESIGN	Suzanne Noli
PHOTO EDITORS	Hilary Newman, Sara Wight
PHOTO RESEARCHER	Teri Stratford
ILLUSTRATION EDITOR	Sigmund Malinowski
ELECTRONIC ILLUSTRATIONS	Rolin Graphics
COVER PHOTOS	*Guitar player,* ©FPG International; *Lighthouse,* ©Chad Ehlers/Stone; *Compact disc,* ©Peter Poulides/Stone

This book was set in Times Roman by PRD Group, Inc. and printed and bound by Von Hoffmann Press. The cover was printed by Von Hoffmann Press.

This book is printed on acid free paper. ∞

Library of Congress Cataloging in Publication Data
Bloomfield, Louis.
  How things work: the physics of everyday life / Louis A. Bloomfield.—2nd ed.
    p. cm.

  Includes index.
  ISBN 0-471-38151-9 (pbk. : alk. paper)
  1. Physics. I. Title.

QC21.2.B59 2001
530—dc21   00-064917

ISBN 0-471-38151-9

Printed in the United States of America

10 9 8 7 5 4 3 2

To Karen, my dearest companion, whose wisdom, compassion, and love have sustained me in this adventure.

To our children, Elana and Aaron, who are growing up to be such wonderful people and such good friends.

And to the students of the University of Virginia, whose curiosity and enthusiasm make teaching fun.

Contents

*All entries in blue are found on the web, www.wiley.com/college/howthingswork, and are accessed by entering the key words at right.

Chapter 5.   Fluids and Motion   145

Chapter 6.   Heat and Thermodynamics   187

Chapter 10. Electronics   327

Chapter 11. Electromagnetic Waves   353

Appendices   469

Preface

This book offers a nonconventional view of physics and science that starts with whole objects and looks inside them to see what makes them work. It's written for liberal arts students who are seeking a connection between science and the world in which they live. While these students may be reluctant to study science as an intellectual exercise, they show remarkable enthusiasm for it when it appears in a useful context. Many of the 5000 students I've taught during the past nine years have been surprised at their own interest in the course, looking forward to classes, asking questions, experimenting on their own, and explaining to friends and family how things they use, and care about, work.

How Things Work brings science to the students rather than the reverse. Like the course on which it's based, this book has always been for nonscientists and is written with their interests in mind. Nonetheless, it has attracted students from the sciences, engineering, architecture, and other technical fields who wish to put familiar scientific concepts into context.

Most physics texts choose to develop the principles of physics first and to delay real-life examples of these principles until later. That teaching method provides few conceptual footholds for the students as they struggle to understand the principles. After all, the comforts of experience and intuition lie in the examples, not the principles. While a methodical and logical development of scientific principles can be satisfying to the seasoned physicist, it's alien to an individual who isn't even familiar with the language being used.

This book conveys an understanding and appreciation for physics by finding its concepts and principles within objects of everyday experience. Because its structure is defined by real-life examples, this book necessarily discusses various physical concepts as they're needed. These concepts are often revisited later on in other objects, thereby showing the universality of the natural laws.

Changes in the Second Edition

Design Changes

- **Color.** The transition to color gives this edition a much richer look and a more relaxed feel and makes the book's artwork far more clear and meaningful. In particular, the color of each vector in the line art indicates its type (e.g., a force, velocity, or acceleration), and pressures in the calculated drawings of fluid flow are indicated dramatically by color. Color also helps to clarify the book's structure and highlights its support features.
- **More open format.** The previous edition had large pages that were densely packed with text and illustrations. This edition follows a smaller and more open design, making the book easier to read and less imposing to the reader. Illustrations have been enlarged when necessary to make them clearer.
- **Linkage to the web.** Icons now direct the reader to the extensive supplementary materials available on the book's website.

Content Changes

- **Fewer, more focused presentations.** Like many texts, the previous edition tried to cover too much material and often overwhelmed both the students and the instructor. To alleviate this problem, a number of sections have been shifted to the web and the remaining sections have been carefully streamlined to make them more accessible to the reader. That streamlining was done with a scalpel, not an ax, and many of the sections were extensively rewritten in the process of giving them a sharper focus. Off-topic diversions have been reduced to a minimum, and the central theme of each section shines through more clearly than before.

- **Two opening chapters instead of one.** The previous edition packed too much into the four sections of the opening chapter. That material is now spread over six sections in two opening chapters, one primarily on translational motion and the other primarily on rotational method. In the process, two new topics have entered the book—skating and bumper cars.

- **Regrouping of sections.** With the streamlining of the book and the transfer of some topics to the web came the need to rethink the overall organization of the sections and their associated objects. In this edition, all of the sophisticated mechanical objects are grouped together in Chapter 3, thermal objects are in Chapter 6, objects involving resonance or mechanical waves are in Chapter 7, and objects involving modern physics are in Chapter 14.

- **Newtonian view of fluid flow.** The sections on fluid flow now treat the changes in pressure in a flowing fluid as the consequences of acceleration and flow deflection rather than the other way around (the Bernoullian view). This shift in emphasis makes the origins of lift and drag much more transparent and provides a stronger and more intuitive connection between fluid-based objects and the mechanical ones that precede them in the text.

- **Improved development of solid-state physics.** This edition consistently uses a theater analogy to aid the student's understanding of metals, insulators, semiconductors, and semiconductor devices.

- **Keeping up with new technology.** While physics at the introductory level changes fairly slowly, the objects that use physics change almost daily. This edition reflects those developments, particularly in the sections on television and optical recording and communication.

- **Fixes to conceptual flaws.** Trying to understand exactly how such a wide range of objects work is a daunting task, and I made a number of embarrassing mistakes in the previous edition. In the intervening four years, many readers have helped me learn what I did not know, and this book reflects my improved understanding of the objects and their physics.

Feature Changes

- **Rearrangement of support features.** To reduce the book's encyclopedic feel, all of the section ending features of the previous edition have been moved to the ends of chapters or to the web. This change allows the exercises and problems to be ordered by physics concept rather than by the section in which they appear. Moving the review questions and physics concept list to the web allows them to become interactive tools to help students assess their understanding and prepare for examinations.

- **Single glossary.** The glossary has been brought together at the end of the book, where it serves as a convenient reference for students who are struggling with the language of physics.

The Goals of This Book

As they read this book, students should:

1. **Begin to see science in everyday life.** This goal is the central focus of the book. Science is all around us; we only need to keep our eyes open to see it. We're surrounded by things that can be understood at various levels and it's within a student's reach to see science in each of those things. That doesn't mean that they should look at an oil painting and see only preferential absorption of incident light by organic and inorganic molecules. Rather, they should realize that there's a beauty to science that can supplement aesthetic beauty. They can learn to look at a glorious red sunset and appreciate both its appearance *and* why it exists.

2. **Learn that science isn't frightening.** The increased technological complexity of our world has done more than just stop people from looking for science in their environment. It has instilled within them a significant fear of science. One of the unintended consequences of modern technology is that few people tinker with anything anymore. Many devices are just disposable, being too complicated to modify or repair, and the resulting unfamiliarity breeds anxiety. To combat that anxiety, this book shows students that most objects can be examined and understood, and that the science behind them isn't so scary.

3. **Learn to think logically in order to solve problems.** Because the universe obeys a system of well-defined rules, it permits a logical understanding of its behaviors. Like mathematics and computer science, physics offers a field of study where logic reigns supreme. Having learned two or three simple rules, students can combine them logically to obtain more complicated rules and be certain that those new rules are true. So the study of physical systems is a good place to practice logical thinking.

4. **Develop and expand their physical intuition.** When you're exiting from a highway, you don't have to consider velocity, acceleration, and inertia to know that you should brake gradually—you already have physical intuition that tells you the consequences of doing otherwise. Such physical intuition is wonderfully useful in everyday life, but it takes time and experience to acquire. This book is intended to broaden a student's physical intuition to situations normally avoided and to ones never even encountered. That is, after all, one of the purposes of reading books and studying other people's scholarship: to learn from other people's experiences.

5. **Learn how things work.** This book is about objects from everyday life. Its sections deal with specific objects and with the mechanisms that permit those objects to work. As it explores the objects, it exposes bits and pieces of the overall physical laws that govern the universe. It approaches those laws as they were originally discovered: while trying to understand real objects. Although each of the objects in this book may only touch on one or two physical concepts, as a whole, this book presents many or most of the laws of

physics. Students should begin to see the similarities between objects, shared mechanisms, and recurring themes that are reused by nature or by people. This book tries to remind students of these connections and is ordered so that later objects build on their understanding of concepts encountered earlier.

6. **Begin to understand that the universe is predictable rather than magical.** One of the most fundamental principles of science, not described by equations or complex diagrams, is the notion that every effect has its cause. Things don't just occur willy-nilly. Whatever happens, we can look backward in time to find its real causes. We can also predict the future to some extent, based on insight acquired from the past and a knowledge of the present. What distinguishes the physical sciences and mathematics from other fields is that there are often absolute answers, free from inconsistency, contraindication, or paradox. Once students understand how the physical laws govern the universe, they can start to appreciate how orderly it is. They can replace a sense of magic at seeing a certain behavior with a sense of structure and understanding.

7. **Obtain a perspective on the history of science and technology.** None of the objects that this book examines appeared suddenly and spontaneously in the workshop of a single individual who was oblivious to what had been done before. These objects were developed in the context of history by people who were fairly aware of what they were doing and usually familiar with any similar objects that already existed. Nearly everything is discovered or developed when related activities make their discoveries or developments inevitable and timely. To establish that historical context, this book describes some of the history behind the objects it discusses.

Artwork

Because this book is about real things, its illustrations and photographs are about real things, too. Whenever possible, artwork is built around familiar objects so that the concepts the artwork is meant to convey become associated with objects students already know. Many students are visual learners—if they see it, they can learn it. By superimposing the abstract concepts of physics onto simple realistic artwork, this book attempts to connect physics with everyday life. That idea is particularly evident at the opening of each section, where the object examined in that section appears in a carefully rendered drawing. This drawing provides students with something concrete to keep in mind as they encounter the more abstract physical concepts that appear in that section. By lowering the boundaries between what the students see in the book and what they see in their environment, the artwork of this book makes science a part of their world.

Features

This printed book contains 40 sections, each of which discusses how something works. The sections are grouped together in 14 chapters according to the major physical themes developed. In addition to the discussion itself, the sections and chapters include a number of features intended to strengthen the educational value of this book. Among these features are:

- **Chapter introductions, experiments, and itineraries.** Each of the 14 chapters begins with a brief introduction to the principal theme underlying that chapter. It then presents an experiment that students can do with household items to observe firsthand some of the issues associated with that physical theme. Lastly, it presents a general itinerary for the chapter, identifying some of the physical issues that will come up as the objects in the chapter are discussed.

- **Section introductions, questions, and experiments**. Each of the 40 sections explains how something works. Often that something is a specific object or group of objects, but it can be more general. A section begins by introducing the object and then asks a number of questions about it, questions that might occur to students as they think about the object and that are answered by the section. Lastly, it suggests some simple experiments that students can do to observe some of the physical concepts that are involved in the object.

- **Check your understanding and check your figures.** Sections are divided into a number of brief segments, each of which ends with a "Check Your Understanding" question. These questions apply the physics of the preceding segment to new situations and are answered and explained in the summary material at the end of this chapter. Segments that introduce important equations also end with a "Check Your Figures" question. These questions show how the equations can be applied and are also answered and explained in the summary material.

- **Chapter epilogue and explanation of experiment.** Each chapter ends with an epilogue that reminds the students of how the objects they studied in that chapter fit within the chapter's physical theme. Following the epilogue is a brief explanation of the experiment suggested at the beginning of the chapter, using physical concepts that were developed in the intervening sections.

- **Chapter summary and laws and equations.** The sections covered in each chapter are summarized briefly at the end of the chapter, with an emphasis on how the objects work. These summaries are followed by a restatement of the important physical laws and equations encountered within the chapter.

- **Chapter exercises and problems.** Following the chapter summary material is a collection of questions dealing with the physics concepts in that chapter. Exercises ask the students to apply those concepts to new situations. Problems ask the students to apply the equations in that chapter and to obtain quantitative results.

- **Chapter cases.** The final items in each chapter are cases, extended exercises that apply the physical concepts from that chapter and sometimes previous chapters to new circumstances. Each case asks a series of questions and leads the students through an examination of the physical principles involved in a new object or situation. Quantitative questions are marked with an asterisk (*). In addition to tying the chapters together, these cases introduce many interesting objects, worthy of sections of their own. By studying these cases, students learn to explain how new objects work.

- **Three-way approach to the equation of physics.** The laws and equations of physics are the groundwork on which everything else is built. But because each student responds differently to the equations, this book presents them carefully and in context. Rather than making one size fit all, these equations are presented in three different forms. The first is a word equation, identifying each physical quantity by name to avoid any ambiguities. The second is a symbolic equation, using the standard format and notation. The third is a sentence that conveys the meaning of the equation in simple terms and often by example. Each student is likely to find one of these three forms more comfortable, meaningful, and memorable than the others.

- **Glossary.** The key physics terms are assembled into a glossary at the end of the book. Each glossary term is also marked in bold in the text when it first appears together with its contextual definition.
- **Historical, technical, and biographical asides.** To show how issues discussed in this book fit into the real world of people, places, and things, a number of brief asides have been placed in the margins of the text. An appropriate time at which to read a particular aside is marked in the text by a ❏.

Organization

The 40 sections that make up this book are ordered so that they follow a familiar path through physics: mechanics, heat, resonance and waves, electricity and magnetism, light and optics, and modern physics. Because there are too many topics here to cover in a single semester, the book is designed to allow shortcuts through the material. In general, the final sections in each chapter and the final chapters in each of the main groups mentioned above can be omitted without serious impact on the material that follows. The only exceptions to that rule are the first two chapters, which should be covered in their entirety as the introduction to any course taught from this book. The book also divides neatly in half so that it can be used for two independent one-semester courses—the first covering Chapters 1–7 and the second covering Chapters 1, 2, and 8–14. A detailed guide to shortcuts appears in the Instructor's Manual.

Student Website

This book is supported by an ever-increasing array of web-based student supplements. Starting from the book's student entry URL,

www.wiley.com/college/
howthingswork

a student has free access to resources that include:

- **Review questions and physics concept summaries.** Tied to each of the book's 40 sections are a collection of review questions and a list of the important physical and scientific concepts covered in those sections. Answers are available for the review questions so that students can see if they understand the material.
- **Additional material for each section.** Many of the subjects discussed in this book have so many interesting facets that they could serve as the bases for hundred-page dissertations. However, to keep the printed sections at a reasonable length, some of their content has been placed on the web. When the ⊕ icon appears next to a paragraph, additional discussion of the current topic can be found on the web. When the ⊕ appears at the end of a section, additional topics associated with that section are available on the web. Simply go to the student web page and enter the keyword or phrase accompanying the ⊕ icon.
- **Additional exercises, problems, and cases.** These questions provide alternatives to those that appear in the printed book.
- **Additional sections and chapters.** More than 20 complete sections are available online to supplement the 40 that appear in this book. These include several special opening sections that can be used to get the course off to an exciting start.
- **Links to related websites for each section.**

Instructor Website and Instructor's Manual

The instructor's website, located at URL

www.wiley.com/college/
howthingswork/instructor

provides everything described above, plus access to:

- **Additional homework questions and solutions**
- **Test questions and solutions**
- **Organizational ideas for designing a course**
- **Demonstration ideas for each section**
- **Lecture slides for each section**
- **Artwork to use in presentations**
- **Resource lists**

Many of these items also appear in printed form in the Instructor's Manual.

Acknowledgments

There are a great many people who have contributed to this second edition and to whom I am enormously grateful. First among them is my editor, Stuart Johnson, whose enthusiasm for this project has kept things moving forward all these years. It has also been a pleasure working with Anne Smith, Deanna Campbell, and Tom Hempsted during the development process and with Barbara Russiello during production. Harry Nolan has orchestrated a beautiful design for the book, Hilary Newman has been wonderful at finding interesting photographs, Sigmund Malinowski has done a terrific job coordinating the development of the line art, Geraldine Osnato has been very helpful at finding existing line art, and Christina della Bartolomea has done a spectacular job of line editing the manuscript. Thanks to Bob Smith and Sue Lyons for helping learn what instructors and students want from this book and to Mark Gerber for helping plan the website for this book. I also appreciate the continued interest and support from *How Things Work* veterans Barbara Heaney and Maddy Lesure.

I continue to enjoy tremendous assistance from colleagues here and elsewhere who have supported the *How Things Work* concept and have often taught the course themselves. These people include Rob Watkins, Mike Noel, Bascom Deaver, Vittorio Celli, and Michael Fowler at the University of Virginia, John Krupczak at Hope College, Laura Lising at NIST, William McNairy at Duke University, Martin Mason at College of the Desert, Robert Reynolds at Reed College, Maarten Rutgers at Ohio State University, Richard Superfine at the University of North Carolina at Chapel Hill, and Robert Welsh at College of William and Mary.

Because this book is about real objects, the course I teach from it is enriched by countless demonstrations. I couldn't have done those demonstrations, nor taken many of the photographs for this book, without the vast experience and enormous enthusiasm of Mike Timmins, John Malone, and Roger Staton. Together with several talented student assistants, they make as fine a lecture-demonstration group as one could ever want.

I am also extremely grateful to the students of the University of Virginia for being so eager and interested, and also so tolerant of my endless experimenting with the content and delivery of the course. Many students have contributed directly or indirectly to the book itself and are now too numerous to name individually. To all of you who have talked to me or written to me about the course, voiced your concerns, ideas, and observations, and urged me onward, my sincerest thanks.

Thanks also to my family, Karen, Elana, and Aaron Bloomfield, for helping me with everything from editing sections of text to photographing whirling wineglasses.

Lastly, this book has benefited more than most from the constructive criticism of a number of talented reviewers. In addition to getting a better sense of how to present the material in this book, I have learned a great deal of physics from their reviews. My deepest thanks to all of these fine people:

Charles Ardary
Edmonds Community College

Ali Badakhshan
University of Northern Iowa

Keith Bonin
Wak Forest University

Edward R. Borchardt
Mankato State University

Robert Boughton
Bowling Green State University

Arthur J. Braundmeier
Southern Illinois University

David Buckley
East Stroudsburg University

Neal Cason
University of Notre Dame

Joshua Cohn
University of Miami

James J. Donaghy
Washington & Lee University

Gordon Donaldson
University of Strathclyde, Scotland

Sherman Frye
Northern Virginia
    Community College

Bob Hallock
University of Massachusetts—
    Amherst

Glenn M. Julian
Miami University

Mary Lu Larsen
Towson State University

David Markowitz
University of Connecticut

Martin Mason
College of the Desert

David C. McKenna
Rensselaer Polytechnic Institute

William McNairy
Virginia Military Institute

Jon Nadler
Richland Community College

Adam Niculescu
Virginia Commonwealth University

David Raffaelle
Glendale Community College

John C. Raich
Colorado State University

Paul Schuyler
Spokane Community College

Romeo A. Segnan
American University

Peter Sheldon
Randolph-Macon Woman's College

Stanley J. Sobolewski
Indiana University of Pennsylvania

Robert Tremblay
Southern Connecticut
    State University

Tim Usher
California State University—
    San Bernardino

Terrance Vacha
Cuyahoga Community College

David Wagner
Edinboro University
    of Pennsylvania

Bob Welsh
The College of William and Mary

John Yelton
University of Florida

I have always felt that the real test of this book, and of any course taught from it, is its impact on students' lives long after final grades have been recorded. It is my sincere hope that many of these students will find themselves looking at objects in the world around them years later with understanding and insight that they would not have had were it not for their encounter with this book.

Louis A. Bloomfield
Charlottesville, Virginia
lab3e@virginia.edu

The Laws of Motion Part I

The purpose of this book is to broaden your perspectives on familiar objects and situations by helping you understand the physical processes that make them work. While science is part of our daily existence—not some special activity we do only occasionally, if at all—most of us ignore it or take it for granted. In this book we'll counter that tendency by seeking out science in the world around us, in the objects we encounter every day. We'll see that seemingly magical objects and effects are really very straightforward once we know a few of the physical concepts that make them possible. In short, we'll learn about *physics*—the study of the material world and the rules that govern its behavior.

To help us get started, this first pair of chapters will do two main things: introduce the language of physics, which we'll be using throughout the book, and present the basic laws of motion on which everything else will rest. In later chapters, we'll explore objects that are interesting and important, both in their own right and because of the scientific issues they raise. Most of these objects, as we'll see, involve many different aspects of physics and thus bring variety to each section and chapter. But these first two chapters are special because they must provide an orderly introduction to the discipline of physics itself.

EXPERIMENT: Removing a Tablecloth from a Table

One famous "magical" effect is the feat of removing a tablecloth from a set table without breaking the dishes on top. The person performing this stunt pulls the tablecloth out from under the place settings in one lightning-swift motion. With a lit-

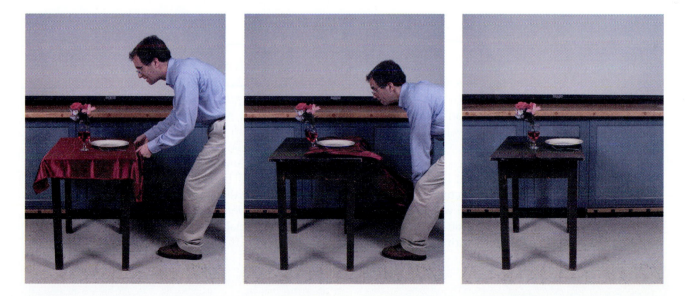

tle luck—not to mention a smooth, slippery tablecloth—the covering slides off the table abruptly, leaving the dishes behind and virtually unaffected.

With a little practice, you too can do this stunt. Choose a slick, unhemmed tablecloth, one with no projections that might catch on the dishes. A soft, flexible material such as silk helps because you can then pull the cloth downward and over the edge of the table. When you finally get up the nerve to try this stunt—with unbreakable dishes, of course—make sure you pull as abruptly as possible, keeping the time you spend moving the cloth out from under the dishes to an absolute minimum. It helps to hold the cloth with your palms downward and to let the cloth hang loosely between each hand and the table so that you can get your hands moving before the cloth snaps taut and begins to slide off the table. Don't make the mistake of starting slowly or the dishes will end up on the floor.

Before you perform this stunt, try to *predict* what will happen when you pull the cloth. How far will the dishes move—or will they not move at all? How important is the speed with which you pull the cloth? How does the weight of each dish affect its movement or lack of movement? Is the surface texture of each dish important? How would rubbing each dish with wax paper alter the results? Write down your predictions.

Now give the tablecloth a pull and *observe* what happens. Hopefully, the table will remain set. If it doesn't, try again, but this time vary the tablecloth's speed or the types of dishes or the way you pull the cloth. See if you can *measure* the effects of these variations on the dishes. Does everything work as you expected? Do the results *verify* your predictions, or were those predictions wrong?

If you don't have a suitable tablecloth, or any dishes you care to risk, there are many similar experiments you can try. Put several coins on a sheet of paper and whisk that sheet out from under them. Or stack several books on a table and use a stiff ruler to knock out the bottom one. Most impressive of all is to balance a short eraser-less pencil on top of a wooden needlepoint ring that is itself balanced on the open mouth of a glass bottle. If you yank the ring away quickly enough, the pencil will be left behind and will drop right into the bottle.

We'll return to the tablecloth stunt at the end of this chapter. In the meantime, we'll explore some of the physics concepts that help explain why your stunt worked—or, if it didn't, why the floor is now covered with dishes.

Chapter Itinerary

To examine these concepts, we'll look carefully at three kinds of everyday activities and objects: (1) *skating,* (2) *falling balls,* and (3) *ramps.* In *skating,* we'll see how objects move when nothing pushes on them. In *falling balls,* we'll see how that movement can be influenced by gravity. In *ramps,* we'll explore mechanical advantage and how gradual inclines make it possible to lift heavy objects without pushing very hard.

These activities may seem mundane, but understanding them in terms of basic physical laws requires considerable thought. These two introductory chapters will be like climbing up the edge of a high plateau: the ascent won't be easy, and our destination will be hidden from view. But once we arrive at the top, with the language and basic concepts of physics in place, we'll be able to explain a broad variety of objects with only a small amount of additional effort. And so we begin the ascent.

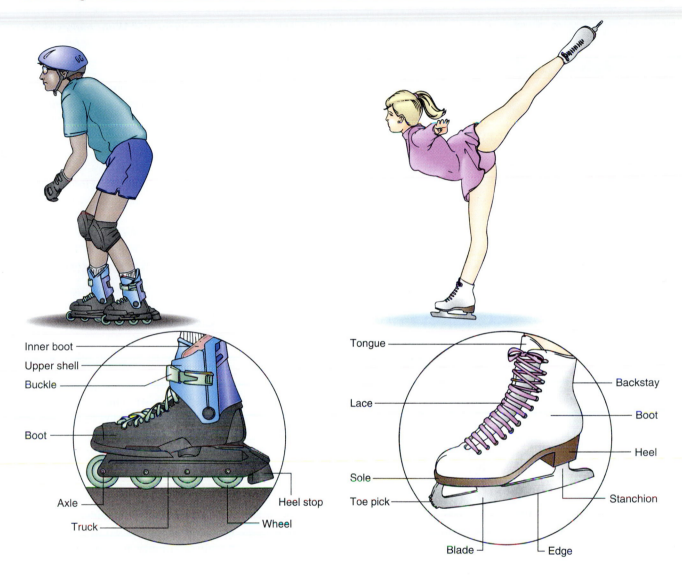

Inner boot
Upper shell
Buckle
Boot
Axle
Truck
Heel stop
Wheel

Tongue
Lace
Sole
Toe pick
Backstay
Boot
Heel
Stanchion
Blade
Edge

SECTION 1.1 # Skating

Like many sports, skating is trickier than it appears. As a first-time skater, you're likely to find yourself getting up repeatedly from the ground or ice, and it takes some practice before you can glide smoothly forward or come gracefully to a stop. But whether you're wearing ice skates or Rollerblades, the physics of your motion is surprisingly simple. When you're on a level surface with your skates pointing forward, you coast!

Coasting is one of the most basic concepts in physics and our starting point in this book. Joining it in this section will be starting, stopping, and turning, which together will help us understand the first few laws of motion. We'll leave sloping surfaces for the section on ramps and won't have time to teach you how to do spins or win a race. Nonetheless, our exploration of skating will get us well on the road to an understanding of the fundamental principles that govern all motion.

Fig. 1.1.1 Skater Michelle Kwan glides without any horizontal influences. If she's stationary, she'll tend to remain stationary; if she's moving, she'll tend to continue moving.

❑ **Aristotle (Greek philosopher, 384–322 BC)** theorized that objects' velocities were proportional to the forces exerted on them. While this theory correctly predicted the behavior of a sliding object, it incorrectly predicted that heavier objects should fall faster than lighter objects. Nonetheless, Aristotle's theory was retained for a long time, in part because finding the simpler and more complete theory was hard and in part because the scientific method of relating theory and observation took time to develop.

examples of inertia

Questions to Think About: *What do we mean by "movement"? What makes skaters move and, once they're moving, what keeps them in motion? What does it take to stop a moving skater or turn that skater in another direction?*

Experiments to Do: *An hour or two on the ice or roller rink would be ideal, but if you don't have skates then try a skateboard or a chair with wheels. Get yourself moving forward on a level surface and then let yourself coast. What's propelling you forward? Are you being pushed forward by anything? In what direction do you head as you coast? How would you describe where you are at a given moment? How would you measure your speed?*

Before you run into a wall or tree, slow yourself to a stop. What was it that slowed you down? Were you still coasting as you stopped? Did anything push on you as you slowed yourself?

Get yourself moving again. What caused you to speed up? How quickly can you pick up speed and what do you do differently to speed up quickly? Now turn to one side or the other. Did anything push on you as you turned? What happened to your speed? What happened to your direction of travel?

Gliding Forward: Inertia and Coasting

While you're putting on your skates, let's take a moment to think about what happens to a person who has nothing pushing on her at all. When she's completely free of outside influences (Fig. 1.1.1), free of pushes and pulls, does she stand still? Does she move? Does she speed up? Does she slow down? In short, what does she do?

The correct answer to that apparently simple question eluded people for thousands of years; even Aristotle, perhaps the most learned philosopher of the classical world, was mistaken about it (see ❑). What makes this question so tricky is that no objects on earth are truly free of outside influences; instead, they all push on, rub against, or interact with one another in some way or other.

As a result, it took the remarkable Italian astronomer, mathematician, and physicist Galileo Galilei many years of careful observation and logical analysis to answer the question ❑. The solution he came up with, like the question itself, sounds simple: if the person is stationary, he/she will remain stationary; if the person is moving in some particular direction, he/she will continue moving in that direction at a steady pace, following a straight-line path. This property of steady motion in the absence of any outside influence is called **inertia.**

> **Inertia**
> A body in motion tends to remain in motion; a body at rest tends to remain at rest.

The main reason why Aristotle failed to discover inertia, and why we often overlook inertia ourselves, is friction. When you slide across the floor in your shoes,

This icon indicates a web link that you can follow to obtain further information about inertia. To access this material, go to the book web site at
www.wiley.com/college/howthingswork
and enter the phrase "examples of inertia" in the box asking for a keyword or phrase.

friction quickly slows you to a stop and hides your inertia. To make inertia more obvious, we have to get rid of friction. That's why you're wearing skates.

Skates almost completely eliminate friction, at least in one direction, so that you can glide effortlessly across the ice or roller rink and experience your own inertia. For simplicity, let's imagine that your skates are perfect and experience no friction at all as you glide. Also, for this and the next couple of sections, we'll forget not only about friction but also about air resistance. As long as the air is calm and you're not moving too fast, air resistance isn't all that important to skating anyway.

Now that you're ready to skate, we'll begin to examine five important physical quantities relating to motion and look at their relationships to one another. These quantities are position, velocity, mass, acceleration, and force.

Let's begin by describing your location. At any particular moment, you're located at a **position**—that is, at a specific point in space. Whenever we report your position, it's always as a **distance** and **direction** from some reference point: how many meters north of the refreshment stand or how many kilometers west of Cleveland.

Position is an example of a vector quantity. A **vector quantity** consists of both a magnitude and a direction; the **magnitude** tells you how much of the quantity there is, while the direction tells you which way the quantity is pointing. Vector quantities are common in nature. When you encounter one, pay attention to the direction part; if you're looking for buried treasure 30 paces from the old tree but forget that it's due east of that tree, you'll have a lot of digging ahead of you.

You're on your feet and beginning to skate. If you're moving, then your position is changing. In other words, you have a velocity. **Velocity** measures how quickly your position changes; it's our second vector quantity and consists of the speed at which you're moving and the direction in which you're heading. Your **speed** is the distance you travel in a certain amount of time,

$$\text{speed} = \frac{\text{distance}}{\text{time}},$$

and your direction might be east, north, or down—if you're taking a spill.

But when you're gliding freely, with nothing pushing you horizontally, you have a particularly simple velocity. Since you travel at a steady pace, along a straight-line path, your velocity is constant and never changes. If you're heading west at a speed of 10 meters-per-second, you'll continue to travel with that velocity indefinitely. A speed of 10 meters-per-second means that, if you travel for 1 second at your present speed, you'll cover a distance of 10 meters. Since your velocity is constant, you'll travel 100 meters in 10 seconds, 1000 meters in 100 seconds, and so on. Furthermore, the path you'll take is a straight line. In a word, you coast.

Thanks to your skates, we can now restate the previous description of inertia in terms of velocity: an object that is not subject to any outside influences moves at a constant velocity, covering equal distances in equal times along a straight-line path. This statement is frequently identified as **Newton's first law of motion,** after its discoverer, the English mathematician and physicist Sir Isaac Newton ❏. The outside influences referred to in this law are called **forces,** a technical term for pushes and pulls.

> **Newton's First Law of Motion**
> An object that is not subject to any outside forces moves at a constant velocity, covering equal distances in equal times along a straight-line path.

❏ While a professor in Pisa, **Galileo Galilei (Italian scientist, 1564–1642)** was obliged to teach the natural philosophy of Aristotle. Troubled with the conflict between Aristotle's theory and observations of the world around him, Galileo devised experiments that measured the speeds at which objects fall and determined that all falling objects fall at the same rate.

vectors

speed versus velocity

❏ In 1664, while **Sir Isaac Newton (English scientist and mathematician, 1642–1727)** was a student at Cambridge University, the university was forced to close for 18 months because of the plague. Newton retreated to the country and discovered the laws of motion and gravitation and invented the mathematical basis of calculus. These discoveries, along with his observation that celestial objects such as the moon obey the same simple physical laws as terrestrial objects such as an apple (a new idea for the time), are recorded in his *Philosophiae Naturalis Principia Mathematica,* first published in 1687. This book is perhaps the most important and influential scientific and mathematical work of all time.

Check Your Understanding #1: A Puck on Ice
Why does a moving hockey puck continue to slide across an ice rink even though no one is pushing on it?

The Alternative to Coasting: Acceleration

As you glide forward with nothing pushing you horizontally, what prevents your speed and direction from changing? The answer is your mass. **Mass** is the measure of your inertia, your resistance to changes in velocity. Almost everything in the universe has mass. Because you have mass, your velocity will change only if something pushes on you—that is, only if you experience a force. You'll keep moving steadily in a straight path until something pushes on you with a force to stop you or send you in another direction. Force is our third vector quantity, having both a magnitude and a direction. After all, a push to the right is different from a push to the left.

When something pushes on you, your velocity changes; in other words, you accelerate. **Acceleration,** our fourth vector quantity, measures how quickly your velocity changes. *Any* change in your velocity is acceleration, whether you're speeding up, slowing down, or even turning. If either your speed or direction of travel is changing, you're accelerating!

Like any vector quantity, acceleration has a magnitude and a direction. To see how these two parts of acceleration work, imagine that you're at the starting line of a race, waiting for it to begin. The starting buzzer sounds and you're off! You begin to accelerate—your speed increases and you cover ground more and more quickly. The magnitude of your acceleration depends on how hard the surface you're skating on pushes you forward. If it's a long race and you're not in a hurry, you obtain a modest push from the ground and the magnitude of your acceleration is small. Your velocity changes slowly. But if the race is a sprint and you haven't a moment to spare, you spring forward hard and the ground exerts an enormous forward force on you. The magnitude of your acceleration is large and your velocity changes rapidly. In this case, you can actually feel your inertia opposing your efforts to pick up speed.

But acceleration has more than just a *magnitude*. When you start the race, you also select a *direction* for your acceleration—the direction toward which your velocity is shifting with time. This acceleration is in the same direction as the force causing it. If you obtain a forward force from the ground, you'll accelerate forward—your velocity will shift more and more forward. If you obtain a sideways force from the ground, well. . . . The other racers will have to jump out of your way as you careen into the wall. They'll laugh all the way to the finish line at your failure to recognize the importance of direction in the definitions of force and acceleration.

Once you're going fast enough, you can stop fighting inertia and begin to glide. You coast forward at a constant velocity. Now inertia is helping you; it keeps you moving steadily along even though nothing is pushing you forward. (Recall that we're neglecting friction and air resistance. In reality, those effects push you backward and gradually slow you down as you glide. However, since we're ignoring them in this section, your motion is smooth and steady.)

But even when you're not trying to speed up or slow down, you can still accelerate. As you steer your skates or go over a bump, you experience sideways or up–down forces that change your *direction of travel* and thus cause you to accelerate.

Finally the race is over and you skid to a stop. You're accelerating again. This time you're accelerating backward, in the direction opposite your forward velocity. While we often call this process *deceleration,* it's just a special type of acceleration. Your forward velocity gradually diminishes until you come to rest.

To help you recognize acceleration, here are some accelerating objects:

1. A runner who's leaping forward at the start of a race—the runner's velocity is changing from zero to forward so the runner is accelerating *forward*.
2. A bicycle that's stopping at a crosswalk—its velocity is changing from forward to zero so it's accelerating *backward* (it's decelerating).
3. An elevator that's just starting upward from the first floor to the fifth floor—its velocity is changing from zero to upward so it's accelerating *upward*.
4. An elevator that's stopping at the fifth floor after coming from the first floor—its velocity is changing from upward to zero so it's accelerating *downward*.
5. A car that's beginning to shift left to pass another car—its velocity is changing from forward to forward and leftward so it's accelerating toward *the left*.
6. An airplane that's just beginning its descent—its velocity is changing from level forward to descending forward so it's accelerating mostly *downward*.
7. Children riding a carousel around in a circle—while their speeds are constant, their directions of travel are always changing. We'll discuss the directions in which they're accelerating in Section 3.3.

Here are some objects that are *not* accelerating:

1. A parked car—its velocity is always zero.
2. A car traveling straight forward on a level road at a steady speed—no change in its speed or direction of travel.
3. A bicycle that's climbing up a smooth, straight hill at a steady speed—no change in its speed or direction of travel.
4. An elevator that's moving straight upward at a steady pace, halfway between the first floor and the fifth floor—no change in its speed or direction of travel.

Seeing acceleration isn't as easy as seeing velocity. You have to watch skaters closely for some time to see whether or not they're accelerating. If their paths aren't straight or if their speeds aren't steady, then they're accelerating.

acceleration versus velocity

Check Your Understanding #2: Changing Trains
Trains spend much of their time coasting along at constant velocity. When does a train accelerate forward? backward? leftward? downward?

How Forces Affect Skaters

Now that we've learned what acceleration is, let's see how you accelerate in response to a particular force. Your acceleration depends on your mass: the more massive you are, the less you accelerate. For example, it's easier to change your velocity before you eat Thanksgiving dinner than afterward. Your acceleration also depends on the force exerted on you: the stronger the force, the more you accelerate.

There is a simple relationship between your acceleration, your mass, and the force exerted on you. The force exerted on you equals the product of your mass times your acceleration. The acceleration, as we've seen, is in the same direction as the force. This relationship was deduced by Newton from his observations of motion and is referred to as **Newton's second law of motion.**

interpreting symbolic equations

Equation 1.1.1 can be rearranged algebraically as

$$\text{acceleration} = \frac{\text{force}}{\text{mass}}.$$

This form makes it clear that a skater's acceleration depends both on the force causing that acceleration and on the skater's mass. Increasing the force increases the acceleration, while increasing the skater's mass with pads and a helmet decreases the person's acceleration.

Fig. 1.1.2 A baseball accelerates easily because of its small mass. A bowling ball has a large mass and is harder to accelerate.

Newton's Second Law of Motion
The force exerted on an object is equal to the product of that object's mass times its acceleration. The acceleration is in the same direction as the force.

This law can be written in a word equation:

$$\text{force} = \text{mass} \cdot \text{acceleration}, \tag{1.1.1}$$

in symbols:

$$\mathbf{F} = m \cdot \mathbf{a},$$

and in everyday language:

Throwing a baseball is much easier than throwing a bowling ball (Fig. 1.1.2).

Remember that in Eq. 1.1.1 the direction of the acceleration is the same as the direction of the force.

Because it's an equation, the two sides of Eq. 1.1.1 are equal. The force on you equals the product of your mass times your acceleration. Since your mass can't change without a visit to the snack bar, Eq. 1.1.1 indicates that an increase in the force on you is accompanied by a similar increase in your acceleration. That way, as the left side of the equation increases, the right side increases to keep the two sides equal. Thus the harder something pushes on you, the more rapidly your velocity changes. (For another useful form of Eq. 1.1.1, see ❏.)

We can also compare the effects of a specific force on two different masses, for example, you and the former sumo wrestler to your left. Since each of you experiences the same force, the left side of Eq. 1.1.1 is the same for both of you. Since the right side must also stay the same, the equation indicates that an increase in mass must be accompanied by a corresponding decrease in acceleration. Sure enough, your velocity changes more rapidly than the velocity of the sumo wrestler when the two of you are subjected to identical forces (Fig. 1.1.3).

So far we've explored five principles:

1. Your position indicates exactly where you're located.
2. Your velocity measures how quickly your position changes.
3. Your acceleration measures how quickly your velocity changes.
4. In order for you to accelerate, something must exert a force on you.
5. The more mass you have, the less acceleration you experience for a given force.

We've also encountered five important physical quantities—mass, force, acceleration, velocity, and position—as well as some of the rules that relate them to one another. Much of the groundwork of physics rests on these five quantities and on their interrelationships.

Skating certainly depends on these quantities. We can now see that, in the absence of any horizontal forces, you either remain stationary or coast along at a constant velocity. To start, stop, or turn, something must push you horizontally and that something is the ice or pavement. We haven't talked about how you obtain horizontal forces from the ice or pavement and we'll leave that problem for later sections. But as you skate, you should be aware of these forces and notice how they change your speed, direction of travel, or both. Learn to watch yourself accelerate.

► Check Your Understanding #3: Hard to Stop
It's much easier to stop a bicycle traveling toward you at 5 kilometers-per-hour than an automobile traveling toward you at the same velocity. What accounts for this difference?

► Check Your Figures #1: At the Bowling Alley
Bowling balls come in various masses. Suppose that you try bowling two different balls, one with twice the mass of the other. If you push on them with equal forces, which one will accelerate faster and how much faster?

Measure for Measure: The Importance of Units

If you went to the grocery store and asked for "6 sugar," the clerk wouldn't know how much sugar you wanted. The number 6 wouldn't be enough information; you'd need to specify the units you were referring to—cups or pounds or cubes or tons. This need to specify units applies to almost all physical quantities—velocity, force, mass, and so on—and it has led our society to develop units that everyone agrees on and can understand, also known as **standard units.**

For example, if a skater has a velocity of 20 miles-per-hour toward the east, the units "miles-per-hour" are essential to give meaning to the number 20. This same velocity can be specified using different units—for example, as 32.2 kilometers-per-hour eastward. In fact, we can use many different units to specify velocity: feet-per-second, yards-per-day, and inches-per-century, to name only a few. We may use any units that are convenient, and we can always find a simple relationship to convert one unit into another. For example, to convert miles-per-hour into kilometers-per-hour, we'd simply multiply by 1.609 kilometers/mile.

The value of standard units comes from a general agreement about what they mean. Everyone agrees how far a mile is and how long an hour is. When combined, these two basic units form a composite unit for measuring speed, the mile-per-hour, that everyone also agrees on.

Many of the common units in the United States come from the old **English System of Units,** which most of the world has abandoned in favor of the **SI Units** (Systéme Internationale d'Unités). The continued use of English units in the United States often makes life difficult. If you have to triple a cake recipe that calls for $\frac{3}{4}$ cup of milk, you must work hard to calculate that you need $2\frac{1}{4}$ cups. Then you go to buy $2\frac{1}{4}$ cups of milk, which is slightly more than half a quart, but end up buying two pints instead. You now have 14 ounces of milk more than you need. But is that 14 fluid ounces or 14 ounces of weight? And so it goes.

The SI system has two important characteristics that distinguish it from the English system and make it easier to use:

1. Different units for the same physical quantity are related by factors of 10.
2. Most of the units are constructed out of a few basic units: the meter, the kilogram, and the second.

Let's take the first characteristic first: different units for the same physical quantity are related by factors of 10. When measuring volume, 1000 milliliters is exactly 1 liter and 1000 liters is exactly 1 meter^3. When measuring mass, 1000 grams is exactly 1 kilogram and 1000 kilograms is exactly 1 metric ton. Because of this consistent relationship, enlarging a recipe that's based on the SI system is as simple as mul-

Fig. 1.1.3 If you give these two skaters identical pushes, the boy will accelerate more quickly than the girl. That's because the girl has more mass than the boy.

tiplying a few numbers. You never have to think about converting pints into quarts, teaspoons into tablespoons, or ounces into pounds. Instead, if you want to triple a recipe that calls for 500 milliliters of sugar, you just multiply the recipe by 3 to obtain 1500 milliliters of sugar. Since 1000 milliliters is 1 liter, you'll need 1.5 liters of sugar. Converting milliliters to liters is as simple as multiplying by 0.001 liter/milliliter. (See Appendix B for more conversion factors.)

how to change units

SI units remain somewhat mysterious to many U.S. residents, even though some of the basic units are slowly appearing on our grocery shelves and highways. As a result, while the SI system really is more sensible than the old English system, developing a feel for some SI units is still difficult. How many of us know our heights in meters (the SI unit of length) or our masses in kilograms (the SI unit of mass)? If your car is traveling 200 kilometers-per-hour and you pass a police car, are you in trouble? Yes, because 200 kilometers-per-hour is about 125 miles-per-hour. Actually, the hour is not an SI unit—the SI unit of time is the second—but it remains customary for describing long periods of time. Thus the kilometer-per-hour is a unit that is half SI (the kilometer part) and half customary (the hour part).

The second characteristic of the SI system is its relatively small number of basic units. So far, we've noted the SI units of mass (the **kilogram,** abbreviated kg), length (the **meter,** abbreviated m), and time (the **second,** abbreviated s). One kilogram is about the mass of a liter of water or soda; one meter is about the length of a long stride; one second is about the time it takes to say "one banana." From these three basic units, we can create several others, such as the SI units of velocity (the **meter-per-second,** abbreviated m/s) and acceleration (the **meter-per-second²,** abbreviated m/s²). One meter-per-second is a healthy walking speed; one meter-per-second² is about the acceleration of an elevator after the door closes and it begins to head upward. This conviction that many units are best constructed out of other, more basic units dramatically simplifies the SI system; the English system doesn't usually suffer from such sensibility.

The SI unit of force is also constructed out of the basic units of mass, length, and time. If we choose a 1-kilogram object and ask just how much force is needed to make that object accelerate at 1 meter-per-second², we define a specific amount of force. Since 1 kilogram is the SI unit of mass and 1 meter-per-second² is the SI unit of acceleration, it's only reasonable to let the force that causes this acceleration be the SI unit of force: the *kilogram-meter-per-second²*. Since this composite unit is pretty unwieldy but very important, it has been given its own name: the **newton** (abbreviated N)—after, of course, Sir Isaac, whose second law defines the relationship among mass, length, and time that the unit expresses. One newton is about the weight of 10 U.S. quarter dollars; if you hold 10 quarters steady in your hand, you'll feel a downward force of about 1 newton.

Because of our slow but steady move to the SI system, we'll be using SI units for most of the book. But whenever we might benefit from an intuitive feel for a physical quantity, we'll examine it in English and customary units as well. A bullet train traveling "67 meters-per-second" won't mean much to most of us, while one moving "150 miles-per-hour" (150 mph) or "241 kilometers-per-hour" (241 km/h) might elicit our well-deserved respect.

Check Your Understanding #4: Going for a Walk

If you're walking at a pace of 1 meter-per-second, how many miles will you travel in an hour?

SECTION 1.2   # Falling Balls

We've all dropped balls from our hands or seen them arc gracefully through the air after being thrown. These motions seem like simplicity itself and, not surprisingly, they're governed by only a few universal rules. We encountered several of those rules in the previous section, but we're about to examine our first important type of force: gravity. Like Newton, who reportedly began his investigations after seeing an apple fall from a tree, we'll start simply by exploring gravity and its effects on motion in falling objects.

Questions to Think About: What do we mean by "falling," and why do balls fall? Which falls faster: a heavy ball or a light ball? Can a ball that's heading upward still be falling? How does gravity affect a ball that's thrown sideways?

Experiments to Do: A few seconds with a baseball will help you see some of the behaviors that we'll be exploring. Toss the ball into the air to various heights, catching it in your hand as it returns. Work with a friend to help you time the flight of the ball. As you toss the ball higher, how much more time does it spend in the air? How does it feel coming back to your hands? Is there any difference in the impact it makes? Which takes the ball longer: rising from your hand to its peak height or returning from its peak height back to your hand?

Now drop two different balls—a baseball, say, and a golf ball. If you drop them simultaneously, without pushing either one up or down, does one ball strike the ground first, or do they arrive together? Now throw one ball horizontally while dropping the second. If they both leave your hands at the same time and the first one's initial motion is truly horizontal, which one reaches the ground first?

Weight and Gravity

Like everything else around us, a ball has a weight. For example, a golf ball weighs about 0.5 N (0.5 newton). But what is weight? Evidently it's a force, since the newton is a unit of force. But to understand what weight really is—and, in particular, where it comes from—we need to look at gravity.

Gravity is a physical phenomenon that produces an attractive force between every pair of objects in the universe. In our daily lives, however, the only object large enough and near enough to have an obvious direct effect on us is our planet, the earth. While the moon and the sun are also important, their weak gravitational attractions cause subtler effects, such as the ocean tides.

Acceleration

Weight
(Force of gravity)

Fig. 1.2.1 A ball experiencing the force of gravity. It accelerates downward.

On the earth's surface, gravity creates a downward force on any object, attracting that object directly toward the center of the earth. We call this downward gravitational force the object's **weight** (Fig. 1.2.1). A baseball has a weight of about 1.4 N (0.32 pound). An automobile has a weight of about 10,000 N (2200 pounds). A fly has a weight of about 0.001 N (0.0002 pound).

One remarkable characteristic of gravity is that the force it exerts on an object is exactly proportional to the object's mass—if one ball has twice the mass of another ball, it also has twice the weight. This relationship between weight and mass is astonishing because they're very different things: weight is how hard gravity pulls on a ball, and mass is how difficult the ball is to accelerate. Because of their proportionality, a ball that's heavy is also hard to shake!

An object's weight is also proportional to the local strength of gravity, which is measured by a quantity called the **acceleration due to gravity.** We can express these observations in a word equation:

$$\text{weight} = \text{mass} \cdot \text{acceleration due to gravity}, \tag{1.2.1}$$

in symbols:

$$\mathbf{w} = m \cdot \mathbf{g},$$

and in everyday language:

> *You can lose weight either by reducing your mass or by going someplace, like a small planet, where the gravity is weaker.*

mass versus weight

The acceleration due to gravity is determined by various properties of the earth and doesn't depend on the object being considered. At the surface of the earth, the acceleration due to gravity is about 9.8 m/s² (9.8 meters-per-second², or 32 feet-per-second²) toward the center of the earth. While this quantity doesn't appear to involve force, its units can be rewritten in terms of newtons as 9.8 newtons-per-kilogram; thus a ball (or any other object, for that matter) weighs 9.8 newtons for each kilogram of its mass.

If the only force on a ball is its weight, the ball accelerates downward; in other words, it falls. While a ball moving through the earth's atmosphere encounters additional forces due to air resistance, we'll ignore those forces for the time being. Doing so will cost us only a little in terms of accuracy—the effects of air resistance are negligible as long as the ball is dense and its speed relatively small—and it will allow us to focus exclusively on the effects of gravity.

How much does the falling ball accelerate? According to Eq. 1.1.1, the force on the ball is equal to the ball's mass times its acceleration. According to Eq. 1.2.1, the ball's weight is equal to the ball's mass times the acceleration due to gravity. But because the ball is *falling,* the only force on it is its own weight so these two independent equations can be brought together. Equating the force on the ball with the ball's weight, we get

$$m \cdot \mathbf{a} = \mathbf{F} = \mathbf{w} = m \cdot \mathbf{g}$$
$$\mathbf{a} = \mathbf{g}.$$

As you can see, the ball's acceleration is equal to the acceleration due to gravity!

Thus a ball falling near the earth's surface experiences a downward acceleration of 9.8 m/s^2, regardless of its mass. This downward acceleration is substantially more than that of an elevator starting its descent. When you drop a ball, it picks up speed very quickly in the downward direction.

Because all falling objects at the earth's surface accelerate downward at exactly the same rate, a billiard ball and a bowling ball dropped simultaneously from the same height will reach the ground together. (Remember that we're not considering air resistance yet.) Although the bowling ball weighs more than the billiard ball, it also has more mass; so while the bowling ball experiences a larger downward force, its larger mass ensures that its downward acceleration is equal to that of the lighter and less massive billiard ball.

▶ Check Your Understanding #1: Weight and Mass

Out in deep space, far from any celestial object that exerts significant gravity, would an astronaut weigh anything? Would that astronaut have a mass?

▶ Check Your Figures #1: Weighing In on the Moon

You're in your spacecraft on the surface of the moon. Before getting into your suit, you weigh yourself and find that your moon weight is almost exactly one-sixth your earth weight. What is the moon's acceleration due to gravity?

The Velocity of a Falling Ball

We're now ready to examine the motion of a falling ball near the earth's surface. A falling ball is one that has only the force of gravity acting on it; and gravity, as we've seen, causes any falling object to accelerate downward at a constant rate. But we're usually less interested in the falling object's acceleration than we are in its position and velocity. Where will the object be in 3 seconds, and what will its velocity be then? When you're trying to summon up the courage to jump off the high dive, you want to know how long it'll take you to reach the water and how fast you'll be going when you hit.

The first step in answering these questions is to look at how a ball's velocity is related to the time you've been watching it fall. To do that, you'll need to know the ball's *initial velocity*—that is, its speed and direction at the moment you start watching it. If you drop the ball from rest, its initial velocity is zero.

You can then describe the ball's present velocity in terms of its initial velocity, its acceleration, and the time that has passed since you started watching it. Because a constant acceleration causes the ball's velocity to change by the same amount each second, the ball's present velocity differs from its initial velocity by the product of the acceleration times the time over which you've been watching it. We can relate these quantities as a word equation:

$$\text{present velocity} = \text{initial velocity} + \text{acceleration} \cdot \text{time,} \qquad \textbf{(1.2.2)}$$

in symbols:

$$\mathbf{v} = \mathbf{v_0} + \mathbf{a} \cdot t,$$

and in everyday language:

> *A stone dropped from rest descends faster with each passing second, but you can give it a boost by throwing it downward instead of just letting go.*

For a ball falling from rest, the initial velocity is zero, the acceleration is downward at 9.8 m/s², and the time you've been watching it is simply the time since it started to drop (Fig. 1.2.2). After one second, the ball has a downward velocity of 9.8 m/s (9.8 meters-per-second, or 32 feet-per-second). After two seconds, the ball has a downward velocity of 19.6 m/s. After three seconds, its downward velocity is 29.4 m/s, and so on. Since the ball's motion is strictly vertical, we often put a negative sign in front of the acceleration to indicate the direction. By convention, we say that a negative sign means "down."

Check Your Understanding #2: Half a Fall

You drop a marble from rest and after 1 s, its velocity is 9.8 m/s in the downward direction. What was its velocity after only 0.5 s of falling?

Check Your Figures #2: The High Dive

If it takes you about 1.4 s to reach the water from the 10-meter diving platform, how fast are you going just before you enter the water?

The Position of a Falling Ball

The ball's velocity continues to increase as it falls, but where exactly is the ball located? To answer that question, you need to know the ball's *initial position*—that is,

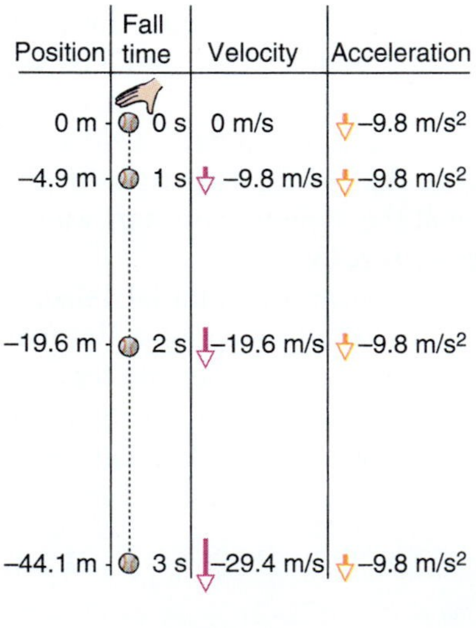

Position	Fall time	Velocity	Acceleration
0 m	0 s	0 m/s	−9.8 m/s²
−4.9 m	1 s	−9.8 m/s	−9.8 m/s²
−19.6 m	2 s	−19.6 m/s	−9.8 m/s²
−44.1 m	3 s	−29.4 m/s	−9.8 m/s²

Fig. 1.2.2 The moment you let go of a ball that was resting in your hand, it begins to fall. Its weight causes it to accelerate downward. After 1 second, it has fallen 4.9 m and has a velocity of 9.8 m/s downward. After 2 seconds, it has fallen 19.6 m and has a velocity of 19.6 m/s downward, and so on. As the ball continues to accelerate downward, its velocity continues to increase downward. Negative values for the position and velocity are meant to indicate downward movement, caused by a negative or downward acceleration.

where it was when you started watching it fall. If you drop the ball from rest, the initial position is your hand and you can define that spot as 0.

You can then describe the ball's present position in terms of its initial position, its initial velocity, its acceleration, and the time that has passed since you started watching it. However, because the ball's velocity is changing, you can't simply multiply its present velocity by the time that it's been falling to determine how much the ball's present position differs from its initial position. Instead, you must use the ball's average velocity during the whole period you've been watching it. Since the ball's velocity has been changing uniformly from its initial velocity to its present velocity, the ball's average velocity is exactly halfway in between the two:

$$\text{average velocity} = \text{initial velocity} + \tfrac{1}{2} \cdot \text{acceleration} \cdot \text{time}.$$

The ball's present position differs from its initial position by the product of this average velocity times the time over which you've been watching it. We can relate these quantities as a word equation:

$$\text{present position} = \text{initial position} + \text{initial velocity} \cdot \text{time}$$
$$+ \tfrac{1}{2} \cdot \text{acceleration} \cdot \text{time}^2, \tag{1.2.3}$$

in symbols:

$$\mathbf{x} = \mathbf{x_0} + \mathbf{v_0} \cdot t + \tfrac{1}{2} \cdot \mathbf{a} \cdot t^2,$$

and in everyday language:

The longer a stone has been falling, the more its height diminishes with each passing second. However, it won't overtake a stone that was dropped next to it at an earlier time or dropped from beneath it at the same time.

For a ball falling from rest, the initial velocity is zero, the acceleration is downward at 9.8 m/s^2, and the time you've been watching it is simply the time since it started to drop (Fig. 1.2.2). After one second, the ball has fallen 4.9 m (16 feet). After two seconds, the ball has fallen downward a total of 19.6 m. After three seconds, the ball has fallen a total of 44.1 m, and so on.

Equations 1.2.2 and 1.2.3 depend on the definition of acceleration as the measure of how quickly *velocity* changes and the definition of velocity as the measure of how quickly *position* changes. Because the acceleration of a falling ball doesn't change with time, the two equations can be derived using algebra. But in more complicated situations, where an object's acceleration changes with time, predicting its position and velocity usually requires the use of *calculus*. Calculus is the mathematics of change, invented by Newton to address just these sorts of problems.

We've been discussing what happens to a falling ball, but we could have chosen another object instead. Everything falls the same way; heavy or light, large or small, all objects take the same amount of time to fall some distance near the earth's surface, as long as they're dense enough to overcome air resistance. If there were no air, this statement would be exactly true for any object; a feather and a lead brick would plummet downward together if you dropped them simultaneously.

Now that we've explored acceleration due to gravity, we can see why a ball dropped from a tall ladder is more dangerous than the same ball dropped from a short stool. The farther the ball has to fall, the longer it takes to reach the ground and the more time it has to accelerate. During its long fall from the tall ladder, the ball acquires a large downward velocity and becomes very hard to stop. If you try to

❏ In 1782, William Watts, a plumber from Bristol, England, patented a technique for forming perfectly spherical, seamless lead shot for use in guns. His idea was to pour molten lead through a sieve suspended high above a pool of water. The lead droplets cool in the air as they fall, solidifying into perfect spheres before reaching the water. Shot towers based on this idea soon appeared throughout Europe and eventually in the United States. Nowadays, iron shot has all but replaced environmentally dangerous lead shot. Iron shot is cast, rather than dropped, because the longer cooling time needed to solidify molten iron would require impractically tall shot towers.

catch it, you'll have to exert a very large upward force on it to accelerate it upward and bring it to rest quickly. Exerting that large upward force may hurt your hand.

The same notion holds if you're the falling object. If you leap off a tall ladder, a substantial amount of time will pass before you reach the ground. By the time you hit the ground, you will have acquired considerable downward velocity. The ground will then accelerate you upward and bring you to rest with a very large and unpleasant upward force. (For an interesting and less painful application of long falls, see ❏.)

Check Your Understanding #3: Half a Fall Again

You drop a marble from rest and after 1 s, it has fallen downward a distance of 4.9 m. How far had it fallen after only 0.5 s?

Check Your Figures #3: Extreme Physics

You're planning to construct a bungee-jumping amusement at the local shopping center. If you want your customers to have a 5-second free-fall experience, how tall will you need to build the tower from which they'll jump? (Don't worry about the extra height needed to stop people after the bungee pulls taut.)

How a Thrown Ball Moves: Projectile Motion

If the only force acting on an object is its weight, then the object is falling. So far, we've explored this principle only as it pertains to balls dropped from rest. However a thrown ball is falling, too; once it leaves your hand, it's subject only to the downward force of gravity and it falls. It may seem odd but even though it's initially traveling upward, a ball tossed upward is accelerating downward at 9.8 m/s². As a result, the tossed ball's upward velocity diminishes, it stops rising, its velocity becomes downward, and it eventually returns to the ground.

Equation 1.2.2 still describes how the ball's velocity depends on the fall time, but now the initial velocity isn't zero; it points in the upward direction! If you toss a ball straight up in the air, it leaves your hand with a large upward velocity (Fig. 1.2.3). As soon as you let go of it, it begins to accelerate downward. If the ball's initial upward velocity is 29.4 m/s, then after one second its upward velocity is 19.6 m/s. After another second, its upward velocity is only 9.8 m/s. After a third second, the ball momentarily comes to a complete stop with a velocity of zero. It then descends from this peak height, falling just as it did when you dropped it from rest.

The ball's flight before and after its peak is symmetrical. It travels upward quickly at first, since it has a large upward velocity. As its upward velocity diminishes, it travels more and more slowly until it comes to a stop. It then begins to descend, slowly at first and then faster and faster as it continues its constant downward acceleration. The time the ball takes to rise from its initial position in your hand to its peak height is exactly equal to the time it takes to descend back down from that peak to your hand. Equation 1.2.3 indicates how the position of the ball depends on the fall time, with the initial velocity being the upward velocity of the ball as it leaves your hand.

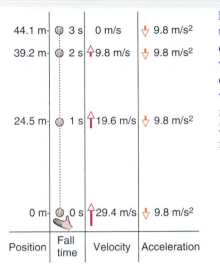

Position	Fall time	Velocity	Acceleration
44.1 m	3 s	0 m/s	9.8 m/s²
39.2 m	2 s	9.8 m/s	9.8 m/s²
24.5 m	1 s	19.6 m/s	9.8 m/s²
0 m	0 s	29.4 m/s	9.8 m/s²

Fig. 1.2.3 The moment you let go of a ball thrown straight upward, it begins to accelerate downward at 9.8 m/s². The ball rises but its upward velocity diminishes steadily until it momentarily comes to a stop. It then descends with its downward velocity increasing steadily. In this example, the ball rises for 3 s and comes to rest. It then descends for 3 s before returning to your hand in a very symmetrical flight.

The larger the initial upward velocity of the ball, the longer it rises and the higher it goes before its velocity is reduced to zero. It then descends for the same amount of time it spent rising. The higher the ball goes before it begins to descend, the longer it takes to return to the ground and the faster it's traveling when it arrives. That's why catching a high fly ball with your bare hands stings so much: the ball is traveling very, very fast when it hits your hands, and a large force is required to bring the ball to rest.

What happens if you don't toss the ball exactly straight up? Suppose you throw the ball upward at some angle. The ball still rises to a peak height and then begins to descend; but as it rises and descends, it also travels away from you horizontally so that it strikes the ground at some distance from your feet. How much does this horizontal travel complicate the motion of a falling ball?

The answer is not very much. One of the beautiful simplifications of physics is that we can often treat an object's vertical motion independently of its horizontal motion. This technique involves separating the vector quantities—acceleration, velocity, and position—into **components,** those portions of the quantities that lie along specific directions (Fig. 1.2.4). For example, the vertical component of an object's position is that object's altitude.

If you know a ball's altitude, you know only part of its position; you still need to know where it is to your left or right and to your front or back. In fact, you can completely specify its position (or any other vector quantity) in terms of three components along three directions that are at right angles to one another. This means that you can completely specify the ball's position by its vertical altitude, its horizontal distance to your left or right, and its horizontal distance in front or in back of you. For example, the ball might be 10 m above you, 3 m to your left, and 2 m behind you. These three distances indicate precisely where the ball is located.

Up until now, we've actually been examining only the vertical components of position, velocity, and acceleration. But now we're going to let the ball move horizontally so that we also have to consider what happens to the two horizontal components of each vector quantity. Keeping track of all three components of each quantity is difficult. Since we're interested in the motion of a tossed ball, we can eliminate the left–right component by throwing the ball forward; that way, the ball will move only in a vertical plane that extends directly in front of us. We can then specify the ball's position as its altitude and its horizontal distance in front of us.

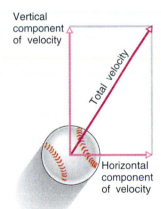

Vertical component of velocity

Total velocity

Horizontal component of velocity

Fig. 1.2.4 Even if the ball has a velocity that is neither purely vertical nor purely horizontal, its velocity may nonetheless be viewed as having a vertical component and a horizontal component. Part of its total velocity acts to move this ball upward, and part of its total velocity acts to move this ball in a horizontal direction.

Because the falling ball's acceleration is constant and doesn't depend on where the ball is near the earth's surface, the ball's horizontal motion is independent of its vertical motion. We already know about the falling ball's vertical motion, but what is its horizontal motion? It appears we need some new relationships to describe how the horizontal components of position, velocity, and acceleration change with time. Fortunately, we can reuse Eqs. 1.2.2 and 1.2.3.

While Eqs. 1.2.2 and 1.2.3 were introduced to describe a ball's vertical motion, they're actually more general. They relate three vector quantities—position, velocity, and constant acceleration—to one another. They describe how an object's position and velocity change with time when the object undergoes constant acceleration, regardless of the direction of that constant acceleration.

These equations also apply to the components of position, velocity, and constant acceleration. If you add the words "vertical component of" in front of each vector quantity in Eqs. 1.2.2 and 1.2.3, the equations correctly describe the object's vertical motion. The same can be done with the object's horizontal motion. Our previous look at vertical motion implicitly inserted the words "vertical component of" into the equations. Now we'll try inserting the words "horizontal component of" to understand the tossed ball's horizontal motion.

As soon as the ball leaves your hand, its motion can be broken into two parts: a vertical motion and a horizontal motion (Fig. 1.2.4). Part of the ball's initial velocity is in the upward direction, and that vertical velocity component determines the object's ascent and descent. Part of the ball's initial velocity is in the horizontal or downfield direction, and that horizontal velocity component determines the ball's drift downfield.

Because gravity is the only force on the ball and it acts only in the downward vertical direction, there is no horizontal component of acceleration. The horizontal velocity component therefore remains constant, and the ball travels downfield at a steady rate throughout its flight (Fig. 1.2.5). Overall, the upward vertical component of the ball's initial velocity determines how high the ball goes and how long it stays aloft before striking the ground, while the horizontal component of the initial velocity determines how quickly the ball travels downfield during its time aloft (Fig. 1.2.6).

When the ball hits the ground it still has its original horizontal velocity component, but its vertical velocity component is now in the downward direction. The total velocity of the ball is composed of these two components. The ball starts with its velocity up and forward and ends with its velocity down and forward.

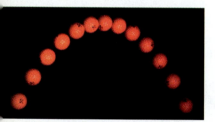

Fig. 1.2.5 This golf ball drifts steadily to the right after being thrown because gravity only affects the ball's vertical component of velocity.

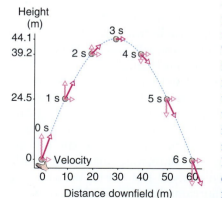

Height (m)

44.1

39.2

24.5

0 s

0   Velocity

0   10   20   30   40   50   60

1 s   2 s   3 s   4 s   5 s

6 s

Distance downfield (m)

Fig. 1.2.6 If you throw a ball upward, at an angle, part of the initial velocity will be in the vertical direction and part will be in the horizontal direction. The vertical and horizontal motions will take place independently of one another. The ball will rise and fall just as it did in Figs. 1.2.2 and 1.2.3; at the same time, however, it will move downfield. Because there is no horizontal acceleration (gravity acts only in the vertical direction), the horizontal velocity remains constant, as indicated by the velocity arrows. In this example, the ball travels 10 m downfield each second and strikes the ground 60 m from your feet after a 6-s flight through the air.

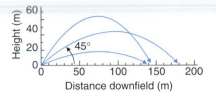

Fig. 1.2.7 If you want the ball to hit the ground as far from your feet as possible, given a certain initial speed, throw the ball at 45°. Halfway between horizontal and vertical, such a throw gives the ball equal initial vertical and horizontal components of velocity. The ball then stays aloft for a relatively long time and makes good use of that flight time to travel downfield.

If you want a ball or shot put to hit the ground as far from your feet as possible, you should keep it aloft for a long time *and* give it a sizable horizontal component of velocity; in other words, you must achieve a good balance between the vertical and horizontal velocity components (Fig. 1.2.7). These components of velocity together determine the ball's flight path, its **trajectory.** If you throw the ball straight up, it will stay aloft for a long time but will not travel downfield at all (and you will need to wear a helmet). If you throw the ball directly downfield, it will have a large initial horizontal velocity component but will hit the ground almost immediately.

Neglecting air resistance and the altitude difference between your throwing arm and the ground that the ball will eventually hit, your best choice is to throw the ball at an angle of 45° above horizontal. At that angle, the initial upward velocity component will be the same as the initial downfield velocity component. The ball will stay aloft for a reasonably long time and will make good use of that time to move downfield. Other angles won't make such good use of the initial speed to move the ball downfield. (A discussion of how to determine the horizontal and vertical components of a velocity appears in Appendix A.)

These same ideas apply to two baseballs, one dropped from a cliff and the other thrown horizontally from that same cliff. If both leave your hands at the same time, they will both hit the ground below at the same time (Fig. 1.2.8). The fact that the second ball has an initial horizontal velocity doesn't affect the time it takes to descend to the ground, because the horizontal and vertical motions are independent. Of course, the ball thrown horizontally will strike the ground far from the base of the cliff, while the dropped ball will land directly below your hand.

Check Your Understanding #4: Aim High

Why must a sharpshooter or an archer aim somewhat above her target? Why can't she simply aim directly at the bull's-eye to hit it?

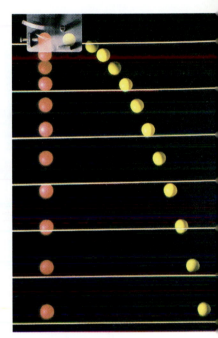

Fig. 1.2.8 When these two balls were dropped, they accelerated downward at the constant rate of 9.8 m/s² and their velocities increased steadily in the downward direction. Even though one of them was initially moving to the right, they descended together.

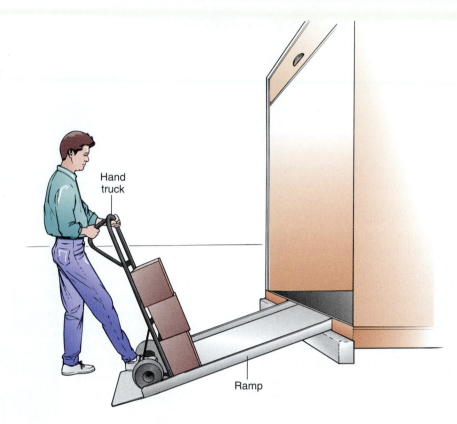

Hand
truck

Ramp

In the previous section, we looked at what happens to an object experiencing only a single force: the downward force of gravity. But what happens to objects that experience two or more forces at the same time? Imagine, for example, an object resting on the floor. That object experiences both the downward force of gravity and an upward force from the floor. If the floor is level, the object doesn't accelerate; but if the floor is tilted, so that it forms a ramp, the object accelerates downhill. In this section, we'll examine the motion of objects traveling along ramps. Also called inclined planes, ramps are common tools that make things easier to move.

Questions to Think About: *How does a ramp make it possible for one person to lift a very heavy object? Why is it so much easier to lower a heavy object than to raise that same object? What is the difference between a heavy object resting on the ground and that same object high up in the air? Why is it scarier to ski or sled down a steep hill than it is to slide down a more gradual slope? Why is it more difficult to ride a bicycle up a steep hill than a gradual hill?*

Experiments to Do: *Place a book on a smooth, flat table or board. Hold the book steady for a second and then let go of it without pushing on it. What happens to the book?*

Now, equip yourself with a pencil. Have a friend tilt the surface of the table or board slightly so that the book begins to slide downhill. Can you stop the book by pushing on it with the pencil? Place the book back on the table and have your friend tilt the table more sharply. Does the tilt of the table affect your ability to stop the book from sliding? Now try to push the book uphill with the pencil (a) when the table is slightly tilted and (b) when it is more sharply tilted. Which task requires more force? Why?

Evidently a gentle push is all that may be needed to raise a relatively heavy object if you use a ramp to help you. To understand why this feat is possible, we need to explore a handful of physical concepts and a few basic laws of motion.

A Piano on the Sidewalk

Imagine that you have a friend who's a talented but undiscovered pianist. She's renting a new apartment, and because she can't afford professional movers (Fig. 1.3.1), she's asked you to help her move her baby grand. Fortunately, her new apartment is only on the second floor. But the two of you still face a difficult challenge: how do you get that heavy piano up there? More important, how do you keep it from falling on you during the move?

The problem is that you can't push upward hard enough to lift the entire piano at once. One solution to this problem, of course, would be to break the piano into pieces and carry them up one by one. But this method has obvious drawbacks: your friend isn't expecting a firewood delivery. A better solution would be to find something else to help you push upward, and one of your best choices would be the simple machine known as a ramp.

Throughout the ages, **ramps,** also known as inclined planes, have made tasks like piano-moving possible. Because ramps can exert the enormous upward forces needed to lift stone and steel, they've been essential building equipment since the days of the pyramids. To see how ramps provide these lifting forces, we'll continue to explore the example of the piano, looking first at the force that the piano experiences when it touches a surface. For the time being, we'll continue to ignore friction and air resistance; they'd needlessly complicate our discussion. Besides, as long as the piano is on wheels, friction is negligible.

Fig. 1.3.1 A ramp would make this move much easier.

Which requires larger forces: lifting a pile of bricks one at a time or lifting them all together?

The Sidewalk Pushes Back: Newton's Third Law

With the piano resting on the sidewalk outside the apartment you make a startling discovery: the piano is *not* falling. Has gravity disappeared? The answer to that question would be painfully obvious if your foot were underneath one of the piano's wheels. No, the piano's weight is still all there. But something is happening at the surface of the sidewalk to keep the piano from falling. Let's take a careful look at the situation.

To begin with, the piano is clearly pushing down hard on the sidewalk. In fact, it's exerting a downward force equal to its entire weight on the sidewalk's surface. That's why you're keeping your toes out of the way. So we now have two forces: the piano being pushed downward by its weight and the sidewalk being pushed downward by a force equal to that weight.

But the presence of a new downward force on the sidewalk still doesn't explain why the piano itself isn't falling. The answer lies in the sidewalk's response to the piano's downward push: the sidewalk pushes back on the piano! You can feel this kind of force by leaning over and pushing downward on the sidewalk with your hand—the sidewalk pushes back. Those two forces, your downward push on the sidewalk and its upward push on your hand, are exactly equal in magnitude but opposite in direction.

This observation, that two things exert equal but opposite forces on one another, isn't unique to sidewalks, pianos, and hands; in fact, it's always true. If you push on any object with a certain amount of force, that object will push on you with exactly the same amount of force in exactly the opposite direction. This rule—often expressed as "for every action, there is an equal but opposite reaction"—is known as **Newton's third law of motion,** the last of his three laws.

why dropped eggs break

> ### Newton's Third Law of Motion
> For every force that one object exerts on a second object, there is an equal but oppositely directed force that the second object exerts on the first object.

escaping from the ice

The universality of this law is astounding. Whether an object is large or small, hard or soft, stationary or faster than a rocket, if you can reach over and push on it, it *will* push back on you with an equal but oppositely directed force.

In the present case, the sidewalk and piano push on one another with equal but oppositely directed forces. Of this equal-but-opposite pair of forces, only one acts on the piano: the sidewalk's upward push. This upward push on the piano is what balances the piano's weight and keeps the piano from falling. We've solved the mystery.

Apart from direction, there's one crucial difference between the force gravity exerts on the piano and the force the sidewalk exerts on it. While the piano's weight is dispersed throughout the piano, the sidewalk's upward force acts only on the bottoms of the piano's wheels. Although these two forces ultimately balance one another, their different locations lead to considerable stress within the piano. If it weren't built so well, the piano might lose a leg or two during the move. And just imagine what would happen if one of its wheels were made from an egg rather than sturdy metal. In that case, the enormous upward force from the sidewalk on the egg-wheel would all act in a small spot on the egg's bottom. Since the shell isn't strong enough to convey the sidewalk's upward force all the way to the piano leg, the inadequately supported piano would fall briefly and smash the egg flat.

> ### Summary of Newton's Laws of Motion
>
> **1.** An object that is not subject to any outside forces moves at a constant velocity, covering equal distances in equal times along a straight-line path.
> **2.** The force exerted on an object is equal to the product of that object's mass times its acceleration. The acceleration is in the same direction as the force.
> **3.** For every force that one object exerts on a second object, there is an equal but oppositely directed force that the second object exerts on the first object.

You are pushing a child on a playground swing. If you exert a 50-N force on him as he is swinging away from you, how much force will he exert back on you?

Adding Up the Forces

So far we've examined two sources of force on pianos: gravity and the sidewalk. A piano near the earth experiences gravity all the time, but it only feels a force from the sidewalk when the two touch. Since the sidewalk won't let a piano pass through it, it exerts an upward **support force** on the piano to keep it away. The piano in Fig. 1.3.2 is being held up by such a support force. A support force is always directed exactly away from the surface that creates it (see ❑). In the absence of friction, for example, a level sidewalk doesn't push you to the left or to the right; it pushes you exactly upward.

Returning to the motionless piano in Fig. 1.3.2, you might wonder why the piano isn't accelerating even a little. Why does the upward force from the sidewalk exactly balance the piano's weight? This perfect balance is reached because the support force from the sidewalk adjusts automatically to the piano's presence. If the sidewalk weren't exerting enough upward force on the piano, the piano would accelerate downward and penetrate the sidewalk's surface; if the sidewalk were exerting too much upward force on the piano, the piano would accelerate upward and move away from the sidewalk. A balance is quickly reached where the piano neither penetrates nor moves away from the sidewalk's surface.

Another way to state that the upward support force on an object exactly cancels the object's downward weight is to say that the **net force** on the object is zero, meaning that the sum of all the forces on the object is zero. Objects often experience more than one force at a time, and it's the net force, together with the object's mass, that determines how it accelerates. When several forces are exerted in the same direction, they add up, assisting one another so that the object accelerates in that direction (Fig. 1.3.3a). When several forces are exerted in opposite directions, they oppose and at least partially cancel one another (Fig. 1.3.3b).

Sometimes forces are exerted in two or more different directions so that the net force is pointing in yet another direction (Fig. 1.3.3c). For example, if you're standing on slippery ice and someone pushes you north while someone else pushes you east, the net force will point somewhere toward the northeast and you'll accelerate in that direction. The precise angle of the net force and your subsequent accelera-

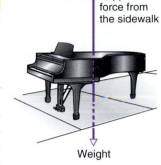

Fig. 1.3.2 A piano resting on the sidewalk. The sidewalk exerts an upward support force that exactly balances the piano's downward weight. The net force on the piano is zero, so the piano doesn't accelerate.

❑ Forces that are directed exactly away from surfaces are called **normal forces,** since the term **normal** is used by mathematicians to describe something that points exactly away from a surface—at right angles or perpendicular to that surface.

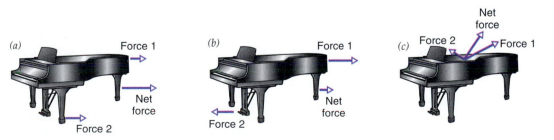

Fig. 1.3.3 When several forces act simultaneously on an object, the object responds to the sum of the forces. This sum is called the net force, and it has both a magnitude and a direction. Here, as elsewhere in this book, the length of each force arrow indicates its magnitude.

tion depend on exactly how hard each person pushes. For most of the following discussion, we'll only need a rough estimate of the net force's magnitude and direction, and we'll obtain that estimate using common sense.

> ### Check Your Understanding #3: Riding the Elevator
>
> As you ride upward in an elevator at a constant velocity, what two forces act on your body and what is the net force on you?

Lifting the Piano: Energy and Work

As you approach the task of moving your friend's piano, you might begin to worry about safety. There is clearly a difference between the piano resting on the sidewalk and the piano suspended on a board just outside the second floor apartment. After all, which one would you rather be sitting beneath? The elevated piano has something that the piano on the sidewalk doesn't have: the ability to produce motion and structural rearrangement (i.e., damage) in itself and the things beneath it. This capacity to make things happen is called *energy,* and the process of making them happen is called *work.*

Energy and work are both important physical *quantities,* meaning that both are measurable. For example, you can measure the amount of energy in the suspended piano, and you can measure the amount of work the piano does when the board breaks and it falls to the sidewalk. As you may suspect, the physical definitions of energy and work are somewhat different from those of common English. Physical **energy** isn't the exuberance of a 5 year old at the amusement park or the contents of a large cup of coffee; instead, it's defined as the capacity to do work. Similarly, physical **work** doesn't refer to activities at the office or in the yard; instead, it refers to the process of transferring energy.

Energy is what's transferred, and work does the transferring. The most important characteristic of energy is that it's conserved. In physics, a **conserved quantity** is one that can't be created or destroyed but can be transferred between objects or converted from one form to another. Conserved quantities are very special in physics; there are only a few of them, and energy is one. An object that has energy can't simply make that energy disappear; it can only get rid of energy by giving it to another object, and it makes this transfer by doing work on that object.

This situation is analogous to money and spending: money is what is transferred, and spending does the transferring. Sensible, law-abiding citizens don't create or destroy money; they simply transfer it among themselves through spending. Just as the most interesting aspect of money is spending it, so the most interesting aspect of energy is doing work with it. We can define money as the capacity to spend, just as we define energy as the capacity to do work.

So far we've been using a circular definition: work is the transfer of energy and energy is the capacity to do work. But what is involved in doing work on an object? You do work on an object by exerting a force on it as it moves in the direction of that force. As you throw a ball, exerting a forward force on the ball and moving the ball forward, you do work on the ball; as you lift a rock, pushing upward and moving the rock upward, you do work on the rock. In both cases, you transfer energy from yourself to an object by doing work on it.

Sometimes transferring energy to an object produces an obvious change in that object. In these situations, the added energy is easy to observe. When you throw a ball, it picks up speed and its energy increases, so that it can then do work on the objects it hits. The moving ball has energy of motion or **kinetic energy.** When you lift a rock, its distance from the earth increases and so does its energy, so that it can then do work on the objects beneath it. Energy stored in the forces between or within objects is called **potential energy,** and since the rock's energy is stored in the gravitational forces between the earth and the rock, its energy is called **gravitational potential energy.**

If you spent the day holding up a heavy anvil, it might seem as though you were doing an awful lot of work on it. However, the anvil's energy isn't changing; you're not transferring any energy to it because you're not doing any work on it. Instead, your muscles are converting useful chemical potential energy from your last meal into useless thermal energy. That thermal energy is staying in your body so it's no wonder that you're getting overheated.

To do work on an object, you have to push on it while it covers some distance in the direction you're pushing. The amount of work you do on the object depends on how hard you push and how much distance the object covers. We can express this relationship as a word equation:

$$\text{work} = \text{force} \cdot \text{distance}, \tag{1.3.1}$$

in symbols:

$$W = \mathbf{F} \cdot \mathbf{d},$$

and in everyday language:

If you're not pushing or it's not moving, then you're not working.

Note that the force and the distance traveled must be *in the same direction.*

Sometimes the force and distance traveled aren't in the same direction or are even in opposite directions. In these cases, the work you do on an object is the product of how far it travels times the component of your force along its direction of motion. If you push more or less in the direction of its motion, you do work on the object. If you push more or less in the direction opposite its motion, the amount of work you do on the object is negative, and the object is doing work on you. If you push at right angles to its motion or it doesn't move, no work is done at all.

examples of work

The fact that when you do negative work on an object, it does work on you is the reason why energy is conserved. When you lift an anvil, you push it up as it moves up and do work on it. At the same time, it pushes your hand down but your hand moves up, so it does negative work on your hand. Overall, the anvil's energy increases by exactly the same amount that your energy decreases! You are transferring energy to the anvil. When you lower the anvil, the process is reversed and it transfers energy to you.

Let's return again to the piano on the sidewalk. Pretend that you're strong enough to carry it slowly up a ladder all by yourself. You do work on the piano as you lift it because you exert an upward force on it and it moves upward. You're transferring energy to the piano, energy that originates in the food you eat, and the piano's gravitational potential energy increases. If you subsequently lower the piano, it does work on you. The piano transfers energy to you, and its gravitational potential energy decreases. While you receive energy from the piano, your body can't turn that energy back into food energy. Instead, you become hotter, as we'll discuss in Section 2.2. But despite your body's inability to reuse work done on it, it's usually easier to have work done on you than to do work on something else. That's why it's harder to lift most objects than to lower them.

> **Summary of Energy and Work**
> *Energy:* The capacity to do work.
> *Kinetic Energy:* The form of energy contained in an object's motion.
> *Potential Energy:* The form of energy stored in the forces between or within objects.
> *Work:* The mechanical means for transferring energy.

Check Your Understanding #4: Pitching

When you throw a baseball horizontally, you're not pushing against gravity. Are you doing any work on the baseball?

Check Your Figures #1: Light Work, Heavy Work

You are moving books to a new shelf, 1.2 m above the old shelf. The books weigh 10 N each and you have 10 of them to move. How much work must you do on them as you move them? Does it matter how many you move at once?

Gravitational Potential Energy

Just how much work do you do in carrying the piano straight up to the apartment? You must exert an upward force on the piano equal to its weight and push it upward from the sidewalk to the second floor. Actually, to get the piano moving, you must push a little harder at the beginning; the net force on the piano then points upward and the piano accelerates upward. Once the piano is rising, however, you only have to support its weight so that the net force on it is zero. The piano then continues to coast upward at constant velocity. Because you're pushing upward and the piano is moving upward, the work you do on the piano is the product of the piano's weight times the distance you lift it.

The piano's gravitational potential energy increases as it rises. The amount of that increase is equal to the work you do on the piano in lifting it. If we agree that the piano has zero gravitational potential energy when it rests on the sidewalk, then the suspended piano's gravitational potential energy is simply its weight times its height above the sidewalk. Since the piano's weight is equal to its mass times the acceleration due to gravity, its gravitational potential energy is its mass times the acceleration due to gravity times its height above the sidewalk.

These ideas are not limited to pianos. You can determine the gravitational potential energy of any object by multiplying its mass times the acceleration due to gravity times its height above the level at which its gravitational potential energy is zero. This relationship can be expressed as a word equation:

gravitational potential energy = mass · acceleration due to gravity · height, **(1.3.2)**

in symbols:

$$U = m \cdot g \cdot h,$$

and in common language:

The higher they were, the harder they hit.

Of course, if you know the object's weight, you can use it in place of the object's mass times the acceleration due to gravity.

So what is the piano's gravitational potential energy when it reaches the second floor? If it weighs 2000 newtons (450 pounds) and the second floor is 5 meters above the sidewalk, you will have done 10,000 newton-meters of work in lifting it up there, and the piano's gravitational potential energy will thus be 10,000 newton-meters. The **newton-meter** is the SI unit of energy and work; it's so important that it has its own name, the **joule** (abbreviated J). At the second floor, the piano's gravitational potential energy is 10,000 J (10,000 joules).

A few everyday examples should give you a feeling for how much energy a joule is. Lifting a liter bottle of water 10 centimeters (4 inches) upward requires about 1 J of work. A 100-watt light bulb needs 100 J every second to operate. Your body is able to extract about 2,000,000 J from a slice of cherry pie. When you're bicycling or rowing hard, your body can do about 1000 J of work each second. A typical flashlight battery has about 10,000 J of stored energy.

Check Your Understanding #5: Mountain Biking

Bicycling to the top of a mountain is much harder than rolling back down to the bottom. At which place do you have the most gravitational potential energy?

Check Your Figures #2: Watch Out Below

If you carry a United States penny (0.0025 kg) to the top of the Empire State Building (449 m), how much gravitational potential energy will it have?

Lifting the Piano with a Ramp

Unfortunately, you probably can't carry a grand piano up a ladder by yourself. You're going to need a ramp. What happens when you place the piano on a ramp? The ramp exerts a support force on the piano to prevent the piano from passing through its surface. However, that support force doesn't point directly upward (Fig. 1.3.4). Except for friction, which we're ignoring for now, the ramp doesn't resist motion along its surface, and it pushes the piano directly away from (or normal to) its surface. Since the ramp isn't exactly horizontal, this support force isn't exactly vertical. Because the piano's weight is directed straight down and the ramp's support force isn't directed straight up, the two forces can't balance one another. As a result, there is a nonzero net force on the piano.

The nearly upward support force of the ramp's surface balances most of the piano's downward weight. The components of force normal to the ramp's surface cancel one another, but the components of force along the ramp's surface don't cancel. Because the small net force that remains points down the ramp, the piano accelerates down the ramp. But because the net force is much smaller than the piano's weight, the acceleration down the ramp is slower than it would be if the piano were falling freely. This effect is familiar to anyone who has bicycled downhill or watched a glass slip slowly off a tilted table. Gravity still accelerates these objects, but more slowly than falling and in the direction of the downward slope.

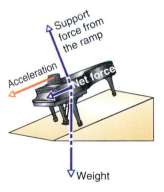

Fig. 1.3.4 A piano sliding on a frictionless ramp while experiencing a force due to gravity. It accelerates down the ramp more slowly than it would if it were falling freely.

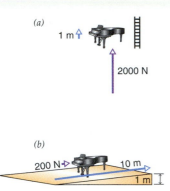

Fig. 1.3.5 To lift a piano weighing 2000 N, you can either (*a*) push it straight up or (*b*) push it along a ramp. To keep the piano moving at a constant velocity, you must make sure it experiences a net force of zero. If you lift it straight up the ladder in (a), you must exert an upward lifting force of 2000 N to balance the piano's downward weight. If you push it up the ramp shown in (*b*), you will only have to push the piano uphill with a force of 200 N in order to give the piano a net force of zero.

Herein lies the beauty of the ramp. By putting the piano on a ramp, you let the ramp supply most of the force needed to keep the piano from accelerating downward. The piano only experiences a small residual net force, which pushes it downhill along the ramp. If you now push uphill on the piano with a force that exactly balances the downhill force, the new net force on the piano is zero and the piano stops accelerating. If you push uphill a little harder, the piano will accelerate up the ramp!

How does the ramp change the job of moving the piano? Suppose that you build a ramp 50 m long that extends from the sidewalk to the apartment's balcony, 5 m above the pavement. This ramp is sloped so that traveling 10 m uphill along its surface actually lifts the piano only 1 m upward (Fig. 1.3.5). You can push the 2000-N piano up this 10 to 1 grade at constant velocity with a force of only 200 N (45 pounds): Most people can push that hard, so the moving job is now realistic. To reach the apartment, you must push the piano 50 m along this ramp with a force of 200 N so that you will do a total of 10,000 J of work.

By pushing the piano up a ramp, you've used physical principles to help you perform a task that would otherwise be nearly impossible. But you didn't get something for nothing. The ramp is much longer than the ladder, and you have had to push the piano for a longer distance in order to raise it to the second floor. Of course, you have had to push with less force.

Remarkably, the amount of work you do in either case is 10,000 J. In carrying the piano up the ladder, the force you exert is large but the distance the piano travels in the direction of that force is small. In pushing the piano up the ramp, the force is small but the distance is large. Either way, the final result is the same: the piano ends up on the second floor with an additional 10,000 J of gravitational potential energy and you have done 10,000 J of work. Expressed graphically in an equation, this relationship would appear as follows:

$$\text{work} = \text{large force} \cdot \text{small distance}$$
$$= \text{small force} \cdot \text{large distance}.$$

In the absence of friction, the amount of work you do on the piano to get it to the second floor doesn't depend on how you raise it. By doing work on the piano, you're increasing its energy, transferring energy to it from yourself. The amount of energy you transfer to the piano depends only on how its situation changes, not on how you achieve this change. No matter how you move that piano up to the second floor, you'll have to do 10,000 J of work on it and its energy will have to increase by 10,000 J. Even if you disassemble the piano into parts, carry them individually up the stairs, and reconstruct the piano in your friend's living room, you will have done 10,000 J of work lifting the piano.

Unless you're an experienced piano tuner, you'll probably be better off sticking with the ramp. It offers an easy method for one person to lift a baby grand piano. The ramp provides **mechanical advantage,** the process whereby a mechanical device redistributes the amounts of force and distance that go into performing a specific amount of mechanical work. In moving the piano with the help of the ramp, you've performed a task that would normally require a large force over a small distance by supplying a small force over a large distance. You might wonder whether the ramp itself does any work on the piano; it doesn't. Although the ramp exerts a support force on the piano and the piano moves along the ramp's surface, this force and the distance traveled are at right angles to one another. The ramp does no work on the piano.

Mechanical advantage occurs in many situations involving ramps. For example, it appears when you ride a bicycle up a hill. Climbing a gradual hill that is 500 m (1640 feet) high takes far less uphill force than climbing a steep hill of the same height. Since your pedaling ultimately provides the uphill force, it's much easier to climb the gradual hill than the steep one. Of course, you must travel a longer distance along the road as you climb the gradual hill than you do on the steep hill, so the work you do is the same in either case. (If you use a mountain bike to pedal up one of these hills, you will also obtain mechanical advantage from gears and levers, as we'll discuss later on in this book.)

Ramps and inclined planes show up in many devices, where they usually increase the forces available and allow us to perform tasks that would otherwise be difficult. They also change the character of certain activities. Skiing wouldn't be very much fun if the only slopes available were horizontal or vertical. By choosing ski slopes of various grades, you can select the net forces that set you in motion. Gentle slopes leave only small net forces and small accelerations; steep slopes produce large net forces and large accelerations. The steeper the slope, the more rapid the downhill acceleration and the larger the maximum downhill velocity. (The existence of a maximum downhill velocity is due to friction and air resistance, which we're ignoring for the moment but will return to later in this book.)

Finally, our observation about mechanical advantage is this: mechanical advantage allows you to do the same work, but you must make a trade-off—you must choose whether you want a large force or a large distance. The product of the two parts, force times distance, remains the same.

Fig. 1.3.6 An access ramp allows this woman to raise herself to the height of the door using modest forces exerted over a long distance.

Check Your Understanding #6: Access Ramps

Ramps for handicap entrances to buildings are often quite long and may even involve several sharp turns (Fig. 1.3.6). A shorter, straighter ramp would seem much more convenient. What consideration leads the engineers designing these ramps to make them so long?

Epilogue for Chapter 1

In this chapter we examined three everyday things and explored the basic physical laws that govern their behaviors. In *skating*, we looked at the concept of inertia and observed that objects accelerate only in response to forces. In *falling balls*, we introduced an important type of force—weight—and saw how weight causes otherwise independent objects to fall with equal downward accelerations.

In *ramps*, we encountered another type of force—support force. We also saw how the work done in changing an object's altitude doesn't depend on how you raise the object, since work is actually the mechanical means for transferring energy from one object to another. Energy, we noted, is one of the conserved physical quantities that govern the motion of objects in our universe. The particular type of energy we studied in *ramps* was gravitational potential energy—the potential energy associated with the force of gravity.

Explanation: Removing a Tablecloth from a Table

The dishes remain in place because of their inertia. As we've seen, an object in motion tends to remain in motion, while an object at rest tends to remain at rest.

Before you pull on the tablecloth, the dishes sit motionless on its surface and they tend to remain that way. By sliding the tablecloth off the table as quickly and smoothly as possible, you ensure that whatever force the tablecloth exerts on each dish occurs only for a very short time. As a result, the dishes undergo only the tiniest changes in velocity and remain essentially stationary on the tabletop.

Chapter Summary

How Skating Works: When you're gliding forward on frictionless skates, you experience no horizontal forces and move at a constant velocity. Inertia alone carries you forward. To change your velocity—to accelerate—something must exert a horizontal force on you. A forward force speeds you up while a backward force slows you down. Since sideways forces change your direction of travel, they, too, make you accelerate.

How Falling Balls Work: Any ball that's subject only to the force of gravity is a falling ball. It accelerates downward at a steady rate. Gravity affects only the ball's vertical motion, causing the ball's vertical component of velocity to increase steadily in the downward direction. If the ball were initially moving horizontally, it would continue that horizontal motion and drift steadily downfield as it falls.

A falling ball that's initially rising soon stops rising and begins to descend. The larger its initial upward component of velocity, the longer the ball rises and the greater its peak height. When the ball then begins to descend, the peak height determines how long it takes for the ball to reach the ground.

When you throw a ball, the vertical component of initial velocity determines how long the ball remains aloft. The horizontal component of initial velocity determines how quickly the ball moves downfield. A thrower intuitively chooses an initial speed and direction for a ball so that it moves just the right distance downfield by the time it descends to the desired height.

How Ramps Work: An object at rest on the floor experiences two forces—its downward weight and an upward support force from the floor that exactly balances that weight. The net force on the object is thus zero. But if the floor is replaced with a ramp, the support force is no longer directly upward and the net force on the object isn't zero. Instead, the net force points downhill along the ramp and is equal to the weight of the object multiplied by the ratio of the ramp rise to the ramp length. If a 10-m long ramp rises 1 m in height, then this ratio is 1 m divided by 10 m or 0.10. The net force downhill along this ramp is thus only 10% of the object's weight.

To stop an object from accelerating down a ramp, you must balance the downhill force by pushing equally hard uphill. In fact, if you exert more force up the ramp than it experiences down the ramp, the object will begin to accelerate up the ramp.

It takes less force to push an object up a ramp than to lift it directly upward, but you must push that object a longer distance along the ramp. Overall, the work you do in raising the object from one height to another remains the same as if you simply lifted it straight up. However, the ramp gives you mechanical advantage, allowing you to do work that would require an unrealistically large force by instead exerting a much smaller force for a much longer distance.

Important Laws and Equations

1. Newton's First Law of Motion: An object that is free from all outside forces travels at a constant velocity, covering equal distances in equal times along a straight-line path.

2. Newton's Second Law of Motion: An object's acceleration is equal to the force exerted on that object divided by the object's mass, or

$$\text{force} = \text{mass} \cdot \text{acceleration.} \qquad (1.1.1)$$

3. Relationship Between Mass and Weight: An object's weight is equal to its mass times the acceleration due to gravity, or

$$\text{weight} = \text{mass} \cdot \text{acceleration due to gravity.} \quad (1.2.1)$$

4. The Velocity of an Object Experiencing Constant Acceleration: The object's present velocity differs from its initial velocity by the product of its acceleration times the time since it was at that initial velocity, or

$$\begin{aligned} \text{present velocity} = {}&\text{initial velocity} \\ &+ \text{acceleration} \cdot \text{time.} \qquad (1.2.2) \end{aligned}$$

5. The Position of an Object Experiencing Constant Acceleration: The object's present position differs from its initial position by the product of its average velocity since it was at that initial position times the time since it was at that initial position, or

$$\begin{aligned} \text{present position} = {}&\text{initial position} + \text{initial velocity} \cdot \text{time} \\ &+ \tfrac{1}{2} \cdot \text{acceleration} \cdot \text{time}^2. \qquad (1.2.3) \end{aligned}$$

6. Newton's Third Law of Motion: For every force that one object exerts on a second object, there is an equal but oppositely directed force that the second object exerts on the first object.

7. The Definition of Work: The work done on an object is equal to the product of the force exerted on that object times the distance that object travels in the direction of the force, or

$$\text{work} = \text{force} \cdot \text{distance.} \qquad (1.3.1)$$

8. Gravitational Potential Energy: An object's gravitational potential energy is its mass times the acceleration due to gravity times its height above a zero level, or

$$\begin{aligned} \text{gravitational potential energy} = {}&\text{mass} \cdot \text{acceleration} \\ &\text{due to gravity} \cdot \text{height.} \quad (1.3.2) \end{aligned}$$

Check Your Understanding—Answers

Section 1.1 SKATING

1. The puck travels at constant velocity across the ice because it has inertia.

Why: A hockey puck resting on the surface of wet ice is almost completely free of horizontal influences. If someone pushes on the puck, so that it begins to travel with some horizontal velocity across the ice, inertia will ensure that the puck continues to slide.

2. The train accelerates forward when starting out from a station, backward when it arrives at the next station, to the left when it turns left, and downward when it begins its descent out of the mountains.

Why: Whenever the train changes its speed or its direction of travel, it is accelerating. When it speeds up on leaving a station, it is accelerating forward (increasing its velocity in the forward direction). When it slows down at the next station, it is accelerating backward (decreasing its velocity in the forward direction). When it turns left, it is accelerating to the left (changing its velocity more toward the left). When it begins to descend, it is accelerating downward (changing its velocity more toward the downward direction).

3. An automobile has a much greater mass than a bicycle.

Why: To stop a moving vehicle, you must exert a force on it in the direction opposite its velocity. The vehicle will then accelerate backward so that it eventually comes to rest. If the vehicle is heading toward you, you must push it away from you. The more mass the vehicle has, the less it will accelerate in response to a certain force and the longer you will have to push on it to stop it completely. While it's easy to stop a bicycle by hand, stopping even a slowly moving automobile by hand requires a large force exerted for a substantial amount of time.

4. About 2.24 miles.

Why: There are many different units in this example, so we must do some converting. First, an hour is 3600 seconds, so that in an hour of walking at 1 meter-per-second, you will have walked 3600 meters. Second, a mile is about 1609 meters so that each time you travel 1609 meters, you have traveled a mile. By walking 3600 meters, you have completed 2 miles and are about one-quarter of the way into your third mile.

Section 1.2 FALLING BALLS

1. The astronaut would have zero weight but would still have a normal mass.

Why: Weight is a measure of the force exerted on the astronaut by gravity. Far from the earth or any other large object, the astronaut would experience virtually no gravitational force and would have zero weight. But mass is a measure of inertia and doesn't depend at all on gravity. Even in deep space, it would be much harder to accelerate a school bus than to accelerate a baseball because the school bus has more mass than the baseball.

2. 4.9 m/s in the downward direction.

Why: A freely falling object accelerates downward at a steady rate. Its velocity changes by 9.8 m/s in the downward direction each and every second. In half a second, the marble's velocity changes by only half that amount or 4.9 m/s.

3. About 1.2 m.

Why: While a freely falling object's velocity changes steadily in the downward direction, its height is more complicated. When you drop the marble from rest, it starts its descent slowly but picks up speed and covers the downward distance faster and faster. In the first 0.5 s, it travels only a quarter of the distance it travels in the first 1 s, or about 1.2 m.

4. The bullet or arrow will fall in flight so she must compensate for its loss of height.

Why: To hit the bull's-eye, the sharpshooter or archer must aim above the bull's-eye because the projectile will fall in flight. The longer the bullet or arrow is in flight, the more it will fall and the higher she must aim. As the distance to the target increases, the flight time increases and her aim must move upward.

Section 1.3 RAMPS

1. Lifting the bricks all together.

Why: To lift a brick, you must exert an upward force on it that is greater than its downward weight so that the brick will begin to accelerate upward. If you must lift several bricks at once, you will have to accelerate all of them upward together. Each one will require an upward force, and the overall upward force you exert will be much larger than for a single brick.

2. 50 N.

Why: If you exert a 50-N force on any object you encounter, whether it's moving or stationary, it will exert a 50-N force back on you. If you push on a friend's hands with 50 N of force, it doesn't matter whether your friend is stationary or moving or wearing roller skates or even sound asleep: she will push back with 50 N of force. She has no choice in the matter. Similarly, if someone pushes on you, you will feel yourself pushing back. That's how Newton's third law works.

3. The two forces are the downward force of gravity (your weight) and an upward support force from the floor. They cancel, so that the net force on you is zero.

Why: Whenever anything is moving with constant velocity, it's not accelerating and thus has zero net force on it. Although the elevator is moving upward, the fact that you are not accelerating means that the car must exert an upward support force on you that exactly balances your weight. You experience zero net force.

4. Yes.

Why: Any time you exert a force on an object and the object moves in the direction of that force, you are doing work on the object. Since gravity doesn't affect horizontal motion, the work you do on the baseball as you throw it ends up in the baseball as kinetic energy (energy of motion). As anyone who has been hit by a pitch can attest, a moving baseball has more energy than a stationary baseball.

5. At the top of the mountain.

Why: Bicycling up the mountain is hard because you must do work against the force of gravity. This work is stored as an increasing gravitational potential energy on your way uphill. Gravity then does work on you as you roll downhill and your gravitational potential energy decreases.

6. The engineers must limit the amount of force needed to propel a wheelchair steadily up the ramp. The steeper the ramp, the more force is required.

Why: A person traveling in a wheelchair on a level surface experiences little horizontal force and can move at constant velocity with very little effort. But climbing a ramp at constant velocity requires a substantial uphill force equal in magnitude to the downhill force from gravity. The steeper the ramp, the more uphill force is needed to maintain constant velocity. A 12 to 1 grade (12 meters of ramp surface for each meter of rise in height) is the accepted limit to how steep such a long ramp can be.

Check Your Figures—Answers

Section 1.1 SKATING

1. The less massive ball will accelerate twice as rapidly.

Why: You can rearrange Eq. 1.1.1 to show that an object's acceleration is inversely proportional to its mass:

$$\text{acceleration} = \frac{\text{force}}{\text{mass}}.$$

If you push on both bowling balls with equal forces, then their accelerations will only depend on their masses. Doubling the mass on the right side of this equation halves the acceleration on the left side. That means that the more massive ball will accelerate only half as quickly as the other ball.

Section 1.2 FALLING BALLS

1. About 1.6 m/s².

Why: You can rearrange Eq. 1.2.1 to show that the acceleration due to gravity is proportional to an object's weight:

$$\text{acceleration due to gravity} = \frac{\text{weight}}{\text{mass}}.$$

Your mass doesn't change in going to the moon, so any change in your weight must be due to a change in the acceleration due to gravity. Since your moon weight is one-sixth of your earth weight, the moon's acceleration due to gravity must be one-sixth that of the earth or about 1.6 m/s².

2. About 14 m/s (50 km/h or 31 mph).

Why: The downward acceleration due to gravity is 9.8 m/s². You fall for 1.4 s, during which time your velocity increases steadily in the downward direction. Since you start with zero velocity, Eq. 1.2.2 gives a final velocity of

$$\text{final velocity} = 9.8 \text{ m/s}^2 \cdot 1.4 \text{ s} = 13.72 \text{ m/s}.$$

Since the time of the fall is only given to two digits of accuracy (1.4 s could really be 1.403 s or 1.385 s), we shouldn't claim that our calculated final velocity is accurate to 4 digits. We should round the value to 14 m/s.

3. About 122 meters (402 feet or a 40-story building).

Why: As they fall, the jumpers will travel downward at ever increasing speeds. Since the jumpers start from rest and fall *downward* for 5 seconds, we can use Eq. 1.2.3 to determine how far they will fall:

$$\begin{aligned}\text{final height} &= \text{initial height} - \tfrac{1}{2} \cdot 9.8 \text{ m/s}^2 \cdot (5 \text{ s})^2 \\ &= \text{initial height} - 122.5 \text{ m}.\end{aligned}$$

The downward acceleration is indicated here by the negative change in height. At the end of 5 seconds, the jumpers will have fallen more than 122 m and will be traveling downward at about 50 m/s. The tower will need additional height to slow the jumpers down and begin bouncing them back upward. Clearly, a 5-second free-fall is pretty unrealistic. Try for a 2- or 3-second free-fall instead.

Section 1.3 RAMPS

1. It takes 120 J, no matter how many you lift at once.

Why: To keep each book from accelerating downward, you must support its weight with an upward force of 10 N. You must then move it upward 1.2 m. The work you do pushing upward on the book as it moves upward is given by Eq. 1.3.1:

$$\text{work} = \text{force} \cdot \text{distance} = 10 \text{ N} \cdot 1.2 \text{ m} = 12 \text{ J}.$$

It takes 12 J of work to lift each book, whether you lift it together with other books or all by itself. The total work you must do on all 10 books is 120 J.

2. About 11 J.

Why: The penny's gravitational potential energy is given by Eq. 1.3.2:

$$\begin{aligned}\text{gravitational potential energy} &= 0.0025 \text{ kg} \cdot 9.8 \text{ N/kg} \cdot 449 \text{ m} \\ &= 11 \text{ N} \cdot \text{m} = 11 \text{ J}\end{aligned}$$

This 11-J increase in energy would be quite evident if you were to drop the penny. In principle, the penny could accelerate to very high speed (up to 340 km/h or 210 mph) and do lots of damage when it hit the ground. Fortunately, a falling penny actually tumbles and is slowed so much by the air that it isn't particularly dangerous.

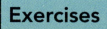

Exercises

1. A dolphin can leap several meters above the ocean's surface. Why doesn't gravity stop the dolphin from leaving the water?

2. As you jump across a small stream, does a horizontal force keep you moving forward? If so, what is that force?

3. Why does stamping your feet clean the snow off them?

4. Why does tapping your toothbrush on the sink dry it off?

5. The back of your car seat has a headrest to protect your neck during a collision. What type of collision causes your head to press against the headrest?

6. An unseatbelted driver can be injured by the steering wheel during a head-on collision. Why does the driver hit the steering wheel when the car suddenly comes to a stop?

7. Why do loose objects on the dashboard slide to the right when the car turns suddenly to the left?

8. Why is your velocity continuously changing as you ride on a carousel?

9. When you apply the brakes on your bicycle, which way do you accelerate?

10. One type of home coffee grinder has a small blade that rotates very rapidly and cuts the coffee beans into powder. Nothing prevents the coffee beans from moving so why don't they get out of the way when the blade begins to push on them?

11. A blacksmith usually hammers hot metal on the surface of a massive steel anvil. Why is this more effective than hammering the hot metal on the surface of a thin steel plate?

12. A sprinter who is running a 200-m race travels the second 100 m in much less time than the first 100 m. Why?

13. If you pull slowly on the top sheet of a pad of paper, the whole pad will move. But if you yank suddenly on that sheet, it will tear away from the pad. What causes these different behaviors?

14. A ball falls from rest for 5 seconds. Neglecting air resistance, during which of the 5 seconds does the ball's speed increase most?

15. If you drop a ball from a height of 4.9 m, it will hit the ground 1 s later. If you fire a bullet exactly horizontally from a height of 4.9 m, it will also hit the ground 1 s later. Explain.

16. An acorn falls from a branch located 9.8 m above the ground. After 1 second of falling, the acorn's velocity will be 9.8 m/s downward. Why hasn't the acorn hit the ground?

17. A diver leaps from a 50-m cliff into the water below. The cliff is not perfectly vertical so the diver must travel forward several meters in order to avoid the rocks be-

neath him. In fact, he leaps directly forward rather than upward. Explain why a forward leap allows him to miss the rocks.

18. The kicker in a sporting event isn't always concerned with how far downfield the ball travels. Sometimes the ball's flight time is more important. If he wants to keep the ball in the air as long as possible, which way should he kick it?

19. The heads of different golf clubs are angled to give the golf ball different initial velocities. The golf ball's speed remains almost constant, but the angle changes with the different clubs. Neglecting any air effects, how does changing the initial angle of the ball affect the distance the ball travels?

20. A speedboat is pulling a water-skier with a rope, exerting a large forward force on her. The skier is traveling forward in a straight line at constant speed. What is the net force she experiences?

21. Your suitcase weighs 50 N. As you ride up an escalator toward the second floor, carrying that suitcase, you are traveling at a constant velocity. How much upward force must you exert on the suitcase to keep it moving with you?

22. What is the net force on (a) the first car, (b) the middle car, and (c) the last car of a metro train traveling at constant velocity?

23. Two teams are having a tug-of-war with a sturdy rope. It has been an even match so far, with neither team moving. What is the net force on the left team?

24. When you kick a soccer ball, which pushes on the other harder: your foot or the soccer ball?

25. The earth exerts a downward force of 850 N on a veteran astronaut as he works outside the space shuttle. What force (if any) does the astronaut exert on the earth?

26. A car passes by, heading to your left, and you reach out and push it toward the left with a force of 50 N. Does this moving car push on you and, if so, with what force?

27. Which is larger: the force the earth exerts on you or the force you exert on the earth?

28. Comic book superheroes often catch a falling person only a hairsbreadth from the ground. Why would this rescue actually be just as fatal for the victim as hitting the ground itself?

29. You accidentally miss the doorway and run into the wall. You suddenly experience a backward force that is several times larger than your weight. What's the origin of this force?

30. When you fly a kite, there is a time when you must do (positive) work on the kite. Is that time when you let the kite out or when you pull it in?

31. Which does more work in lifting a grain of rice over its head: an ant or a person? Use this result to explain how insects can perform seemingly incredible feats of lifting and jumping.

32. Are you doing work while kneading bread? If so, when?

33. While hanging a picture, you accidentally dent the wall with a hammer. Did the hammer do work on the wall?

34. You're cutting wood with a handsaw. You have to push the saw away from you as it moves away from you and pull the saw toward you as it moves toward you. When are you doing work on the saw?

35. The steel ball in a pinball game rolls around a flat, tilted surface. If you flick the ball straight uphill, it gradually slows to a stop and then begins to roll downhill. Which way is the ball accelerating as it rolls uphill? downhill?

36. Why do less snow and other debris accumulate on a steep roof than on a flatter roof?

37. You roll a marble down a playground slide that starts level, then curves downward, and finally curves very gradually upward so that it's level again at the end. Where along its travel does the marble experience its greatest acceleration? its greatest speed?

38. When you're roller skating on level pavement, you can maintain your speed for a long time. But as soon as you start up a gradual hill, you begin to slow down. What slows you?

39. When the brakes on his truck fail, the driver steers it up a runaway truck ramp. As the truck rolls up the ramp, it slows to a stop. What happens to the truck's kinetic energy, its energy of motion?

Problems

1. If your car has a mass of 800 kg, how much force is required to accelerate it forward at 4 m/s^2?

2. If your car accelerates from rest at a steady rate of 4 m/s^2, how soon will it reach 88.5 km/h (55.0 mph or 24.6 m/s)?

3. On Mars, the acceleration due to gravity is 3.71 m/s^2. What would a rock's velocity be 3 s after you dropped it on Mars?

4. How far would a rock fall in 3 s if you dropped it on Mars? (See Problem 3.)

5. How would your Mars weight compare to your earth weight? (See Problem 3.)

6. A basketball player can leap upward 0.5 m. What is his initial velocity at the start of the leap?

7. How long does the basketball player in Problem 6 remain in the air?

8. A sprinter can reach a speed of 10 m/s in 1 s. If the sprinter's acceleration is constant during that time, what is the sprinter's acceleration?

9. If a sprinter's mass is 60 kg, how much forward force must be exerted on the sprinter to make the sprinter accelerate at 0.8 m/s^2?

10. How much does a 60-kg person weigh on earth?

11. If you jump upward with a speed of 2 m/s, how long will it take before you stop rising?

12. How high will you be when you stop rising in Problem 11?

13. How much force must a locomotive exert on a 12,000-kg boxcar to make it accelerate forward at 0.4 m/s^2?

14. How long will it take the boxcar in Problem 13 to reach its cruising speed of 100 km/h (62 mph or 28 m/s)?

15. The builders of the pyramids used a long ramp to lift 20,000-kg (22-ton) blocks. If a block rose 1 m in height while traveling 20 m along the ramp's surface, how much uphill force was needed to push it up the ramp at constant velocity?

16. How much work was done in raising one of the blocks in Problem 15 to a height of 50 m?

17. What is the gravitational potential energy of one of the blocks in Problem 15 if it's now 75 m above the ground?

18. As water descends from the top of a tall hydroelectric dam, its gravitational potential energy is converted to electric energy. How much gravitational potential energy is released when 1000 kg of water descends 200 m to the generators?

19. The tire of your bicycle needs air so you attach a bicycle pump to it and begin to push down on the pump's handle. If you exert a downward force of 25 N on the handle and the handle moves downward 0.5 m, how much work do you do?

20. You're using a wedge to split a log. You are hitting the wedge with a large hammer to drive it into the log. It takes a force of 2000 N to push the wedge into the wood. If the wedge moves 0.2 m into the log, how much work have you done on the wedge?

21. The wedge in Problem 20 acts like a ramp, slowly splitting the wood apart as it enters the log. The work you do on

the wedge, pushing it into the log, is the work it does on the wood, separating its two halves. If the two halves of the log only separate by a distance of 0.05 m while the wedge travels 0.2 m into the log, how much force is the wedge exerting on the two halves of the log to rip them apart?

22. You're sanding a table. You must exert a force of 30 N on the sandpaper to keep it moving steadily across the table's surface. You slide the paper back and forth for 20 minutes, during which time you move it 1000 m. How much work have you done?

Cases

1. You're riding on a playground swing. You're traveling back and forth once every few seconds.
 a. At what point(s) in your motion is your velocity zero?
 b. At what point(s) in your motion is your gravitational potential energy at its maximum?
 c. At what point(s) in your motion is your kinetic energy at its maximum?
 d. As you reach the bottom of a swing, when the swing's ropes are exactly vertical, are you accelerating?
 e. At the moment described in part d, is the force that the swing seat exerts on you more, less, or equal to your weight?

2. Diving boards and platforms offer a nearly ideal opportunity in which to experience the various laws of motion. When you jump off the high diving board, you are a falling object and, if you can keep your presence of mind as you fall, you can learn something about physics. Imagine yourself diving off a platform 10 m above the water below.
 a. If you walk very slowly off the platform, so that you fall directly downward, roughly how long will it take for you to reach the water? Look at Fig. 1.2.2 and make a reasonable estimate.
 b. In the situation described in part a, about how fast will you be traveling downward when you reach the water? Estimate your velocity from Fig. 1.2.2.
 c. If you jump upward as you leave the platform, so that you begin with a modest upward velocity, will your downward velocity when you hit the water be more or less than in part b?
 d. You leave the platform simultaneously with a friend. She walks slowly off the platform and you jump to give yourself a modest initial upward velocity. Who will reach the water first?
 e. You leave the platform simultaneously with a friend. He walks slowly off the platform and you run off the

platform, so that your initial velocity is in the horizontal direction. Who will reach the water first?
 f. In the situation described in part e, who hits the water with the largest speed or are your speeds equal?

3. An escalator is essentially a moving staircase. The individual steps are supported by metal tracks that run on either side of the escalator. These steps follow one another in a complete loop, driven by an electric motor. When you step onto an escalator at the ground floor, it soon begins to carry you upward and forward at a constant velocity toward the second floor.
 a. While you're moving toward the second floor at a constant velocity, what is the net force exerted on you by all outside forces? (Specify the amount and the direction of the net force.)
 b. You know that gravity gives you a weight in the downward direction. What force does the escalator exert on you as you move toward the second floor at a constant velocity? (Specify the amount and the direction of the force.)
 c. Is the escalator doing work on you as you move toward the second floor?
 d. As you first step onto the escalator, you begin to accelerate toward the second floor. Is the net force exerted on you by all outside forces the same as in part a?
 e. If the rapidly moving escalator suddenly stopped moving, you would be thrown forward and might even fall over. What causes you to be thrown forward?
 f. You have more energy when you reach the second floor than you had on the first floor. Why aren't you moving faster as a result?

4. Imagine that you're sledding alone down a steep hill on a toboggan and that you left the top of the hill at the same time as an identical toboggan, loaded with six people.
 a. Neglecting the effects of air resistance and friction, which toboggan will reach the bottom of the hill first?

b. During the descent, your toboggan brushes up against the six-person toboggan. Which toboggan will experience the largest change in velocity as the result of the impact?

c. If you were to take a steeper route down the hill, how would that affect the speed of your descent? Explain.

d. Before each downhill run, you must pull the toboggan back to the top of the hill. Explain how the toboggan's gravitational potential energy changes on the way up the hill and on the way down it.

e. When are you doing (positive) work on the toboggan?

f. When is gravity doing (positive) work on the toboggan?

5. You're a pilot for the Navy. For your airplane to be able to lift itself off the ground, it must be traveling forward with a speed of 208 km/h (130 mph). At this takeoff speed your airplane will have about 50,000,000 N·m (or 50,000,000 J) of kinetic energy.

*a. During takeoff, your airplane's jet engine exerts a force of 250,000 N in the backward direction on the air leaving the engine. What force does that same air exert on the airplane? (Specify the amount and the direction of force.)

*b. The force exerted by the air on the airplane causes it to accelerate down the runway. How long must the runway be for the airplane to reach its takeoff speed?

*c. An aircraft carrier runway is only about 100 m long. As your answer to part b indicates, this distance is not enough for the airplane to reach takeoff speed on its own. The aircraft carrier must assist the airplane by exerting an extra force on it. The aircraft carrier uses a steam-powered catapult to help push the airplane forward. How much additional force must the catapult exert on the airplane to bring the airplane to takeoff speed at the end of the 100-m runway?

d. During an aircraft carrier takeoff, the airplane and the catapult exert forces on one another. Which of these two objects does (positive) work on the other, and which object transfers some of its energy to the other?

e. During an aircraft carrier landing, the airplane hooks onto a cable that slows the airplane to a stop. The airplane and cable exert forces on one another. Which of these two objects does (positive) work on the other, and which object transfers some of its energy to the other?

6. You're about to go on a bicycle trip through the mountains. Being ambitious, you decide to take two children along. The children sit in a trailer that you pull with your bicycle.

a. As you wait to begin your trip, you and your bicycle are motionless. What is the net force on your body?

b. You begin to bicycle into the mountains. You soon find yourself ascending a steep grade. The road rises smoothly uphill and you're traveling up it at a steady pace in a straight line. You're traveling at a constant velocity up the hill. What is the net force on your body?

*c. The road rises 1 m upward for every 10 m you travel along its surface. If the children and their trailer weigh 400 N, how much uphill force must you exert on the trailer to keep it moving uphill at a constant velocity?

*d. When you reach the top of the hill, your altitude has increased by 500 m. How much work did you do on the trailer and children as you pulled them up this hill? Does the amount of work you did on them depend on whether you took the long, gradually sloping road or the short, steep road? (Answer both questions.)

*e. How much work must you do on the trailer and children as you pull them back down the 500-m high hill at constant velocity?

7. You have recently taken up track and field as a way to keep in shape. You soon begin to notice how simple physical laws appear in many of the events.

a. You notice that great sprinters have extremely strong legs. Why is it so important that a sprinter be able to push back hard on the starting blocks at the beginning of a race?

b. You find that throwing a heavy metal shot is far more difficult than throwing a baseball. Weight isn't the whole problem. Even if you try to throw the shot horizontally or downward, so that weight is not an issue, you have great difficulty getting the shot to move quickly. Why?

c. As you land on the soft foam pad beneath the pole vault, you realize that its job is to bring you to rest by accelerating you upward gradually with only modest support forces. If there were no pad there, only concrete, what would the acceleration and support forces be like during your landing?

d. You cross the finish line at the end of a race. The net force on your body points in what direction as you slow down?

e. In the long jump, you run rapidly down a path and then leap into the air. You find that the best distance comes from pushing yourself upward rather than forward during the leap. Why is it so important to have a large upward component of velocity at the start of the leap?

C H A P T E R **2**

The Laws of Motion Part II

In the first chapter, we learned about how things move from place to place and began to develop an understanding of energy, an important conserved quantity. But motion doesn't always involve a change of position, and energy isn't the only conserved quantity that exists in nature. In this chapter, we'll take a look at a second type of motion—rotation—and at two other conserved quantities—momentum and angular momentum. Spinning objects are quite common, and we'll find it useful to understand their laws of motion before proceeding much further. Once we get those additional concepts under our belts, we'll be ready to explore the physics behind a broad assortment of mechanical objects.

EXPERIMENT: A Spinning Pie Dish

Spinning a dish on the top of a narrow post seems like a simple activity. But don't let its uncomplicated appearance deceive you: there is lots of physics involved in keeping the dish turning, in gradually slowing it down, and in preventing it from falling off the post.

An easy way to experiment with a spinning dish is to tape a pencil vertically to the edge of a table or chair so that its eraser projects several inches upward into the air. To avoid wobbling problems, the tape should hold the pencil rigidly in place, and the table or chair should be sturdy and stable.

Now prepare to balance a metal pie dish on the eraser before giving it a twist. If you don't have a pie dish, you can use a Frisbee, a deep plastic plate, or a shallow plastic bowl instead. Be creative. You probably have something that will work; just don't use grandma's heirloom porcelain unless you're willing to face the possible consequences.

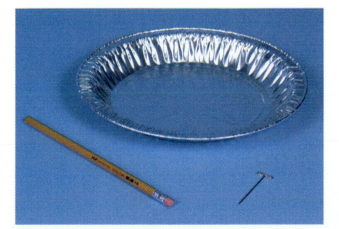

The first thing you'll need to do is to balance the dish on the eraser. Try to *predict* whether this task will be easy or hard. Will it matter whether the dish opens up or down? *Observe* what happens when you set the dish on the eraser. Did you *verify* your predictions?

Find an arrangement in which the dish balances well and then give the dish a gentle spin. What sort of influence do you have to exert on the dish to start it rotating? If the dish doesn't wobble and the pencil remains stationary, the dish should spin for a while before coming to a stop. Once you're no longer touching the dish, what keeps it turning? On the other hand, why doesn't it keep turning forever?

Now use a wire cutter or heavy scissors to cut the head off a pin or to shorten a sewing needle. The goal is to make a 1.25-cm (0.5-inch) long post with a needle-sharp point. Insert the dull end of this post into the pencil eraser, using a coin or a metal spoon handle to push it in so that you don't pierce your finger. Be careful not to dull the point. When you're done, its sharp tip should stick up from the eraser a small fraction of an inch.

What will happen when you place the dish on this tiny point? Will the dish still balance? When you spin the dish, will it turn for a longer or shorter time than on the bare eraser? Try to *measure* how changing the pivot on which the dish spins affects the duration of that spin. If the dish is soft and the point digs into it, protect the dish's bottom by taping a coin to it. How does this improved pivot affect the dish's rotation? Can you prolong the spin by taping weights around the outer edge of the dish? Is there a way to get the dish spinning just by blowing on it?

Chapter Itinerary

We're going to explore the laws of rotational motion and two new conserved quantities in the context of three everyday objects: (1) *seesaws*, (2) *wheels*, and (3) *bumper cars*. In *seesaws*, we'll look at twists and turns, and see how two children manage to rock a seesaw back and forth. In *wheels*, we'll examine how friction affects motion and learn how wheels make a vehicle more mobile. In *bumper cars*, we'll learn the physics behind collisions and how three conserved quantities make it easier to understand otherwise complicated motions.

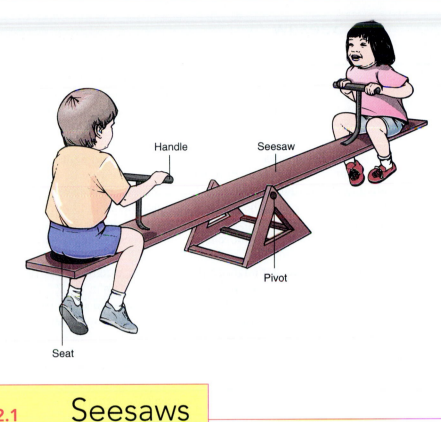

Handle   Seesaw

Pivot

Seat

SECTION 2.1 Seesaws

The ramp that we examined in Section 1.3 is only one tool that provides mechanical advantage. In this section, we'll look at another such device: the type of lever known as a seesaw. As we discuss seesaws, we'll revisit many of the laws of motion that we encountered in the previous chapter. However, we'll see these laws in a new context: rotational motion.

Questions to Think About: *A playground seesaw only balances when the children riding it are properly situated. What do we mean by a* balanced *seesaw? Why does it matter just where the children sit on the board? What are they doing to make the balanced seesaw rock back and forth? Who is doing work on whom as the seesaw rocks back and forth?*

Experiments to Do: *To get a feel for how levers work, find a rigid ruler with a hole in its center—the kind that can be clipped into a three-ring notebook. If you support the ruler by putting the tip of an upright pencil into the central hole, you'll find that the ruler balances; that is, it either remains stationary, at whatever orientation you choose, or rotates steadily about the central hole. (Eventually, the ruler comes to rest because of friction, a detail that we'll continue to ignore for now.) Now, push on one end of the ruler. What happens? Try pushing the ruler's end toward its central hole. What happens then? What is the most effective way to make the ruler spin?*

Now lay the pencil on a table and place the ruler flat on top of it so that the pencil and the ruler are at right angles, or perpendicular, to each other. If you center the ruler on the pencil, the ruler will balance. Load the two ends of the ruler with coins or other small weights, trying as you do to keep the ruler balanced. Try placing the coins at different positions relative to the pencil. Is there any way you can balance a light weight on one end with a heavy weight on the other end?

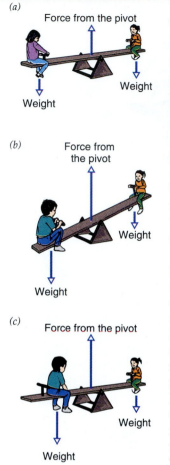

(a)

Force from the pivot

Weight

Weight

(b)

Force from
the pivot

Weight

Weight

(c)

Force from the pivot

Weight

Weight

Fig. 2.1.1 (*a*) When two
children of equal weight sit
at opposite ends of a seesaw, it balances. (*b*) When
their weights are not equal,
the heavy child descends.
(*c*) If the heavy child
moves closer to the pivot,
the seesaw can balance.

The Seesaw

Any child who has played on a seesaw with friends of different sizes knows that the toy works best for two children of roughly the same weight (Fig. 2.1.1*a*). Evenly matched riders balance each other, and this balance allows them to rock back and forth easily. In contrast, when a light child tries to play seesaw with a heavy child, the heavy child's side of the seesaw drops rapidly and hits the ground with a thud (Fig. 2.1.1*b*). The light child is tossed into the air.

There are several solutions to the heavy child/light child problem. Of course, two light children could try to balance one heavy child. But most children eventually figure out that if the heavy child sits closer to the seesaw's pivot, the board will balance (Fig. 2.1.1*c*). The children can then make the board tip back and forth easily, just as it does when two evenly matched children ride at its ends. This is a pretty useful trick, and we'll explore it later in this section. First, though, we'll need to look carefully at the nature of rotational motion.

For simplicity, let's ignore the mass and weight of the board itself. There are then only three forces acting on the seesaw shown in Fig. 2.1.1: two downward forces (the weights of the two children) and one upward force (the support force of the central pivot). Seeing those three forces, we may immediately think about net forces and begin to look for some overall acceleration of this toy and its riders. But we know that the seesaw remains where it is in the playground and isn't likely to head off for Kalamazoo or the center of the earth anytime soon. Because the seesaw's fixed pivot always provides just enough upward or sideways force to keep the board from accelerating as a whole, the board always experiences zero net force and never leaves the playground. Overall movement of an object from one place to another is called **translational motion.** While the board never experiences this kind of motion, it can turn around the pivot, and thus it experiences a different kind of motion. Motion around a fixed point (which prevents translation) is called **rotational motion.** The hands of a clock experience rotational motion as they go around in a circle.

Rotational motion is what makes a seesaw interesting. The whole point of a seesaw is that it can rotate so that one child rises and the other descends. (You may not think of going up and down as rotating, but if the ground weren't there, the seesaw would be able to rotate in a big circle.) What causes the seesaw to rotate, and what observations can we make about the process of rotation?

To answer those questions, we'll need to examine several new physical quantities associated with rotation and explore the laws of rotational motion that relate them to one another. We'll do these things both by studying the workings of seesaws and other rotating objects and by looking for analogies between translational motion and rotational motion.

Imagine holding onto the seesaw in Fig. 2.1.1*a* to keep it level for a moment while the child on the left climbs off the board. Now imagine letting go of the board. As soon as you let go, the board will begin to rotate, and the child on the right will descend toward the ground. The board's motion will be fairly slow at first, but it will move more and more quickly until that child strikes the ground with a teeth-rattling thump.

If we focus only on the rotation itself, we might describe the motion of the seesaw board in the following way:

> "The board starts out not rotating at all. When we release the board, it begins to rotate clockwise. The board's rate of rotation increases steadily in the clockwise direction until the moment the board strikes the ground."

This description sounds a lot like the description of a falling ball released from rest:

> "The ball starts out not moving at all. When we release the ball, it begins to move downward. The ball's rate of translation increases steadily in the downward direction until the moment the ball strikes the ground."

The statement about the seesaw involves rotational motion, while the statement about the ball involves translational motion. The similarities between these two descriptions are not coincidental; they're similar because the concepts and laws of rotational motion have many analogies in the concepts and laws of translational motion. The familiarity that we've acquired with translational motion will help us examine rotational motion.

Check Your Understanding #1: Wheel of Fortune Cookies

The guests at a large table in a Chinese restaurant use a revolving tray, a lazy Susan, to share the food dishes. How does the lazy Susan's motion differ from that of the passing dessert cart?

The Motion of an Isolated Seesaw

In the previous chapter, we looked at the concept of translational inertia, which holds that a body in motion tends to stay in motion and a body at rest tends to stay at rest. This concept led us to Newton's first law of translational motion. Inserting the word "translational" here is a useful revision because we're about to encounter analogous concepts associated with rotational motion. First we'll continue our examination of the rotational motion of an isolated seesaw board; once we've seen how such an isolated seesaw rotates, we can begin to explore how it responds to outside influences such as the pivot or a handful of young riders. Because of the similarities between rotational and translational motion, this section will closely parallel our earlier examination of skating and falling balls.

Imagine an unoccupied seesaw somewhere out in space with nothing pushing on it or twisting it (Fig. 2.1.2). Such an isolated seesaw is free to turn in any direction, even completely upside down. You, the observer, are located near the seesaw and have nothing pushing on you or twisting you. When you look over at the seesaw, what does it do?

If the seesaw is stationary, then it will remain stationary. However, if it's rotating, it will continue rotating at a steady pace, about a fixed line in space. What keeps the seesaw rotating? Its **rotational inertia.** A body that's rotating tends to remain rotating; a body that's not rotating tends to remain not rotating. That's how our universe works.

To describe the seesaw's rotational inertia and rotational motion more accurately, we'll need to identify several physical quantities associated with rotational motion. The first is the seesaw's orientation. At any particular moment, the seesaw is oriented in a certain way—that is, it has a particular angular position. **Angular position** describes the seesaw's orientation relative to some reference orientation; it can be specified by determining how far the seesaw has rotated away from its reference orientation and the pivot line about which that rotation has occurred. The seesaw's angular position points along the pivot line and has a magnitude equal to the amount of rotation. For example, to describe the seesaw's current angular position, we could define the reference orientation as horizontal, from left to

Fig. 2.1.2 A seesaw that is all alone. Since nothing twists it, the seesaw rotates steadily about a fixed line in space.

examples of rotational inertia

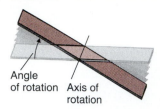

Fig. 2.1.3 You can specify this seesaw's angular position, relative to its horizontal reference orientation, as the axis about which it was rotated to reach its new orientation and the angle through which it was rotated.

right, and then figure out what line we would have to pivot the seesaw about and how far we would have to turn it to reach its current orientation (Fig. 2.1.3). The SI unit of angular position is the **radian,** the natural unit for angles. Since the radian is a natural unit, it is special and can be omitted from calculations and derived units.

If the seesaw is rotating, then its angular position is changing; in other words, it has an angular velocity. **Angular velocity** is our first important vector quantity of rotational motion and measures how quickly the seesaw's angular position changes; it consists of the angular speed at which the seesaw is rotating and the axis or line about which that rotation proceeds. The seesaw's **angular speed** is its change in angle divided by the time required for that change:

$$\text{angular speed} = \frac{\text{change in angle}}{\text{time}}.$$

The SI unit of angular velocity is the **radian-per-second** (abbreviated 1/s).

The seesaw's **axis of rotation** is the line in space about which the seesaw is rotating. But just knowing that line isn't quite enough: is the seesaw rotating about it clockwise or counterclockwise?

To resolve this ambiguity, we take advantage of the fact that any line has two directions to it. Once we have identified the line about which the seesaw is rotating, we can look down that line at the seesaw from either direction. From one direction, the seesaw appears to be rotating clockwise; from the other direction, it appears to be rotating counterclockwise. By convention, we choose the direction in which the seesaw appears to be rotating clockwise and say that the seesaw's axis of rotation points away from our eye toward the seesaw. This convention is called the **right-hand rule** because if the fingers of your right hand are curling around the axis in the way the seesaw is rotating, then your thumb is pointing along the seesaw's axis of rotation (Fig. 2.1.4).

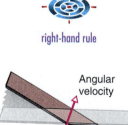

right-hand rule

Fig. 2.1.4 This seesaw is spinning about the rotation axis shown. The direction of the seesaw's angular velocity is defined by the right-hand rule.

Remembering this convention isn't as important as understanding why we must specify the direction about which rotation occurs when describing a rotating object's angular velocity. Just as translational velocity consists of a translational speed and a direction in which the translational motion occurs, so angular velocity consists of a rotational speed and a direction about which the rotational motion occurs.

We're now prepared to describe the rotational motion of an isolated seesaw. Because of its isolation and rotational inertia, its angular velocity is constant. The isolated seesaw just keeps on turning and turning, always at the same angular speed, always about the same axis of rotation.

As you might suspect, this observation isn't unique to seesaws. It is **Newton's first law of rotational motion,** which states that a rigid object that is not wobbling and is not subject to any outside influences rotates at a constant angular velocity, turning equal amounts in equal times about a fixed axis of rotation. The outside influences referred to in this law are called **torques**—a technical term for twists and spins. When you twist off the lid of a jar or spin a top with your fingers, you're exerting a torque. The word "rigid" appears in this law because this law doesn't apply to an object that can change its shape as it rotates.

The phrase "not wobbling" appears in the law because the complicated motions of wobbling objects are governed by a more general principle that we'll learn in the next section. As long as an object is symmetric about its axis of rotation—like a ball or a seesaw board turning end over end—it won't wobble and will obey Newton's first law of rotational motion.

> **Newton's First Law of Rotational Motion**
> A rigid object that is not wobbling and is not subject to any outside torques rotates at a constant angular velocity, turning equal amounts in equal times about a fixed axis of rotation.

Check Your Understanding #2: Going for a Spin
A rubber basketball floats in a swimming pool. It experiences zero torque, no matter which end of it is up. If you spin the basketball and then let go, how will it move?

The Seesaw's Center of Mass

Although you've never seen an isolated seesaw, you've seen other objects that are effectively isolated and nearly free from torques: a baton thrown overhead by a baton twirler, for example, or a juggler's club whirling through the air between two clowns. These motions, however, are complicated because the objects rotate and translate at the same time. The spinning baton travels up and down, and the turning club arcs through the air. How can we distinguish their translational motions from their rotational motions?

Once again, we can make use of a wonderful simplification of physics. There's a special point in or near an isolated object about which all its mass is evenly balanced and about which it naturally spins—its **center of mass.** The axis of rotation always passes right through this point so that, as the object rotates, the center of mass doesn't move unless the object has an overall translational velocity. The center of mass of a typical ball is at its geometrical center, while the center of mass of a less symmetrical object depends on how the mass of that object is distributed. You can begin to find a small object's center of mass by spinning it on a smooth table and looking for the fixed point about which it spins (Fig. 2.1.5).

Center of mass allows us to separate an object's translational motion from its rotational motion. As a juggler's club arcs through space, its center of mass follows the simple path we discussed in Section 1.2 on falling balls (Fig. 2.1.6). At the same time, the club's rotational motion about its center of mass is that of an isolated object: if it's not wobbling, it rotates with a constant angular velocity.

While the seesaw is hardly an isolated object, we'll examine many effectively isolated objects in the course of this book, and it's worth remembering that their translational and rotational motions can be separated if we pay attention to their centers of mass. Even in the case of the seesaw, the pivot is strategically located at or very near the board's center of mass. As a result, the pivot prevents any translational motion of the seesaw while permitting nearly free rotational motion of the board about its center of mass, at least about one axis.

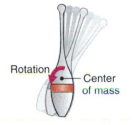

Fig. 2.1.5 This club spins about its center of mass, which remains stationary.

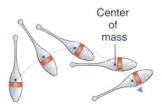

Fig. 2.1.6 A juggler's club that is traveling through space rotates about its center of mass as its center of mass travels in the simple arc associated with a falling object.

Check Your Understanding #3: Tracking the High Dive
When a diver does a rigid, open somersault off a high diving board, his motion appears quite complicated. Can this motion be described simply? How?

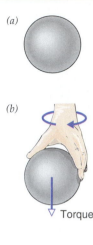

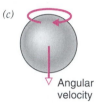

(a)

(b)

Torque

(c)

Angular
velocity

Fig. 2.1.7 If you start
with a ball that's not spinning (*a*), and twist it with a
torque (*b*), the ball will acquire an angular velocity
(*c*) that's in the same direction as that torque.

Fig. 2.1.8 Spinning a
merry-go-round is hard
work because of its large
moment of inertia. Despite
the large torque exerted by
this boy, the merry-go-
round's angular velocity
increases slowly.

How the Seesaw Responds to Torques

Why can't the isolated seesaw board change its rotational speed or axis of rotation? Because it has a moment of inertia. **Moment of inertia** is the measure of an object's *rotational* inertia, its resistance to changes in its *angular* velocity. Moment of inertia is analogous to mass, which is the measure of an object's *translational* inertia, its resistance to changes in its *translational* velocity. An object's moment of inertia depends both on its mass and on how that mass is distributed within the object. The SI unit of moment of inertia is the **kilogram-meter^2** (abbreviated kg·m^2). Because the seesaw has a moment of inertia, its angular velocity will change only if something twists or spins it. In other words, it must experience a torque.

Torque—our second important vector quantity of rotational motion—has both a magnitude and a direction. The more torque you exert on the seesaw, the more rapidly its angular velocity changes. Depending on the direction of the torque, you can make the seesaw turn more rapidly or less rapidly or even rotate about a different axis. But how do you determine the direction of a particular torque? One way is to imagine exerting this torque on a stationary ball floating in water (Fig. 2.1.7*a,b*). The ball will begin to rotate, acquiring a nonzero angular velocity (Fig. 2.1.7*c*). The direction of this angular velocity is that of the torque. The SI unit of torque is the newton-meter (abbreviated N·m).

The larger an object's moment of inertia, the more slowly its angular velocity changes in response to a specific torque (Fig. 2.1.8). You can easily spin a basketball with the tips of your fingers, but it's much harder to spin a bowling ball. The bowling ball's larger moment of inertia comes about primarily because it has a greater mass than the basketball. But moment of inertia also depends on the shape of an object and how far its mass is from the axis of rotation. An object that has most of its mass located near the axis of rotation will have a smaller moment of inertia than an object of the same mass that has most of its mass located far from that axis. Thus a ball of pizza dough has a smaller moment of inertia than the finished pizza. The bigger the pizza gets, the harder it is to spin.

Because an object's moment of inertia depends on how far its mass is from the axis of rotation, changes in the axis of rotation are likely to change its moment of inertia. For example, less torque is required to spin a tennis racket about its handle (Fig. 2.1.9*a*) than to flip the racket head-over-handle (Fig. 2.1.9*b*). When you spin the tennis racket about its handle, the axis of rotation runs right through the handle so that most of the racket's mass is fairly close to the axis and the moment of inertia is small. When you flip the tennis racket head-over-handle, the axis of rotation runs across the handle so that both the head and the handle are far away from the axis and the moment of inertia is large. The tennis racket's moment of inertia becomes even larger when you hold it in your hand and make it rotate about your shoulder rather than its center of mass (Fig. 2.1.9*c*).

When something exerts a torque on the seesaw, its angular velocity changes; in other words, it undergoes angular acceleration, our third important vector quantity of rotational motion. **Angular acceleration** measures how quickly the seesaw's *angular* velocity changes. It's analogous to acceleration, which measures how quickly an object's *translational* velocity changes. Just as with acceleration, angular acceleration involves both a magnitude and a direction. An object undergoes angular acceleration when its angular speed increases or decreases or when its angular velocity changes directions. The SI unit of angular acceleration is the **radian-per-second^2** (abbreviated 1/s^2).

There is a simple relationship between the seesaw's angular acceleration, its moment of inertia, and the torque exerted on it. The torque exerted on the seesaw equals the product of the seesaw's moment of inertia times its angular acceleration. This relationship between torque, moment of inertia, and angular acceleration is **Newton's second law of rotational motion** and can be written in a word equation:

$$\text{torque} = \text{moment of inertia} \cdot \text{angular acceleration}, \qquad \textbf{(2.1.1)}$$

in symbols:

$$\tau = I \cdot \alpha,$$

and in everyday language:

Spinning a marble is much easier than spinning a merry-go-round.

It resembles Newton's second law of translational motion (force = mass · acceleration), except that torque has replaced force, moment of inertia has replaced mass, and angular acceleration has replaced (translational) acceleration. However, this new law doesn't apply to wobbling objects because they're being affected by more than one moment of inertia simultaneously (see the discussion of tennis rackets above) and follow a much more complicated law. (For another useful form of Eq. 2.1.1, see ❏ on page 48.)

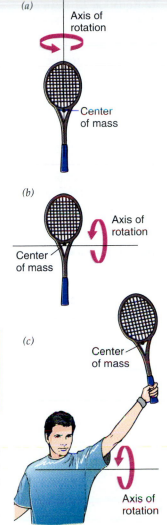

(a) Axis of rotation — Center of mass

(b) Axis of rotation — Center of mass

(c) Center of mass — Axis of rotation

Newton's Second Law of Rotational Motion
The torque exerted on an object that is not wobbling is equal to the product of that object's moment of inertia times its angular acceleration. The angular acceleration points in the same direction as the torque.

Because it's an equation, the two sides of Eq. 2.1.1 are equal. Any change in the torque you exert on a rigid object must be accompanied by a proportional change in its angular acceleration. As a result, the harder you twist or spin a seesaw, the more rapidly its angular velocity changes.

We can also compare the effects of a specific torque on two objects with different moments of inertia. In this case, since the left side of the equation doesn't change, the right side must remain constant. If we replace the playground seesaw with one from a dollhouse, the moment of inertia will decrease and the angular acceleration will have to increase to keep the right side of the equation constant. The angular velocity of a doll's seesaw thus changes more rapidly than the angular velocity of a playground seesaw when the two experience identical torques.

In summary:

1. Your angular position indicates exactly how you're oriented.
2. Your angular velocity measures how quickly your angular position changes.
3. Your angular acceleration measures how quickly your angular velocity changes.
4. In order for you to undergo angular acceleration, something must exert a torque on you.
5. The more moment of inertia you have, the less angular acceleration you experience for a given torque.

Fig. 2.1.9 A tennis racket's moment of inertia depends on the axis about which it rotates. Its moment of inertia is small (*a*) when it rotates about its handle and large (*b*) when it rotates head-over-handle. (*c*) If you make it rotate about your shoulder, its moment of inertia becomes even larger.

This summary of the physical quantities of rotational motion is analogous to the one for translational motion on p. 8. Take a moment to compare the two.

❏ Equation 2.1.1 can be rearranged algebraically as

angular acceleration =

$$\frac{torque}{moment\ of\ inertia}.$$

This form shows that an object's angular acceleration depends both on the torque causing that angular acceleration and on the object's moment of inertia. Increasing the torque increases the angular acceleration while increasing the object's moment of inertia by reshaping it decreases the angular acceleration.

Check Your Understanding #4: The Merry-Go-Round

The merry-go-round is a popular playground toy (see Fig. 2.1.8). Already challenging to spin empty, a merry-go-round is even harder to start or stop when there are lots of children on it. Why is it so difficult to change a full merry-go-round's angular velocity?

Check Your Figures #1: Hard to Turn

Automobile tires are normally hollow and filled with air. If they were made of solid rubber, their moments of inertia would be about 10 times as large. With the wheel lifted off the ground, how much more torque would an automobile have to exert on a solid tire to make it undergo the same angular acceleration as a hollow tire?

Forces, Torques, and Seesaws

We're now ready to mount the seesaw on its pivot. Although the pivot limits the seesaw's motion, it still permits the board to rotate freely about one axis. As we'll see shortly, the board's weight doesn't exert any torque on it about this axis so it obeys Newton's first law of rotational motion. That is, the unoccupied board rotates with constant angular velocity about its pivot.

The unoccupied seesaw is balanced, meaning that it has zero torque on it. As a result, it experiences no angular acceleration. You might think that a balanced seesaw always remains horizontal, but that isn't necessarily so. What is certain is that its angular velocity is constant. If it's rotating, then it continues to rotate steadily about the pivot; if it's stationary, then it remains stationary at its current tilt angle, whether horizontal or not.

To change the seesaw's angular velocity, you must exert a torque on it. But how do you actually exert a torque? You put your hand on one end of the seesaw and push that end down (Fig. 2.1.10a). The seesaw begins to rotate, and your end soon hits the ground. You have exerted a torque on the board.

But you started by exerting a *force* on the board—you pushed on it—so forces and torques must be related somehow. A force can produce a torque and a torque can produce a force. To help us explore that relationship, let's think of all the ways not to exert a torque on a seesaw.

What happens if you push on the seesaw board right where the pivot passes through it (Fig. 2.1.10b)? Nothing—no angular acceleration. If you move a little away from the pivot, you can get the board rotating but you have to push hard. You do much better to push on the end of the board, where even a small force can start the seesaw rotating. The distance from the pivot to the place where you push on the board is called the **lever arm**; in general, the longer the lever arm, the less force it takes to cause a particular angular acceleration. Our first observation about producing a torque with a force is this: you obtain more torque by exerting that force farther from the pivot or axis of rotation. In other words, the torque is proportional to the lever arm.

Another ineffective way to start the seesaw board rotating is to push its end directly toward or away from the pivot (Fig. 2.1.10c). A force exerted toward or away from the axis of rotation doesn't produce any torque about that axis. At least a component of the force you exert must be at right angles to the lever arm, which is actually a vector pointing along the board's surface from the pivot to the place where

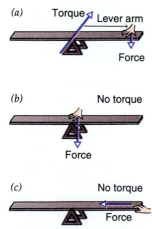

Fig. 2.1.10 (*a*) When you push down on the seesaw, far from its pivot, you exert a torque on it. But when you (*b*) exert your force at the pivot or (*c*) exert your force toward the pivot, you exert no torque.

you push on the board. Our second observation about producing a torque with a force is that you must exert at least a component of that force at right angles to the lever arm. Only that component of force contributes to the torque. To produce the most torque, push at a right angle to the lever arm.

We can summarize these two observations as follows: the torque produced by a force is equal to the product of the lever arm times that force, where we include only the component of the force that is at right angles to the lever arm. This relationship can be written as a word equation:

$$\text{torque} = \text{lever arm} \cdot \text{force,} \qquad (2.1.2)$$

in symbols:

$$\tau = r \times F,$$

and in everyday language:

When twisting an unyielding object, it helps to use a long stick.

The directions of the force and lever arm also determine the direction of the torque. The three directions follow another right-hand rule (Fig. 2.1.11). If you extend the index finger of your right hand in the direction of the lever arm and bend the middle figure of that hand in the direction of the force, then your thumb will point in the direction of the torque. Thus in Fig. 2.1.11a, the lever arm points to the right, the force points downward, and the resulting torque points into the page so that the seesaw undergoes angular acceleration in the clockwise direction. In Fig. 2.1.11b, the lever arm has reversed directions and so has the torque.

What happens if you and a friend push down simultaneously on both seats at once? Then you produce two torques on the seesaw about its pivot, and these torques have opposite directions. The seesaw responds to the **net torque** it experiences, the sum of all the individual torques on the seesaw. Since your two torques oppose one another, they at least partially cancel. If you carefully exert identical downward forces at identical distances from the pivot, the magnitudes of the two torques will be exactly equal and the torques will add to zero. The seesaw will experience zero net torque, and it will be balanced.

This observation explains the need for careful seating of the children on the seesaw. Each child's weight exerts a downward force; by properly distributing those weights on both sides of the pivot, the torques that the weights produce can be made to add to zero. The net torque on the seesaw about its pivot is then zero, and the seesaw balances.

In fact, the weight of the board itself is balanced in this manner. While each end's weight exerts a torque on the board, those two torques sum to zero and have no overall effect on the board's rotation.

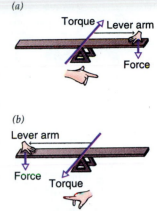

Fig. 2.1.11 The torque on a seesaw obeys a right-hand rule: if your index finger points along the lever arm and your middle finger points along the force, your thumb points along the torque.

Check Your Understanding #5: Cutting Up Cardboard

When you cut cardboard with a pair of scissors, it's best to move the cardboard as close as possible to the scissors' pivot. Explain.

Check Your Figures #2: A Few Loose Screws

You're trying to remove some rusty screws from your refrigerator, using an adjustable wrench with a 0.2-meter (20-centimeter) handle. Although you push as hard as you can on the handle, you can't produce enough torque to loosen one of the screws. You have a 1.0-meter (100-cm) pipe that you can slip over the handle of the wrench to make the wrench effectively 1 meter long. How much more torque will you then be able to exert on the screw?

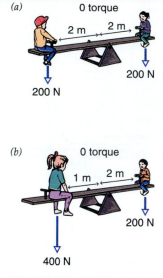

Fig. 2.1.12 (*a*) When two children of equal weight sit at equal distances from the pivot, they produce equal but oppositely directed torques about the pivot. These torques sum to zero so that the seesaw experiences zero net torque. (*b*) When one child weighs twice as much as the other, the seesaw balances when the heavy child sits at half the distance from the pivot.

Net Torque and Mechanical Advantage

The amount of torque that a child's weight produces on a seesaw depends on that child's distance from the pivot. If the child sits on the pivot, the lever arm is zero and she produces no torque; but if she sits at the extreme end of the board, the lever arm is long and she produces a large torque. She can adjust her torque by moving along the board because the seesaw provides her with mechanical advantage. As we saw in Section 1.3, mechanical advantage appears when a device redistributes the amounts of force and distance used to produce a particular amount of work. The seesaw allows a small force exerted at the end of the rotating board to do the same work as a large force exerted near the board's pivot.

To see how mechanical advantage appears in a seesaw, think of what happens when two children sit on its ends. If two 5 year olds, each weighing 200 N, sit at opposite ends of the seesaw, 2 m from the pivot (Fig. 2.1.12*a*), each one exerts a torque of 400 N · m on the seesaw about its pivot (200 N · 2 m = 400 N · m). But because these torques are oppositely directed, they add to zero. The net torque on the seesaw is zero and the seesaw balances.

If you replace one of the 5 year olds with a 400-N teenager, the teenager must sit at half the distance from the pivot (Fig. 2.1.12*b*). Doubling the force while halving the lever arm leaves the torque unchanged at 400 N · m. The two children again produce equal but oppositely directed torques about the pivot, so that the net torque on the seesaw is zero and the seesaw balances. This effect explains how a small child at the end of the seesaw can balance a large child nearer the pivot.

With the seesaw balanced, nothing is accelerating. Each child experiences zero net force; the seesaw pushes up on the small child with a force of 200 N and on the large child with a force of 400 N. Ultimately, it's the small child's 200-N weight that gives rise to the 400-N supporting force experienced by the large child. The seesaw's mechanical advantage allows the small child to support and lift the much heavier child. This effect, where a small force on one part of a rotating system produces a large force elsewhere in that system, is an example of the mechanical advantage associated with levers.

Check Your Understanding #6: Pulling Nails

Some hammers have a special claw designed to remove nails from wood. When you slide the claw under the nail's head and rotate the hammer by pulling on its handle, the claw pulls the nail out of the wood. The hammer's head contacts the wood to form a pivot that's about 10 times closer to the nail than to the handle. The torque you exert on the hammer twists it in one direction, while the torque that the nail exerts on the hammer twists it in the opposite direction. The hammer isn't undergoing any significant angular acceleration, so the torques must be nearly balanced. If you're exerting a force of 100 N on the hammer's handle, how much force is the nail exerting on the hammer's claw?

Riding a Seesaw

Each seesaw in Fig. 2.1.12 is balanced, meaning that the net torque on it is zero. Although each child's weight exerts a torque on the board, the two torques add to zero. Since the seesaw experiences zero net torque and no angular acceleration, it continues rotating at constant angular velocity.

However, as it presently stands, a balanced seesaw should either remain motionless forever or else rotate endlessly in the same direction. Children are unlikely to wait motionless forever, and endless rotation implies that the children will be upside-down periodically. We've obviously neglected a few details.

What do the children do when the board is motionless? To start the board moving, they have to unbalance the seesaw. One of the children must change the torque she exerts on it. She can either change the downward force she exerts on the board or change the distance between that force and the pivot. Actually, children change both the force and the lever arm frequently without even thinking about it. If a child leans inward, toward the pivot, the lever arm decreases and the child exerts less torque on the board; as a result, the board begins to rotate and the child rises. If the child pushes on the ground with his feet, the ground exerts an upward force on him, reducing the force and torque he exerts on the board; again, the board begins to rotate and the child rises.

So either by leaning or by pushing on the ground, the children can start an initially motionless, balanced seesaw rotating. Similarly, when one end of the seesaw hits the ground, the ground exerts a strong, upward support force on it. Located far from the pivot and almost at right angles to the board, this force produces a huge torque on the board and abruptly stops it from rotating. The angular acceleration is so uncomfortably large that most children push on the ground with their feet to cushion the impact. The child on the ground can continue to push down with her feet until the seesaw rotates in the opposite direction. That child begins to rise and the other child descends. When the other end of the seesaw reaches the ground, this cycle begins again.

As they play on a seesaw, the two children frequently change the torques they exert on the board so that the seesaw tips back and forth. During the moments when a child is pushing on the ground or leaning inward or outward to get a stationary seesaw moving, the seesaw is no longer balanced. A balanced seesaw has zero angular acceleration; it's only by unbalancing the seesaw that the children can change the angular velocity of the seesaw.

Check Your Understanding #7: Rocking the Boat

Loading a large container ship requires some care in balancing the cargo and fastening it down firmly. The effective pivot about which the ship can rotate in the water is located roughly along the centerline of the ship, from its bow to its stern. Why is improperly fastened-down cargo so dangerous on such a ship, possibly causing it to capsize during a storm?

seesaws and work, levers

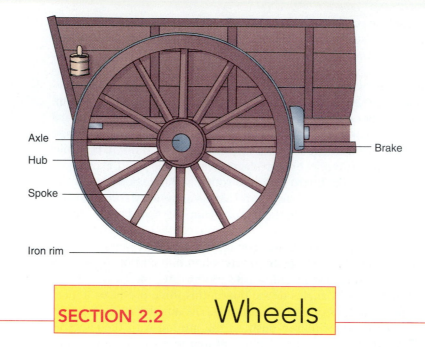

Axle

Hub   Brake

Spoke

Iron rim

SECTION 2.2   # Wheels

Like ramps and levers, wheels are simple tools that make our lives easier. But the wheel's main purpose isn't mechanical advantage, it's overcoming friction. Up until now, we've ignored friction, looking at the laws of motion as they apply only in idealized situations. But our real world does have friction, and an object in motion tends to slow down and stop because of it. One of our first tasks in this section will therefore be to understand friction—though, for the time being, we'll continue to neglect air resistance.

Questions to Think About: *If objects in motion tend to stay in motion, why is it so hard to drag a heavy box across the floor? If objects should accelerate downhill on a ramp, why won't a plate slide off a slightly tilted table? What makes the wheels of a cart turn as you pull the cart forward? How does turning the wheels of an automobile propel the car forward?*

Experiments to Do: *To observe the importance of wheels in eliminating friction, try sliding a book along a flat table. Give the book a push and see how quickly it slows down and stops. Which way is friction pushing on the book? Does the force that friction exerts on the book depend on how fast the book is moving? Let the book come to a stop. Is friction still pushing on the book when it isn't moving? If you push gently on the stationary book, what force does friction exert on it?*

Lay down three or four round pencils, parallel to one another and a few inches apart. Rest the book on top of the pencils and give the book a push in the direction that the pencils can roll. Describe how the book now moves. What do you think has caused the difference?

Moving a File Cabinet: Friction

When we imagined moving your friend's piano into a new apartment back in Section 1.3, we neglected a familiar force—friction. Luckily for us, your friend's piano had wheels on its legs, and wheels facilitate motion by reducing the effects of friction. We'll focus on wheels in this section. But first, to help us understand the relationship between wheels and friction, we'll look at another item that needs to be moved—a file cabinet.

The file cabinet is resting on a smooth and level hardwood floor; it's full of sheet music and weighs about 1000 N (225 pounds). Despite its large mass, you know that it should accelerate in response to a horizontal force, so you give it a gentle push toward the door. Nothing happens. Something else must be pushing on the file cabinet in just the right way to cancel your force and keep it from accelerating. Undaunted, you push harder and harder until finally, with a tremendous shove, you manage to get the file cabinet sliding across the floor. But the cabinet moves slowly even though you continue to push on it. Something else is pushing on the file cabinet, trying to stop it from moving.

That something else is **friction,** a force that opposes the relative motion of two surfaces in contact with one another. Two surfaces that are in **relative motion** are traveling with different velocities so that a person standing still on one surface would observe the other surface as moving. In opposing relative motion, friction exerts forces on both surfaces in directions that tend to bring them to a single velocity.

For example, when the file cabinet slides by itself toward the left, the floor exerts a rightward frictional force on it (Fig. 2.2.1). The frictional force exerted on the file cabinet, *toward the right,* is in the direction opposite the file cabinet's velocity, *toward the left.* Since the file cabinet's acceleration is in the direction opposite its velocity, the file cabinet slows down and eventually comes to a stop.

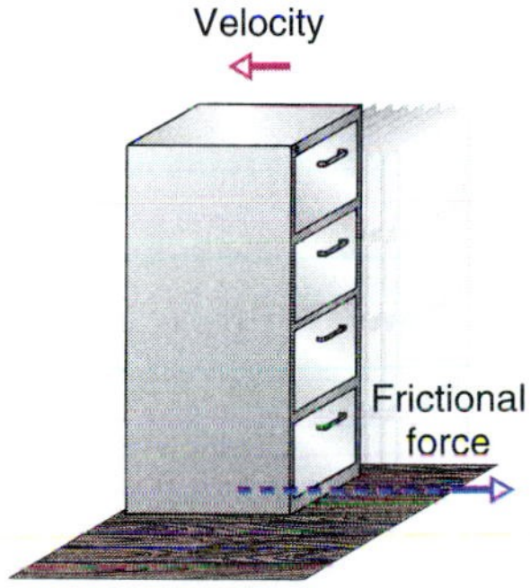

According to Newton's third law of motion, an equal but oppositely directed force must be exerted by the file cabinet on the floor. Sure enough, the file cabinet does exert a leftward frictional force on the floor. However, the floor is rigidly attached to the earth, so it accelerates very little. The file cabinet does almost all the accelerating, and soon the two objects are traveling at the same velocity.

Frictional forces always oppose relative motion, but they vary in strength according to (1) how tightly the two surfaces are pressed against one another, (2) how slippery the surfaces are, and (3) whether or not the surfaces are actually moving relative to one another. In (1), the harder you press two surfaces together, the larger the frictional forces they experience. For example, an empty file cabinet slides more easily than a full one. In (2), roughening the surfaces generally increases friction, while smoothing or lubricating them generally reduces it. Riding a toboggan down the driveway is much more interesting when the driveway is covered with snow or ice than when the driveway is bare asphalt. We'll examine issue (3) later on.

Fig. 2.2.1 A file cabinet sliding to the left across the floor. The file cabinet experiences a frictional force toward the right that gradually brings it to a stop.

Check Your Understanding #1: The One That Got Away

Your table at the restaurant isn't level, and your water glass begins to slide slowly downhill toward the edge. Which way is friction exerting a force on it?

A Microscopic View of Friction

As the file cabinet slides by itself across the floor, it experiences a horizontal frictional force that gradually brings it to a stop. But from where does this frictional force come? The obvious forces on the file cabinet are both vertical, not horizontal: the cabinet's weight is downward and the support force from the floor is upward. How can the floor exert a horizontal force on the file cabinet?

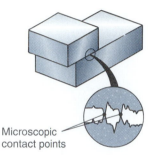

Microscopic
contact points

Fig. 2.2.2 Two surfaces that are pressed against one another actually touch only at specific contact points. When the surfaces slide across one another, these contact points collide, producing sliding friction and wear.

friction and weight

The answer lies in the fact that neither the bottom of the file cabinet nor the top of the floor is perfectly smooth. They both have microscopic hills and valleys of various sizes. The file cabinet is actually supported by thousands of tiny contact points, where the file cabinet directly touches the floor (Fig. 2.2.2). As the file cabinet slides, the microscopic projections on the bottom of the file cabinet pass through similar projections on the top of the floor. Each time two projections collide, they experience horizontal forces. These tiny forces oppose the relative motion and give rise to the overall frictional forces experienced by the file cabinet and floor. Because even an apparently smooth surface still has some microscopic surface structure, all surfaces experience friction as they rub across one another.

Increasing the size or number of these microscopic projections by roughening the surfaces generally leads to more friction. If you put sandpaper on the bottom of the file cabinet, it will experience larger frictional forces as it slides across the floor. On the other hand, a microscopically smoother "nonstick" surface, like that used in modern cookware, would let the file cabinet slide easily.

Increasing the number of contact points by squeezing the two surfaces more tightly together also leads to more friction. The microscopic projections simply collide more often. That's why adding more sheet music to the file cabinet would make it harder to slide. Doubling the file cabinet's weight would roughly double the number of contact points and make it about twice as hard to move across the floor. A useful rule of thumb is that the frictional forces between two surfaces are proportional to the forces pressing those two surfaces together.

Friction also causes wear when the colliding contact points break one another off. With time, this wear can remove large amounts of material so that even seemingly indestructible stone steps are gradually worn away by foot traffic. The best way to reduce wear between two surfaces (other than to insert a lubricant between them) is to polish them so that they are extremely smooth. The smooth surfaces will still touch at contact points and experience friction as they slide across one another, but their contact points will be broad and round and will rarely break one another off during a collision.

Check Your Understanding #2: Weight and Friction

How much harder is it to slide a stack of two identical books across a table than it is to slide just one of those books?

Static and Sliding Friction

There are really two kinds of friction—sliding and static. When two surfaces are moving across one another, they experience **sliding friction,** which acts to stop the surfaces from sliding. But even when two surfaces have the same velocity, they may still experience **static friction,** which acts to prevent two surfaces from starting to slide across one another.

When you try to slide the file cabinet across the floor, you find it particularly hard at the beginning. Because the contact points between the cabinet and floor have settled into one another and are acting to keep the cabinet from moving, a small push does nothing; you need a mighty shove to get the cabinet going. Static friction is exerting a force in the direction opposite to your push, and this frictional force is always exactly equal to the force of your push. The net force on the file cabinet is therefore zero, and it doesn't accelerate.

However, the force that static friction can exert is limited. To get the file cabinet moving, all you have to do is exert more horizontal force on it than static friction can exert. The net force on the file cabinet will then no longer be zero, and it will accelerate.

Once the file cabinet is moving, static friction is replaced by sliding friction. Because sliding friction acts to bring the file cabinet back to rest, you must push on the cabinet to keep it moving. With the file cabinet sliding across the floor, however, the contact points between the surfaces no longer have time to settle into one another, and they consequently experience weaker horizontal forces. That's why the force of sliding friction is generally weaker than that of static friction and why it's easier to keep the file cabinet moving than it is to get it started.

Check Your Understanding #3: Skidding to a Stop

Antilock brakes keep an automobile's wheels from locking and skidding during a sudden stop. Apart from issues of steering, what is the advantage of preventing the wheels from skidding (sliding) on the pavement?

Work, Energy, and Power

There is another difference between static and sliding friction: sliding friction wastes energy. It can't make that energy disappear altogether because energy, as we've seen, is a conserved quantity: it can't be created or destroyed. But energy can be transferred between objects or converted from one form to another. What sliding friction does is convert useful, **ordered energy**—energy that can easily be used to do work—into relatively useless, disordered energy. This disordered energy is called **thermal energy** and is the energy we associate with temperature. It's sometimes called *internal energy* or *heat*. Sliding friction makes things hotter by turning work into thermal energy.

As we saw in Section 1.3, energy is the capacity to do work and is transferred between objects by doing that work. Energy can also change forms, appearing as either kinetic energy in the motions of objects or as potential energy in the forces between or within those objects. With practice, you can "watch" energy flow through a system just as an accountant watches money flow through an economy.

The most obvious form of energy is kinetic energy, the energy of motion. It's easy to see when kinetic energy is transferred into or out of an object. As kinetic energy leaves an object, the object slows down; thus moving water slows down as it turns a gristmill, and a bowling ball slows down as it knocks over bowling pins. Conversely, as kinetic energy enters an object, the object speeds up. A baseball moves faster as you do work on it during a pitch; you're transferring energy from your body into the baseball, where the energy becomes kinetic energy in the baseball's motion.

Potential energy usually isn't as visible as kinetic energy. It can take many different forms, some of which appear in Table 2.2.1. In each case nothing is moving; but because the objects still have a great potential to do work, they contain potential energy.

We measure energy in many common units: joules (J), calories, food Calories (also called kilocalories), and kilowatt-hours, to name only a few. All of these units measure the same thing, and they differ from one another only by numerical conversion factors, some of which can be found in Appendix B. For example, 1 food

Table 2.2.1   Several Forms and Examples of Potential Energy

FORM OF POTENTIAL ENERGY	EXAMPLE
Gravitational potential energy	A bowling ball at the top of a hill
Elastic potential energy	A wound clock spring
Electrostatic potential energy	A cloud in a thunderstorm
Chemical potential energy	A firecracker
Nuclear potential energy	Uranium

Calorie is equal to 1000 calories or 4184 J. Thus a jelly donut with about 250 food Calories contains about 1,000,000 J of energy. Since a joule is the same as a newton-meter, 1,000,000 J is the energy you'd use to lift your friend's file cabinet into the second-floor apartment 200 times (1000 N times 5 m upward is 5000 J of work per trip). No wonder eating donuts is hard on your physique!

Of course, you can eventually use up the energy in a jelly donut; it just takes time. You can only do so much work each second. The measure of how quickly you do work is **power**—the amount of work you do in a certain amount of time, or

$$\text{power} = \frac{\text{work}}{\text{time}}.$$

The SI unit of power is the **joule-per-second,** also called the **watt** (abbreviated W). Other units of power include Calories-per-hour and horsepower; like the units for energy, these units differ only by numerical factors, which are again listed in Appendix B. For example, 1 horsepower is equal to 745.7 W. Since a 1-horsepower motor does 745.7 J of work each second, and since it takes 5000 J of work to move the file cabinet to the second floor, that motor has enough power to do the job in about 6.7 s.

Check Your Understanding #4: Apple Overtures

Trace the flow of energy as an archer shoots an apple off the head of her assistant with an arrow.

Friction and Thermal Energy

So far we've looked at kinetic energy and a variety of different potential energies. But what about the thermal energy produced by sliding friction? Is thermal energy a new kind of potential energy or an alternative to kinetic energy? In truth, it's neither. Thermal energy is actually a mixture of the same kinetic and potential energies that we've seen before. But unlike the kinetic energy in a moving ball or the potential energy in an elevated piano, the kinetic and potential energies in thermal energy are disordered at the atomic and molecular level. Thermal energy makes every microscopic particle in an object jiggle about independently; at any moment, each particle has its own tiny supply of potential and kinetic energies, and this dispersed energy is collectively referred to as thermal energy.

As you push the file cabinet across the floor, you do work on the file cabinet, but it doesn't pick up speed. Instead, sliding friction converts your work into thermal energy, so that the cabinet becomes hotter as the energy you transfer to it is dispersed among its atoms and molecules. But while sliding friction is an easy way to turn work into thermal energy, there's no easy way to turn thermal energy back into work. Disorder makes anything harder to use; while it's easy to disorder something, it's hard to undo that disordering. When you drop your favorite coffee mug on the floor and it shatters into a thousand pieces, the cup is still all there, but it's disordered and thus much less useful. Just as dropping the pieces on the floor a second time isn't likely to reassemble your cup, energy converted into thermal energy can't easily be reassembled into useful, ordered energy.

Unlike sliding friction, static friction doesn't convert work into thermal energy. Since two surfaces experiencing static friction don't move relative to one another, there is no distance traveled and thus no work done. You can push against the stationary file cabinet all day without doing any work on it. Even if you lift the file cab-

(a) Velocity

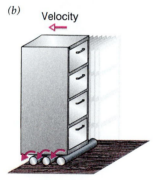

(b) Velocity

Fig. 2.2.3 (*a*) A file cabinet that's supported on turning rollers experiences only static friction. (*b*) Since the top surface of a roller moves forward with the file cabinet, while its bottom surface stays behind with the sidewalk, the roller's center of mass moves only half as fast as the file cabinet. As a result, the rollers are soon left behind.

inet upward with your hands (no easy task), static friction between your hands and the file cabinet's sides merely assists you in doing work on the file cabinet itself. As you lift the file cabinet upward, all of your work goes into increasing the file cabinet's gravitational potential energy.

In contrast, sliding friction converts at least some work into thermal energy. Since two surfaces sliding across one another experience frictional forces that oppose their relative motion, sliding friction does negative work on the surfaces; it extracts energy from a sliding object and converts it into thermal energy. Thus when you do work on the file cabinet by pushing it across the floor, sliding friction does negative work on it. The file cabinet's kinetic energy doesn't change very much, but its thermal energy continues to increase.

Fig. 2.2.4 As this stage-coach rolls forward, sliding friction between its axles and hubs converts some of its kinetic energy into thermal energy. To reduce this wasted energy, the coach has narrow axles that are lubricated with axle grease.

Check Your Understanding #5: Burning Rubber

If you push too hard on your car's accelerator pedal when the traffic light turns green, your wheels will slip and you'll leave a black trail of rubber behind. Such a "jack-rabbit start" can cause as much wear on your tires as 50 km (31 miles) of normal driving. Why is skidding so much more damaging to the tires than normal driving?

Wheels

You've wrestled your friend's file cabinet out the door of the old apartment and are now dragging it along the sidewalk. You're doing work against sliding friction the whole way, producing large amounts of thermal energy in both the bottom of the cabinet and the surface of the sidewalk. You're also damaging both objects, since sliding friction is causing wear on their surfaces. The four-drawer file cabinet may be down to three drawers by the time you arrive at the new apartment.

Fortunately, you can move one object across another object without sliding, and thus without sliding friction, if you use a mechanical system. The classic example is a roller (Fig. 2.2.3). If you place the file cabinet on rollers, you can push it along without any sliding friction; the rollers rotate as the file cabinet moves so that their surfaces never slide across the bottom of the cabinet or the top of the sidewalk. To see how the rollers work, make a fist with one hand and roll it across the palm of your other hand. The skin of one hand doesn't slide across the skin of the other hand; since the motion is silent and doesn't convert work into thermal energy, your skin remains cool. Now slide your two open palms across one another; this time, because there is sliding friction, the motion makes noise and warms your skin.

Although the rollers don't experience sliding friction, they do experience static friction. The top of each roller is touching the bottom of the cabinet, and the two surfaces move along together because of static friction; they grip one another tightly until the roller's rotation pulls them apart. A similar process takes place between the rollers and the top of the sidewalk; static friction exerts torques on the rollers and hence is what makes them rotate in the first place. Again, you can illustrate this behavior with your hands. Try to drag your fist across your open palm. Just before your fist begins to slide, you'll feel a torque on it. Static friction between the skins of your two hands, acting to prevent sliding, causes your fist to begin rotating just like a roller.

Once you get the file cabinet moving on rollers, you can keep it rolling along the level sidewalk indefinitely. Without any sliding friction, the cabinet doesn't lose kinetic energy, so it continues at constant velocity without your having to push it. However, the rollers move out from under the file cabinet as it travels, and you fre-

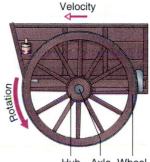

Fig. 2.2.5 A wheel rotates as it travels so that no sliding friction occurs between it and the ground. The load is imparted to the wheel through an axle that enters the central hole of the wheel, the hub. As the wheel turns, the hub rubs against the axle, causing some sliding friction and wear.

carts and energy

roller bearings

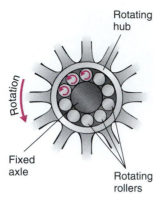

Rotating hub

Rotation

Fixed axle

Rotating rollers

Fig. 2.2.6 In a roller bearing, the hub of the wheel doesn't touch the axle directly. Instead, the two are separated by a set of rollers that turn with the hub. The bottom few rollers bear most of the load since the hub pushes up on them and they push up on the axle. As the wheel turns, the rollers recirculate, traveling up to the right and over the top of the axle before returning down to the left to bear the load once again. The rollers, wheel, and axle experience only static friction, not sliding friction.

ball bearings

powered wheels

quently have to move a roller from the front of the cabinet to the back. In fact, you need at least three rollers to ensure that the file cabinet never falls to the ground when a roller pops out the front. Although the rollers have eliminated sliding friction, they've created another headache—one that makes the prospect of traveling cross-country in a roller-supported, horse-drawn vehicle unappealing. Is there another device that can reduce sliding friction without requiring constant attention?

One alternative would be a four-wheeled cart. The simplest cart rests on fixed poles or axles that pass through central holes or hubs in the four wheels (Fig. 2.2.4, p. 57). The ground exerts upward support forces on the wheels, the wheels exert upward support forces on the axles, and the axles support the cart and its contents. As the cart moves forward, its wheels turn so that their bottom surfaces don't slide or skid across the ground; instead, each wheel lowers a portion of its surface onto the sidewalk, leaves it there briefly to experience static friction, and then raises it back off the sidewalk, with a new portion of wheel surface taking its place. Thus there is only static friction between the cart's wheels and the ground.

Unfortunately, as each wheel rotates, its hub slides across the stationary axle at its center (Fig. 2.2.5, p. 57). This sliding friction wastes energy and causes wear to both hub and axle. However, the narrow hub moves relatively slowly across the axle so that the work and wear done each second are small. Still, this sliding friction is undesirable and can be reduced significantly by lubricating the hub and axle with "axle grease."

A better solution is to insert rollers between the hub and axle (Fig. 2.2.6). The result is a bearing—a mechanical device that minimizes sliding friction between a hub and an axle. A complete bearing consists of two rings separated by rollers that keep those rings from rubbing against one another. In this case, the bearing's inner ring is attached to the stationary axle while its outer ring is attached to the spinning wheel hub. The nondriven wheels of an automobile are supported by such bearings on essentially stationary axles. As the vehicle starts forward, static friction from the ground exerts torques on these free wheels and they begin to turn.

A car's driven wheels are also supported by bearings, but these bearings act somewhat differently. Because the engine must be able to exert a torque on each driven wheel, those wheels are rigidly connected to their axles. As the engine spins one of these axles, the axle spins its wheel. A bearing prevents the spinning axle from rubbing against the car's frame. This bearing's outer ring is attached to the stationary car frame while its inner ring is attached to the spinning axle.

As the driven wheel begins to spin, it experiences static friction with the ground and the ground pushes horizontally on the wheel's bottom to keep it from skidding. Since that is the only horizontal force on the automobile, the automobile accelerates forward.

Recognizing a good idea when you think of it, you load the file cabinet into the back of your Jaguar XJ8 and climb into the driver's seat. The car isn't quite as responsive as usual because of the added mass, but it's still able to accelerate respectably and reach full speed in a reasonable amount of time. In a few minutes, you arrive at your destination and break to a stop. Again, you're aware of the added mass because the car decelerates less quickly than usual. Nonetheless, you've reached your goal safely and are now a hero.

➤ **Check Your Understanding #6: Jewel Movements**

Many antique mechanical watches and clocks proudly proclaim that they have "jewel movements." Gears in these timepieces turn on axles that are pointed at either end and are supported at those ends by very hard, polished gemstones. What is the advantage of having needlelike ends on an axle and supporting those needles with smooth, hard jewels?

Contact

Power source grid

Riser

Shoulder harness

Steering wheel

Headlight

Bumper

Neutral power return (the floor)

Bumper Cars

While car crashes normally aren't much fun, unless you're watching movies or television, there is one delightful exception: bumper cars. For a few minutes, drivers in this amusement park ride race madly about an oval track, deliberately crashing their vehicles into one another and laughing hysterically at the violent impacts. Jolts, jerks, and spins are par for the course, and it's a wonder that no one gets whiplash. But hidden in the fun are several important physics concepts that influence everything from tennis to billiards.

Questions to Think About: *Why does a stationary car begin rolling forward after being struck by a moving car? What aspects of motion are passed between cars as they collide? Why does your car jolt more when the car that hits it contains two big adults rather than one small child? What would happen if the bumper cars had hard steel bumpers rather than soft rubber ones? Why is your car often set spinning by collisions, and what keeps it spinning?*

Experiments to Do: *Place a coin on a smooth table and flick a second, identical coin so that it slides along the table and strikes the stationary coin squarely. What happens? Try this experiment again, but now use two coins with different masses. How is the collision different? Does it matter which coin you crash into the other?*

Now line up several identical coins so that they touch and slide another coin into one end of this line. How does the collision affect the coin that was originally moving? How does it affect the line of coins? What was transferred among the coins by the collision?

Now stand a coin on its edge and flick it so that it spins rapidly. Did you give it something that keeps it spinning? Why does the coin eventually stop spinning?

Coasting Forward: Linear Momentum

Bumper cars are small, electrically powered vehicles that can turn on a dime and are protected on all sides by oval-shaped rubber bumpers. Each car has only two controls: a pedal that activates its motor and a steering wheel that controls the direction in which the motor pushes the car. Since the car itself is so small, its occupants account for much of the car's total mass and moment of inertia.

Imagine that you have just sat down in one of these cars and put on your safety strap. The other people also climb into their cars, usually one person per car, and the ride begins.

With your car free to move or turn, you quickly become aware of its translational and rotational inertias. The car's translational inertia makes it hard to start or stop, and its rotational inertia makes it difficult to spin or stop from spinning. While we've seen these two types of inertia before, let's take another look at them and at how they affect your bumper car. This time, we'll see that they're associated with two new conserved quantities—linear momentum and angular momentum. It turns out that energy isn't the only conserved quantity in nature!

Linear momentum, usually just called **"momentum,"** is the measure of an object's translational motion—its tendency to continue moving in a particular direction. Roughly speaking, your car's momentum indicates which way it's heading and just how difficult it was to get the car moving with its current velocity. The car's momentum is its mass times its velocity and can be written as a word equation:

$$\text{momentum} = \text{mass} \cdot \text{velocity}, \tag{2.3.1}$$

in symbols:

$$\mathbf{p} = m \cdot \mathbf{v},$$

and in everyday language:

It's hard to stop a fast-moving truck.

Note that the direction of momentum, a vector quantity, is the same as the direction of velocity. As we might expect, the faster your car is moving or the more mass it has, the more momentum it has in the direction of its motion.

To physicists, conserved quantities are rare treasures that make it easier to understand otherwise complicated motions. Like all conserved quantities, momentum can't be created or destroyed. It can only be transferred between objects. Momentum plays a very basic role in bumper cars: the whole point of crashing them into one another is to enjoy the momentum transfers. During each collision, momentum shifts from one car to the other so that they abruptly change their speeds or directions or both. As long as these momentum transfers aren't too jarring, everyone has a good time.

You've stopped your car, so it has zero velocity and zero momentum. To begin moving again, something must transfer momentum to your car. While you could press the pedal and let the motor gradually transfer momentum from the ground to your car, that's not much fun. Instead, you let two grinning couch potatoes in an overloaded green car slam into you at breakneck speed (Fig. 2.3.1).

The green car was heading westward, and in a few moments your car is moving westward, too, while the green car has slowed significantly. Before you recover from the jolt, your car pounds a child's car westward and your car slows down abruptly. Finally, its impact with a wall stops the child's car. Despite disapproving looks from the child's parents, there's no harm done. Overall, westward momentum has flowed from the spud-

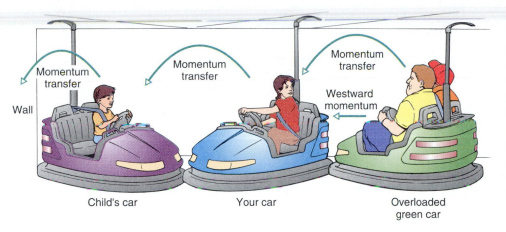

Momentum transfer

Momentum transfer

Momentum transfer

Wall

Westward momentum

Child's car

Your car

Overloaded green car

mobile to your car, to the child's car, and into the wall. No momentum has been created or destroyed; you've all simply enjoyed passing it along from car to car.

Check Your Understanding #1: Stuck on the Ice

Suppose you're stuck in the middle of a frozen lake, with a surface so slippery that you can't get any traction. You take off a shoe and throw it toward the southern shore. You find yourself coasting toward the northern shore and soon escape from the lake. Why did this scheme work?

Check Your Figures #1: Follow That Train!

The bad guys are getting away in a four-car train and you're trying to catch them. The train has a mass of 20,000 kg and it's rolling forward at 22 m/s (80 km/h or 50 mph). What is the train's momentum?

Exchanging Momentum in a Collision: Impulses

Momentum is transferred to a car by giving it an **impulse**—that is, a force exerted on it for a certain amount of time. When the motor and floor push your car forward for a few seconds, they give your car an impulse and transfer momentum to it. This impulse is the change in your car's momentum and is equal to the product of the force exerted on the car times the duration of that force. This relationship can be written as a word equation:

$$\text{impulse} = \text{force} \cdot \text{time}, \tag{2.3.2}$$

in symbols:

$$\Delta \mathbf{p} = \mathbf{F} \cdot t,$$

and in everyday language:

The harder and longer you push a bobsled forward at the start of a race, the more momentum it will have when it starts down the hill.

The more force or the longer that force is exerted, the larger the impulse and the more your car's momentum changes.

Different forces exerted for different amounts of time can transfer the same momentum to a car:

$$impulse = \text{large force} \cdot \text{short time}$$
$$= \text{small force} \cdot \text{long time}. \qquad (2.3.3)$$

Thus you can get your car moving with a certain forward momentum either by letting the motor and floor push on it with a small forward force of long duration or by letting the colliding green car push on it with a large forward force of short duration.

force versus momentum

We can now explain why bumper cars have soft rubber bumpers. If the bumpers were hard steel, the collision between the green car and your car would last only an instant and would involve an enormous forward force. You'd be in need of a neck brace and the services of a personal injury lawyer. However, amusement parks don't like lawsuits and sensibly limit the car impact forces. To do this, they use rubber bumpers and rather slow-moving cars.

Nonetheless, you can get a pretty good jolt when you collide head-on with another car. Your two cars then start with oppositely directed momenta, and the collision roughly exchanges those momenta between cars. In almost no time, you go from heading forward to heading backward. The impulse that causes this reversal of motion is especially large because it not only stops your forward motion, it also causes you to begin heading backward.

Why should momentum be a conserved quantity? It's conserved because of Newton's third law of motion. When one car exerts a force on a second car for a certain amount of time, the second car exerts an equal but oppositely directed force on the first car for exactly the same time. Because of the equal but oppositely directed nature of the two forces, cars that push on one another receive impulses that are equal in amount but opposite in direction. Since the momentum gained by one car is exactly equal to the momentum lost by the other car, we say that momentum is transferred from one car to the other.

The more mass a car has, the less its velocity changes as a consequence of a momentum transfer. That's why the green car doesn't stop completely when it crashes into your car, while your car speeds up dramatically. The green car has so much forward momentum that transferring a fraction of it to your car causes a large change in your car's velocity. Like a bug being hit by a car windshield, your car does most of the accelerating.

► Check Your Understanding #2: Bowling Them Over

When a beanbag hits the wall, it transfers all of its forward momentum to the wall and comes to a stop. When a rubber ball hits the wall, it transfers all of its forward momentum, comes to a stop, and then rebounds. During the rebound it transfers still more forward momentum to the wall. If you wanted to knock over a weighted bowling pin at the county fair, which would be the most effective projectile: the rubber ball or the beanbag, assuming they have identical masses and you throw them with identical velocities?

► Check Your Figures #2: Stop That Train!

The engine of the train you're chasing (see Check Your Figures #1) has broken down, but it's still rolling forward. To stop it, you grab onto the last car and begin to drag your boot heels on the ground. The backward force on the train is 200 N. How long will it take you to stop the train?

Spinning in Circles: Angular Momentum

Spinning is another important feature of bumper cars, and it involves angular momentum. **Angular momentum** is the measure of an object's rotational motion—its tendency to continue spinning about a particular axis. Simply put, your car's angular momentum indicates the direction of its rotation and just how difficult it was to get it spinning with its current angular velocity. The car's angular momentum is its moment of inertia times its angular velocity and can be written as a word equation:

$$\text{angular momentum} = \text{moment of inertia} \cdot \text{angular velocity}, \quad (2.3.4)$$

in symbols:

$$\mathbf{L} = I \cdot \boldsymbol{\omega},$$

and in everyday language:

It's hard to stop a spinning carousel.

Note that the direction of angular momentum, a vector quantity, is the same as the direction of angular velocity. The faster your car is spinning or the larger its moment of inertia, the more angular momentum it has in the direction of its angular velocity.

Angular momentum is another conserved quantity and hence can only be transferred between objects. For your car to begin spinning, something must transfer angular momentum to it, and your car will then continue to spin until it transfers this angular momentum elsewhere. But to study angular momentum properly, we must pick the pivot about which all the spinning will occur. In the present situation, a good choice for this pivot is your car's initial center of mass.

Your car is stationary again, so it has zero angular velocity and zero angular momentum. Suddenly, a purple car sweeps by and strikes your car a glancing blow (Fig. 2.3.2). Because the purple car was circling your car counterclockwise, it had counterclockwise angular momentum about the pivot. Its impact transfers some of this angular momentum to your car, which begins spinning counterclockwise itself. Since it has given up some of its angular momentum, the purple car circles your car less rapidly. Your car gradually stops spinning as its wheels and friction transfer the angular momentum to the ground and earth. Overall, no angular momentum was created or destroyed during the collision. Instead, it was transferred from the purple car to your car to the earth.

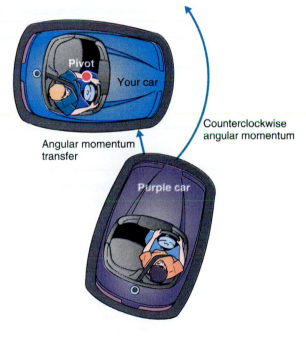

Fig. 2.3.2 Since the purple car is circling your car counterclockwise, it has counterclockwise angular momentum. When it hits your car, it transfers some of that angular momentum to your car. Because of this transfer, the purple car stops circling quickly as your car begins to spin counterclockwise.

Pivot

Your car

Counterclockwise angular momentum

Angular momentum transfer

Purple car

Check Your Understanding #3: Many Happy Re-Turns
Satellites are often set spinning during launch in order to give them added stability. When astronauts visit these satellites years later, they find them still spinning. Why don't the satellites stop spinning?

Check Your Figures #3: Want to Go for a Spin?
Spinning satellites are particularly stable. Suppose that the astronauts launching a particular satellite decide to increase its angular velocity by a factor of 5. How will that change affect the satellite's angular momentum?

Angular Impulses

Angular momentum is transferred to a car by giving it an **angular impulse**—that is, a torque exerted on it for a certain amount of time. When the purple car hits your car and exerts a torque on it briefly, it gives your car an angular impulse and transfers angular momentum to it. This angular impulse is the change in your car's angular momentum and is equal to the product of the torque exerted on your car times the duration of that torque. This relationship can be written as a word equation:

$$\text{angular impulse} \ = \ \text{torque} \cdot \text{time}, \qquad\qquad \textbf{(2.3.5)}$$

in symbols:

$$\Delta \mathbf{L} = \boldsymbol{\tau} \cdot t,$$

and in everyday language:

> *To get a merry-go-round spinning rapidly, you must twist it hard and for a long time.*

The more torque or the longer that torque is exerted, the larger the angular impulse and the more your car's angular momentum changes.

Different torques exerted for different amounts of time can transfer the same angular momentum to a car:

$$\text{angular impulse} = \text{large torque} \cdot \text{short time}$$
$$= \text{small torque} \cdot \text{long time}. \qquad \textbf{(2.3.6)}$$

Thus you can get your car spinning with a certain angular momentum either by letting the motor and floor twist it with a small torque of long duration or by letting the colliding purple car twist it with a large torque of short duration. As with linear momentum, sudden transfers of angular momentum can break things, so the cars are designed to limit their impact torques to reasonable levels. Even so, you may find yourself reaching for the motion sickness bag after a few spinning collisions.

Why should angular momentum be a conserved quantity? Like linear momentum, angular momentum is conserved because of Newton's third law of motion. In this case, we are referring to **Newton's third law of rotational motion:** if one object exerts a torque on a second object, then the second object will exert an equal but oppositely directed torque on the first object.

> **Newton's Third Law of Rotational Motion**
> For every torque that one object exerts on a second object, there is an equal but oppositely directed torque that the second object exerts on the first object.

When one car exerts a torque on a second car for a certain amount of time, the second car exerts an equal but oppositely directed torque on the first car for exactly the same amount of time. Because of the equal but oppositely directed nature of the two torques, cars that exert torques on one another receive angular impulses that are equal in amount but opposite in direction. Since the angular momentum gained by one car is exactly equal to the angular momentum lost by the other car, we say that angular momentum is transferred from one car to the other.

Because a car's angular momentum depends on its moment of inertia, two different cars may end up rotating at different angular velocities even though they have identical angular momenta. For example, when the purple car hits the overloaded green car and transfers angular momentum to it, the green car's enormous moment of inertia makes it spin relatively slowly. The same sort of behavior occurs with linear momentum, where a car's mass affects how fast it travels when it's given a certain amount of linear momentum. But while a bumper car can't change its mass, it can change its moment of inertia. If it does so while it's spinning, its angular *momentum* won't change, but its angular *velocity* will!

To see this change in angular velocity, consider the overloaded green car. Its two large occupants are disappointed with the ride because their huge mass and moment of inertia prevent them from experiencing the intense jolts and spins that you've been enjoying. Suddenly they get a wonderful idea. As their car slowly spins, one of them climbs into the other's lap and the two sit very close to the car's center of mass. By rearranging the car's mass this way, they have reduced the car's overall moment of inertia and the car actually begins to spin faster than before.

Since the green car's mass has been redistributed, it's no longer a freely turning rigid object covered by Newton's first law of rotational motion. However, its angular momentum must remain constant as long as it doesn't transfer that angular momentum elsewhere. Since the car's moment of inertia has become smaller, its angular velocity must increase in order to keep its angular momentum constant. That's just what happens. This effect of changing one's moment of inertia explains how an ice skater can achieve an enormous angular velocity by pulling herself into a thin, spinning object on ice (Fig. 2.3.3).

Fig. 2.3.3 When Kristi Yamaguchi pulls in her arms, she reduces her moment of inertia. She begins to spin more rapidly so as to maintain the same angular momentum.

Check Your Understanding #4: Spinning the Merry-Go-Round

A person who is initially motionless starts a merry-go-round spinning and then returns to being motionless. If angular momentum is truly conserved, what is the source of the angular momentum that the spinning merry-go-round now has?

Check Your Figures #4: Spin Away!

How much longer will it take the astronauts launching the satellite in Check Your Figures #3 to bring it to the faster angular velocity, if they use the initially planned torque?

The Conserved Quantities and Potential Energy

As you drive your bumper car around the oval, its motion is governed in large part by three conserved quantities: energy, linear momentum, and angular momentum (Table 2.3.1). While you can exchange those quantities with the earth and the power company by steering your car or switching on its motor, most of the interesting exchanges involve collisions.

Each time your car pushes another car forward, your car does work on that other car and transfers energy to it. Each time your car pushes a car northward briefly, your car gives a northward impulse to the other car and transfers northward momentum to it. And each time your car twists a car clockwise about its center of mass, your car gives a clockwise angular impulse to the other car and transfers clockwise angular momentum to it. These exchanges of energy, momentum, and angular momentum are fast and furious and make for an exciting ride.

| Table 2.3.1 | The Three Conserved Quantities of Motion and the Mechanism by Which They Are Transferred Between Objects |

CONSERVED QUANTITY	TRANSFER MECHANISM
Energy	Work
Linear momentum	Impulse
Angular momentum	Angular impulse

Shortly before the ride stops, you notice that there is a low point in the floor. After years of use, its metal surface has dented into a bowl-shaped depression and you observe that cars naturally tend to roll into this bowl and accelerate toward its bottom. We've seen this tendency to accelerate downhill before with ramps, but now let's look at it in terms of energy: a car always accelerates in the direction that reduces its total potential energy as quickly as possible. Since a lone car's only potential energy is gravitational potential energy, it accelerates in such a way as to reduce its gravitational potential energy as quickly as possible: down the steepest route to the bottom of the bowl.

potential energy and force

This behavior of accelerating in the direction that reduces total potential energy as quickly as possible is universal. Potential energy and forces are related to one another, so this rule is really just a way to determine the direction of the net force on an object or its parts. An object accelerates in the direction of the net force on it, which is also the direction that will reduce its total potential energy as quickly as possible. This rule is a useful way to determine how motion will proceed: which way a spring will leap, a chair will tip, or a bumper car will roll. We'll use it frequently in this book.

→ Check Your Understanding #5: Heading Down

When you pull a child back on a playground swing and let go, which way does that child accelerate?

Kinetic Energy

Before leaving bumper cars, there's one more question to answer: how do you determine your car's kinetic energy, its energy of motion? One way is to calculate just how much work its motor would have to do on it to bring it from rest to its current speed. The result of such a calculation is that a moving car's kinetic energy is equal to one-half of its mass times the square of its speed. This relationship can be written as a word equation:

$$\text{kinetic energy} = \tfrac{1}{2} \cdot \text{mass} \cdot \text{speed}^2, \qquad (2.3.7)$$

in symbols:

$$K = \tfrac{1}{2} \cdot m \cdot v^2,$$

and in everyday language:

Racing around at twice the speed takes four times the energy.

An automobile traveling 100 km/h (62 mph) has four times the kinetic energy of one traveling 50 km/h (31 mph). This huge increase in energy with a modest increase in speed explains why high-speed crashes are far more deadly than those at lower speeds. It also explains why your bumper car has a top speed of roughly 15 km/h (10 mph).

A rotating car also has kinetic energy. Like the kinetic energy of translational motion, the kinetic energy of rotational motion depends on the car's inertia and speed. But for a spinning car, it's the rotational inertia and rotational speed that matter. The car's kinetic energy is equal to one-half of its moment of inertia times the square of its angular speed. This relationship can be written as a word equation:

$$\text{kinetic energy} = \tfrac{1}{2} \cdot \text{moment of inertia} \cdot \text{angular speed}^2, \qquad \textbf{(2.3.8)}$$

in symbols:

$$K = \tfrac{1}{2} \cdot I \cdot \omega^2,$$

and in everyday language:

It takes a very energetic person to spin his wheels twice as fast.

The power is turned off and the bumper cars slow to a stop. In the final moments of the ride, you find that your car is rotating and translating at the same time. Its total kinetic energy is then simply the sum of the translational and rotational parts. Its translational kinetic energy depends on the speed of the car's center of mass and its rotational kinetic energy depends on the angular speed at which the car spins about that center of mass. Friction gradually turns all of this kinetic energy into thermal energy, and it's time to head for the exit.

➤ Check Your Understanding #6: Throwing a Fastball
A typical grade-school pitcher can throw a baseball at 80 km/h (50 mph), but only a few professional athletes have the extraordinary strength needed to throw a baseball at twice that speed. Why is it so much harder to throw the baseball only twice as fast?

➤ Check Your Figures #5: Blowing in the Wind
The air in a hurricane travels at 200 km/h (124 mph). How much more kinetic energy does 1 kg of this air have than 1 kg of air moving at only 20 km/h?

➤ Check Your Figures #6: Playing Around at the Playground
When children climb onto a playground merry-go-round, they increase its rotational inertia. If the children triple the merry-go-round's moment of inertia, how will they alter the kinetic energy it has when it spins at a certain angular speed?

Epilogue for Chapter 2

In this chapter we looked at rotating and colliding objects and studied the physical laws that describe their motions. In *seesaws,* we examined rotational inertia and saw how torques cause angular accelerations. We also noticed how useful it can be to separate an object's rotational motion from its translational motion. In *wheels,* we discussed another important type of force, friction, as well as a new type of energy, thermal energy—the energy associated with heat and temperature. In *bumper cars,* we introduced two more conserved physical quantities: momentum and angular momentum. As we'll see, following the flows of energy, momentum, and angular momentum between objects often helps in understanding how those objects work.

Explanation: Spinning a Pie Dish

Because the balanced dish has rotational inertia, torques are required both to start it spinning and to stop it from doing so. When you twist the dish with your hand, the torque you exert gives the dish an angular impulse and sets it spinning with a certain amount of angular momentum. If the pivot were truly frictionless and there were no air resistance, the dish would spin indefinitely because it would be unable to get rid of its angular momentum. However, friction in the pivot exerts a small but significant torque that opposes the dish's motion and gradually slows it down. As this frictional torque transfers angular momentum out of the dish and into the pen, chair, and earth, the dish turns more and more slowly until it finally comes to a stop. The sharper the pivot and the smaller the contact area between point and dish, the less frictional torque the dish experiences and the longer it spins.

Balancing the dish is easy as long as it's upside-down. With its edge drooping downward, the dish has relatively little gravitational potential energy and is surprisingly stable. If it begins to tip to one side, the dish's average height rises and so does its gravitational potential energy. Since objects naturally accelerate in whatever direction lowers their potential energy as quickly as possible, the upside-down dish quickly tips back toward level after being disturbed. In contrast, an upright dish is virtually impossible to balance on a point because any tip will lower its gravitational potential energy and lead quickly to catastrophe. We'll look at these stabilizing/destabilizing effects more carefully later on in this book.

Chapter Summary

How Seesaws Work: A seesaw is a rotating toy that works best when it's almost perfectly balanced, meaning that it experiences zero net torque. The seesaw's pivot usually passes through the board's center of mass so that the board balances when it's not occupied. The riders arrange themselves so that the torques they exert on the seesaw cancel one another completely. The board then experiences zero net torque and zero angular acceleration, and it rotates with constant angular velocity. It either remains motionless or turns steadily in one direction or the other.

To make the seesaw tip back and forth, the riders subtly adjust the torques they exert on the seesaw. They do this either by leaning, thus varying their distances from the pivot, or by pushing against the ground with their feet, thus varying the forces they exert on the board. In either case, they unbalance the seesaw, and it experiences both a net torque and an angular acceleration. By rhythmically changing the net torque on the seesaw, the riders cause it to rotate back and forth.

How Wheels Work: Wheels facilitate motion by eliminating or reducing sliding friction between an object and a surface. The wheels convey the support forces needed to hold the object up but allow the object to move without sliding. As a cart with freely turning wheels moves along the surface, static friction between each wheel and the surface exerts a torque on that wheel and causes it to turn. However, rubbing may occur between the wheel's hub and the axle, where sliding friction can waste energy and cause wear. To eliminate this sliding friction, roller or ball bearings are often used.

The torque that causes a powered wheel on a vehicle to turn comes from an engine by way of an axle. In this case, static friction between the outside of the wheel and the ground exerts a torque on the wheel that opposes the torque from the engine. This static frictional force also contributes to the net force on the vehicle and causes it to accelerate.

Once supported on wheels and bearings, objects can move freely and can retain linear momentum, angular momentum, and energy for long periods of time. By eliminating sliding friction, wheels can also keep objects from converting their ordered energies, either kinetic or potential, into thermal energy. Wheels allow vehicles to hold onto these conserved quantities for extended periods and make transportation far more practical.

How Bumper Cars Work: Since they start from rest, bumper cars must obtain their initial momenta and angular momenta from the ground and their initial kinetic energies from the power company. They do this with the help of motors and wheels, which gradually transfer energy, momentum, and angular momentum into the cars.

Once the cars are moving, they can begin to exchange those conserved quantities by way of collisions. Each impact usually changes the cars' speeds and directions of travel in a manner that may seem rather complicated. However, following the exchanges of momentum, angular momentum, and energy often makes it easier to understand these collisions.

Cars containing massive riders respond weakly when collisions transfer momentum and angular momentum to them. That's because their large masses and moments of inertia minimize their changes in velocity and angular velocity. Because of their small masses, children experience the wildest rides.

Important Laws and Equations

1. **Newton's First Law of Rotational Motion:** A rigid object that is not wobbling and is not subject to any outside torques rotates at a constant angular velocity, turning equal amounts in equal times about a fixed axis of rotation.

2. **Newton's Second Law of Rotational Motion:** The torque exerted on an object is equal to the product of that object's moment of inertia times its angular acceleration, or

$$\text{torque} = \text{moment of inertia} \cdot \text{angular acceleration.} \quad (2.1.1)$$

The angular acceleration points in the same direction as the torque. This law doesn't apply to objects that are wobbling.

3. **Relationship Between Force and Torque:** The torque produced by a force is equal to the product of the lever arm times that force, or

$$\text{torque} = \text{lever arm} \cdot \text{force,} \quad (2.1.2)$$

where we include only the component of the force that's at right angles to the lever arm.

4. **Linear Momentum:** An object's linear momentum is its mass times its velocity, or

$$\text{linear momentum} = \text{mass} \cdot \text{velocity.} \quad (2.2.1)$$

5. **The Definition of Impulse:** The impulse given to an object is equal to the product of the force exerted on that object times the length of time that force is exerted, or

$$\text{impulse} = \text{force} \cdot \text{time.} \quad (2.2.2)$$

6. **Angular Momentum:** An object's angular momentum is its moment of inertia times its angular velocity, or

$$\text{angular momentum} = \text{moment of inertia} \cdot \text{angular velocity.} \quad (2.2.4)$$

7. **The Definition of Angular Impulse:** The angular impulse given to an object is equal to the product of the torque exerted on that object times the length of time that torque is exerted, or

$$\text{angular impulse} = \text{torque} \cdot \text{time.} \quad (2.2.5)$$

8. Newton's Third Law of Rotational Motion: For every torque that one object exerts on a second object, there is an equal but oppositely directed torque that the second object exerts on the first object.

9. Kinetic Energy: An object's translational kinetic energy is one-half of its mass times the square of its speed, or

$$\text{kinetic energy} = \tfrac{1}{2} \cdot \text{mass} \cdot \text{speed}^2. \qquad (2.3.7)$$

An object's rotational kinetic energy is one-half of its moment of inertia times the square of its angular speed, or

$$\text{kinetic energy} =$$
$$\tfrac{1}{2} \cdot \text{moment of inertia} \cdot \text{angular speed}^2. \qquad (2.3.8)$$

10. Potential Energy and Acceleration: An object accelerates in the direction that reduces its total potential energy as quickly as possible.

Check Your Understanding—Answers

Section 2.1 SEESAWS

1. The lazy Susan undergoes rotational motion while the dessert cart undergoes translational motion.

Why: The lazy Susan has a fixed pivot at its center. This pivot never goes anywhere, regardless of how much you turn the lazy Susan. The tray simply rotates about this fixed pivot. In contrast, the dessert cart moves about the room and has no fixed point. The server can rotate the dessert cart when necessary, but its principal motion is translational.

2. It will continue to spin at a steady pace about a fixed rotational axis (although friction with the water will gradually slow the ball's rotation).

Why: Because the basketball is free of torques, the outside influences that affect rotational motion, it has a constant angular velocity. If you spin the basketball, it will continue to spin about whatever axis you chose. If you don't spin the basketball, its angular velocity will be zero and it will remain stationary.

3. Yes. His center of mass falls smoothly, obeying the rules governing falling objects. As he falls, his body rotates at constant angular velocity about his center of mass.

Why: Like a thrown football or tossed baton, the diver is a rigid, rotating object. His motion can be separated into translational motion of his center of mass (it falls) and rotational motion about his center of mass (he rotates about it at constant angular velocity). While the diver may never think of his motion in these terms, he is aware intuitively of the need to handle both his rotational and translational motions carefully. Hitting the water with his chest because he did not rotate properly isn't much more fun than hitting the board because he didn't handle translation properly.

4. The full merry-go-round has a huge moment of inertia.

Why: Starting or stopping a merry-go-round involves angular acceleration. As the pusher, you exert a torque on the merry-go-round and it undergoes angular acceleration. But this angular acceleration depends on the merry-go-round's moment of inertia, which in turn depends on how much mass it has and how far that mass is from the axis of rota-

tion. With many children adding to the merry-go-round's moment of inertia, its angular acceleration tends to be small.

5. The closer the cardboard is to the pivot, the more force it must exert on the scissors to produce enough torque to keep the scissors from rotating closed. When the cardboard is unable to produce enough torque, the scissors cut through it.

Why: When you place paper close to the pivot of a pair of scissors, you are requiring that paper to exert enormous forces on the scissors to keep them from rotating closed. Rotations are started and stopped by torques, and forces exerted close to the pivot exert relatively small torques.

6. About 1000 N.

Why: Since the nail is 10 times closer to the pivot, the nail must exert 10 times the force on the hammer to create the same magnitude of torque as you do pulling on the handle. As the nail pulls on the hammer, the hammer pulls on the nail. Although the wood exerts frictional forces on the nail to keep it from moving, the extracting force overwhelms this friction and the nail slides slowly out of the wood.

7. A substantial shift in the cargo's position during a storm can unbalance the ship, creating a net torque on the ship, and cause it to begin rotating about the effective pivot. The ship may then capsize.

Why: Although most boats can compensate for some amount of cargo imbalance, shifting cargo can easily flip even a fairly stable boat. It happens frequently in real life, often with fatal consequences. Some boats, particularly canoes and racing shells, are notoriously sensitive to unbalanced loading and are easily flipped by careless or moving occupants.

Section 2.2 WHEELS

1. Friction is pushing the glass uphill.

Why: The glass is sliding downhill across the top of the stationary table. Since friction always opposes relative motion, it pushes the glass uphill, in the direction opposite its motion.

2. About twice as hard.

Why: The frictional forces between the table and books are roughly proportional to the force pressing them together. The book's weight is what pushes them together. When you effectively double the book's weight, by stacking a second book on top of it, you double the frictional forces between the table and the books.

3. If the wheels continue to turn, they experience static friction with the pavement. If they lock and begin to skid, they experience sliding friction. The maximum force of static friction is larger than the force of sliding friction, so the car will decelerate faster if the wheels don't skid.

Why: For a rapid stop, the car needs the maximum possible force in the direction opposite its velocity. The most effective way to obtain that stopping force from the road is with static friction between the turning wheels and the pavement. Sliding friction, the result of skidding tires, is much less effective at stopping the car, wears out the tires, and diminishes the driver's ability to steer the vehicle.

4. The archer does work on the string and bow as she draws the arrow back. (Chemical energy from her body is transferred to the bow, where it is stored as elastic potential energy.) As she releases the arrow, the string and bow do work on the arrow. (Elastic potential energy is transferred to the arrow, where it becomes kinetic energy.) Finally, the arrow does work on the apple, knocking it off the head of the assistant. (Kinetic energy is transferred from the arrow to the apple.)

Why: Because energy is conserved, we could in principle follow it back to the origins of the universe. Whatever energy we see around us now was somewhere in our universe yesterday, last week, and a million years ago, although its form may have changed. It will still be in our universe next year, too, but maybe not in as useful a form.

5. Normal driving involves mostly static friction because the surfaces of the tires don't slide across the pavement. Skidding involves sliding friction as the tire surfaces move independently of the pavement. Because it involves sliding friction, skidding creates thermal energy and damages the tires.

Why: The expression "burn rubber" is an appropriate name for skidding during a "jack-rabbit start." Substantial thermal energy is produced, and a trail of hot rubber is left on the pavement behind the car. At drag races, the frictional heating that results from skidding at the start can be so severe that the tires actually catch on fire.

6. Because all the supporting forces are very close to the axis of rotation, the jewels exert almost zero torque on the axle. The axle turns remarkably freely.

Why: Mechanical timepieces need almost ideal motion to keep accurate time. One of the best ways to allow a rotating

object free movement is to support it exactly on the axis of rotation, where the support can't exert torque on the object.

Section 2.3 BUMPER CARS

1. By transferring southward momentum to the shoe, you would obtain northward momentum.

Why: Initially, both you and your shoe have zero momentum. But when you throw the shoe southward, you give it southward momentum. Since the only source of that southward momentum is you, you must have lost southward momentum. A negative amount of southward momentum is actually northward momentum, and thus you coast northward. Interestingly enough, the total momentum of you and the shoe hasn't changed. It's still zero, as it must be because momentum is conserved. It has simply been redistributed.

2. The bouncy rubber ball would be more effective.

Why: Either projectile will transfer all of its original momentum to the bowling pin while coming to a stop. But then the bouncy rubber ball will bounce back and continue to exert a force on the bowling pin. The impulse (force · time) delivered by the rubber ball will be greater than that delivered by the beanbag because the ball will exert its forward force for a longer time (during stopping *and* rebounding). The ball will rebound with its momentum reversed, having transferred roughly twice its original momentum to the pin.

3. The satellites are unable to get rid of their angular momentum.

Why: Because of their extreme isolation, orbiting satellites have nothing with which to exchange angular momentum. The angular momentum given to them at launch stays with them indefinitely, so they continue to spin for decades.

4. It came from the entire earth.

Why: Because the person stood on the earth as he started the merry-go-round spinning, he transferred angular momentum from the earth to the merry-go-round. The merry-go-round spins in one direction, and the earth's rotation changes ever so slightly in the other direction. Because the earth is so enormous and has such an enormous moment of inertia, its slight change in rotation is undetectable.

5. The child accelerates forward, in the direction that will reduce the potential energy as quickly as possible.

Why: The child has only one form of potential energy—gravitational potential energy. This gravitational potential energy is lowest when the child is directly below the swing's supporting bar. The child accelerates forward because that will put the child below the support as quickly as possible.

6. Doubling the speed of the baseball requires quadrupling the energy transferred to it by the pitcher.

Why: To throw a 160-km/h (100-mph) fastball, a major league pitcher must put four times as much kinetic energy into both the ball and his arm as when pitching an 80-km/h slow ball. He also pitches the fastball in half the time needed to pitch the slow ball. Overall, he must do four times as much work throwing a fastball and he must do that work in one-half the time. That means the pitcher produces eight times as much power while throwing a fastball as when throwing a slow ball. No wonder amateurs have trouble duplicating that feat.

Check Your Figures—Answers

Section 2.1 SEESAWS

1. About 10 times as much torque.

Why: To keep the angular acceleration in Eq. 2.1.1 unchanged while increasing the moment of inertia by a factor of 10, the torque must also increase by a factor of 10. Solid tires are extremely difficult to spin or to stop from spinning, which is why automobiles use hollow tires.

2. Five times as much torque as before.

Why: The pipe increases the wrench's lever arm by a factor of 5, from 0.2 meter to 1.0 meter. According to Eq. 2.1.2, the same force exerted five times as far from the pivot will produce five times as much torque about that pivot. Extending the handle of a leverlike tool is a common technique to increase the available torque, although it can be hazardous for both the tool and its user. Some tools that are designed for such extreme use come with removable handle extensions.

Section 2.3 BUMPER CARS

1. 440,000 kg·m/s.

Why: You can use Eq. 2.2.1 to calculate the train's momentum from its mass and velocity:

$$\text{linear momentum} = 20,000 \text{ kg} \cdot 22 \text{ m/s}$$
$$= 440,000 \text{ kg} \cdot \text{m/s}.$$

That momentum is in the forward direction.

2. 2200 s, so kiss your boot heels goodbye!

Why: To stop the train, you must give it a backward impulse that completely cancels its forward momentum. Since its forward momentum is 440,000 kg·m/s, the backward impulse must be 440,000 kg·m/s. Since 200 N can also be written as 200 kg·m/s^2, we can use Eq. 2.2.2 to find the time:

$$\text{time} = \frac{440,000 \text{ kg} \cdot \text{m/s}}{200 \text{ kg} \cdot \text{m/s}^2}$$
$$= 2200 \text{ s}.$$

3. The angular momentum will increase by a factor of 5.

Why: Because the satellite's angular momentum is proportional to its angular velocity, spinning it five times faster will increase its angular momentum by that same factor.

4. It will take them five times as long.

Why: To reach the new, faster angular velocity, the astronauts will need an angular impulse that's five times as large as originally planned. Since they will be using the same torque, they will have to exert that torque for five times as long.

5. One hundred times as much kinetic energy.

Why: Because kinetic energy is proportional to speed squared, the kilogram of air in the hurricane moves 10 times as fast but has 100 times as much kinetic energy as the slower moving air. This enormous increase in energy is what makes a hurricane's wind dangerous. The air's terrific speed also brings large quantities of it to you quickly, so that the wind power arriving each second is overwhelming.

6. The children will triple the merry-go-round's kinetic energy.

Why: The kinetic energy of a spinning object is proportional to its moment of inertia. Since the children triple the merry-go-round's moment of inertia, they triple its kinetic energy.

Exercises

1. The chairs in an auditorium aren't all facing the same direction. How could you describe their angular positions in terms of a reference orientation and a rotation?

2. When an airplane starts its propellers, they spin slowly at first and gradually pick up speed. Why does it take so long for them to reach their full rotational speed?

3. A mechanic balances the wheels of your car to make sure that their centers of mass are located exactly at their geometrical centers. Neglecting friction and air resistance, how would an improperly balanced wheel behave if it were rotating all by itself?

4. An object's center of mass isn't always inside the object, as you can see by spinning it. Where is the center of mass of a boomerang or a horseshoe?

5. Why is it hard to start the wheel of a roulette table spinning, and what keeps it spinning once it's started?

6. Why can't you open a door by pushing its doorknob directly toward or away from its hinges?

7. Why can't you open a door by pushing on its hinged side?

8. It's much easier to carry a weight in your hand when your arm is at your side than it is when your arm is pointing straight out in front of you. Use the concept of torque to explain this effect.

9. A gristmill is powered by falling water, which pours into buckets on the outer edge of a giant wheel. The weight of the water turns the wheel. Why is it important that those buckets be on the wheel's outer edge?

10. How does the string of a yo-yo get the yo-yo spinning?

11. One way to crack open a walnut is to put it in the hinged side of a door and then begin to close the door. Why does a small force on the door produce a large force on the shell?

12. A common pair of pliers has a place for cutting wires, bolts, or nails. Why is it so important that this cutter be located very near the pliers' pivot?

13. You can do push-ups with either your toes or your knees acting as the pivot about which your body rotates. When you pivot about your knees, your feet actually help you to lift your head and chest. Explain.

14. Tightrope walkers often use long poles for balance. Although the poles don't weight much, they can exert substantial torques on the walkers to keep them from tipping and falling off the ropes. Why are the poles so long?

15. Some racing cars are designed so that their massive engines are near their geometrical centers. Why does this design make it easier for these cars to turn quickly?

16. How does a bottle opener use mechanical advantage to pry the top off a soda bottle?

17. A jar-opening tool grabs onto a jar's lid and then provides a long handle for you to turn. Why does this handle's length help you to open the jar?

18. When you climb out on a thin tree limb, there's a chance that the limb will break off near the trunk. Why is this disaster most likely to occur when you're as far out on the limb as possible?

19. How does a crowbar make it easier to lift the edge of a heavy box a few centimeters off the ground?

20. The basket of a wheelbarrow is located in between its wheel and its handles. How does this arrangement make it relatively easy for you to lift a heavy load in the basket?

21. Skiers often stop by turning their skis sideways and skidding them across the snow. How does this trick remove energy from a skier, and what happens to that energy?

22. A horse does work on a cart it's pulling along a straight, level road at a constant speed. The horse is transferring energy to the cart, so why doesn't the cart go faster and faster? Where is the energy going?

23. Explain why a rolling pin flattens a piecrust without encountering very much sliding friction as it moves.

24. Professional sprinters wear spikes on their shoes to prevent them from sliding on the track at the start of a race. Why is energy wasted whenever a sprinter's foot slides backward along the track?

25. A yo-yo is a spool-shaped toy that spins on a string. In a sophisticated yo-yo, the end of the string forms a loop around the yo-yo's central rod so that the yo-yo can spin almost freely at the end of the string. Why does the yo-yo spin longest if the central rod is very thin and very slippery?

26. As you begin pedaling your bicycle and it accelerates forward, what is exerting the forward force that the bicycle needs to accelerate?

27. When you begin to walk forward, what exerts the force that allows you to accelerate?

28. If you are pulling a sled along a level field at constant velocity, how does the force you are exerting on the sled compare to the force of sliding friction on its runners?

29. Why does putting sand in the trunk of a car help to keep the rear wheels from skidding on an icy road?

30. When you're driving on a level road and there's ice on the pavement, you hardly notice that ice while you're heading straight at a constant speed. Why is it that you only notice how slippery the road is when you try to turn left or right, or to speed up or slow down?

31. Describe the process of writing with chalk on a blackboard in terms of friction and wear.

32. Falling into a leaf pile is much more comfortable than falling onto the bare ground. In both cases you come to a complete stop, so why does the leaf pile feel so much better?

33. In countless movie and television scenes, the hero punches a brawny villain who doesn't even flinch at the impact. Why is the immovable villain a Hollywood fantasy?

34. Why can't an acrobat stop himself from spinning while he is in midair?

35. While a gymnast is in the air during a leap, which of the following quantities must remain constant for her: velocity, momentum, angular velocity, or angular momentum?

36. If you sit in a good swivel chair with your feet off the floor, the chair will turn slightly as you move about but will immediately stop moving when you do. Why can't you make the chair spin without touching something?

37. When a star runs out of nuclear fuel, gravity may crush it into a neutron star about 20 km (12 miles) in diameter. While the star may have taken a year or so to rotate once before its collapse, the neutron star rotates several times a second. Explain this dramatic increase in angular velocity.

38. A toy top spins for a very long time on its sharp point. Why does it take so long for friction to slow the top's rotation?

39. It's easier to injure your knees and legs while hiking downhill than while hiking uphill. Use the concept of energy to explain this observation.

40. When you first let go of a bowling ball, it's not rotating. But as it slides down the alley, it begins to rotate. Use the concept of energy to explain why the ball's forward speed decreases as it begins to spin.

41. Firefighters slide down a pole to get to their trucks quickly. What happens to their gravitational potential energy, and how does it depend on the slipperiness of the pole?

Problems

1. When you ride a bicycle, your foot pushes down on a pedal that's 17.5 cm (0.175 m) from the axis of rotation. Your force produces a torque on the crank attached to the pedal. Suppose that you weigh 700 N. If you put all your weight on the pedal while it's directly in front of the crank's axis of rotation, what torque do you exert on the crank?

2. An antique carousel that's powered by a large electric motor undergoes constant angular acceleration from rest to full rotational speed in 5 seconds. When the ride ends, a brake causes it to decelerate steadily from full rotational speed to rest in 10 seconds. Compare the torque that starts the carousel to the torque that stops it.

3. When you start your computer, the hard disk begins to spin. It takes 6 seconds of constant angular acceleration to reach full speed, at which time the computer can begin to access it. If you wanted the disk drive to reach full speed in only 2 seconds, how much more torque would the disk drive's motor have to exert on it during the starting process?

4. An electric saw uses a circular spinning blade to slice through wood. When you start the saw, the motor needs 2 seconds of constant angular acceleration to bring the blade to its full angular velocity. If you change the blade so that the rotating portion of the saw now has three times its original moment of inertia, how long will the motor need to bring the blade to its full angular velocity?

5. When the saw in Problem 4 slices wood, the wood exerts a 100-N force on the blade, 0.125 m from the blade's axis of rotation. If that force is at right angles to the lever arm, how much torque does the wood exert on the blade? Does this torque make the blade turn faster or slower?

6. When you push down on the handle of a doll-like wooden nutcracker, its jaw pivots upward and cracks a nut. If the point at which you push down on the handle is five times as far from the pivot as the point at which the jaw pushes on the nut, how much force will the jaw exert on the nut if you exert a force of 20 N on the handle? (Assume all forces are at right angles to the lever arms involved.)

7. Some special vehicles have spinning disks (flywheels) to store energy while they roll downhill. They use that stored energy to lift themselves uphill later on. Their flywheels have relatively small moments of inertia but spin at enormous angular speeds. How would a flywheel's kinetic energy change if its moment of inertia were five times larger but its angular speed were five times smaller?

8. What is the momentum of a fly if it's traveling 1 m/s and has a mass of 0.0001 kg?

9. Your car is broken, so you're pushing it. If your car has a mass of 800 kg, how much momentum does it have when it's moving forward at 3 m/s (11 km/h)?

10. You begin pushing the car forward from rest (see Problem 9). Neglecting friction, how long will it take you to push your car up to a speed of 3 m/s on a level surface if you exert a constant force of 200 N on it?

11. When your car is moving at 3 m/s (see Problems 9 and 10), how much translational kinetic energy does it have?

12. No one is driving your car (see Problems 9, 10, and 11) and it crashes into a parked car at 3 m/s. Your car comes to a stop in just 0.1 s. What force did the parked car exert on it to stop it that quickly?

13. You're at the roller-skating rink with a friend who weighs twice as much as you do. The two of you stand motionless in the middle of the rink so that your combined momentum is zero. You then push on one another and begin to roll apart. If your momentum is then 450 kg · m/s to the left, what is your friend's momentum?

Cases

1. You're seated in a crowded bus that has stopped at a bus stop. Two people board the bus, one wearing rubber-soled shoes and one wearing inline skates (roller skates—i.e., no friction). They both have to stand in the middle of the aisle and neither one holds onto anything.

   a. The bus driver is new and the bus lurches forward from the bus stop. Which way does the person wearing inline skates move in relationship to the bus?

   b. The bus is accelerating forward as it pulls away from the bus stop. Why doesn't the person wearing inline skates accelerate with the bus?

   c. The person wearing rubber-soled shoes remains in place as the bus starts moving. What provides the force that causes that person to accelerate with the bus?

   d. Once the bus has reached a constant velocity (it's traveling along a straight, level road at a steady pace), the person wearing inline skates is able to stand comfortably in the aisle without rolling anywhere. What is the net force on the person wearing skates?

   e. The driver accidentally runs into a curb and the bus stops so abruptly that everything slides forward, including the person wearing rubber-soled shoes. In stopping quickly, the bus experiences an enormous backward acceleration. Why doesn't the person wearing rubber-soled shoes accelerate backward with the bus?

2. Revolving doors are popular in northern hotels and office buildings as a way to prevent cold outside air from blowing directly into the lobby. Most revolving doors have four panels arranged in a cross, as viewed from above. You step in between two panels of the door and push on the panel in front of you. The revolving door begins to rotate and once you reach the inside of the building, you step out into the lobby.

   a. It is much easier to make the revolving door rotate by pushing on the panel far from the central pivot than it is by pushing near the pivot. Why?

   b. As you push on the door, it begins to turn more and more quickly. What is your pushing doing to the door?

   c. One of the dangers of revolving doors is being hit by the panel behind you as you step out of the door. The door tends to keep on turning after you stop pushing, and it can bump you if you're not careful. Why does it keep on turning after you stop pushing?

   d. What eventually stops the revolving door when no one uses it for a minute or two?

   e. The people who built the revolving door wanted it to be easy to start and stop. They had to be most careful to minimize the mass of which edge of each door panel: the upper edge, lower edge, inside edge (nearest the pivot), or outside edge (farthest from the pivot)?

3. A 100,000-kg train is traveling north along a straight, level track at a constant speed of 30 m/s (108 km/h or 67 mph).

   a. As it moves forward at constant velocity, what is the net force on the train?

   b. What is the train's momentum?

   c. To stop the train, how much momentum must you transfer to it?

   *d. If you exert a 500-N southward force on the coasting train, how long will it take to stop the train?

4. In bowling, you roll a heavy plastic ball down a smooth wooden lane and try to knock over 10 upright pins.

   a. When you first let go of the ball, it is not rotating and skids along the wooden surface. What is happening to its kinetic energy as it skids?

   b. The ball gradually starts to rotate. What provides the torque that causes it to spin?

   c. When the ball strikes the pins, they go flying while the ball continues forward. Why doesn't the ball simply bounce off the pins?

   d. Bowling balls come in a variety of different weights. How does the ball's mass affect its behavior when it reaches the pins?

5. A popular exercise machine in a modern weight room simulates the act of climbing stairs. You stand on two pedals and move your feet up and down. The pedals move with your feet, and hidden machinery inside the device makes a whirring sound. It takes a lot of force to push each pedal downward, but that pedal rises upward easily as you lift your foot. As long as you keep moving your feet up and down at the right pace, you remain stationary. If you slow your pace, you begin to sink downward. If you speed up, you begin to rise upward.

   a. When you're moving your feet up and down at the right pace, so that you remain stationary, how much total downward force are you exerting on the two pedals?

   b. As you push your foot and the pedal downward, you exert a large downward force on the pedal. Between your foot and the pedal, which is doing (positive) work on which?

   c. As you lift your foot upward, the pedal exerts a small upward force on your foot. Between your foot and the pedal, which is doing (positive) work on which?

   d. On the average, what is doing (positive) work on what (you and the machine)?

   e. Clearly, energy is being transferred out of you and into the machine. The machine converts that energy into a form that can be carried away in the air. What is that final form of energy?

6. Local fairs and amusement parks usually offer games in which you can win a large prize by performing a seemingly easy task. In many cases, these tasks are made surprisingly difficult by simple physical principles and few people receive prizes. Here are several of those games.

a. A pitching game requires that you knock over three milk bottles with a baseball. The bottles are filled with sand. Why does filling the bottles with sand make them so hard to knock over with the baseball?

b. A tossing game requires that you throw a coin forward and have it come to a stop on a smooth glass plate. Why doesn't the coin stop when it hits the smooth glass plate?

c. Another game requires that you knock over a miniature bowling pin with a ball hanging from a string. The string is suspended from a point directly above the pin. To win, you must swing the ball past the pin and have the ball knock over the pin on its return swing. This feat can't be done. The ball keeps circling around the pin at a relatively constant distance. The ball can't stop this circling to hit the pin and knock it over. Why does the ball keep circling the pin?

d. Still another tossing game requires that you throw a basketball into a shallow basket that is tipped toward you. Whenever you throw the ball into the basket, it bounces back out of the basket and falls onto the floor. What conserved physical quantity is the basketball unable to get rid of in time to remain in the basket?

7. Advanced skiers turn by sliding the backs of their skis across the snow. Since the fronts of their skis don't move much, the skis end up pointed in a new direction.

a. The amount of sideways force that a skier must exert on the skis to slide them sideways is proportional to how hard the skis press down on the snow beneath them. Why?

b. To make it easier to slide the skis sideways, the skier "unweights"—reduces the force pressing the skis downward against the snow. The skier does this by jumping upward. How can the downward force that the skier exerts on the snow be less than the skier's weight?

c. When during the jump is it easiest to slide the skis sideways?

d. Rather than jumping, some racers simply pull their legs upward suddenly. Why does this action reduce the force pressing their skis against the snow?

e. Less skilled skiers sometimes turn without unweighting—they push their skis sideways so hard that the skis slide anyway. This technique is exhausting. Why does it require so much work?

Mechanical Objects

Now that we've surveyed the laws of motion, we can begin using those laws to explain the behavior of everyday objects. But while we can already understand some of the central features at work in a toy wagon, a weight machine, or a ski lift, we're still missing a number of mechanical concepts that are important in the world around us. In this chapter, we'll look at some of those additional concepts.

One of the most important will be acceleration. If we treat acceleration passively, it can be fairly uninteresting: we push on the cart and the cart accelerates. But if we think of the concept more actively—for example, if we envision ourselves on a roller coaster as it plummets down that first big hill—then acceleration becomes much more intriguing. In fact, we might even need to hold on to our hats.

EXPERIMENT: Swinging Water Overhead

To examine some of the novel effects of acceleration, try experimenting with a bucket of water. If you're careful, you can swing this bucket over your head and up-

side-down without spilling a drop. In the process, you'll be demonstrating a number of important physical concepts.

To do this experiment, you'll need a bucket with a handle. (You might substitute some equivalent container; even a plastic cup will do in a pinch.) Fill the bucket part way full of water and then hold it by the handle so that it hangs down by your side.

Now swing the bucket backward about an eighth of a turn and bring it forward rapidly. In one smooth, fluid motion, swing it forward, up, and over your head. Continue this motion all the way around behind you and then bring the bucket forward again. You'll need to swing the bucket quickly to avoid getting wet. As you swing the bucket around and around, you'll notice that the water stays in it even when it travels over your head and is upside down. Why doesn't the water fall out?

You can carry this experiment a step further by swinging the bucket at various speeds—that is, if you don't mind getting wet. First, try to *predict* what will happen if you swing the bucket less rapidly or more rapidly. Now do the experiments and *observe* what happens. Did the experiments *verify* your predictions?

As you swing the bucket over your head, try to *measure* how strongly the bucket pulls on your hand. Is the pull stronger or weaker when you swing less rapidly? more rapidly? Is there any relationship between the upward pull you feel from the inverted bucket and the water's tendency to remain inside?

You might vary this experiment in several ways. Try swinging a plastic cup held in your fingers, or placing a full wine glass in the bucket and swinging the two objects together. In the latter case, you'll find that the wine will stay in the glass and the glass will stay at the bottom of the bucket, even when the bucket is upside-down.

By the way, the hardest part of all these tricks is stopping. To avoid a catastrophe, you'll need to do the same thing you did to get started, only in reverse. Come to a smooth, gradual stop about an eighth of a turn in front of you, and then let the bucket loosely drop back to your side. If you stop moving the bucket too abruptly, the water, wine, or glass will spill or smash. Why do you suppose that happens?

Chapter Itinerary

In this chapter, we'll examine four types of everyday objects: (1) *spring scales,* (2) *bouncing balls,* (3) *carousels and roller coasters,* and (4) *bicycles.* In *spring scales,* we'll review the relationship between mass and weight and explore how the distortion of a spring can be used to measure an object's weight. In *bouncing balls,* we'll study how balls store and return energy and how their bouncing depends both on their own characteristics and on those of the objects they hit. In *carousels and roller coasters,* we'll look at how acceleration gives rise to gravity-like apparent forces that can make us scream with delight at the amusement park. And in *bicycles,* we'll see how motion and clever design can make an apparently unstable object stable enough to ride without hands on the handlebars.

The concepts these objects illustrate can explain other phenomena as well. Almost any solid object, from a mattress to a diving board to a tire, behaves like a spring scale's spring when you push on it. Bouncing balls offer a view of collisions that will help you comprehend what happens when two cars crash or when a hammer hits a nail. The sensations associated with acceleration that you experience on a roller coaster are also present when you ride in airplanes, on subways, or on swing sets. And the stability issues facing a bicyclist are the same as those that waiters, water-skiers, and warehouse clerks must deal with all the time. When it comes to the physics of everyday objects, there really is nothing new under the sun.

Spring Scales

How much of you is there? From day to day, depending on how much you eat, the amount of you stays approximately the same. But how can you tell how much that is? The best measure of quantity is mass: kilograms of gold, beef, grain, or you. Mass is the measure of an object's inertia and, as we saw in Section 1.1, doesn't depend on the object's environment or on gravity. A kilogram box of cookies always has a mass of 1 kilogram, no matter where in the universe you take it.

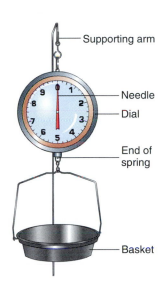

Supporting arm

Needle

Dial

End of spring

Basket

But mass is difficult to measure directly. Moreover, the very concept of mass is only about 300 years old. Consequently, people began quantifying the material in an object by measuring its weight. Spring scales eventually became one of the simplest and most practical tools for accomplishing this task, and they are still found in bathrooms and grocery stores today. They really do contain springs, although these are normally hidden from view.

Questions to Think About: *How is your weight related to your mass? If the earth's gravity became twice as strong, how would your mass be affected? What about your weight? Does jumping up and down change either your mass or your weight? If you stand on a strong spring, how does your weight affect the shape of the spring? Why should there be a relationship between your weight and how much the spring bends?*

Experiments to Do: *Find a hanging spring scale of the type used in the produce section of a grocery store and watch the scale's basket and weight indicator as you put objects in the basket. What happens to the basket as you fill it up? Can you find a relationship between the basket's height and the weight reported by the scale? If you drop something into the basket, instead of lowering it gently, how does the scale respond? Why does the weight indicator bounce back and forth rhythmically? What happens to the gravitational potential energy of the dropped item?*

Now find a spring bathroom scale—the short, flat kind with a rotating dial. Stand on it. Why does it only read your correct weight when you are standing still? If you jump upward, how does the scale's reading change? What about if you let yourself drop? Bounce up and down gently. How does the scale's average reading compare with your normal weight? Does bouncing really change your weight? You can also change the scale's reading by pushing on the floor, wall, or other nearby objects. Which way must you push to increase the scale's reading? to decrease it? When you change the reading in these ways, are you actually changing your weight?

Why You Must Stand Still on a Scale

Whenever you stand on a scale in your bathroom or place a melon on a scale at the grocery store, you are measuring weight. An object's weight is the force exerted on it by gravity, usually the earth's gravity. When you stand on a bathroom scale, the scale measures just how much upward force it must exert on you in order to keep you from moving downward toward the earth's center. As in most scales you'll encounter, the bathroom scale uses a spring to provide this upward support. If you are stationary, you are not accelerating, so your downward weight and the upward force from the spring must cancel one another; that is, they must be equal in magnitude but opposite in direction so that they sum to zero net force. Consequently, although the scale actually displays how much upward force it's exerting on you, that amount is also an accurate measure of your weight.

This subtle difference between your actual weight and what the scale is reporting is important. While an object's weight depends only on its gravitational environment, not on its motion, the weighing process itself is extremely sensitive to motion. If anything accelerates during the weighing process, the scale may not report the object's true weight. For example, if you jump up and down while you're standing on a scale, the scale's reading will vary wildly because you're accelerating. Since you're no longer experiencing zero net force, your downward weight and the upward force from the scale no longer cancel. If you want an accurate measurement of your weight, therefore, you have to stand still.

But even if you stand still, weighing is not a perfect way to quantify the amount of material in your body. That's because your weight depends on your environment. If you always weigh yourself in the same place, the readings will be pretty consistent, as long as you don't routinely eat a dozen jelly doughnuts for lunch. But if you moved to the moon, where gravity is weaker, you'd weigh only about one-sixth as much as on earth. Even a move elsewhere on earth will affect your weight: the earth bulges outward slightly at the equator, and gravity there is about 0.5% weaker than at the poles. That change, together with a small acceleration effect due to the earth's rotation, means that a scale will read 1.0% less when you move from the north pole to the equator. Obviously, moving south is not a useful weight-loss plan.

Check Your Understanding #1: Space Merchants

You're opening a company that will export gourmet food from the earth to the moon. You want the package labels to be accurate at either location. How should you label the amount of food in each package—by mass or by weight?

Stretching a Spring

So you know now that when you put a melon in the basket of a scale at the grocery store and read its weight from the scale's dial, the scale is actually reporting just how much upward force its spring is exerting on the melon. While your shopping cart could support the melon equally well, there's no simple way to determine just how much upward force the cart exerts on the melon. Therein lies the beauty, and the utility, of a spring: a simple relationship exists between its length and the forces it's exerting on its ends. The spring scale can therefore determine how much force it's exerting on the melon by measuring the length of its spring.

The spring shown in Fig. 3.1.1 consists of a wire coil that pulls inward on its ends when it's stretched and pushes outward on them when it's compressed. If a coil spring is neither stretched nor compressed, it exerts no forces on its ends.

The top spring (Fig. 3.1.1a) is neither stretched nor compressed, so that when it lies on a table to keep it from falling, its ends remain motionless. Those ends are in **equilibrium**—experiencing zero net force. As the phrase "zero net force" suggests, equilibrium occurs whenever the forces acting on an object sum perfectly to zero so that the object doesn't accelerate. When you sit still in a chair, for example, you are in equilibrium.

The spring in Fig. 3.1.1a is also at its equilibrium length, its natural length when you leave it alone. No matter how you distort this spring, it tries to return to this equilibrium length. If you stretch it so that it's longer than its equilibrium length, it

local weight variations

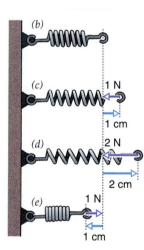

Fig. 3.1.1 Five identical springs. The ends of spring (*a*) are free so that it can adopt its equilibrium length. The left ends of the other springs are fixed so that only their right ends can move. When the free end of a spring (*b*) is moved away from its original equilibrium position (*c*, *d*, and *e*), the spring exerts a restoring force on that end that is proportional to the distance between its new position and the original equilibrium position.

will pull inward on its end. If you compress it so that it's shorter than its equilibrium length, it will push outward on its ends.

Let's attach the left end of our spring to a post (Fig. 3.1.1*b*) and look at the behavior of its free right end. With nothing pushing or pulling on the spring, this free end will be in equilibrium at a particular location—its **equilibrium position.** Since the spring's end naturally returns to this equilibrium position if we stretch or compress it and then let go, the end is in a **stable equilibrium.**

But what happens if we pull the free end to the right and don't let go? The spring now exerts a steady inward force on that end, trying to return it to its original equilibrium position. The more we stretch the spring, the more inward force it exerts on the end. This inward force is exactly proportional to how far we stretch the end away from its original equilibrium position. Since the end of the spring in Fig. 3.1.1*c* has been pulled 1 cm to the right of its original equilibrium position, the spring now pulls this end to the left with a force of 1 N; if the end is instead pulled 2 cm to the right, as it has been in Fig. 3.1.1*d*, the spring pulls it to the left with a force of 2 N. This proportionality continues to work even when we compress the spring: in Fig. 3.1.1*e*, the end has been pushed 1 cm to the left, and the spring is pushing it to the right with a force of 1 N.

In summary, the force exerted by a coil spring has two interesting properties. First, this force is always directed so as to return the spring to its equilibrium length. We call this kind of force a **restoring force** because it acts to restore the spring to its equilibrium length. Second, the spring's restoring force is proportional to how far it has been distorted (stretched or compressed) from its equilibrium length.

These two observations are expressed in **Hooke's law,** named after the Englishman Robert Hooke, who first demonstrated it in the late 17th century. This law can be written in a word equation:

$$\text{restoring force} = -\text{spring constant} \cdot \text{distortion,} \qquad (3.1.1)$$

in symbols:

$$\mathbf{F} = -k \cdot \mathbf{x},$$

and in everyday language:

The farther you squeeze a roll of paper towels, the harder it pushes back.

Here the **spring constant,** k, is a measure of the spring's stiffness. The larger the spring constant—that is, the stiffer the spring—the larger the restoring forces the spring exerts. The negative signs in these equations indicate that a restoring force always points in the direction opposite the distortion.

> **Hooke's Law**
> The restoring force exerted by an elastic object is proportional to how far it has been distorted from its equilibrium shape.

Springs are distinguished by their stiffness, as measured by their spring constants. Some springs are very flexible and have small spring constants—for example, the one that pops the toast out of your toaster, which you can easily compress with your hand. Others, like the large springs that suspend an automobile chassis above the wheels, are very stiff and have large spring constants. But no matter the stiffness, all springs obey Hooke's law.

Hooke's law is remarkably general and isn't limited to the behavior of coil springs. Almost anything you push or pull on will pull or push back with a force that's proportional to how far you have distorted it away from its equilibrium length—or, in the case of a complicated object, its equilibrium shape. Equilibrium shape is the shape an object adopts when it's not subject to any outside forces. If you bend a tree, it will push back with a force proportional to how far it has been bent. If you pull on a rubber band, it will pull back with a force proportional to how far it has been stretched, up to a point. If you squeeze a ball, it will push outward with a force proportional to how far it has been compressed. If a heavy truck bends a bridge downward, the bridge will push upward with a force proportional to how far it has been bent (Fig. 3.1.2).

There is a limit to Hooke's law, however. If you distort an object too far, it will usually begin to exert less force than Hooke's law demands. This is because you will have exceeded the **elastic limit** of the object and will probably have permanently deformed it in the process. If you pull on a spring too hard, you'll stretch it forever; if you push on a tree too hard, you'll break it. But as long as you stay within the elastic limit, almost everything obeys Hooke's law—a rope, a ruler, an orange, and a trampoline.

Distorting a spring requires work. When you stretch a spring with your hand, pulling its end outward, you transfer some of your energy to the spring. The spring stores this energy as **elastic potential energy.** If you reverse the motion, the spring returns most of this energy to your hand, while a small amount is converted to thermal energy by frictional effects inside the spring itself. Work is also required to compress, bend, or twist a spring. In short, a spring that is distorted away from its equilibrium shape always contains elastic potential energy.

Check Your Understanding #2: Going Down Anyone?

As you watch people walk off the diving board at a pool, you notice that it bends downward by an amount proportional to each diver's weight. Explain.

Check Your Figures #1: A Sinking Sensation

You're hosting a party in your third floor apartment. When the first 10 guests begin standing in your living room, you notice that the floor has sagged 1 centimeter in the middle. How far will the floor sag when 20 guests are standing on it? when 100 guests are standing on it?

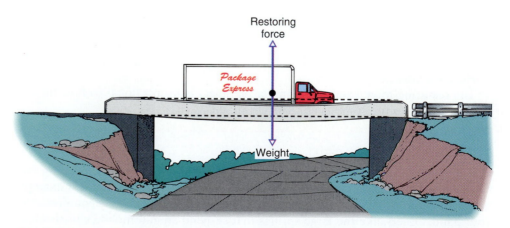

Fig. 3.1.2 A steel bridge sags under the weight of a truck. The bridge bends downward until the upward restoring force it exerts on the truck exactly balances the truck's weight.

How a Hanging Grocery Scale Measures Weight

We're now ready to understand how spring scales work. Imagine a hanging spring scale of the kind used to weigh produce. Inside this scale is a coil spring, suspended from the ceiling by its upper end (Fig. 3.1.3). Hanging from its lower end is a basket. For the sake of simplicity, imagine that this basket has little or no weight. With no force pulling down on it, the scale's spring adopts its equilibrium length, and the basket, experiencing zero net force, is in a position of stable equilibrium. If you shift the basket up or down and then let go of it, the spring will push it back to this position.

When you place a melon in the basket, the melon's weight pushes it downward. The basket starts descending and as it does, the spring stretches and begins to exert an upward force on the basket. The more the spring stretches, the greater this upward force so that eventually the spring is stretched just enough so that its upward force exactly supports the melon's weight. The basket is now in a new stable equilibrium position—again experiencing zero net force.

But how does the scale determine the melon's weight? It uses Hooke's law. Once the basket has adopted its new equilibrium position, where the melon's weight is exactly balanced by the upward force of the spring, the amount the spring has stretched is an accurate measure of the melon's weight.

The scales in Fig. 3.1.3 differ only in the way they measure how far the spring has stretched beyond its equilibrium length. The top scale uses a pointer attached to the end of the spring, while the bottom scale uses a "rack and pinion" gear system that converts the small linear motion of the stretching spring into a much more visible rotary motion of the dial needle. The rack is the series of evenly spaced teeth attached to the lower end of the spring; the pinion is the toothed wheel attached to the dial needle. As the spring stretches, the rack moves downward, and its teeth cause the pinion to rotate. The farther the rack moves, the more the pinion turns, and the higher the weight reported by the needle.

Each of these spring scales reports a number for the weight of the melon you put in the basket. In order for that number to mean something, the scale has to be calibrated. **Calibration** is the process of comparing a local device or reference to a generally accepted standard to ensure accuracy. To calibrate a spring scale, the device or its reference components must be compared against standard weights. Someone must put a standard weight in the basket and measure just how far the spring stretches. Each spring is different, although spring manufacturers try to make all their springs as identical as possible.

Check Your Understanding #3: Scaling Down

If you pull the basket of a hanging grocery store scale downward 1 centimeter, it reports a weight of 5 N (about 1.1 pounds) for the contents of its basket. If you pull the basket downward 3 centimeters, what weight will it report?

Bouncing Bathroom Scales

As we noted earlier, the most common type of bathroom scale is also a spring scale (Fig. 3.1.4). When you step on this kind of scale, you depress its surface and levers inside it pull on a hidden spring. That spring stretches until it exerts, through the levers, an upward force on you that is equal to your weight. At the same time, a rack and pinion mechanism inside the scale turns a wheel with numbers printed on it.

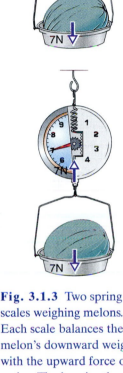

Fig. 3.1.3 Two spring scales weighing melons. Each scale balances the melon's downward weight with the upward force of a spring. The heavier the melon, the more the spring will stretch before it exerts enough upward force to balance the melon's weight. The top scale has a pointer to indicate how far the spring has stretched and thus how much the melon weighs. The bottom scale has a rack and pinion gear that turns a needle on a dial. As the comblike rack moves up and down, it turns the toothed pinion gear.

problems with scales

Fig. 3.1.4 When you step on this bathroom scale, its surface moves downward slightly and compresses a stiff spring. The extent of this compression is proportional to your weight, which is reported by the dial on the left. Levers inside the scale make it insensitive to exactly where you stand.

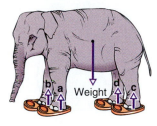

Fig. 3.1.5 You can weigh a baby elephant by placing a spring scale under each of its four feet. These scales exert upward forces **a, b, c,** and **d** to support the elephant. The elephant's weight is equal to the sum of those four forces, as measured by the scales.

When the wheel stops moving, you can read one of these numbers through a window in the scale. Because which number you see depends on how far the spring has stretched, this number indicates your weight.

However, the wheel rocks briefly back and forth around your actual weight before it settles down. The wheel moves because you're bouncing up and down as the scale gradually gets rid of excess energy. When you first step on the scale's surface, its spring is not stretched and it isn't pushing up on you at all. You begin to fall. As you descend, the spring stretches and the scale begins to push up on your feet. But by the time you reach the equilibrium height, where the scale is exactly supporting your weight, you are traveling downward quickly and coast right past that equilibrium. The scale begins to read more than your weight.

The scale now accelerates you upward. Your descent slows, and you soon begin to rise back toward equilibrium. Again you overshoot the proper height, and now the scale begins to read less than your weight. You are bouncing up and down because you have excess energy that is shifting back and forth between gravitational potential energy, kinetic energy, and elastic potential energy. This bouncing will continue until sliding friction in the scale has converted it all into thermal energy. Only then will the bouncing stop and the scale read your correct weight.

One scale is enough for you, but how can you weigh your pet baby elephant? It's too heavy for a single scale, but four scales will do the trick. If you put one scale under each of the elephant's feet, the scales will work together to support its weight. Each scale will report just how much upward force it's exerting on the elephant, so the sum of the four measurements will equal the animal's overall weight (Fig. 3.1.5).

The specific readings of the four scales will depend on the position of the elephant's center of gravity. Its **center of gravity** is the effective location of its weight and coincides with its center of mass. If the elephant's center of gravity is closer to one scale than the others, that scale will bear more of its weight and will report a higher value than the other scales. To keep from breaking any of the scales, you should position the elephant's center of gravity about midway between all of them.

Check Your Understanding #4: Weighed Down

When you step on the surface of a spring bathroom scale, you can feel it move downward slightly. How is the distance that the scale's surface moves downward when you step on it related to the weight it reports?

other scales, vibrations, weighing astronauts, balances, improved balances

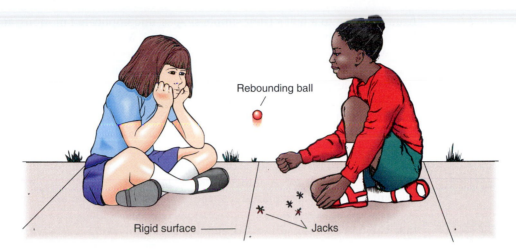

Rebounding ball

Rigid surface ——————   Jacks

SECTION 3.2 — Bouncing Balls

If you visit a toy or sporting goods store, you'll find a lot of different balls—almost a unique ball for every sport or ball game. These balls differ in more than just size and weight. Some are very hard, others very soft; some are smooth, others rough or ridged.

In this section we'll focus primarily on another difference: the ability to bounce. A superball, for example, bounces extraordinarily well, while a foam rubber ball hardly bounces at all. Even balls that appear identical can be very different; a new tennis ball bounces much better than an old one. We'll begin this section by exploring these differences.

Questions to Think About: Is it possible for a ball to bounce higher than the height from which it was dropped? Where does a ball's kinetic energy go as it bounces, and what happens to the energy that doesn't reappear after the bounce? What happens when a ball bounces off a moving object, such as a baseball bat? What role does the baseball bat's structure have in the bouncing process? Does it matter which part of a baseball bat hits the ball?

Experiments to Do: Drop a ball on a hard surface and watch it bounce. What happens to the ball's shape during the bounce? Hold the ball in your hands and push its surface inward with your fingers. What is the relationship between the force it exerts on your fingers and how far inward you dent it? Denting the ball takes work. Why? How does the ball's energy change as it dents? What happens to the ball's energy as its shape returns to normal (to equilibrium)?

Drop the ball from various heights and see if you can find a simple relationship between the ball's initial height and the height to which it rebounds. Now drop the ball on a soft surface, such as a pillow, or on a lively surface, such as an inflated balloon. Why does the surface it hits change the way the ball bounces?

The Way the Ball Bounces: Balls as Springs

In many ways, balls are perfect objects. What would most sports be like without them? How would industrial machines function without ball bearings to keep them

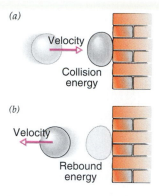

(a)

Velocity

Collision
energy

(b)

Velocity

Rebound
energy

Fig. 3.2.1 A bounce from a wall has two halves— *(a)* the collision and *(b)* the rebound. During the collision between the ball and the wall, some of their kinetic energy is transformed into other forms—an amount called the collision energy. During the rebound, some stored energy is released as kinetic energy—an amount called the rebound energy. The rebound energy is always less than the collision energy because some energy is lost as thermal energy. However, a lively ball wastes less energy than a dead one.

from grinding to a halt? Their simple shapes, uncomplicated motions, and ability to bounce make balls both fascinating and useful. Most balls are spherical, meaning that when no outside forces act on them they adopt spherical equilibrium shapes. But some balls, such as those used in U.S. football and rugby, have equilibrium shapes that are not spherical but oblong.

The term "equilibrium shape," of course, is one we've seen already: the previous section used it to describe springs. This is no coincidence, for a spherical ball behaves like a spherical spring. In fact, everything we associate with springs has some place in the behavior of balls. For example, when you push a ball's surface inward, it exerts an outward restoring force on you. When you do work on the ball as you distort its surface, it stores some of this work as elastic potential energy and when you let the ball return to its equilibrium shape, it releases that stored energy.

This springlike behavior is evident when a ball collides with the floor or with a bat. The ball's surface distorts during the collision, giving it elastic potential energy that is released when the ball rebounds. If the ball is moving, then some of this stored energy comes from the ball's kinetic energy. If the ball hits a moving surface, then some of this stored energy comes from the surface's kinetic energy. Much of the stored energy reappears as kinetic energy in the ball and surface as the ball rebounds.

Some balls bounce better than others. We frequently call a very bouncy ball "lively" and a ball that doesn't bounce well "dead." One way to look at a ball's liveliness is to compare kinetic energies before and after the bounce. We can do that by dividing the bounce into two halves—the collision and the rebound (Fig. 3.2.1). During the collision, the ball and surface convert some of their overall kinetic energy into elastic potential energy and thermal energy. The amount of kinetic energy transformed at impact is called the **collision energy.** A lively ball does a good job of converting the collision energy into elastic potential energy, while a dead ball converts most of it into thermal energy (Fig. 3.2.2).

During the rebound, the ball and surface push away from one another, converting elastic potential energy back into kinetic energy. The total amount of kinetic energy released as the surface and ball push apart is the **rebound energy.** Collision energy that doesn't reappear as rebound energy has been transformed into thermal energy.

The ratio of rebound energy to collision energy (Table 3.2.1) determines how high a ball will bounce when you drop it on a hard floor (Fig. 3.2.3). An ideally elastic ball would have a ratio of 1.00 and would rebound to its initial height. But a real ball wastes some of the collision energy, so it rebounds to a lesser height. The height from which you drop the ball is proportional to its initial gravitational potential energy and therefore to its collision energy. The height to which it rebounds is similarly proportional to its rebound energy. So the ratio of these two heights is a good measure of the ratio of the rebound energy to the collision energy. The smaller this ratio, the less kinetic energy the ball receives during the rebound and the weaker the bounce.

While this energy ratio is often useful, a ball is usually characterized by its **coefficient of restitution**—the ratio of its rebound *speed* to its collision *speed* when it bounces off a hard, immobile surface:

$$\text{coefficient of restitution} = \frac{\text{outgoing speed of ball}}{\text{incoming speed of ball}}. \qquad (3.2.1)$$

Scientists have found that, for most balls, this speed ratio remains constant over a wide range of collision speeds (Table 3.2.1). A ball that rebounds with the same speed that it had when it collided with the surface has a coefficient of restitution of 1.00. A superball is almost this lively, with a coefficient of restitution of about 0.90. Thus, when a superball traveling at 10 km/h collides with a concrete wall, it rebounds

Table 3.2.1 **Approximate Energy Ratios and Coefficients of Restitution for a Variety of Balls**

TYPE OF BALL	REBOUND ENERGY COLLISION ENERGY	COEFFICIENT OF RESTITUTION
Superball	0.81	0.90
Racquet ball	0.72	0.85
Golf ball	0.67	0.82
Tennis ball	0.56	0.75
Steel ball bearing	0.42	0.65
Baseball	0.30	0.55
Foam rubber ball	0.09	0.30
"Unhappy" ball	0.01	0.10
Beanbag	0.002	0.05

Fig. 3.2.2 When a tennis ball hits the floor, it dents inward to store energy and then rebounds somewhat more slowly than it arrived. These images show the ball's position at twelve equally spaced times. Is the ball bouncing to the left or the right? How can you tell?

at about 9 km/h. In contrast, a foam rubber ball's coefficient of restitution is about 0.30, while that of a beanbag is almost zero.

Actually, the energy ratio and the coefficient of restitution are directly related. Remember, a ball's kinetic energy equals half its mass times its velocity squared, or $\frac{1}{2}mv^2$. Even if we don't know the masses of the balls, we know that the energy ratio is equal to the square of the speed ratio. Thus if a foam rubber ball rebounds at only 0.30 times its collision speed, it retains only 0.30^2—0.09 times, or 9%—of its original kinetic energy, and the remaining 91% is converted to thermal energy in the rubber and air that make up the ball. A superball, in contrast, retains about 81% of its original kinetic energy after a bounce.

Balls bounce best when they store energy through compression rather than through surface bending. That's because bending can lead to lots of wasteful internal friction. Since solid balls involve compression, they usually bounce well, whether they're made of rubber, wood, plastic, or metal. But air-filled balls bounce well only when properly inflated. A normal basketball, which stores most of its energy in its compressed air, has a high coefficient of restitution. In contrast, an underinflated basketball, which experiences lots of surface bending during a collision, barely bounces at all. Similarly, a tennis ball bounces best when new; after a while, the compressed air inside leaks out and the ball's coefficient of restitution drops.

Check Your Understanding #1: A Game of Marbles
As you head into the park to play a game of marbles with your friends, several of the glass marbles fall through a hole in your marble bag and bounce nicely on the granite walkway. How can a marble bounce?

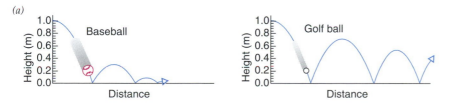
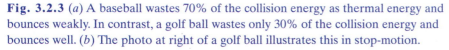

Fig. 3.2.3 (*a*) A baseball wastes 70% of the collision energy as thermal energy and bounces weakly. In contrast, a golf ball wastes only 30% of the collision energy and bounces well. (*b*) The photo at right of a golf ball illustrates this in stop-motion.

How the Surface Affects the Bounce

Fig. 3.2.4 When a tennis ball bounces from a moving racket, both the ball and racket dent inward. The ball and racket share the collision energy almost evenly.

If the surface on which a ball bounces is not perfectly hard, that surface will contribute to the bouncing process. It will distort and store energy when the ball hits it and will return some of this stored energy to the ball as it rebounds. Overall, the collision energy is shared between the ball and the surface, and each provides part of the rebound energy.

Just how the collision energy is distributed between the surface and ball depends on how stiff each one is. During the bounce, the surface and the ball both behave as springs, pushing on one another with equal but oppositely directed forces. Since the forces denting them inward are equal, the work done in distorting each object is proportional to how far inward it dents. Whichever object dents the furthest receives the most collision energy.

Since the ball usually distorts more than the surface it hits, most of the collision energy normally goes into the ball. As a result, you might expect the ball to provide most of the rebound energy, too. However, that's not always true. Some lively elastic surfaces store collision energy very efficiently and return almost all of it as rebound energy. Since a relatively dead ball wastes most of the collision energy it receives, a lively surface's contribution to the rebound energy can be very important to the bounce. A lively racket is critical to the game of tennis because the racket's strings provide much of the rebound energy as the ball bounces off the racket (Fig. 3.2.4). Trampolines and springboards are even more extreme examples, with surfaces so lively that they can even make people bounce. People, like beanbags, have coefficients of restitution near zero; when you land on a trampoline, it receives and stores most of the collision energy and then provides most of the rebound energy.

The stiffnesses of the ball and surface also determine how much force each object exerts on the other and thus how quickly the collision proceeds. Stiff objects resist denting much more strongly than soft objects. When both objects are very stiff, the forces involved are large and the acceleration is rapid. Thus a steel ball rebounds very quickly from a concrete floor because the two exert enormous forces on one another. If the ball and/or surface are relatively soft, the forces are weaker and the acceleration is slower.

What if the surface that a ball hits isn't very massive? In that case, the surface may do part or all of the "bouncing." During the bounce, the ball and the surface accelerate in opposite directions and share the rebound energy. Massive surfaces, such as floors and walls, accelerate little and receive almost none of the rebound energy. But when the surface a ball hits is not very massive, you may see it accelerate. When a ball hits a lamp on the coffee table, the ball will do most of the accelerating, but the lamp is likely to fall over, too.

Similarly, when a baseball strikes a baseball bat, the ball and bat accelerate in opposite directions. The more massive the bat, the less it accelerates. To ensure that most of the rebound energy went to the baseball, the legendary hitters of the early 20th century used massive bats. Such bats are no longer in vogue because they are too hard to swing. But in the early days of baseball, when pitchers were less highly skilled, massive bats drove many long home runs.

Check Your Understanding #2: The Game Begins

You are playing the game of marbles on a soft dirt field. The goal is to knock glass marbles out of a circle by hitting them with other marbles. You initially drop several marbles onto the ground inside the circle and they hardly bounce at all. What prevents them from bouncing well here?

How a Moving Surface Affects the Bounce

The last paragraph describes the act of hitting a baseball with a moving baseball bat as though it were a case of "bouncing." That might sound a little strange. When a baseball hits a stationary bat, the ball bounces. But if a moving bat hits a stationary baseball, is it proper to say that the ball bounces?

The answer is yes. In fact, which object is moving and which is stationary depends on your point of view—your **frame of reference.** A fly resting on the baseball will claim that the baseball is stationary and that it's being struck by a moving bat. Another fly resting on the baseball bat will claim that the bat is stationary and that it's being struck by a moving baseball. Which fly has the correct frame of reference?

Both frames of reference are equally valid. One of the remarkable discoveries of Galileo and Newton is that the laws of physics work perfectly in any inertial frame of reference. An **inertial frame of reference** is one that's not accelerating and is thus either stationary or moving at a constant velocity. As long as you view the world around you from an inertial frame of reference, the laws of motion will accurately describe what you see, and energy, momentum, and angular momentum will all be conserved.

But frames of reference aren't the first things you think of during a baseball game. When you swing a bat and drive the pitch toward center field, your main concern is how fast the outgoing ball is traveling toward the bleachers. A speedy ball will be a home run, while a slower ball will be an out. What determines the ball's outgoing speed?

With both bat and ball moving relative to the playing field, there are several useful frames of reference from which to study the collision. However, we'll find it easier to focus on how quickly the bat and ball move toward or away from one another. This relative motion is what matters most in a collision. After all, whether a rock hits a bottle or a bottle hits a rock, it's going to be bad for the bottle.

When a ball bounces off a *moving* surface that's rigid and very massive, the ball's coefficient of restitution still applies. But now we must use a more general form of that speed ratio. This improved version compares the speed at which the ball and surface separate after the bounce to the speed at which they approach before the bounce:

$$\text{coefficient of restitution} = \frac{\text{speed of separation}}{\text{speed of approach}}. \qquad \textbf{(3.2.2)}$$

When the surface is stationary, Eq. 3.2.2 is equivalent to Eq. 3.2.1.

To see how this generalization allows us to explain why the baseball you hit is now traveling over the centerfielder's head, let's examine the collision between bat and ball. Let's suppose that, just before the collision, the pitched baseball was approaching home plate at 100 km/h and that, as you swung to meet the ball, your bat was moving toward the pitcher at 100 km/h (Fig. 3.2.5a). Since each object is moving toward the other, their speed of approach is the sum of their individual speeds, or 200 km/h.

The baseball's coefficient of restitution is 0.55, so after the collision (Fig. 3.2.5b) the speed of separation will be only 0.55 times the speed of approach, or 110 km/h. The outgoing ball and the swinging bat separate from one another at 110 km/h. Since the bat is still moving toward the pitcher at 100 km/h, the ball must be traveling toward the pitcher even faster: at 100 km/h plus 110 km/h or a total speed of 210 km/h! That's why it flies past everyone and into the stands.

(a)

(a)   (b)

Fig. 3.2.5 (*a*) Before they collide, the bat and ball are approaching one another at an overall speed of 200 km/h. (*b*) After the collision, the two are separating from one another at a speed of 110 km/h. However, because the bat is moving toward the pitcher at 100 km/h, the outgoing ball is traveling at 210 km/h in that same direction.

(b)

> **Check Your Understanding #3: Marble Frames of Reference**
> Two of you flick your marbles into the circle simultaneously from opposite sides of the circle and they collide head on. Each marble was traveling forward at 1 m/s. From the inertial reference frame of your marble, what was the velocity of the other person's marble just before the two marbles hit?

(c)

Fig. 3.2.6 When a ball hits a bat, the bat experiences both acceleration and angular acceleration. (*a*) If the ball hits near the bat's center, the angular acceleration is small and the bat's handle accelerates backward. (*b*) If the ball hits near the bat's end, the angular acceleration is large and the bat's handle accelerates forward. (*c*) But if the ball hits the bat's center of percussion, the angular acceleration is just right to keep the handle from accelerating.

Surfaces Also Bounce . . . and Twist and Bend

As you can see, a surface's motion has a large effect on a ball bouncing off it. A surface that moves toward an incoming ball will strengthen the rebound, while one that moves away from the ball will weaken it. But we're neglecting the ball's effect on the surface itself. Sometimes that surface bounces, too.

A baseball bat is a case in point. When you swing your bat into a pitched ball, the bat doesn't continue on exactly as before. The ball pushes on the bat during the collision and the bat responds in a number of interesting ways.

First, as we noted before, the bat rebounds from the ball. The bat decelerates slightly during the collision so that its speed after the impact is a little less than before it. Since the ball's final speed depends on the bat's final speed, a slower bat means a slower ball. Thus we've slightly overestimated the ball's speed as it heads toward the bleachers.

Second, the ball's impact sets the bat spinning. When the ball pushes on the bat and makes it accelerate backward, it also exerts a torque on the bat about its center of mass and makes it undergo angular acceleration (Fig. 3.2.6). While these two types of acceleration might seem inconsequential, your hands notice their effects. The bat's acceleration tends to yank its handle toward the catcher, while its angular acceleration tends to twist its handle toward the pitcher. The extent of these two motions depends on just where the ball hits the bat. If the ball hits a point known as the **center of percussion**, the handle experiences no overall acceleration (Fig. 3.2.6c). The smooth feel of such a collision explains why the center of percussion, located about 7 inches from the end of the bat, is known as a "sweet spot."

Finally, the collision often causes the bat to vibrate. Like a xylophone bar struck by a mallet (Fig. 3.2.7a), the bat bends back and forth rapidly with its ends and center moving in opposite directions (Fig. 3.2.7b). These vibrations sting your hands and

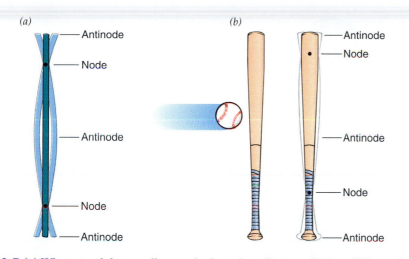

Fig. 3.2.7 (*a*) When struck by a mallet, a xylophone bar vibrates with its middle and ends moving back and forth in opposite directions. The parts that move farthest are antinodes, while the points that don't move at all are nodes. (*b*) When struck by a ball, a baseball bat vibrates in a similar fashion. However, an impact at one of the bat's nodes causes no vibration.

can even break a wooden bat. However, near each end of the bat, there's a point that doesn't move when the bat vibrates—a **vibrational node.** When the ball hits the bat at its node, no vibration occurs. Instead, the bat emits a crisp, clear "crack" and the ball travels farther. Fortunately, the bat's vibrational node and its center of percussion almost coincide so you can hit the ball with both sweet spots at once.

Check Your Understanding #4: Mass and Marbles

The marbles you're playing with are not all the same size and mass. You notice that larger marbles are particularly effective at knocking other marbles out of the circle. You decide to use a 10-cm-diameter glass ball as a marble, expecting to clean out the entire circle. But when you flick it with your thumb, your thumb merely bounces off. Why doesn't the glass ball move forward quickly?

frames of reference, hitting bats, sweet spot, hitting balls, hitting surfaces, measuring the c.o.r.

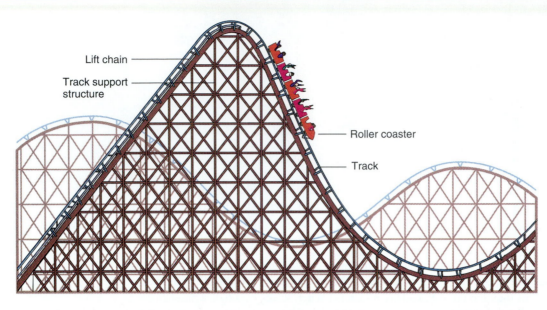

Lift chain

Track support
structure

Roller coaster

Track

Carousels and Roller Coasters

As your sports car leaps forward at a green light, you're pressed firmly back against your seat. It's as though gravity were somehow pulling you both down and backward at the same time. But it's not gravity that pulls back on you; it's your own inertia preventing you from accelerating forward with the car.

When this happens, you're experiencing the sensation of acceleration. We feel this sensation many times each day, whether through turning in an automobile or riding up several floors in a fast elevator. But nowhere is our sensation of acceleration more acute than at the amusement park. We accelerate up, down, and around on the carousel, back and forth in the bumper cars, and left and right in the scrambler. The ultimate ride, of course, is the roller coaster, which is one big, wild experience of acceleration. When you close your eyes on a straight stretch of highway, you can hardly tell the automobile is moving. But when you close your eyes on a roller coaster, you have no trouble feeling every last turn in the track. It's not the speed you feel, but the acceleration. What is often called motion sickness should really be called acceleration sickness.

Questions to Think About: *How does your body feel its own weight? When you swing a bucket full of water around in a circle, as we asked you to do in the chapter opening, why do you feel such a strong outward force from the bucket? Why can you swing that bucket completely over your head without spilling the water inside it? What keeps you from falling out of a roller coaster as it goes over the top of a loop-the-loop? Which car of a roller coaster should you sit in to experience the best ride?*

Experiments to Do: *To begin associating the familiar sensations of motion with the physics of acceleration, travel as a passenger in a vehicle that makes lots of turns and stops. Close your eyes and see whether you can tell which way the vehicle is turning and when it's starting or stopping. Which way do you feel pulled when the vehicle turns left? turns right? starts? stops? How is this sensation related to the direction of* *the vehicle's acceleration? Now find a time when the vehicle is traveling at constant velocity on a level path and see if you feel any sensations that tell you which way it's heading. Try to convince yourself that it's heading backward or sideways rather than forward.*

The Experience of Acceleration

Nothing is more central to the laws of motion than the relationship between force and acceleration. Up until now, we've looked at forces and noticed that they can produce accelerations; in this section we'll reverse that process, looking at accelerations and noticing that they require forces. For you to accelerate, something must push or pull on you. Just where and how that force is exerted on you determines what you feel when you accelerate.

The backward "force" you feel as your car accelerates forward is caused by your body's inertia, its tendency not to accelerate (Fig. 3.3.1). The car and your seat are accelerating forward, and since the seat acts to keep you from traveling through its surface, it exerts a forward support force on you that causes you to accelerate forward. But the seat can't exert a force uniformly throughout your body. Instead, it pushes only on your back, and your back then pushes on your bones, tissues, and internal organs to make them accelerate forward. Each piece of tissue or bone is responsible for the forward force needed to accelerate forward the tissue in front of it. A whole chain of forces, starting from your back and working forward toward your front, causes your entire body to accelerate forward.

Let's compare this situation with what happens when you're standing motionless on the floor. Since gravity exerts a downward force on you that's distributed uniformly throughout your body, each part of your body has its own independent weight; these individual weights, taken together, add up to your total weight. The floor, for its part, is exerting an upward support force on you that keeps you from accelerating downward through its surface. But the floor can't exert a force uniformly throughout your body. Instead, it pushes only on your feet, and your feet then push on your bones, tissues, and internal organs to keep them from accelerating downward. Each piece of tissue or bone is responsible for the upward force needed to keep the tissue above it from accelerating downward. A whole chain of forces, starting from your feet and working upward toward your head, keeps your entire body from accelerating downward.

As you probably noticed, the two previous paragraphs are very similar. But so are the sensations of gravity and acceleration. When the ground is preventing you from falling, you feel "heavy"; your body senses all the internal forces needed to support its pieces so that they don't accelerate, and you interpret these sensations as weight. When the car seat is causing you to accelerate forward, you also feel "heavy"; your body senses all the internal forces needed to accelerate its pieces forward, and you interpret these sensations as weight. This time the "weight" is experienced toward the back of the car.

Try as you may, you can't distinguish the gravity-like "force" that you experience as you accelerate from the true force of gravity. And you're not the only one fooled by acceleration. Even the most sophisticated laboratory instruments can't determine directly whether they are experiencing gravity or are simply accelerating. However, despite the convincing sensations, the backward heavy feeling in your gut as you accelerate forward is the result of inertia and is not due to a real backward force. This experience of acceleration is called a **fictitious force.** It always points in

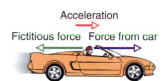

Fig. 3.3.1 As you accelerate forward in a car, you feel a gravity-like fictitious force in the direction opposite to the acceleration. This fictitious force is really the mass of your body resisting acceleration.

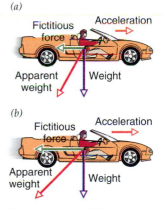

Fig. 3.3.2 (*a*) When you accelerate forward gently, the backward fictitious force you feel is small and your apparent weight is mostly downward. (*b*) When you accelerate forward quickly, you experience a strong backward fictitious force and your apparent weight is backward and down.

the direction opposite the acceleration that causes it, and its strength is proportional to that acceleration.

If you accelerate forward quickly, the backward fictitious force you experience can be quite large. However, you don't experience this fictitious force all by itself; you also experience your downward weight, and together these effects feel like an especially strong gravitational force at an angle somewhere between straight down and the back of the car. We will call the combined effects of gravitational and fictitious forces **apparent weight.** The faster you accelerate, the stronger the backward fictitious force, and the more your apparent weight points toward the back of the car (Fig. 3.3.2).

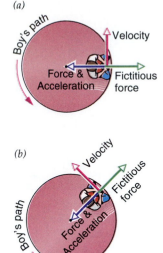

(a)

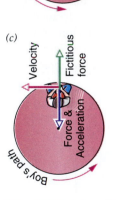

(b)

Check Your Understanding #1: The Feel of a Tight Turn

You're sitting in the passenger seat of a racing car that is moving rapidly along a level track. The track takes a sharp turn to the left and you find yourself thrown against the door to your right. What horizontal forces, real and fictitious, are acting on you?

Carousels

When you ride on a carousel, you travel in a circle around a central pivot. That's an unusual motion. If you were experiencing zero net force, you would travel in a straight line at a steady pace in accordance with Newton's first law. But since your path is circular instead of straight, you must be experiencing a nonzero net force and you must be accelerating.

Which way are you accelerating? Remarkably, you are always accelerating toward the center of the circle. To see why that's so, let's look down on a simple carousel that's turning counterclockwise at a steady pace (Fig. 3.3.3). At first, the boy riding the carousel is directly east of its central pivot and is moving northward (Fig. 3.3.3a). If nothing were pulling on the boy, he would continue northward and fly off the carousel. Instead, he follows a circular path by accelerating toward the pivot—that is, toward the west. As a result, his velocity turns toward the northwest and he heads in that direction. To keep from flying off the carousel, he must continue to accelerate toward the pivot, which is now southwest of him (Fig. 3.3.3b). His velocity turns toward the west, and he follows the circle in that direction. And so it goes (Fig. 3.3.3c).

The boy's body is always trying to go in a straight line, but the carousel keeps pulling him inward so that he accelerates toward the central pivot. The boy is experiencing **uniform circular motion.** "Uniform" means that the boy is always moving at the same speed, although his direction keeps changing. "Circular" describes the path the boy follows as he moves, his trajectory.

Like any object undergoing uniform circular motion, the boy is always accelerating toward the center of the circle. An acceleration of this type, toward the center of a circle, is called a **centripetal acceleration** and is caused by a centrally directed force, a **centripetal force.** A centripetal force is not a new, independent type of force like gravity, but the net result of whatever forces act on the object. Centripetal means "center-seeking" and a centripetal force pushes the object toward that center. The carousel uses support forces and friction to exert a centripetal force on the boy, and he experiences a centripetal acceleration. Amusement park rides often involve centripetal acceleration (Fig. 3.3.4).

Fig. 3.3.3 A boy riding on a turning carousel is always accelerating toward the central pivot. His velocity vector shows that he is moving in a circle but his acceleration vector points toward the pivot. When he is heading north (a), he is accelerating toward the west. His velocity gradually changes direction until he is heading northwest (b), at which time he is accelerating toward the southwest. He turns further until he is heading west (c) and is then accelerating toward the south. (North is upward.)

The amount of acceleration the boy experiences depends on his speed and the radius of the carousel. The faster the boy is moving and the smaller the radius of his circular trajectory, the more he accelerates. His acceleration is equal to the square of his speed divided by the radius of his circular trajectory. We can express this relationship as a word equation:

$$\text{acceleration} = \frac{\text{speed}^2}{\text{radius of circular trajectory}}, \quad \textbf{(3.3.1)}$$

in symbols:

$$a = \frac{v^2}{r},$$

and in everyday language:

Making a tight, high-speed turn involves lots of acceleration.

Fig. 3.3.4 The people on this ride travel in a circle, pulled inward by cords so that they accelerate toward the center of the circle.

Since the boy is accelerating inward, toward the center of the circle, he feels a fictitious force outward, away from the center of the circle. This fictitious force is often called **"centrifugal force."** The quotation marks indicate that "centrifugal force" is not a force at all but an outward fictitious force due to inward acceleration. Fictitious forces such as "centrifugal force" don't contribute to the net force that acts on an object. Thus, if the centripetal force on an object is removed, the object will travel in a straight line with the velocity it had at the moment the centripetal force stopped. But despite its fictitious nature, "centrifugal force" creates a compelling sensation of gravity-like force. The boy on the carousel feels as though gravity is pulling him outward as well as down and clings tightly to the carousel to keep from falling off.

Gravitational force and "centrifugal force" can differ significantly in their amounts. While the gravitational force on a particular object has one set value on the earth's surface, the amount of "centrifugal force" that an object can experience has no limit. For an object going very rapidly around a small circle, the fictitious force can easily exceed the force of gravity.

Fictitious force is often measured relative to the earth's gravity. On the earth's surface, the acceleration due to gravity is 9.8 m/s^2 or 9.8 N/kg. Thus a 100-kg man weighs 980 N, an amount we can call 1 gravity or 1g, for short. On the scrambler or on an airplane maneuvering sharply, that same man may experience a fictitious force five times his weight, or 4900 N. We describe this fictitious force as 5g's.

centrifuges and spin-dryers

▶ Check Your Understanding #2: Banking on a Curve

Your racing car comes to a banked, left-hand turn, which it completes easily. Why is it essential that a racetrack turn be banked so that it slopes downhill toward the center of the turn?

▶ Check Your Figures #1: Going for a Spin

Some children are riding on a playground carousel with a radius of 1.5 m. The carousel turns once every two seconds. How quickly are the children accelerating?

Fig. 3.3.5 As this roller coaster plunges down the first hill, its last car is pulled over the edge by the cars in front of it and its riders feel almost weightless.

Fig. 3.3.6 As it goes over the loop-the-loop, this roller coaster is accelerating rapidly toward the center of the circle. When it reaches the top of the loop, the track pushes the roller coaster downward and the riders feel pressed into their seats. If they close their eyes, they won't even be able to tell that they're upside-down.

Roller Coasters

While roller coasters offer interesting visual effects, such as narrowly missing obstacles, and strange orientations, such as upside-down, the real thrill of roller coasters comes from their accelerations. Plenty of other amusement park rides suspend you sideways or upside-down, so that you feel ordinary gravity pulling at you from an unusual angle. But why pay for these when you can stand on your head for free? For a *real* thrill, you need acceleration to give you the weightless feeling you experience as a roller coaster dives over its first big hill or the several-g fictitious force you feel as you go around a sharp corner. A change in the amount of "gravity" you feel is much more exciting than a change in its direction. We're now prepared to look at a roller coaster and understand what you feel as you go over hills (Fig. 3.3.5) and loop-the-loops (Fig. 3.3.6).

Every time the roller coaster accelerates, you feel a fictitious force in the direction opposite your acceleration. That fictitious force gives you an apparent weight that's different from your true weight. As we saw with a car, rapid forward acceleration tips your apparent weight backward toward the rear of the vehicle, while rapid deceleration tips it toward the front. But a roller coaster can do something a car can't: it can accelerate downward rapidly! In that case, the fictitious force you experience is upward and opposes the downward force of gravity so that they partially cancel. As a result, your apparent weight is less than your real weight and points downward or perhaps, if the downward acceleration is fast enough, points upward.

Consider that last possibility: if you accelerate downward at just the right rate, the upward fictitious force will exactly cancel your downward weight. You will feel perfectly weightless, as though gravity didn't exist at all. The rate of downward acceleration that causes this perfect cancellation is that of a freely falling object. Your weight won't have changed, but the roller coaster will no longer be supporting you and you will be accelerating downward at 9.8 m/s². You will experience the same sensations as a sky diver who has just stepped out of an airplane.

Since freely falling objects are subject only to the forces of gravity, they don't have to push on one another to keep their relative positions. As you fall, your hat and sunglasses will fall with you and won't require any support forces from your head. Even if your sunglasses come off, they will hover in front of you as the two of you accelerate downward together. Similarly, your internal organs don't need to support one another, and the absence of internal support forces gives rise to the exhilarating sensation of free fall.

However, a roller coaster is attached to a track, and its rate of downward acceleration can actually exceed that of a freely falling object. In those special

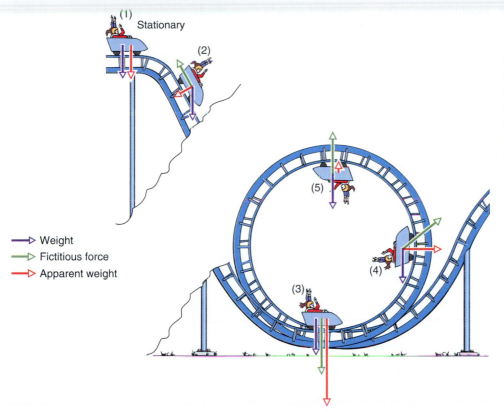

(1) Stationary

(2)

(5)

(3)

(4)

Weight
Fictitious force
Apparent weight

Fig. 3.3.7 A single-car roller coaster going over the first hill and a loop-the-loop. At each point along the track, the car experiences its weight, a fictitious force due to its current acceleration, and an apparent weight that is the sum of those two. The apparent weight always points toward the track, which is why the car doesn't fall off it.

situations, the track will be assisting gravity in pushing the roller coaster downward. As a rider, you will feel less than weightless. The upward fictitious force will be so large that your apparent weight will be in the upward direction, as though the world had turned upside-down!

Figure 3.3.7 shows a single-car roller coaster at various points along a simple track with one hill and one loop-the-loop. Weight, fictitious force, and apparent weight are all vector quantities, as are the car's velocity and acceleration. These quantities are indicated with arrows of varying lengths that show each vector's direction and magnitude. The longer the arrow, the greater the magnitude of the quantity it represents.

At the top of the first hill of Fig. 3.3.7, the single-car roller coaster is almost stationary (1). You, the rider, feel only your weight, straight down—nothing exciting yet. But as soon as the car begins its descent, accelerating down the track, a fictitious force appears pointing up the track (2). The combination of this fictitious force and your weight gives you an apparent weight that is very small and points down and into the track. Most people find this sudden reduction in apparent weight terrifying. Our bodies are very sensitive to partial weightlessness, and this falling sensation is half the fun of a roller coaster. An astronaut, weightless in space, has this disquieting feeling for days on end. No wonder astronauts have such frequent troubles with motion (or rather acceleration) sickness.

On earth, the weightless feeling can't last. It occurs only during downward acceleration and disappears as your car levels off near the bottom of the hill. By the time the car begins its rise into the loop-the-loop, it is traveling at maximum speed and has begun to accelerate upward (3). This upward acceleration creates a downward ficti-tious force so that your apparent weight is huge and downward. You feel pressed into your seat as you experience about 2 or 3g's.

The trip through the loop-the-loop is almost uniform circular motion. In effect, you are taking a single turn around a vertical carousel. However, as the car rises into the loop-the-loop, some of its kinetic energy becomes gravitational potential energy and the car slows down. As the car descends out of the loop-the-loop, this gravita-tional potential energy returns to kinetic form and the car speeds up. As a result of these speed changes, your acceleration is not exactly inward toward the center of the loop-the-loop and the fictitious force you experience is not exactly outward. Still, an inward acceleration and an outward fictitious force are good approxima-tions for what occurs.

Halfway up the loop-the-loop, your true acceleration is inward and downward, so the fictitious force you experience is outward and upward (4). Your apparent weight is still much more than your weight and is directed outward. You feel pressed into your seat, and the car itself is pressed against the track (Fig. 3.3.8).

Finally, you reach the top of the loop-the-loop (5). The car has slowed somewhat as the result of its climb against the force of gravity. But it's still accelerating toward the center of the circle, and you still experience a fictitious force outward and, in this case, upward. Your weight is downward, but the upward fictitious force exceeds your weight. Your apparent weight is upward (Fig. 3.3.6)!

Not only does the inverted car stay on its track, but you feel as though a weak gravity-like force is pressing you into your seat. Actually, the car is pushing on you to help gravity accelerate you around the loop. If your hat were to come off at the top of the loop, it would land in your seat, even though that seems to involve some sort of upward movement. In fact, the car is accelerating almost directly downward at a rate that's faster than that of your freely falling hat. Gravity and the track to-gether push the car downward so fast that it overtakes the hat—your hat is really falling but the car is plummeting even faster.

Many roller coaster tracks are designed so that acceleration leaves the riders pressed into their seats even when the cars go upside-down. In principle, roller coasters that travel on such tracks don't need seat belts to prevent riders from falling out (although seat belts are comforting to the passengers and insurance com-panies). But some roller coaster tracks have special cars and seat belts that per-mit them to have apparent weights away from the track. These roller coasters can and do go upside-down without enough downward acceleration to keep the riders in their seats. In such roller coasters, the riders feel like they are hanging and if one of them were to lose a hat, it would fall to the ground rather than into the car.

What about a roller coaster with more than one car? For the most part, the same rules apply. How-ever, a new source of force now acts on each car: the other cars. The effects of these cars are most pro-nounced at the top of the very first hill. As the train disconnects from the lift chain and approaches the descent, it is rolling forward slowly and the first cars are well over the crest of the hill before they pick up

Fig. 3.3.8 The roller coaster pushes these riders inward as it travels around the loop-the-loop. The rid-ers feel a strong fictitious force outward and can hardly tell that they're turning upside-down.

much speed (Fig. 3.3.9a). They're pulling hard on the cars behind them, and those cars are pulling back, slowing their descent. By the time the train is moving fast, the first car is well down the hill. By then, the track is beginning to turn upward and the first car's riders feel mostly upward acceleration and downward fictitious force. That's why riders in the first few cars of a roller coaster don't feel much weightlessness.

In contrast, the last car is moving at high speed early in its descent. It undergoes a dramatic downward acceleration as it's yanked over the crest of the hill by the cars in front of it (Fig. 3.3.9b). As a result, its riders feel large upward fictitious forces and quite extreme weightlessness. In fact, the designers of the track must be careful not to make the downward accelerations too rapid, or the roller coaster will flick the riders in the last car right out of their seats.

Obviously, it does matter where you sit on a roller coaster. The first seat offers the most exciting view, but less than spectacular fictitious forces. The last car almost always offers the best fictitious forces. Probably the dullest seat in the roller coaster is the second; it offers a relatively tame ride and a view of the people in the front seat.

Check Your Understanding #3: Taking to the Air

Your racing car travels over a bump in the track and suddenly becomes airborne. What keeps you in the air? Has gravity disappeared?

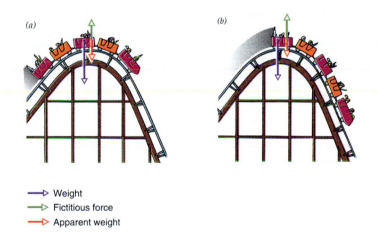

(a) *(b)*

→ Weight
→ Fictitious force
→ Apparent weight

Fig. 3.3.9 When a multicar roller coaster descends the first hill, the ride experienced in the first cars is different from that in the last cars. (*a*) The first cars travel over the crest of the hill slowly and reach high speed only well down the hill. The cars behind them slow their descent. (*b*) The last cars are whipped over the top and are traveling very rapidly early on. The last cars accelerate downward dramatically as they go over the first hill and their riders experience a strong feeling of weightlessness.

Bicycles

SECTION 3.4

A bicycle is a wonderfully energy efficient human-powered vehicle. Its wheels allow its rider to make full use of inertia and gravitational potential energy, rolling forward easily on a level surface and accelerating down hills. Compare the easy motion of a bicycle to that of walking, which requires effort every step of the way. Bicycles are very simple machines, and most of their moving parts are quite visible: the pedals, gears, brakes, and steering mechanisms, to name a few. Their simplicity and visibility make bicycles relatively easy to fix, even for a novice.

Questions to Think About: Why is a two-wheeled bicycle preferable to the apparently more stable three-wheeled tricycle? Why do you lean a bicycle in the direction that you are turning? How is it possible to ride a bicycle without hands on the handlebar? How does pedaling a bicycle make it move forward? Why does a bicycle have gears? Why can you pedal backward on a multispeed bicycle, and what is the clicking sound you hear when you do?

Experiments to Do: If you know how to ride a bicycle, pay attention to its stability as you ride it. Notice how you lean the bicycle in the direction that you are turning and how the bicycle naturally steers into the turn as you begin to lean. This automatic steering is part of the bicycle's stabilizing behavior. While you pedal, observe which part of the chain becomes taut and the relative rates at which the pedals and the rear wheel turn. How does this relative rate depend on your choice of gears? How does the ease of pedaling depend on this choice? Notice also that when you stop pedaling, the gear on the rear wheel stops turning but the wheel itself keeps going. Something inside the gear or wheel hub is only connecting the gear and wheel some of the time. How does that device determine when to connect the gear to the wheel?

Static Stability and Tricycles

Bicycles have a stability problem. With only two wheels to support them, stationary bicycles tip over easily. No matter how hard you try, you can't get a riderless bicycle to stand still on its wheels for more than a few seconds. The slightest breeze and over it goes. Why then do we use two-wheeled bicycles for transportation?

We can begin to answer that question by looking at **static stability,** the stability of an object that's not in motion. An object that is **statically stable** is in a stable equilibrium when at rest: if it's disturbed from that equilibrium, it experiences restoring forces or torques that push it back toward equilibrium. A stool has this behavior and so does a table, but a balancing bicycle most definitely does not. If you disturb the bicycle, no forces or torques keep it upright; instead, it crashes to the ground. However, there's another pedal-powered vehicle that is statically stable—a tricycle (Fig. 3.4.1).

A tricycle is statically stable because it spontaneously recovers from a tip (Fig. 3.4.2a). Like any object, the tricycle naturally accelerates in whatever direction causes its total potential energy to decrease as rapidly as possible. Since tipping the tricycle raises its center of gravity and thus its gravitational potential energy, the tricycle spontaneously accelerates in the direction opposite any tip. If it isn't tipping yet, the tricycle doesn't start and if it is tipping, it accelerates so as to return to upright. No wonder children love tricycles.

Fig. 3.4.1 A tricycle is very stable when it's standing still. However, it tips over easily during a high-speed turn because the rider can't lean in the direction that the tricycle is turning. The rider must also pedal furiously to move at a reasonable speed.

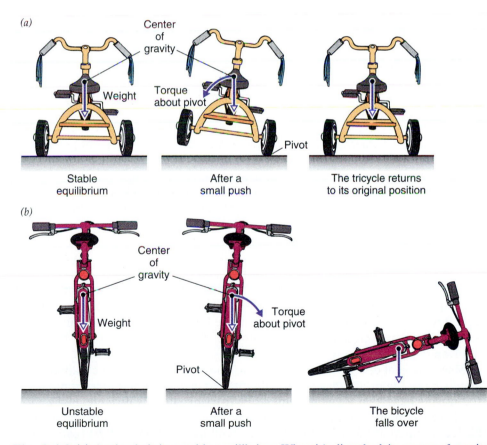

Fig. 3.4.2 (*a*) A tricycle is in a stable equilibrium. When it's disturbed, its center of gravity and gravitational potential energy rise, and it experiences restoring forces that return it to upright. (*b*) A bicycle is in an unstable equilibrium. Any disturbance causes it to fall.

The tricycle establishes a **base of support,** a polygon defined by its contact points with the ground. As long as the tricycle's center of gravity stays above this base of support, tipping it will raise its center of gravity and its gravitational potential energy, and the tricycle will experience restoring forces and torques that return it to upright. The tricycle is in a stable equilibrium.

But a bicycle has no base of support (Fig. 3.4.2*b*). With just two wheels touching the ground, the only polygon that forms is a straight line. If the bicycle tips to one side, its center of gravity will descend and its gravitational potential energy will decrease. The bicycle will accelerate *away* from equilibrium! The upright bicycle is thus in an **unstable equilibrium:** instead of experiencing restoring forces or torques, the disturbed bicycle will tip farther and faster until it hits the ground.

(a)

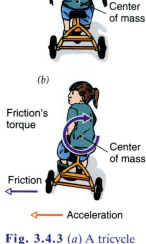

Check Your Understanding #1: Keeping Coffee in Its Place

Some travel mugs taper outward at the bottom so that they have very wide bases. Why does this shape make such a mug particularly stable and keep it from tipping over during the morning commute?

Dynamic Stability and Bicycles

Static stability matters most to people who have difficulty balancing. When it's not moving, a tricycle is hard to tip over, which is why children learn to ride tricycles first (Fig. 3.4.1). But when a tricycle is moving, static stability doesn't guarantee safety. If a child rolls down a steep hill and then suddenly makes a sharp turn, he or she will probably flip over. What has gone wrong?

A tricycle is stable only as long as the rider makes no sudden accelerations, such as a turn at high speed. When a child makes such a turn, he needs a force to produce the necessary sideways acceleration, and that force comes from friction between the wheels and the pavement. As the child steers the tricycle, the pavement pushes horizontally on its wheels. If the turn is slow enough, the wheels convey this force to the child and the tricycle turns safely. But if the turn is too abrupt, the child's inertia keeps her body going straight while the wheels move to one side. Crash. As you can see, a tricycle has good static stability but poor **dynamic stability**—stability in motion.

The tricycle's problem is that it can't handle the torque that friction exerts on it during a turn. The horizontal frictional force that causes the tricycle to accelerate during a turn is exerted well below the tricycle and rider's combined center of mass (Fig. 3.4.3). This frictional force produces a torque about that center of mass, tending to tip the tricycle over. The tricycle's static stability resists this tipping, up to a point. If the turn is too fast, the frictional torque flips the tricycle over. During high-speed turns, the tricycle is dynamically *unstable*. As we'll soon see, that's because it can't lean during turns.

Since the goal of a wheeled vehicle is to go somewhere, dynamic stability is ultimately more important than static stability. And while a bicycle lacks static stability, a moving bicycle is remarkably stable. Its dynamic stability is so great that it's almost hard to tip over. It can even be ridden without any hands on the handlebars. This feat is a popular daredevil stunt among children who haven't yet realized how easy it is.

As British physicist David Jones discovered (Fig. 3.4.4), the bicycle's incredible dynamic stability comes from *two* sources. First, the wheels behave as gyroscopes.

Fig. 3.4.3 (*a*) A tricycle that is heading straight is stable because any tip causes its center of gravity to move upward. (*b*) However, during a fast left turn, the tricycle accelerates left and friction exerts a large leftward force on the wheels. This frictional force produces a torque about the tricycle and rider's combined center of mass and can cause the tricycle to tip over.

gyroscopes

Because of their rotation, they have angular momentum and tend to continue spinning at a constant rate about a fixed axis in space. A wheel's angular momentum can be changed only by a torque so it naturally tends to keep its upright orientation.

But angular momentum alone doesn't keep the bicycle from tipping over. Instead, the front wheel's gyroscopic action automatically steers the bicycle in whatever direction the bicycle is leaning. If the bicycle's frame begins to lean to the left, the front wheel will automatically steer toward the left so as to return the bicycle to an upright position. While a stationary bicycle falls over when it's disturbed from its unstable equilibrium, a forward-moving bicycle naturally drives under the combined center of mass and returns to equilibrium.

The front wheel steers in the direction that the bicycle is leaning because of gyroscopic **precession.** When you exert a torque on a spinning gyroscope, you change the amount and/or direction of its angular momentum. In most cases, the gyroscope begins to spin about a new axis of rotation. When the bicycle begins to lean to the left, the pavement exerts a torque on the front wheel and that spinning wheel undergoes precession. It steers left!

Although gyroscopic precession can keep a forward-moving bicycle from tipping over all by itself, it's assisted by a second effect. The front wheel also steers in the direction that the bicycle is leaning because of the shape and angle of the fork that supports it. The bicycle is always acting to lower its center of gravity and thus its gravitational potential energy, which is why it falls over when stationary. But a bicycle isn't a rigid object; its front wheel can swivel. Because of the shape and angle of the fork, a leaning bicycle can lower its center of gravity still farther by steering its front wheel in the direction that the bicycle is leaning. When the bicycle is leaning toward the left, its front wheel steers toward the left. Once again the bicycle automatically steers in the direction that it is leaning and avoids falling over. These self-correcting effects explain why a riderless bicycle stays up so long when you roll it forward all by itself.

There is little a bicycle designer can do to change the gyroscopic stabilizing effect, but the fork shape and angle can be varied. To be stable, the front wheel must touch the ground behind the steering axis (Fig. 3.4.5). If the fork is bent or improperly made so that the wheel touches the ground ahead of the steering axis, the bicycle will steer in the direction that's opposite its lean and be virtually unridable. Modern bicycles have front forks that arc forward so that the wheel touches the ground just behind the steering axis. This situation leaves the bicycle **dynamically stable** enough to ride easily yet very maneuverable. Some children's bicycles are built with straighter forks so that the wheel touches the ground far behind the steering axis. These bicycles are more stable than adult bicycles but not as easy to turn.

So why does a bicycle rider lean during a turn? The lean prevents the road from exerting a torque on the bicycle and rider about their combined center of mass. A torque would tip them over.

The rider's goal is to tip himself and/or the bicycle so that the total force from the pavement points directly at their combined center of mass. The pavement will then exert no torque on the two, and they won't experience any angular acceleration. Gravity itself produces no torque because it effectively acts at the center of gravity/mass and thus has no lever arm.

Fig. 3.4.4 In 1970, British physicist David Jones investigated the origins of a bicycle's dynamic stability. He built a series of "unridable" bicycles, including one that had a front wheel so small that it often skidded and became almost red hot as the bicycle was ridden. Jones found that gyroscopic effects alone did not account for a bicycle's stability. He discovered that this stability also comes from the shape of the front fork.

Steering axis

Fig. 3.4.5 A bicycle is stable when moving forward in part because its front wheel touches the ground behind the steering axis. As a result, the front wheel naturally steers in the direction that the bicycle is leaning and returns the bicycle to an upright position.

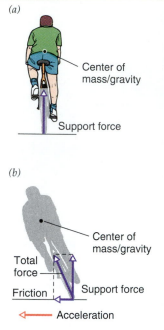

(a)

Center of
mass/gravity

Support force

(b)

Center of
mass/gravity

Total
force

Friction

Support force

Acceleration

Fig. 3.4.6 (*a*) A bicycle
that is heading straight is in
rotational equilibrium when
it's perfectly upright. The
support force from the road
produces no torque about
its center of mass. (*b*) A bicycle that is turning left is in
rotational equilibrium when
it's tilted to the left. Together, the support and frictional forces from the road
produce no torque about its
center of mass.

Fig. 3.4.7 As they round a
turn during a race, these cy-
clists lean toward the inside
of the turn. This leaning prevents the pavement from
producing torques on them
and tipping them over.

When the bicycle is heading straight, the rider keeps the bicycle upright. That
way, the support force from the road points directly toward their combined center
of mass and produces no torque about it (Fig. 3.4.6*a*). But when the rider wants to
turn left, he leans himself and/or the bicycle to the left. Since the bicycle is accelerating to the left during the turn, the road exerts both an upward support force and
a leftward frictional force on the wheel (Fig. 3.4.6*b*). Although the road's total force
on the wheel now points to one side, the lean ensures that this force points directly
toward the pair's center of mass and produces no torque about it. The bicycle
doesn't tip over (Fig. 3.4.7).

With the correct amount of leaning during turns, the bicycle remains in *rotational equilibrium:* it has zero net torque on it. After a while, this habit of leaning the
bike becomes so automatic, you don't even think about it. As they say, you never
forget how to ride a bicycle.

With the proper lean, you can always prevent a turning vehicle from tipping
over. Even if you turn a bicycle or motorcycle too hard and begin to skid, you can
still keep it in equilibrium. But leaning in the direction of a turn is possible only
with vehicles that are statically unstable. Vehicles that can't lean can and do tip
over. Even a car can tip over on a high-speed turn, and some cars or modified
trucks are particularly prone to flipping. The higher a car's center of mass, the
easier it is for frictional forces to tip it over during a turn. There have been several recent production cars that have been found unsafe because their centers of
mass are too high and their bases of support too narrow. Small trucks that have
been lifted up to resemble tractor-trailer cabs may look cool, but they also tip
over easily.

Check Your Understanding #2: Cutting Corners

Why does a skier lean in the direction of a turn during a downhill run?

Pedaling Bicycles

So far, we have only discussed stability. But once we've settled on a bicycle as the
most likely configuration for a useful person-powered vehicle, we need to figure out
how to person-power it. The rider could push his feet on the ground, but that would
be pretty inconvenient and even dangerous at high speeds. Instead, we use foot pedals to produce a torque on one of the wheels. But which wheel, and how do we produce that torque?

The original answer was to power the front wheel by putting pedals and a crank
directly on its axle. The axle was suspended with bearings so that exerting forces on
the pedals produced a torque on the front wheel and it began to turn. Friction between the turning wheel and the ground then pushed the bicycle forward. This
method is still used in children's tricycles, but it has two drawbacks. First, you must
keep pedaling as long as you're moving. If the front wheel is turning, so are the pedals. The second drawback is more serious: you can produce more than enough
torque on the wheels of a tricycle, but you have trouble moving your legs quickly
enough to get a tricycle going very fast. When you ride on level ground, you find
yourself pedaling furiously to move at any decent pace and feel very little resistance
from the pedals.

An early solution to the frantic pedaling problem was to use a gigantic front
wheel. In such a configuration, one turn of the wheel would take you a considerable

distance, so that you no longer had to pedal so fast. The pennyfarthing of the middle 19th century was this sort of bicycle (Fig. 3.4.8). But you still had to keep pedaling this bicycle and it had a new problem: you couldn't push hard enough with your feet to keep its front wheel turning on uphill stretches.

The uphill problem was solved long ago by removing the pedals from the wheel axle and using an indirect drive scheme for the rear wheel. This change makes it possible to use mechanical advantage, when necessary, to reduce the required pedaling force. The problem of the perpetually turning pedals was solved by incorporating a one-way drive or freewheel in the hub of the rear wheel. The freewheel allows the wheel to turn freely when you stop pedaling. A modern bicycle with these improvements appears in Fig. 3.4.9.

Fig. 3.4.8 The pennyfarthing used a large directly driven front wheel to permit the rider to travel at a reasonable speed without having to pedal very rapidly. Its name came from its resemblance to two coins, the large English penny and the smaller farthing.

> **Check Your Understanding #3: Bigger Isn't Always Better**
> Why is it so difficult to climb a hill on a pennyfarthing?

Hills and Straightaways: Using Gears

A bicycle uses a metal chain to convey torque from its pedals to its rear wheel. As your feet push down on the pedals, they produce a torque on the crank and its toothed crank sprocket (Fig. 3.4.10). The crank and sprocket would simply undergo angular acceleration were it not for the chain wrapped around the crank sprocket. The top portion of this chain extends from the crank sprocket to the freewheel sprocket on the rear wheel, so that the crank sprocket can't undergo angular acceleration without dragging the rear wheel along with it. Tension in the top portion of the chain pulls backward on the crank sprocket, opposing the pedaling torque and slowing the crank's angular acceleration. At the same time, this tension pulls forward on the freewheel sprocket, producing a torque on the rear wheel and encouraging it to spin faster.

Fig. 3.4.9 A modern bicycle. The rear wheel is driven by a chain that allows the rider to vary the mechanical advantage between the pedals and the rear wheel. A freewheel in the hub of the rear wheel lets that wheel turn freely in one direction. This free motion allows the bicycle to coast forward, even when the pedals are stationary.

However, there's one additional source of torque in this complicated arrangement—friction. The pavement pushes on the bottom of the rear wheel with whatever static frictional force is needed to prevent skidding and this force produces a torque on the rear wheel. Your pedaling makes the rear wheel push the pavement backward, so the pavement responds by pushing the rear wheel forward. This forward force from the ground on the wheel slows the wheel's angular acceleration, but it also pushes the whole bicycle forward. That's why your pedaling has the overall effect of propelling the bicycle.

The amount of forward force you need as you pedal depends on what you're doing. If you're rolling downhill or going slowly on level ground, you barely need to be pushed forward at all. But if you're climbing a hill, fighting a headwind, or struggling with air resistance at high speeds, the pavement must push forward strongly on the bicycle to prevent it from slowing down. Here is where gears come into play.

Each time you push the pedals in a complete circle, the rear wheel moves a certain distance along the pavement. But that distance depends on which gear you've selected. If the crank sprocket is much larger than the freewheel sprocket, then the rear wheel spins rapidly and the distance traveled is large. But if you select a smaller crank sprocket and a larger freewheel sprocket, then the distance traveled will be small. It might seem best to travel a large distance, but there's a catch: you have to push harder on the pedals when the gears are turning the rear wheel relatively quickly.

gears and belts

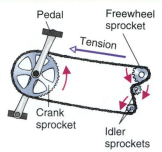

Fig. 3.4.10 The drive system for a modern bicycle. Forces exerted on the pedals produce a torque that turns the crank sprocket. The crank sprocket produces tension in the chain, which exerts a torque on the freewheel sprocket and the rear wheel. The idler sprockets handle the extra chain.

It's a matter of work. To see this clearly, imagine that you are riding a bicycle forward at a steady pace. From your perspective, the rear wheel is pushing the pavement backward and thus doing work on it. That work isn't being wasted; it is actually what keeps you moving forward relative to the ground. It's just that in a bicyclist's frame of reference, the ground is what moves.

With each complete turn of the crank, you do a certain amount of work on the pedals and the wheel does a certain amount of work on the pavement. But the farther the pavement moves under the bicycle and the harder the bicycle pushes on the pavement, the more work the wheel does. Since your legs are providing that work, something about your pedaling has to change. Your feet move a fixed distance around in a circle with each turn of the crank, so the only way you can do more work is to push the pedals harder; more force over the same distance results in more work.

There's a limit to how hard you can push with your legs, so you normally select a gear in which the forces required to turn the pedals are reasonable. The gearing accommodates the fact that sometimes the rear wheel only pushes backward lightly on the pavement, in which case it can move long distances with each turn of the crank, and sometimes the wheel pushes strongly on the pavement, in which case it must move only a short distance with each turn of the crank. Whether you're traveling quickly on a straightaway or slowly up a hill, gears let you pedal at a sensible rate with comfortable forces.

> **Check Your Understanding #4: Racing Down a Hill**
> If you wish to go particularly fast downhill, which sprockets should the chain be on?

Fig. 3.4.11 The ratchet in a bicycle freewheel. If the relative rotation of the inner and outer parts is in the correct direction, the pawls transmit torque from the outer part to the inner part. If the relative rotation direction is reversed, the pawls compress the springs and skip along the teeth on the inside of the outer part. No torque is transmitted.

The Freewheel

Another feature of modern bicycles is the freewheel itself. The freewheel sprocket doesn't drive the rear wheel directly. Instead, it twists a device that transmits torque to the rear wheel in only one direction. When you pedal forward, the chain exerts torque on the rear wheel. When you stop pedaling, or pedal backward, the freewheel allows the rear wheel to turn freely. As a result, you don't have to keep pedaling whenever the bicycle is moving.

The actual freewheel mechanism is called a *ratchet* (Fig. 3.4.11). The outer part of the ratchet pushes on the inner part by way of several *pawls*. These pawls are mounted with springs so that they connect the inner and outer parts of the ratchet only when they try to rotate one direction relative to another. When they try to rotate the other way, the pawls compress their springs and skip freely along the teeth of the outer part. You hear the sound of this skipping when you stop pedaling. Other devices that use ratchets include socket wrenches, pull-starters on gasoline engines, window shades, winding stems on wristwatches, and automobile seat belts.

> **Check Your Understanding #5: Tickets Please**
> Turnstiles in sports stadiums, subways, stores, and amusement parks turn only one way, so that you can't return the way you came. What mechanism prevents them from turning backward?

bike brakes, tires, bike energy

Epilogue for Chapter 3

In this chapter we looked at the physical concepts involved in four types of simple machines. In *spring scales,* we explored the relationship between the force acting on a spring and its distortion, and we examined how this distortion can be used to measure an object's weight. In *bouncing balls,* we examined the process of storing and releasing kinetic energy during a collision. As we saw, both the ball and the object it hits contribute to the bounce.

In *carousels and roller coasters,* we explored both the sensations we feel as we accelerate and why those sensations occur. We saw that circular motion involves a centrally directed acceleration that gives rise to an outwardly directed apparent force. And in *bicycles,* we investigated the concept of dynamic stability, where an object that falls over when stationary becomes remarkably stable in motion. We also saw how the need for mechanical advantage between the turning of the pedals and the rotation of the wheels led to the development of the modern chain-driven, multispeed bicycle.

Explanation: Swinging Water Overhead

As you swing the bucket over your head, you are pulling downward on it and causing it to accelerate downward very rapidly. The water remains in the inverted bucket because the bucket is accelerating downward faster than gravity alone can accelerate the water. Although the water is free to fall, the bucket overtakes the falling water. As a result, the water is pressed into the bottom of the bucket. The same effect occurs when you push a book downward rapidly with your open palm. Although the book is free to fall, it remains pressed into your palm because your hand is accelerating it downward faster than gravity. Finally, if you stop swinging the bucket too abruptly, the bucket's contents will not decelerate with it. Instead, they will spill or smash.

Chapter Summary

How Spring Scales Work: A spring scale measures an object's weight by supporting that object with a spring. When the object is at rest, the spring's upward force exactly balances the object's downward weight; the scale then reports the upward restoring force its spring is exerting on the object. As Hooke's law describes, that restoring force is proportional to the spring's distortion, so the scale can determine it by measuring how far the spring has bent. This measurement is often done mechanically and is reported with a needle or dial.

How Bouncing Balls Work: A ball behaves like a spherical or oblong spring. It stores elastic potential energy as it distorts away from its equilibrium shape and releases some of that energy as it returns to normal. When a ball strikes a surface, some kinetic energy is removed from the ball and the surface and is either stored within those objects as elastic potential energy or lost as thermal energy. As they rebound, some of the stored energy becomes kinetic energy again. The kinetic energy returned—the rebound energy—is always less than the kinetic energy initially removed from the objects—the collision energy. The missing energy has been converted into thermal energy.

How Carousels and Roller Coasters Work: A carousel uses centripetal acceleration to produce an outward fictitious force on each rider. Combined with weight, this gravity-like sensation gives the rider an apparent weight that points downward and outward.

A roller coaster also uses rapid acceleration to create unusual apparent weights for its riders. Each time the coaster accelerates on a hill or turn, the rider experiences a fictitious force in the direction opposite the acceleration. This fictitious force, combined with the rider's true weight, creates an apparent weight that varies dramatically in amount and direction throughout the ride. It's this fluctuating apparent weight, particularly its near approaches to zero, that make riding a roller coaster so exciting.

How Bicycles Work: While a stationary bicycle tips over easily, a moving bicycle is remarkably stable. It remains upright with the help of two stabilizing effects of motion—one due to gyroscopic precession and the other due to the shape and angle of the front fork. Both effects work together to steer the bicycle in the direction that it's leaning. It drives under its center of mass and thus remains upright.

The rider propels the bicycle by exerting a torque on the pedals and crank. The crank's rotary motion is conveyed to the rear wheel by a chain and sprockets, usually with mechanical advantage. As the rider pedals forward, the rear wheel turns, and friction between the tire and the pavement pushes the bicycle forward. A freewheel at the hub of the rear wheel disconnects the rear sprocket from the wheel whenever the wheel is turning forward faster than the sprocket. This disconnection makes it possible to stop pedaling while the rear wheel is still turning forward.

Important Laws and Equations

1. Hooke's Law: The restoring force exerted by an elastic object is proportional to how far it is from its equilibrium shape, or

restoring force = −spring constant · distortion.   (3.1.1)

2. Acceleration of an Object in Uniform Circular Motion:

An object in uniform circular motion has a centripetal acceleration equal to the square of its speed divided by the radius of its circular trajectory, or

$$\text{acceleration} = \frac{\text{speed}^2}{\text{radius of circular trajectory}}. \quad (3.3.1)$$

Check Your Understanding—Answers

Section 3.1 SPRING SCALES

1. You should sell by mass and label your foods in kilograms, not pounds.

Why: Packages that are labeled in pounds are specifying their contents by weight. Pounds are a unit of force, in this case the gravitational force the earth exerts on the package's contents. A 1-pound container of oatmeal will only weigh 1 pound at the surface of the earth. When it's exported to the moon, it will weigh just $\frac{1}{6}$ of a pound, and your company may be fined for selling underweight groceries. If you label the packages according to their masses, that labeling will remain correct no matter where you ship the packages. Mass is the measure of inertia and depends only on the object, not on its environment.

2. The diving board is behaving as a spring, bending downward in proportion to the weight of each diver.

Why: The heavier the diver, the more the board bends downward before exerting enough upward force on the diver to balance the diver's weight.

3. 15 N (about 3.3 pounds).

Why: The scale's dial is simply reporting the position of its basket. The dial is calibrated so that a 1-centimeter drop in the basket indicates that the spring is pulling up on it with a force of 5 N. Since the spring's restoring force is described by Hooke's law, a 3-centimeter drop in the basket means that the spring is exerting an upward force of 15 N on the basket.

4. The distance the scale's surface moves downward is proportional to the weight it reports.

Why: The scale's spring is connected to its surface by levers so that as the surface moves downward, the spring distorts by a proportional amount. The spring's distortion is reported on the dial. Thus the dial's reading is proportional to the surface's downward movement.

Section 3.2 BOUNCING BALLS

1. When the marble collides with the hard granite surface, it dents inward and converts most of its kinetic energy into elastic potential energy. It then rebounds, converting this stored energy back into kinetic energy as it bounces from the surface.

Why: An elastic marble has a very high coefficient of restitution and bounces well from a hard surface.

2. The soft dirt distorts more than a marble's glass surface and receives most of the collision energy. It converts most of that energy into thermal energy so the marble rebounds weakly.

Why: While a marble bounces nicely on a hard surface, the dirt is soft and receives virtually all of the collision energy when the hard marble hits it. The dirt distorts during the impact but stores little energy because it's not very elastic. The marble doesn't rebound much.

3. 2 m/s toward you.

Why: The velocities reported in the question are those observed by people sitting still with respect to the circle. From the inertial reference frame of your marble, the circle itself is heading in your direction at 1 m/s. Since the other person's marble is moving in your direction at 1 m/s faster than the circle, its total velocity according to your marble is 2 m/s in your direction.

4. The glass ball's mass is so much larger than that of your thumb that your thumb receives almost all the rebound energy. Your thumb bounces, not the glass ball.

Why: In any collision, it's the least massive object that experiences the greatest acceleration and that receives the largest share of the rebound energy. The effect is similar to what would happen if you swung a light aluminum baseball bat at a pitched bowling ball. The bat would rebound wildly but the bowling ball would continue to travel toward the catcher. This same effect is true in automobile collisions, where a massive sedan is much less disturbed than the tiny, subcompact with which it collides.

Section 3.3 CAROUSELS AND ROLLER COASTERS

1. The car seat and door are exerting a real leftward force on you, causing you to accelerate leftward with the car. You also experience a rightward fictitious force, as your inertia acts to keep you from accelerating leftward.

Why: As the car turns left, it accelerates toward the left. The car seat and right door together exert a real leftward force on you, to prevent the car from driving out from under you. Because it has mass, your body resists this leftward acceleration. You feel your body trying to go in a straight line, which would carry it out the right door of the car as the car accelerates toward the left.

2. So that the support force exerted by the racetrack on the car's wheels can provide at least some of the centripetal force needed to accelerate the car around the turn.

Why: As a car and driver travel around a circular turn, they are accelerating toward the center of the track and require a huge centripetal force inward. On a level track, the only horizontal force available is static friction between the ground and the car's tires. If static friction is unable to provide enough inward force, the car will skid off the track, following a straight-line path. This type of accident is typical of a highway curve on an icy day and is why designers bank the curves. The banks are ramps sloping down toward the center so that the horizontal component of the support force exerted by the track on the car's wheels provides an additional, inward, centripetal force to help that car accelerate around the curve.

3. Gravity is still present, but your inertia prevents you from following a rapid downturn in the surface you are traveling along.

Why: If the road you are traveling along suddenly turns downward, you must accelerate downward to stay in contact with its surface. The steeper and more abrupt the descent, the more downward acceleration you need. The only downward force you experience is your weight, which can cause a downward acceleration of no more than 9.8 m/s^2. If the surface drops out from under you faster than that, you will become airborne. You will then be falling freely and accelerating downward as fast as gravity will permit. Eventually, you will fall to the surface. In many sports, including skiing, motorcycle racing, and skateboarding, a person traveling along an uneven surface becomes airborne after passing over a bump.

Section 3.4 BICYCLES

1. The mug's wide base ensures that its center of gravity will stay above its base even during a severe tip and it will naturally return to its stable equilibrium.

Why: A mug with a narrow base is stable, but only if it never tips very far. As soon as its center of gravity is no longer above its base, over it will go. But a mug with a wide base has a broad base of support and will recover even from severe tips.

2. To make sure that the force on her skis is directed toward her center of mass.

Why: There are many sports and situations where a person must lean in the direction of a turn. Since a turn always involves horizontal acceleration and horizontal force, the person must shift her center of mass toward the inside of the turn. When she leans just the right amount, the net force on her feet points directly toward her center of mass and ex-

erts no torque on her about her center of mass. She doesn't begin to rotate and doesn't tip over.

3. Each turn of the huge wheel takes you a long distance up the hill and requires lots of work. To do this much work in a single turn of the pedals, you must push on the pedals very hard.

Why: As you pedal a bicycle, you are doing work on the pedals; you push the pedals as they move in the direction of that push. If too much work is required, as is the case when you're going uphill on a pennyfarthing, the force involved becomes excessive.

4. The largest crank sprocket and the smallest rear wheel sprocket.

Why: On a downhill run, you want the rear wheel to turn as fast as possible. Torque is unlikely to be a problem. If you select the largest crank sprocket, the chain will be drawn forward quickly as you pedal. If you select the smallest rear sprocket, the movement of the chain will cause the rear wheel to turn as quickly as possible. With this arrangement, each turn of the crank will cause the rear wheel to turn about five times. This way you can pedal slowly even as the bicycle races down the hill.

5. A ratchet.

Why: As you walk through a turnstile in the permitted direction, pawls in the ratchet skip and it turns easily. You can often hear the pawls clicking. But when you try to go in the other direction, the pawls engage and prevent the turnstile from rotating.

Check Your Figures—Answers

Section 3.1 SPRING SCALES

1. 2 centimeters and 10 centimeters (assuming that the floor doesn't break).

Why: A floor, like most suspended surfaces, behaves like a spring. Your floor distorts 1 centimeter before it exerts an upward restoring force equal to the weight of 10 guests. It will thus distort 2 centimeters before supporting 20 guests and 10 centimeters before supporting 100 guests. While this distortion should be within the elastic limit of the floor beams, it may cause the plaster and paint to crack. If the beams break, the floor will collapse.

Section 3.3 CAROUSELS AND ROLLER COASTERS

1. About 15 m/s^2.

Why: The children are in uniform circular motion so their acceleration is given by Eq. 3.3.1. To use this equation, we need to know the children's speed and the radius of their circular trajectory. Since they complete one trip around the circumference of a 1.5-m radius circle every 2 seconds, their speed is

$$\text{children's speed} = \frac{\text{distance}}{\text{time}}$$
$$= \frac{2\pi \cdot \text{radius}}{\text{time}} = \frac{2\pi \cdot 1.5 \text{ m}}{2 \text{ s}} = 4.7 \text{ m/s}.$$

We can now obtain their acceleration from Eq. 3.3.1:

$$\text{children's acceleration} = \frac{(4.7 \text{ m/s})^2}{1.5 \text{ m}} = 14.7 \text{ m/s}^2.$$

Since our measurements of the carousel's radius and its turning time are only accurate to about 10%, our calculation of the children's acceleration is only accurate to about 10%. We report it as 15 m/s^2.

Exercises

1. In what way does the string of a bow and arrow behave like a spring?

2. As you wind the mainspring of a mechanical watch or clock, why does the knob get harder and harder to turn?

3. Curly hair behaves like a weak spring that can stretch under its own weight. Why is a hanging curl straighter at the top than at the bottom?

4. When you lie on a spring mattress, it pushes most strongly on the parts of you that stick into it the farthest. Why doesn't it push up evenly on your entire body?

5. If you pull down on the basket of a hanging grocery store scale so that it reads 15 N, how much downward force are you exerting on the basket?

6. People often remark that a particular scale "reads heavy," meaning that it reports more than a person's real weight. What is wrong with the scale's spring?

7. If you were to step on several different bathroom scales, one after the next, chances are that they would report slightly different weights. Explain this result in terms of calibration.

8. While you're weighing yourself on a bathroom scale, you reach out and push downward on a nearby table. Is the weight reported by the scale high, low, or correct?

9. There's a bathroom scale on your kitchen table and your friend climbs up to weigh himself on it. One of the table's legs is weak and you're afraid that he'll break it, so you hold up that corner of the table. The table remains level as you push upward on the corner with a force of 100 N. Is the weight reported by the scale high, low, or correct?

10. If you put your bathroom scale on a ramp and stand on it, will the weight it reports be high, low, or correct?

11. When you step on a scale, it reads your weight plus the weight of your clothes. Only your shoes are touching the scale, so how does the weight of the rest of your clothes contribute to the weight reported by the scale?

12. To weigh an infant you can step on a scale once with the infant and then again without the infant. Why is the difference between the scale's two readings equal to the weight of the infant?

13. Why doesn't a marshmallow bounce well when you drop it on the floor?

14. If you drop a ball onto the floor from a height of 1 m, it will rebound to a height of less than 1 m. Why can't it bounce higher than 1 m? Why doesn't it even reach 1 m?

15. An elastic ball that wastes 30% of the collision energy as heat when it bounces on a hard floor will rebound to 70% of the height from which it was dropped. Explain the 30% loss in height.

16. The best running tracks have firm but elastic rubber surfaces. How does a lively surface assist a runner?

17. Why is it so exhausting to run on soft sand?

18. Steep mountain roads often have emergency ramps for trucks with failed brakes. Why are these ramps most effective when they are covered with deep, soft sand?

19. There have been baseball seasons in which so many home runs were hit that people began to suspect that something was wrong with the baseballs. What change in the baseballs would account for them traveling farther than normal?

20. During rehabilitation after hand surgery, patients are often asked to squeeze and knead putty to strengthen their muscles. How does the energy transfer in squeezing putty differ from that in squeezing a rubber ball?

21. Your car is on a crowded highway with everyone heading south at about 100 km/h (62 mph). The car ahead of you slows down slightly and your car bumps into it gently. Why is the impact so gentle?

22. Bumper cars are an amusement park ride in which people drive small electric vehicles around a rink and intentionally bump them into one another. All of the cars travel at about the same speed. Why are head-on collisions more jarring than other types of collisions?

23. When two trains are traveling side by side at breakneck speed, it's still possible for people to jump from one train to the other. Explain why this can be done safely.

24. If you drop a steel marble on a wooden floor, why does the floor receive most of the collision energy and contribute most of the rebound energy?

25. A RIF (Reduced Injury Factor) baseball has the same coefficient of restitution as a normal baseball except that it deforms more severely during a collision. Why does this increased deformability lessen the forces exerted by the ball during a bounce and reduce the chances of it causing injury?

26. Padded soles in running shoes soften the blow of hitting the pavement. Why does padding reduce the forces involved in bringing your foot to rest?

27. Some athletic shoes have inflatable air pockets inside them. These air pockets act like springs that become stiffer

as you pump up the air pressure. High pressure also makes you bounce back up off the floor sooner. Why does high pressure shorten the bounce time?

28. Why does it hurt less to land on a soft foam pad than on bare concrete after completing a high jump?

29. Why must the surface of a hammer be very hard and stiff for it to drive a nail into wood?

30. Some amusement park rides move you back and forth in a horizontal direction. Why is this motion so much more disturbing to your body than cruising at a high speed in a jet airplane?

31. You are traveling in a subway along a straight, level track at a constant velocity. If you close your eyes, you can't tell which way you're heading. Why not?

32. Moving a can of spray paint rapidly in one direction will not mix it nearly as well as shaking it back and forth. Why is it so important to change directions as you mix the paint?

33. Why does a baby's rattle only make noise when the baby moves it back and forth and not when the baby moves it steadily in one direction?

34. In some roller coasters, the cars travel through a smooth tube that bends left and right in a series of complicated turns. Why does the car always roll up the right-hand wall of the tube during a sharp left-hand turn?

35. Railroad tracks must make only gradual curves to prevent trains from derailing at high speeds. Why is a train likely to derail if it encounters a sharp turn while it's traveling fast?

36. What keeps the ball pressed against the outside rim of a spinning roulette wheel? Why does the ball roll inward off the rim as the wheel slows down?

37. As a grand prix car wends its way around the curving streets of a European city, the driver must slow down to avoid sliding off the road. What limits the speed at which a level curve can be negotiated without sliding off the road?

38. Police sometimes use metal battering rams to knock down doors. They hold the ram in their hands and swing it into a door from about 1 m away. How does the battering ram increase the amount of force the police can exert on the door?

39. When a moving hammer hits a nail, it exerts the enormous force needed to push the nail into wood. This force is far greater than the hammer's weight. How is it produced?

40. A hammer's weight is downward, so how can a hammer push a nail upward into the ceiling?

41. As you swing back and forth on a playground swing, your apparent weight changes. At what point do you feel the heaviest?

42. Some stores have coin-operated toy cars that jiggle back and forth on a fixed base. Why can't these cars give you the feeling of actually driving in a drag race?

43. A salad spinner is a rotating basket that dries salad after washing. How does the spinner extract the water?

44. People falling from a high diving board feel weightless. Has gravity stopped exerting a force on them? If not, why don't they feel it?

45. When your car travels rapidly over a bump in the road, you suddenly feel weightless. Explain.

46. Astronauts learn to tolerate weightlessness by riding in an airplane (nicknamed the "vomit comet") that follows an unusual trajectory. How does the pilot direct the plane in order to make its occupants feel weightless?

47. You board an elevator with a large briefcase in your hand. Why does that briefcase suddenly feel particularly heavy when the elevator begins to move upward?

48. As your car reaches the top in a smoothly turning Ferris wheel, which way are you accelerating?

49. Why does shaking your wet hands remove the water?

50. A rodeo rider must hold tight to a bucking bull to avoid being thrown off. The bull contorts its body so that its back accelerates downward faster than the acceleration due to gravity. Why does that movement tend to lose the rider?

51. Sprinters start their races from a crouched position with their bodies well forward of their feet. This position allows them to accelerate quickly without tipping over backward. Explain this effect in terms of torque and center of mass.

52. If the bottom of your bicycle's front wheel becomes caught in a storm drain, your bicycle may flip over so that you travel forward over the front wheel. Explain this effect in terms of rotation, torque, and center of mass.

53. When you turn while running, you must lean in the direction of a turn or risk falling over. If you lean left as you turn left, why don't you fall over to the left?

54. If a motorcycle accelerates too rapidly, its front wheel will rise up off the pavement. During this stunt the pavement is exerting a forward frictional force on the rear wheel. How does that frictional force cause the front wheel to rise?

55. As a skateboard rider performs stunts on the inside of a U-shaped surface, he often leans inward toward the middle of the U. Why does leaning keep him from falling over?

56. Most racing cars are built very low to the ground. While this design reduces air resistance, it also gives the cars better dynamic stability on turns. Why are these low cars more stable than taller cars with similar wheel spacings?

57. The crank of a hand-operated kitchen mixer connects to a large gear. This gear meshes with smaller gears attached to the mixing blades. Since each turn of the crank makes the blades spin several times, how is the force you exert on the crank handle related to the forces the mixing blades exert on the batter around them?

58. The starter motor in your car is attached to a small gear. This gear meshes with a large gear that's attached to the engine's crankshaft. The starter motor must turn many times to make the crankshaft turn just once. How does this gearing allow modest forces inside the starter motor to turn the entire crankshaft?

59. A bread-making machine uses gears to reduce the rotational speed of its mixing blade. While its motor spins about 50 times each second, the blade spins only once each second. The motor provides a certain amount of work each second, so why does this arrangement of gears allow the machine to exert enormous forces on the bread dough?

60. The chain of a motorcycle must be quite strong. Since the motorcycle has only one sprocket on its rear wheel, the top of the chain must pull forward hard to keep the rear wheel turning as the motorcycle climbs a hill. Why would replacing the motorcycle's rear sprocket for one with more teeth make it easier for the chain to keep the rear wheel turning?

Problems

1. Your new designer chair has an S-shaped tubular metal frame that behaves just like a spring. When your friend, who weighs 600 N, sits on the chair, it bends downward 4 cm. What is the spring constant for this chair?

2. You have another friend who weighs 1000 N. When this friend sits on the chair from Problem 1, how far does it bend?

3. You're squeezing a springy rubber ball in your hand. If you push inward on it with a force of 1 N, it dents inward 2 mm. How far must you dent it before it pushes outward with a force of 5 N?

4. When you stand on a particular trampoline, its springy surface shifts downward 0.12 m. If you bounce on it so that its surface shifts downward 0.30 m, how hard is it pushing up on you?

5. Engineers are trying to create artificial "gravity" in a ring-shaped space station by spinning it like a centrifuge.

The ring is 100 m in radius. How quickly must the space station turn in order to give the astronauts inside it apparent weights equal to their real weights at the earth's surface?

6. A satellite is orbiting the earth just above its surface. The centripetal force making the satellite follow a circular trajectory is just its weight, so its centripetal acceleration is about 9.8 m/s^2 (the acceleration due to gravity near the earth's surface). If the earth's radius is about 6375 km, how fast must the satellite be moving? How long will it take for the satellite to complete one trip around the earth?

7. When you put water in a kitchen blender, it begins to travel in a 5-cm radius circle at a speed of 1 m/s. How quickly is the water accelerating?

8. In Problem 7, how hard must the sides of the blender push inward on 0.001 kg of the spinning water?

Cases

1. A spring bathroom scale is designed to report the amount of upward force it's exerting on the objects touching its surface.
   a. When you first step on the scale, you usually have some downward velocity because you "land" on the scale. As it slows your motion to bring you to rest, does the scale report your correct weight, or does it report more than your weight or less than your weight?
   b. If you stand motionless on one foot, rather than two feet, what fraction of your weight does the scale report?
   c. If you jump upward, what does the scale report as you push yourself upward?

   d. You stand motionless on two identical bathroom scales, one foot on the left scale and one foot on the right scale. What can you say about the weights that the two scales report?
   *e. You stand motionless on two identical bathroom scales stacked on top of each other. Each scale weighs 10 N. What weights do the two scales report?

2. You are seated in a subway car, facing forward with your eyes closed. The only way that you can tell where you are going is by feeling the effects of motion on your body.

a. When the car is traveling at a steady speed on a straight section of track, what is the net force on your body?

b. Why is it very difficult to feel which way the car is going when it's traveling at a constant velocity?

c. Explain the sensations you experience as the car increases or decreases its forward speed.

d. How can you tell when the car turns to the left or the right?

3. A trampoline has an elastic surface, supported at the edges by springs or elastic bands so that it's normally flat. This surface stores energy during a bounce so that you can jump very high on it.

a. When you get on a particular trampoline and stand in its center, its surface distorts downward 10 cm. It's behaving like a spring that's distorted away from its equilibrium shape. The distorted trampoline surface is exerting a restoring force on you in which direction?

*b. If someone with twice your weight climbed into the middle of the empty trampoline, how far downward would its surface distort?

c. You begin bouncing up and down on the trampoline. As you land on the trampoline during one of the bounces, its surface distorts downward 30 cm. Your weight hasn't changed, so how can you make it distort so far downward?

d. While you are in the air above the trampoline, nothing is pushing on you except gravity. On both your way upward and your way downward (while in the air), do you feel weightless, your normal weight, or particularly heavy?

e. While you are touching the trampoline during a bounce, its surface is pushing upward on you. Near the bottom of the bounce, you distort the trampoline's surface downward more than 10 cm. On both your way downward and your way upward (while the surface is distorted downward more than 10 cm), do you feel weightless, your normal weight, or particularly heavy?

4. Bumper cars are a popular ride at many amusement parks. You drive about an enclosed area in a small, electrically powered vehicle. This vehicle has a large rubber bumper wrapped all the way around its exterior.

a. Half the fun of driving a bumper car is crashing into other people's cars. When you drive forward at high speed and slam into the car in front of you, you find yourself thrown forward in your car. Which way is your car accelerating?

b. When you are stopped and someone else slams into the front of your car, you find yourself thrown forward in your car. Which way is your car accelerating?

c. What would happen to the support forces between cars and to the accelerations those cars experience if the soft rubber bumpers were replaced by hard, steel bumpers?

d. Suppose that all of the cars are traveling at 10 km/h in various directions, as viewed by the people waiting in line for a turn. You crash first into a car that's heading in about the same direction as yours and feel a gentle thud. You then crash into a car that's heading toward you and experience a tremendous jolt. Explain briefly why these two collisions are different.

e. If you were to fill your car with metal bars so that its mass were 10 times that of any other bumper car, how would it affect the jolts you experience during crashes, as compared to riding in a normal car?

5. The body of a car is suspended above the four wheels on four huge coil springs. The road supports the wheels, the wheels support the springs, and the springs support the body.

*a. When the manufacturer placed the body of the car on the springs, the springs were compressed downward 10 cm. The body weighs 10,000 N. When you and your friends climb into the car, the body descends an additional 2 cm. How much do you and your friends weigh?

*b. While you and your friends are driving down a straight, level highway at a constant speed of 100 km/h, what is the net force on the car body and which way (if any) is it accelerating?

*c. The car now passes over a very sudden rise in the pavement. The wheels move upward about 5 cm, instantly compressing the springs by that same amount. Which way does the car body now accelerate, and how large is the acceleration compared to the acceleration due to gravity?

*d. If the car body were attached directly to the wheels, with no springs in between, and it passed over the 5-cm rise in the pavement, which way would the car body accelerate and how large would that acceleration be relative to the acceleration due to gravity? (You must make some assumptions here; justify those assumptions.)

e. Use your answers to parts c and d to explain why cars have spring suspensions.

6. You're taking a step aerobics class. In front of you is a small platform that you step onto and off of during the course of the exercises. Most of the time, one foot remains stationary on the platform and you use it to lift your body up and down.

a. As you step up onto the platform, your leg does work on your body. What characteristics of you and the platform determine how much work your leg must do?

b. How much work is your leg doing on your body as it lowers you gently back down to the ground?

c. If you let yourself drop back down to the ground, rather than lowering yourself gently, you could injure the leg you land on. Why does wearing padded shoes reduce your risk of injury?

*d. The platforms are all identical and are made of a sturdy plastic that acts like a stiff spring. When you step up onto your platform it distorts downward by about 4 mm. You're curious about the weight of the person to your right, so you watch the platform as that person steps onto it. It distorts downward 8 mm. How much does that person weigh?

7. If you attach a string to a ball and begin swinging it around your head, the ball will travel in a circle. Let's suppose that the ball is traveling counterclockwise, as viewed from overhead.

    a. Although the ball is traveling with a constant speed, it is accelerating. Which way is it accelerating?

    b. In which direction is the net force on the ball?

    c. In which direction is the ball traveling at the moment it passes directly in front of you?

    d. If you let go of the string at the moment the ball passes directly in front of you, which way will the ball travel? Why?

    e. You feel an outward tug on the string, yet there is no outward force on the ball. The only way that the ball can tug on the string is if the string is tugging on the ball. What effect does the string's pull have on the ball if there isn't any outward force on the ball?

8. A one-wheeled cycle, or unicycle, is the ultimate in statically unstable vehicles.

    a. Why does a unicycle always fall over when the rider doesn't try to keep it upright?

    b. To keep the unicycle from falling over, the rider continually tries to position the wheel so that the force the ground exerts on the wheel points toward a particular point. What is that point?

    c. A person who is riding forward on a two-wheeled bicycle must lean left while making a turn toward the left. Does a person who is riding forward on a unicycle also have to lean left during a left turn?

    d. While a unicycle doesn't exhibit dynamic stability the way a two-wheeled bicycle does, there is a way to give it dynamic stability. If you spin the unicycle extremely rapidly about its vertical axis, it will act like a toy top and won't fall over for a very long time. (Unfortunately, it's hard to ride this way.) While gravity will exert a torque on this spinning unicycle about its point of contact with the ground, if its axis isn't perfectly vertical, the unicycle doesn't simply fall over. Instead, its axis of rotation changes directions. This behavior is an example of what?

4

Fluids

So far all of the everyday objects we've examined have been solids. But since gases and liquids are also important parts of the world around us—as the air we breathe, the water we swim in, and even the blood we pump through our veins—we'll now turn to objects that, unlike solids, do not have well-defined shapes. These objects are called fluids, and the study of fluid behavior and motion is a broad field, extending across the sciences and engineering. Fluid dynamics, or hydrodynamics as it's often called, is as important to an oil-well engineer as to an animal physiologist or a stellar astrophysicist. The tools used to analyze fluids are somewhat more complicated than for solids because fluids themselves are more complicated: it's hard to exert a force directly on them, and, even if we could, they usually don't move as a single rigid object. In this chapter, we'll look at some of the concepts and tools needed to understand their complex behaviors.

EXPERIMENT: A Cartesian Diver

One of these concepts is buoyancy. An object immersed in a fluid experiences an upward force from that fluid. This buoyant force is what lifts a helium balloon in

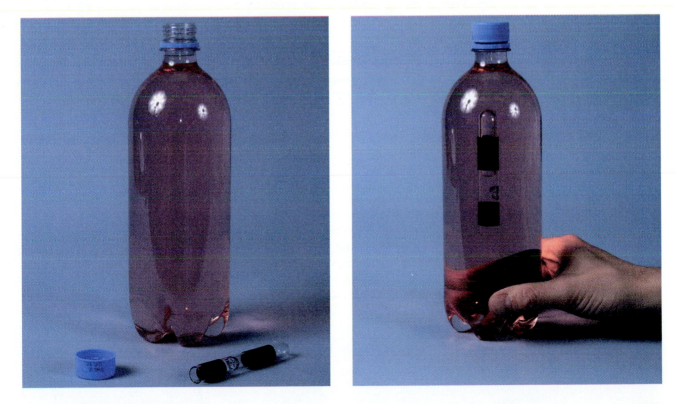

the air and suspends a boat on the surface of water; it depends on the relative densities of the object and the fluid in which it's immersed, where density is the ratio of the object's or the fluid's mass to its volume. As we'll see in this chapter, an object that's more dense than the fluid around it sinks, while one that's less dense than the surrounding fluid floats.

To see the importance of density in determining whether an object floats or sinks, you can construct a simple toy called a Cartesian Diver. This once-popular parlor gadget consists of a small air-filled vial floating in a sealed container of water. Normally, the air bubble inside the vial keeps it floating at the surface of the water, but whenever you squeeze the container, the vial sinks.

To make a Cartesian Diver, you'll need only a plastic soda bottle and a small vial that's open at one end. The vial can be made of almost anything—plastic, metal, or glass—as long as it's dense enough to sink in water. Fill the plastic soda bottle full of water and float the vial in it upside-down; air trapped inside the vial should keep the vial afloat. Now slowly reduce the size of the air bubble inside the vial until the vial barely floats. You can make this adjustment by tipping the vial to let some of its air escape or by removing it from the bottle and pouring water into it. Once you have the vial floating only a few millimeters out of the water at the top of the soda bottle, cap the bottle and prepare to test your diver.

Before you squeeze the soda bottle, try to *predict* what will happen when you do. How will squeezing the bottle affect the air bubble and the vial's position in the water? Now squeeze the bottle hard and *observe* the results. Did they *verify* your predictions? *Measure* the air bubble's size as you squeeze the bottle. How does the bubble's size depend on how hard you squeeze the bottle? Why should the two be related? What is the relationship between the size of the air bubble and the diver's height in the water?

As you release the pressure on the bottle, the diver will float back up to the surface. Why does the sunken diver suddenly become buoyant again? By carefully squeezing the bottle, you can even make the diver hover in the middle of the bottle. Try making the diver hover while your eyes are closed. Why is achieving the hovering state so difficult? Why must you be watching the diver to make it hover?

Chapter Itinerary

We'll return to the diver at the end of the chapter. But first we'll examine two things from the world around us: (1) *balloons* and (2) *water distribution*. In *balloons,* we'll explore how the concepts of pressure and buoyancy help explain how the earth's atmosphere keeps hot air and helium balloons from falling to the ground. In *water distribution,* we'll see both how pressure propels water through plumbing and the ways in which water can contain energy.

The issues we'll be looking at crop up frequently in our everyday experiences. Pressure plays an important role in aerosol cans, steam engines, firecrackers, and even the weather; buoyancy supports ships on water and keeps oil above vinegar in a bottle of salad dressing. Just as important, these concepts will lay the groundwork for Chapter 5, where we'll examine objects in which motion affects the behavior of fluids.

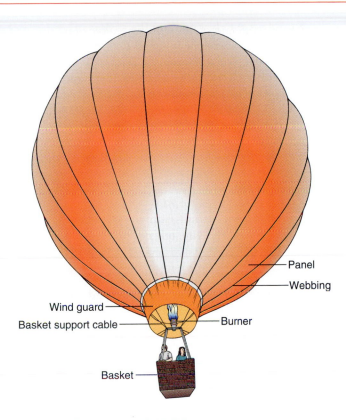

Panel

Webbing

Wind guard

Basket support cable

Burner

Basket

Because gravity gives every object near the earth's surface a weight proportional to its mass, objects fall when you drop them. Why then does a helium-filled balloon—which, after all, is just another object with a mass and a weight—sail upward into the sky when we let go of it? Does the balloon have a negative mass and a negative weight, or are we forgetting something?

We're forgetting air—specifically, the layer of air that sits atop the earth's surface and is held in place by gravity. Since this air is difficult to see and moves out of our way so easily, we often forget that it's there. But it sometimes makes itself noticeable. When we ride a bicycle, we feel its forces; when we blow up a beach ball, we see that it takes up space. And when we release a helium balloon, it's air that lifts the balloon upward.

Questions to Think About: *Since most objects fall to the ground through the atmosphere, why doesn't the atmosphere itself fall downward? Why is the air "thinner" in the mountains than at sea level? If we suck all of the air out of a plastic bag, what squeezes the bag into a thin sheet? Why does blowing air into the bag make it inflate? What happens to the bag's total mass when we fill it with air? with hot air? with helium? If we took a sturdy helium balloon to the moon, where there is no air, and then released it, which way would it move?*

Experiments to Do: *Pull on the string of a helium balloon to get a feel for how it behaves. If it pulls upward on your fingers, does that mean its weight (and mass) is neg-*

ative? How would an object with a negative mass respond to a force? Pull on the balloon's string and convince yourself that the balloon's mass is positive. Do you think there are any objects with negative masses?

Since the balloon's mass and weight are both positive, gravity must be pulling the balloon downward. But how can the balloon pull upward on your finger without accelerating downward? What other forces might be pushing the balloon upward? You can enhance these upward forces by partially submerging the balloon in a container of water. Where else do similar upward forces appear in everyday life?

Take the balloon for a ride in a car. Which way does the balloon move when you start suddenly? when you stop suddenly? Again, it seems as though the balloon's mass is negative. What is pushing on the balloon to make it move in this counterintuitive way?

Air and Air Pressure

Hot-air and helium balloons are supported by the air around them. Although these balloons have positive masses and downward weights, the surrounding air pushes upward on them hard enough to balance their weights so that they float. Thus, if we want to understand balloons, we must start by understanding air.

Like the objects we've already studied, air has mass and weight. Unlike these objects, however, it has no fixed shape or size. If we collected 1 kg of air, we could shape it any way we liked. We could also make it occupy many different **volumes.** Since air is **compressible**—that is, since we can change the volume that a specific mass of it occupies—1 kg could fill a single scuba tank or a whole basketball arena.

This flexibility of size and shape originates in the microscopic nature of air. Air is a **gas,** a substance that consists of tiny, individual particles that travel around independently. These individual particles are atoms and molecules. An **atom** is the smallest portion of an element that retains all of the chemical characteristics of that element; a **molecule,** an assembly of two or more atoms, is the smallest portion of a chemical compound that retains all of the characteristics of that compound. The atoms in each molecule are held together by **chemical bonds,** linkages formed by electromagnetic forces between the atoms.

atoms versus molecules

The particles in air are extremely small, less than a millionth of a millimeter in diameter. Most are nitrogen and oxygen molecules, but others include carbon dioxide, water, methane, and hydrogen molecules, and argon, neon, helium, krypton, and xenon atoms. The atoms, which don't normally form molecules or make strong chemical bonds with other atoms, are called **inert gases** because of their chemical inactivity.

To see how these particles function, think of each one as a tiny marble: each has a size, mass, and weight, and each falls because of gravity whenever it's free of other forces. But this comparison leads to a puzzling observation. When you pour marbles out of a bag, they fall and quickly settle to the ground; but when you pour air molecules out of a cup, they don't seem to fall at all. If air molecules are like tiny marbles, why don't they pile up on the earth's surface?

The answer has to do with the air's **internal kinetic energy.** This energy—the fraction of the air's sun-derived thermal energy that's contained in the motions of its molecules—keeps the tiny air molecules moving, spinning, and away from the earth's surface. In contrast, real marbles are too massive and heavy to be moved noticeably by thermal energy. Since the weak forces between gas molecules allow only a small amount of the air's thermal energy to exist as **internal potential energy,**

energy stored in the forces between particles in the air, almost all of it instead takes the form of internal kinetic energy. This energy per molecule is measured as its **temperature;** the more internal kinetic energy per air molecule, the hotter that air is.

If you could observe the microscopic structure of air, you'd see countless individual molecules in frenetic **thermal motion** (Fig. 4.1.1a). At room temperature, these molecules travel at bulletlike speeds of roughly 500 m/s, but they collide so often that they make little progress in any particular direction. Between collisions, they travel in nearly straight-line paths because gravity doesn't have enough time to make them fall very far.

For the moment, let's ignore gravity and examine what happens to 1 kg of air molecules as they whiz around the inside of a box. Each time a molecule bounces off a wall of the box, it exerts a force on that wall. Although the individual forces are tiny, the number of molecules is not, and together they produce a large average force. The size of this total force depends on the wall's **surface area;** the larger its surface area, the more average force it experiences. In order to characterize the air, however, we don't really need to know the wall's surface area; instead, we can refer to the average force the air exerts on each unit of surface area, a quantity called **pressure.**

Pressure is measured in units of force-per-area. Since the SI unit of surface area is the **meter^2** (abbreviated m^2 and often referred to as the square meter), the SI unit of pressure is the **newton-per-meter^2.** This unit is also called the **pascal** (abbreviated Pa), after French mathematician and physicist Blaise Pascal. One pascal is a small pressure; in contrast, the air around you has a pressure of about 100,000 Pa, so that it exerts a force of about 100,000 N on a 1-m^2 surface. Since 100,000 N is about the weight of a city bus, air pressure can exert enormous forces on large surfaces.

Besides pushing on the outer walls of our hypothetical box, air also pushes on any object contained within that box. The molecules bounce off the object's surfaces, pushing them inward. As long as the object can withstand these compression forces, the air won't greatly affect it since the uniform air pressure ensures that the forces on all sides of the object cancel one another perfectly. A sheet of paper, for example, will experience zero net force because the forces exerted on either side of it will add to zero.

Air molecules also bounce off each other, so that air pressure exerts forces on air, too. A cube of air inserted into the box experiences all of the inward forces that a cube of metal would experience. The air around the cube pushes inward on it, and the cube pushes outward on the air around it. Since the net force on the cube of air is zero, the cube doesn't accelerate.

Check Your Understanding #1: Getting a Grip on Suction

After you push a suction cup against a smooth wall, the elastic cup bends back and a small, empty space is created between the cup and the wall. What keeps the suction cup against the wall?

Pressure, Density, and Temperature

Since air pressure is produced by bouncing air molecules, it depends on how often, and how hard, those molecules hit a particular region of surface. The more frequent or harder the impacts, the greater the air pressure.

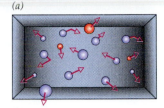
(a)

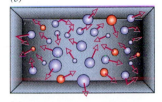
(b)

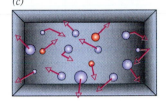
(c)

Fig. 4.1.1 (a) As air molecules bounce off surfaces, they exert pressure on those surfaces; the amount of pressure depends on the air's temperature and on how densely its molecules are packed. (b) Packing the air molecules more densely increases the number of molecules that hit the surfaces each second. (c) Increasing the temperature of the air increases the speed of the molecules (shown by the arrows) so that they hit the surfaces harder and more frequently. Either change, in speed or collision frequency, increases the air's pressure.

To increase the rate at which air molecules hit a surface, we can pack them more tightly. If we add another 1 kg of air to our hypothetical box, we double the number of air molecules in the same volume, which doubles the rate at which they hit each surface and therefore doubles the pressure (Fig. 4.1.1*b*). Thus air's pressure is proportional to its **density,** to how much mass is contained in each unit of volume. Since the SI unit of volume is the **meter³** (abbreviated m³ and often referred to as the cubic meter), the SI unit of density is the **kilogram-per-meter³** (abbreviated kg/m³). The air around you has a density of about 1.25 kg/m³. Water, in contrast, is much denser, with a density of about 1000 kg/m³.

We can also increase the rate at which air molecules hit a surface by speeding them up (Fig. 4.1.1*c*). If we double the internal kinetic energy of the air in our box, we double the average kinetic energy of each molecule. Because a molecule's kinetic energy depends on the square of its speed, doubling its kinetic energy increases its speed by a factor of $\sqrt{2}$. As a result, each molecule hits the surface $\sqrt{2}$ times as often and exerts $\sqrt{2}$ times as much force when it hits. With each molecule exerting $\sqrt{2} \cdot \sqrt{2}$ or two times as much average force, the pressure doubles. Thus air's pressure is proportional to the average kinetic energy of its molecules—to their average internal kinetic energies.

This average kinetic energy per molecule is measured by the air's temperature; the hotter the air, the larger the average kinetic energy per molecule and the higher the air's pressure. But the most convenient scale for relating the temperature of air to its pressure isn't the common **Celsius** (°C) or **Fahrenheit** (°F) scale; instead, it's a special **absolute temperature scale.** The SI scale of absolute temperature is the **Kelvin** scale (K). When the air's temperature is 0 K (−273.15 °C or −459.67 °F), it contains no internal kinetic energy at all and has no pressure; this temperature is called **absolute zero.** The Kelvin scale is identical to the Celsius scale, except that it's shifted so that 0 K is equal to −273.15 °C. In addition to associating the zero of temperature with the zero of internal kinetic energy, the Kelvin scale avoids the need for negative temperatures. Room temperature is about 293 K.

Since air pressure is proportional to both the air's density and its absolute temperature, we can express the relationship among these things in the following way:

$$\text{pressure} \propto \text{density} \cdot \text{absolute temperature.} \tag{4.1.1}$$

This formula is useful, since it allows us to predict what will happen if we change the temperature or density of a specific gas, such as air. But it has its limitations; in particular, it doesn't work if we compare the pressures of two different gases, such as air and helium, which differ in their chemical compositions. To make such a comparison, we'll need to improve on Eq. 4.1.1. We'll do that later when we examine helium balloons.

Even in describing a specific gas, Eq. 4.1.1 has other shortcomings. The main problem is that real gas molecules are not completely independent of one another. If the temperature drops too low, the molecules begin to stick together to form a **liquid,** and Eq. 4.1.1 becomes invalid. At still lower temperatures, the molecules can no longer flow and form a **solid.** Still, despite its limitations, this simple relationship between pressure, density, and temperature will prove useful in understanding how hot-air balloons float: it will help us understand the basic structure of the earth's atmosphere, the origins of the upward force that keeps a hot-air balloon aloft, and the reason why hot air rises.

If you remove a partially filled container of food from the refrigerator and allow it to warm to room temperature, the lid will often bow outward and may even pop off. What has happened?

The Earth's Atmosphere

Most of the mass of the earth's atmosphere is contained in a layer less than 6 km (4 miles) thick. Since the earth is 12,700 km in diameter, this layer is relatively thin—so thin that, if the earth were the size of a basketball, it would be no thicker than a sheet of paper.

The atmosphere stays on the earth's surface because of gravity. Every air molecule, as we've seen, has a weight. Just as a marble thrown upward eventually falls back to the ground, so the molecules of air keep returning toward the earth's surface. Although the molecules are moving too fast for gravity to affect their motions significantly over the short term, gravity works slowly to keep them relatively near the earth's surface. A molecule, like a rapidly moving marble, may appear to travel in a straight line at first, but it will arc over and begin to fall downward eventually. Only the lightest and fastest moving particles in the atmosphere, hydrogen molecules and helium atoms, occasionally manage to escape from earth's gravity and drift off into space.

While gravity pulls the atmosphere downward, air pressure pushes the atmosphere upward. The air molecules try to fall to the earth's surface, but as they do, their density grows higher and higher. As more air molecules are compressed into the same volume, the air pressure increases. It's this air pressure that supports the atmosphere and prevents it from collapsing into a thin pile on the ground.

To understand how gravity and air pressure structure the atmosphere, think of a 1-m square column of the atmosphere as though it were a tall stack of air blocks (Fig. 4.1.2). The 1-kg blocks support one another with air pressure to form a stack of about 10,000 blocks. The bottom block must support the weight of all the blocks above it and is tightly compressed, with a height of about 0.8 m, a density of about 1.25 kg/m^3, and a pressure of about 100,000 Pa. A block farther up in the stack has less weight to support and is less tightly compressed. The higher in the stack you look, the lower the density of the air and the less the air pressure. After all, blocks near the top of the stack have less weight above them to support.

The atmosphere has essentially the same structure as this stack of blocks. The air near the ground supports the weight of several kilometers of air above it, giving it a density of about 1.25 kg/m^3 and a pressure of about 100,000 Pa; at higher altitudes, however, the air's density and pressure are reduced, since there is less atmosphere overhead and the air doesn't have to support as much weight. High-altitude air is thus "thinner" than low-altitude air. Whatever the altitude, the pressure of the surrounding air is referred to as **atmospheric pressure.**

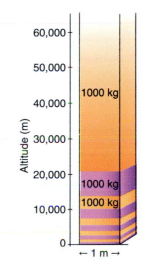

Fig. 4.1.2 The air in a 1-m square column of atmosphere has a mass of about 10,000 kg. The bottom 1000 kg is the most tightly compressed, because it supports the most weight above it. At higher altitudes, the air is less tightly compressed because it has less weight above it.

As you drive up and down in the mountains, you may feel a popping in your ears as air moves to equalize the pressures inside and outside your eardrum. What causes these pressure changes?

pressure on a person

The Lifting Force on a Balloon: Buoyancy

So far we've examined air, air pressure, and the atmosphere. While it may seem that we've avoided dealing with balloons, these topics really are involved in keeping a hot-air or helium balloon aloft. As we've seen, the air in the earth's atmosphere is a **fluid,** a shapeless substance with mass and weight. This air has a pressure and exerts forces on the surfaces it touches; that pressure is greatest near the ground and decreases with increasing altitude. Air pressure and its variation with altitude allow air to lift a hot-air or helium balloon through an effect known as buoyancy.

Buoyancy was first described more than two thousand years ago by the Greek mathematician Archimedes (287–222 BC). Archimedes realized that an object partially or wholly immersed in a fluid is acted upon by an upward **buoyant force** equal to the weight of the fluid it displaces. **Archimedes' principle** is actually very general and applies to objects floating or submerged in any fluid, including air, water, or oil. The buoyant force originates in the forces a fluid exerts on the surfaces of an object. We've seen that such forces can be quite large but tend to cancel one another. How then can pressure create a nonzero total force on an object, and why should that force be in the upward direction?

> **Archimedes' Principle**
> An object partially or wholly immersed in a fluid is acted upon by an upward buoyant force equal to the weight of the fluid it displaces.

Without gravity the forces would cancel each other perfectly because the pressure of a stationary fluid would be uniform throughout. But gravity causes a stationary fluid's pressure to decrease with altitude. For example, when nothing is moving, the air pressure beneath an object is always higher than the air pressure above it. Thus air pushes upward on the object's bottom more strongly than it pushes downward on the object's top, and the object consequently experiences an upward force from the air—a buoyant force.

How large is the buoyant force on this object? It's equal in magnitude to the weight of the fluid that the object displaces. To understand this simple result, imagine replacing the object with a similarly shaped portion of the fluid itself (Fig. 4.1.3a). Since the buoyant force is exerted by the surrounding fluid, not the object, it doesn't depend on the object's composition. A balloon filled with helium will experience the same buoyant force as a similar balloon filled with water or lead or even air. So replacing the object with a similarly shaped portion of fluid will leave the buoyant force on it unchanged.

But a portion of fluid suspended in more of the same fluid doesn't accelerate anywhere; it just sits there, so the net force on it is zero. It has a downward weight, but that weight must be canceled by some upward force that can only come from the surrounding fluid. This upward force is the buoyant force, and it's always equal in magnitude to the weight of the object-shaped portion of fluid, the fluid displaced by the object.

This buoyant principle explains why one object floats upward in a fluid while another sinks downward. An object placed in a fluid experiences two forces: its downward weight and an upward buoyant force. If its weight is more than the buoy-

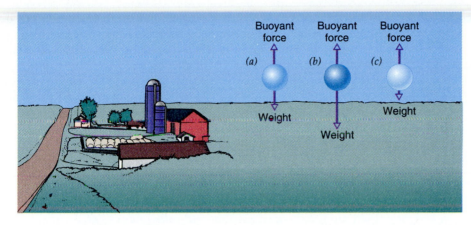

Fig. 4.1.3 (*a*) A portion of air immersed in that same air experiences an upward buoyant force equal to its weight and doesn't accelerate. (*b*) An object that is heavier than the air it displaces sinks, while (*c*) another object that is lighter than the air it displaces floats.

ant force, the object will accelerate downward (Fig. 4.1.3*b*); if its weight is less than the buoyant force, it will accelerate upward (Fig. 4.1.3*c*). And if the two forces are equal, the object won't accelerate at all and will maintain a constant velocity.

Whether or not an object will float in a fluid can also be viewed in terms of density. An object that has an average density greater than that of the surrounding fluid sinks, while one that has a lower average density floats. A water-filled balloon, for example, will sink in air because water and rubber are more dense than air. If you double the volume of the balloon, you double both its weight and the buoyant force on it, so it still sinks. The total volume of an object is not as important as how its density compares to that of the surrounding fluid.

> **Check Your Understanding #4: Why People Don't Float in Air**
> If a person displaces 0.08 m³ of air, what is the buoyant force he experiences?

Hot-Air Balloons

Since air is very light, with a density of only 1.25 kg/m³, few objects float in it. One of these rare objects is a balloon with absolutely nothing inside. Assuming that the balloon has a very thin outer shell or envelope, the whole object will weigh almost nothing and will have an average density of almost zero. It will experience a buoyant force far greater than its negligible weight, and it will float upward nicely.

Unfortunately, this empty balloon will be crushed by the air surrounding it. With an atmospheric pressure of about 100,000 Pa outside, each square meter of surface area on the envelope will experience an inward force of 100,000 N. Because there is nothing inside to support the walls against this enormous force, the balloon will be flattened immediately. A thick, rigid envelope is needed to withstand the crushing pressure of the air outside, but then the average density of the balloon would be large and it would weigh too much to float. So an empty balloon won't work.

What will work is a balloon filled with something that exerts an outward pressure on the envelope equal to the inward pressure of the surrounding air. Then each

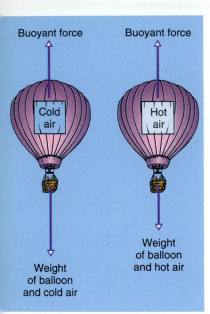

Buoyant force   Buoyant force

Cold air   Hot air

Weight of balloon and cold air   Weight of balloon and hot air

Fig. 4.1.4 A balloon filled with hot air contains fewer air molecules and weighs less than a balloon filled with cold air. If the balloon's weight is small enough, the net force on the balloon will be in the upward direction and the balloon will accelerate upward.

region of the envelope will experience zero net force and the balloon will not be crushed. We could fill the balloon with outside air, but that would make its average density too large. Instead, we need a gas that has the same pressure as the surrounding air but a smaller density.

One gas that has a low density at atmospheric pressure is hot air. Filling our balloon with hot air takes fewer molecules than filling it with cold air, since each hot-air molecule is moving faster and contributes more to the overall pressure than does a cold-air molecule. A hot-air balloon contains fewer molecules, has less mass, and weighs less than it would if it contained cold air. Now we have a practical balloon with an average density less than that of the surrounding air. The buoyant force it experiences is larger than its weight, and up it goes (Fig. 4.1.4).

Because the air pressure inside a hot-air balloon is the same as the air pressure outside the balloon, the air has no tendency to move in or out, and the balloon doesn't need to be sealed (Fig. 4.1.5). A large propane burner, located at the balloon's open end, heats the air that fills the envelope. The hotter the air in the envelope, the lower its density and the less the balloon weighs. The balloon's pilot controls the flame so that the balloon's weight is very nearly equal to the buoyant force on the balloon. If the pilot raises the air's temperature, molecules leave the envelope, the balloon's weight decreases, and the balloon rises. If the pilot allows the air to cool, molecules enter the envelope, the balloon's weight increases, and the balloon descends.

But even if the pilot heats the air very hot, the balloon will not rise upward forever. As the balloon ascends, the air becomes thinner and the pressure decreases both inside and outside the envelope. Although the balloon's weight decreases as the air thins out, the buoyant force on it decreases even more rapidly, and it becomes less effective at lifting its cargo. When the air becomes too thin to lift the balloon any higher, the balloon reaches a *flight ceiling* above which it can't rise, even if the pilot turns the flame on full blast. For each hot-air temperature, then, there is a cruising altitude at which the balloon will hover. When the balloon reaches that altitude, it's in a stable equilibrium. If the balloon shifts downward for some reason, the net force on it will be upward; if it shifts upward, the net force on it will be downward.

hot air rises

> **Check Your Understanding #5: Ballooning Weather**
> Can a hot-air balloon lift more on a hot day or a cold day?

Helium Balloons

Although the molecules in hot and cold air are similar, there are fewer of them in each cubic meter of hot air than in each cubic meter of cold air. We call the number of molecules per unit of volume **particle density,** and hot air has a smaller particle density than cold air (Fig. 4.1.6). Because they contain similar molecules, hot air also has a smaller density than cold air and is lifted upward by the buoyant force.

But there's another way to make one gas float in another: use a gas consisting of very light particles. Helium atoms, for example, are much lighter than air molecules. When they have equal pressures and temperatures, helium gas and air also have equal particle densities. Since each helium atom weighs on average 14% as much as the average air molecule, 1 m³ of helium weighs only 14% as much as 1 m³ of air. Thus a helium-filled balloon has only a fraction of the weight of the air it displaces, and the buoyant force carries it upward easily.

Why should air and helium have the same particle densities whenever their pressures and temperatures are equal? Because a gas particle's contribution to the pressure doesn't depend on its mass (or weight). At a particular temperature, each particle in a gas has the same average kinetic energy in its translational motion, regardless of its mass. Although a helium atom is much less massive than a typical air molecule, the average helium atom moves much faster and bounces more often. As a result, lighter but faster-moving helium atoms are just as effective at creating pressure as heavier but slower-moving air molecules.

Thus, if you allow the helium atoms inside a balloon to spread out until the pressures and temperatures inside and outside the balloon are equal, the particle densities inside and outside the balloon will also be equal (Fig. 4.1.7, p. 128). Since the helium atoms inside the balloon are lighter than the air molecules outside it, the balloon weighs less than the air it displaces, and it will be lifted upward by the buoyant force.

The pressure of a gas is proportional to the product of its particle density and its absolute temperature, as the following formula indicates:

$$\text{pressure} \propto \text{particle density} \cdot \text{absolute temperature.} \qquad (4.1.2)$$

This relationship holds regardless of the gas's chemical composition. Our previous relationship, Eq. 4.1.1, worked only as long as the gas's composition didn't change, so that density and particle density remained proportional to one another. But now we have a relationship with a wider applicability.

Equation 4.1.2, with an associated constant of proportionality, is called the **ideal gas law.** This law relates pressure, particle density, and absolute temperature

Fig. 4.1.5 The bottom of a hot-air balloon is open so that heated air can flow in and cold air can flow out. The heated air displaces more than its weight in cold air and makes the balloon lighter.

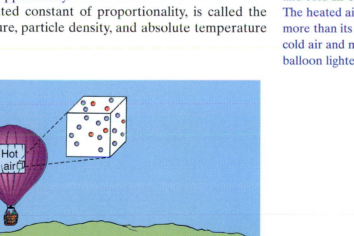

Cold air

Hot air

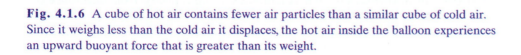

Fig. 4.1.6 A cube of hot air contains fewer air particles than a similar cube of cold air. Since it weighs less than the cold air it displaces, the hot air inside the balloon experiences an upward buoyant force that is greater than its weight.

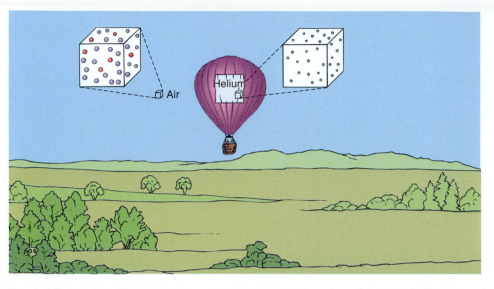

Fig. 4.1.7 A cube of helium gas contains the same number of particles as a similar cube of air, but each helium particle weighs less than the average air particle. Since it weighs less than the air it displaces, the helium inside the balloon experiences an upward buoyant force that is greater than its weight.

Fig. 4.1.8 Rigid airships were appropriately named because they were truly ships that floated through the air. Unfortunately, the hydrogen gas that filled most airships, making them light enough that air could lift them, is very flammable. The *Hindenburg* burned on May 6, 1937, while trying to land at Lakehurst, New Jersey. Because hydrogen is so buoyant in air, most of the combustion occurred above the airship and many passengers survived.

for a gas in which the particles are perfectly independent. It's also fairly accurate for real gases in which the particles do interact somewhat. All that remains to complete the ideal gas law is to replace the proportionality with an equality. The constant of proportionality is the **Boltzmann constant,** with a measured value of 1.381×10^{-23} Pa·m³/(particle·K). Using the Boltzmann constant, the ideal gas law can be written as a word equation:

pressure = Boltzmann constant · particle density · absolute temperature,   **(4.1.3)**

in symbols:

$$p = k \cdot \rho_{particle} \cdot T,$$

and in everyday language:

Don't incinerate a spray can. A hot, dense gas tends to burst its container.

> **The Ideal Gas Law**
> The pressure of a gas is equal to the product of the Boltzmann constant times the particle density times the absolute temperature.

Helium is not the only "lighter-than-air" gas. Hydrogen gas, which is half as dense as helium, is also used to make balloons float. But don't expect hydrogen to lift twice as much weight as helium. A balloon's lifting capacity is the difference between the upward buoyant force it experiences and its downward weight. Although the gas in a hydrogen balloon weighs half that in a similar helium balloon, the balloons experience the same buoyant force. Thus the hydrogen balloon's lifting capacity is only slightly more than that of the helium balloon. Hydrogen's main advantage is that it's cheap and plentiful, while helium is scarce (see ❏). But because hydrogen is also dangerously flammable, it's avoided in situations where safety is important (Fig. 4.1.8). However, even helium-filled airships can have problems (see ❏).

Check Your Understanding #6: What Not to Put in a Balloon

A carbon dioxide molecule is heavier than an average air molecule. If you pour carbon dioxide gas from a cup, which way does it flow in air, up or down?

Check Your Figures #1: Popped Out of the Refrigerator

When you take an air-filled plastic container out of the refrigerator, it warms from 2 °C to 25 °C. How much does the pressure of the air inside it change?

❏ Helium gas is obtained as a by-product of natural gas production from underground reservoirs in the United States. While some of this gas is saved for industrial and commercial use, much is simply released into the atmosphere. The only other source of helium is the atmosphere, where helium is present at a level of 5 parts per million. Once the underground stores are consumed, helium will become a relatively rare and expensive gas.

❏ Even helium-filled airships were easily destroyed by bad weather. The *Shenandoah,* one of two U.S. airships based on the German designs, was destroyed by air turbulence on September 3, 1925, near Ava, Ohio. Crowds from a local fair immediately poured over the wreckage, collecting souvenirs.

elastic balloons

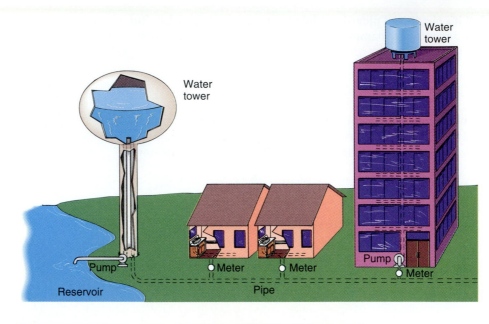

Water Distribution

Now that we've explored the behavior of objects contained in fluids, let's turn to the behavior of fluids contained in objects. In this section, as we examine how plumbing distributes water, we'll see that pressure, density, and weight are just as important in plumbing as they are in ballooning. To keep things simple, we'll focus on the causes of water's motion through plumbing, leaving most of the complications associated with the motion itself for the next chapter. For example, we'll temporarily ignore drag and viscosity and the fascinating pressure changes that accompany fluid motion.

Questions to Think About: *Why is the water pressure higher in the basement than it is in the attic? Why does a deep water well require a pump at the bottom? If a water tower is only a storage device, why is it so tall? Why do skyscrapers have complicated plumbing systems that include reservoirs at various levels in the building? What causes water to flow up through a drinking straw to your mouth?*

Experiments to Do: *To see the effects of pressure and weight on water, try these simple experiments with a drinking straw. First, "suck" water up the straw from a glass to your mouth. Are you exerting an attractive force on the water, or is some other force pushing it upward toward your mouth? With the straw full of water, seal the top with your finger; keeping the seal tight, remove the straw from the glass. What happens to the water inside? What happens to the water when you release the seal? Now blow into one end of a straw full of water while sealing the far end with your finger. When you release the seal on the far end, what happens to the water? What forces are responsible for this effect?*

Water Pressure

Water distribution systems require two things: plumbing and water pressure. Plumbing is what delivers the water, and water pressure is what makes that water flow. Water pressure is important because, like everything else, water has inertia and accelerates only when pushed. If nothing pushed on the water when you opened a faucet, the water simply wouldn't start flowing. Since the pushes that send water through pipes come principally from differences in water pressure, we need to look carefully at how such pressure is created and controlled.

For the present, we'll ignore gravity. As we've seen with the atmosphere, gravity creates **pressure gradients** in fluids—distributions of pressures that vary continuously with position. On earth, pressure decreases with altitude and increases with depth, creating vertical pressure gradients that complicate plumbing in hilly cities and skyscrapers. But if all of our plumbing is in a level region—for example, a single-story house in a very flat city—our job is much simpler. With no significant changes in height, we can safely ignore gravity, since no water is supporting the weight of water above it and gravity's effects are minimal.

In this simplified situation, water accelerates only in response to unbalanced pressures. Just as unbalanced forces make a solid object accelerate, so unbalanced pressures make a fluid accelerate. If the water inside a pipe is exposed to a uniform pressure throughout, then each portion of the water feels no net force and doesn't accelerate; it either remains stationary or coasts in a straight line at a steady pace (Fig. 4.2.1). But if the pressure is out of balance, the water accelerates toward the region of lowest pressure.

This acceleration doesn't mean that the water will instantly begin moving toward the lowest pressure. Because it has momentum, the water changes velocity gradually: it speeds up, slows down, or turns to the side, depending on where the lowest pressure is located. A complicated arrangement of high and low pressures can steer water through an intricate maze of pipes, and that is exactly how water reaches your home from the city pumping station. Every change in its velocity during its trip through the plumbing is caused by a pressure imbalance.

How are these imbalances in pressure created? One method is to push inward on a portion of the water. In that case, the water passively experiences a local increase in pressure and responds by accelerating away from the squeezed region. Because the local increase in pressure isn't caused by the water's motion, it's a **static** variation in pressure. But the motion of water itself can also cause variations in its pressure. These **dynamic** variations in pressure are complicated and fascinating and contribute to such diverse effects as the lift on an airplane's wing, the curve of a curve ball, and the thrust of a rocket engine. We'll examine these dynamic effects of fluid motion in the next chapter.

Check Your Understanding #1: Under Pressure in the Garden

With the water faucet open and the nozzle at the end of your garden hose shut tightly, the hose is full of high-pressure water. Why doesn't this water accelerate?

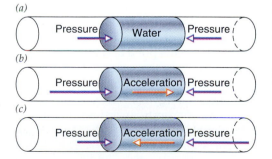

(a)

(b)

(c)

Fig. 4.2.1 (*a*) If the water in a horizontal pipe is exposed to a uniform pressure, then it will not accelerate. (*b,c*) However, if the pressure along the pipe is not uniform, the imbalance will create a net force on each portion of the water and the water will accelerate toward the side with the lower pressure.

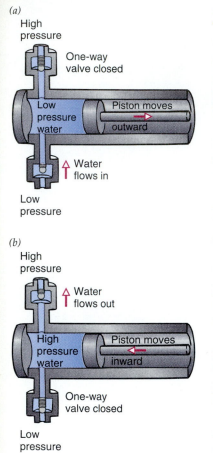

(a)

High pressure

One-way valve closed

Low pressure water

Piston moves outward

Water flows in

Low pressure

(b)

High pressure

Water flows out

High pressure water

Piston moves inward

One-way valve closed

Low pressure

Fig. 4.2.2 Water is pumped from a region of low pressure to a region of high pressure by a reciprocating piston pump. (*a*) As the piston is drawn outward, water flows into the cylinder from the low-pressure region. (*b*) As the piston is pushed inward, the inlet one-way valve closes and water is driven out of the cylinder and into the high-pressure region.

more about pumps

Creating Water Pressure with Water Pumps

To start water flowing through the plumbing in a level house or city, you need a water pump—a device that uses mechanical work to deliver pressurized water through a pipe. At its most basic level, a water pump squeezes a portion of water to raise its local pressure and keeps squeezing as that water accelerates and flows out toward regions of lower pressure elsewhere in the plumbing.

To visualize a simple pump, imagine holding an open plastic bottle full of water. If you don't squeeze the bottle, the pressure inside it is uniform (remember we're neglecting gravity) and the motionless water remains motionless. But if you squeeze the sides of the bottle and push inward on the water, the water responds by pushing outward on you—Newton's third law—and it does this by increasing its local pressure. As its pressure rises inside the bottle, the water begins to accelerate toward the lower pressure above the bottle's open top and the result is a fountain. You are pumping water.

You are also doing work: as the water flows out of the bottle, your hands move inward. Since you are pushing inward on the water and the water is moving inward, you are doing work on the water. Pumps always do work when they deliver pressurized water, and pressurized water always carries with it the energy associated with that work.

While a water bottle can act as a pump briefly, it soon runs out of water. A more practical pump appears in Fig. 4.2.2. In this pump, a piston slides back and forth in the open end of a hollow cylinder, making a watertight seal. Pushing inward on that piston squeezes any water in the cylinder and raises the local water pressure. Water begins to flow.

But what distinguishes this pump from our simple bottle is that its cylinder is easy to refill. The cylinder actually has two openings, each of which has a valve that permits water to flow in only one direction. Water can only leave the cylinder through the top opening and can only enter through the bottom opening. As the pump's piston is pushed into the water-filled cylinder, the water pressure in the cylinder rises and water accelerates and flows out through the top valve. As the pump's piston is pulled out of the water-filled cylinder, the water pressure inside the cylinder drops and water accelerates and flows in through the bottom valve. In fact, as the piston is withdrawn, the pressure inside the cylinder drops below atmospheric pressure, so that even water in an open reservoir nearby will accelerate toward the partial vacuum in the cylinder and refill it.

▶ Check Your Understanding #2: Working a Water Pump

Which normally requires more work: pulling the piston of a water pump out of the cylinder or pushing it back in?

Moving Water: Pressure and Energy

The pump of Fig. 4.2.2 can draw low-pressure water from a pond and fill a hose with high-pressure water. If the other end of the hose is open, the water will accelerate toward lower pressure at that end and will have considerable kinetic energy as it sprays out of the hose. From where does this kinetic energy come?

The energy comes from you and the pump. As you push inward on the piston, pressurizing the water and squeezing it out through the top valve, you're doing work on the water because you're exerting an inward force on the water's surface and the water is moving inward. The amount of work you do is equal to the product of the water pressure times the volume of water you pump. This simple relationship between work, pressure, and volume comes about because the inward *force* the piston exerts on the water is equal to the water pressure times the surface area of the piston, and because the inward *distance* the piston travels is equal to the volume of water being pumped divided by the surface area of the piston. *Force* times *distance* equals *work*.

As you pump the water, it sprays out of the open hose. The energy that makes the water accelerate out of the hose actually travels through the water directly from the pump to the end of the hose. Like all liquids, water is **incompressible**—its volume doesn't change as its pressure increases—so each time a liter of water leaves the pump, a liter of water also leaves the hose. While the water never really stores any energy, the pump gives each liter of water a certain amount of energy as it leaves the hose so we can imagine that this energy is associated with the water and not with the pump. We create a useful fiction: pressure potential energy. Water that's under pressure has a **pressure potential energy** equal to the product of the water's volume times its pressure.

Because pressure potential energy actually comes from the pump, it vanishes as soon as you break the link between the water and the pump: you can't save a bottle of high-pressure water and expect it to retain this potential energy. The concept of pressure potential energy is only meaningful if the water is flowing freely so that water leaving the plumbing is immediately replaced by the pump; then whatever energy leaves the plumbing as kinetic energy in the water is put back into the plumbing by the pump. Actually, the details of the pump aren't as important as the idea that any water moving through the plumbing is immediately replaced by more water with the same pressure. As long as the water is flowing steadily, you can safely use the concept of pressure potential energy, even if you don't know where the pump is or whether there even is one.

Pressure potential energy is most meaningful in **steady-state flow**—a situation in which fluid flows continuously and steadily through a stationary environment, without starting or stopping or otherwise changing its characteristics anywhere. Water spraying steadily out of a hose, wind blowing smoothly across your motionless face, and a gentle current flowing in a quiet river are all cases of steady-state flow in fluids.

Without gravity, the energy in a certain volume of water in steady-state flow is equal to the sum of its pressure potential energy and its kinetic energy. We've already seen that the pressure potential energy is the product of the water's volume times its pressure. The water's kinetic energy is given by Eq. 2.3.7 as one-half the product of its mass times the square of its speed. Since water's mass is its density times its volume, this sum is

pumps require energy

$$
\begin{aligned}
\text{energy} &= \text{pressure potential energy} + \text{kinetic energy} \\
&= \text{pressure} \cdot \text{volume} + \tfrac{1}{2} \cdot \text{density} \cdot \text{volume} \cdot \text{speed}^2,
\end{aligned}
\qquad \textbf{(4.2.1)}
$$

If we divide both sides of this expression by the volume involved, we can obtain another useful form of this relationship:

$$
\begin{aligned}
\frac{\text{energy}}{\text{volume}} &= \frac{\text{pressure potential energy}}{\text{volume}} + \frac{\text{kinetic energy}}{\text{volume}} \\
&= \text{pressure} + \tfrac{1}{2} \cdot \text{density} \cdot \text{speed}^2.
\end{aligned}
\qquad \textbf{(4.2.2)}
$$

As each volume of water moves along with the flow, it is soon replaced by a new volume of water. Because the flow is steady-state, the energy in the new volume of

❑ As a professor in Basel, **Daniel Bernoulli (Swiss mathematician, 1700–1782)** taught not only physics, but also botany, anatomy, and physiology. He correctly proposed that the pressure a gas exerts on the walls of its container results from the countless impacts of tiny particles that make up the gas. He also derived an important relationship between the pressure, motion, and height of a fluid—Bernoulli's equation.

water must be exactly the same as in the volume that preceded it; thus the energy in each volume of water that flows along a particular path must be identical. The particular path that a volume of water takes is called a **streamline,** and the energy-per-volume of fluid along a streamline is constant:

$$\frac{\text{energy}}{\text{volume}} = \frac{\text{pressure potential energy}}{\text{volume}} + \frac{\text{kinetic energy}}{\text{volume}}$$
$$= \text{pressure} + \tfrac{1}{2} \cdot \text{density} \cdot \text{speed}^2$$
$$= \text{constant } (\textit{along a streamline}). \qquad \textbf{(4.2.3)}$$

Equation 4.2.3 is called **Bernoulli's equation,** after Swiss mathematician Daniel Bernoulli ❑, whose work led to its development, although Swiss mathematician Leonhard Euler (1707–1783) actually completed it.

Because energy is conserved, an incompressible fluid such as water that's in steady-state flow can exchange pressure for speed or speed for pressure as it flows along a streamline. As water accelerates out of a hose with a nozzle, for example, its pressure drops but its speed increases because it's converting pressure potential energy into kinetic energy. As the moving water sprays against the car you're washing, it slows down but its pressure increases because it's converting kinetic energy back into pressure potential energy. In both cases, the water's energy is conserved.

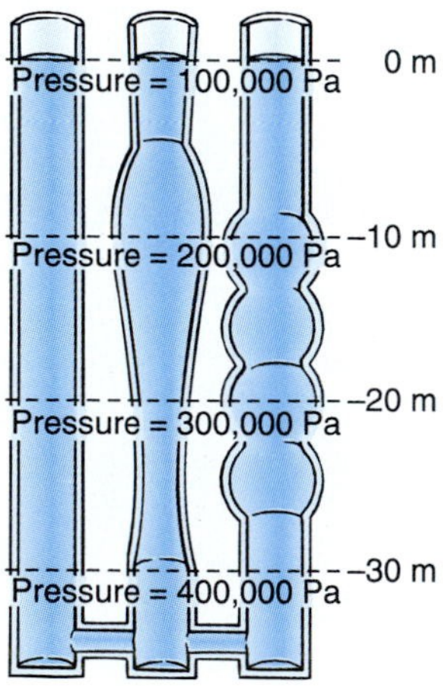

> ➤ Check Your Understanding #3: How Does Your Garden Grow?
> Water in your garden hose has considerable pressure and arcs several meters through the air as you water your plants. What is the water pressure in the falling water once it leaves the end of the hose?

Gravity and Water Pressure

Gravity creates a pressure gradient in water: the deeper the water, the more weight there is overhead, and the greater the pressure. Since water is much denser than air, water pressure increases rapidly with depth. In a vertical pipe that's open on top, the water's surface is at atmospheric pressure (about 100,000 Pa), but only 10 m (33 feet) below the water's surface, the pressure has already risen to 200,000 Pa. At that depth, the weight of water overhead is equal to the weight of air overhead, even though the atmosphere is several kilometers thick.

The shape of the pipe doesn't affect the relationship between pressure and depth. No matter how complicated the plumbing, the pressure of stationary water inside it increases with depth by 10,000 Pa per meter or 10,000 Pa/m (Fig 4.2.3). This uniform pressure gradient creates an upward buoyant force on each volume of water in the pipe and prevents that water from falling (Fig. 4.2.4).

Because the pressure depends only on height, water in a complicated system of plumbing, connected near the bottom and open to the air on top, will flow until it has filled that plumbing to a uniform height. If water in one region of the plumbing is not as high as it is elsewhere, the pressures in that region will be relatively low and water will accelerate toward it. This natural flow quickly equalizes the heights of water in the plumbing and is the origin of the expression, "Water seeks its level." Natural flow is often used in water delivery (see ❑).

The dependence of water pressure on depth has a number of important implications for water distribution. First, water pressure at the bottom of a tall pipe is substantially higher than at the top of that same pipe. Consequently, if only a single

Pressure = 100,000 Pa   0 m

Pressure = 200,000 Pa   —10 m

Pressure = 300,000 Pa   —20 m

Pressure = 400,000 Pa   —30 m

Fig. 4.2.3 The pressure of stationary water in pipes increases with depth by about 10,000 Pa per meter of depth. *The shape of the pipes doesn't matter.* For plumbing that's open on top and connected near the bottom, as shown here, water will tend to flow until its height is uniform throughout the plumbing.

pressure and height

pipe supplies water to a skyscraper, then the water pressure on the ground floor will be dangerously high while the pressure in the penthouse will be barely enough for a decent shower. Tall buildings must therefore handle water pressure very carefully; they can't supply water to every floor directly from the same pipe.

Second, pressure in a city water main does more than simply accelerate water out of a showerhead; it also supports water in the pipes of multistory buildings. Lifting water to the third floor against the downward force of gravity requires a large upward force, and that force is provided by water pressure. The higher you want to lift the water, the more water pressure you need at the bottom of the plumbing. Lifting the water also requires energy, which is often provided by a water pump.

Third, as water travels up and down the streets of a hilly city, its pressure varies with height. In the valleys the pressure can be very large, and at the tops of hills the pressure can be very small. Water mains in valleys must therefore be particularly strong to keep from bursting. The large pressure in a valley is very useful because it helps push the water back uphill on the other side of the valley (Fig. 4.2.5, p. 136). Nonetheless, a very hilly city must have pumping stations and other water pressure control systems located throughout in order to provide reasonable water pressures to all the buildings, regardless of their altitudes.

The notion that water seeks its level is only true in pipes that are open on top, so that the water's surface is at atmospheric pressure. If you seal off part of a plumbing system and reduce the air pressure inside, the water's top surface will rise higher there than elsewhere. It rises until the added weight of the water column compensates for the missing air pressure. This effect of reduced air pressure explains why water rises in a drinking straw and how water can travel between two open containers through an elevated pipe known as a siphon.

Removing all the air pressure from a long straw or siphon will raise the water level inside just 10 m above normal. Even with no pressure above it, this 10-m column of elevated water produces atmospheric pressure at its bottom and prevents water in the rest of the plumbing from lifting it any higher. It's thus impossible to draw water from a deep well simply by lowering a pipe into that well and reducing the air pressure in the pipe: the water will rise no more than 10 m upward. Instead, a pump must be attached to the bottom of the pipe to pressurize the water and push it all the way to the top of the pipe.

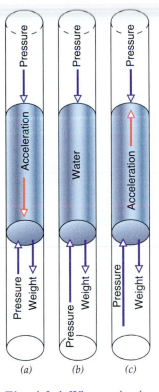

(a)   (b)   (c)

Fig. 4.2.4 When a pipe is oriented vertically, gravity affects the motion of water in the pipe. (a) If the water's pressure doesn't change with depth, the water will accelerate downward (fall) because of its weight. (b) If the water's pressure increases with depth by 10,000 Pa/m, the water won't accelerate. (c) If the water's pressure increases with depth by more than that amount, the water will accelerate upward.

> **Check Your Understanding #4: What's the Water Pressure?**
> If all of the water to a 400-m tall skyscraper were delivered from a single pipe, how much higher would the water pressure be on the ground floor than on the top floor?

Moving Water Again: Gravity

As we've seen, it takes pressure and energy to lift water to the third floor of a building. We can now expand our statement of energy conservation in fluids to include gravity and gravitational potential energy.

Water's gravitational potential energy is equal to its weight times its height (the force required to lift it times the distance it has been lifted), and its gravitational potential energy-per-volume is its weight-per-volume times its height. Since its weight-per-volume is its density times the acceleration due to gravity, water's gravitational potential energy-per-volume is its density times the acceleration due to gravity times its height.

❏ The Romans used gravity to convey water to Rome from sources up to 90 km away. A very gradual slope in the aqueducts kept the water moving in spite of frictional effects that opposed the water's progress. Poisoning from the lead pipes used in some of the aqueducts is blamed in part for the decay of the Roman Empire.

Fig. 4.2.5 Los Angeles receives much of its water from Owens Valley, 300 km north. The water negotiates the mountains and valleys in between, driven by gravity alone. Giant pipes allow pressure to build during downhill stretches in order to push the water back uphill later on. Parts of the 1913 aqueduct support so much pressure that the steel pipe used in them has to be more than an inch thick.

compressible fluids

Fig. 4.2.6 Many buildings in New York City have water towers on their roofs. These towers maintain water pressure in the plumbing and help in fire fighting.

If we include gravitational potential energy in Eq. 4.2.2 and recognize that, for fluid in steady-state flow along a streamline, the energy-per-volume is constant, we obtain a relationship that can be written as a word equation:

$$\frac{\text{energy}}{\text{volume}} = \frac{\text{pressure potential energy}}{\text{volume}} + \frac{\text{kinetic energy}}{\text{volume}}$$
$$+ \frac{\text{gravitational potential energy}}{\text{volume}}$$
$$= \text{pressure} + \tfrac{1}{2} \cdot \text{density} \cdot \text{speed}^2$$
$$+ \text{density} \cdot \text{acceleration due to gravity} \cdot \text{height}$$
$$= \text{constant } (\textit{along a streamline}), \tag{4.2.4}$$

in symbols:

$$p + \tfrac{1}{2} \cdot \rho \cdot v^2 + \rho \cdot g \cdot h = \text{constant } (\textit{along a streamline}),$$

and in everyday language:

When a stream of water speeds up in a nozzle or flows uphill in a pipe, its pressure drops.

This is a revised version of Bernoulli's equation, one that includes gravity. It correctly describes steady-state flow in streamlines that change height.

> **Bernoulli's Equation**
> For an incompressible fluid in steady-state flow, the sum of its pressure potential energy, its kinetic energy, and its gravitational potential energy is constant along a streamline. Equation 4.2.4 expresses this law as a formula.

Because energy is conserved, an incompressible fluid such as water that's in steady-state flow can exchange its speed, pressure, and height for one another. Thus as water flows downhill, its speed or pressure or both increase; if it falls from an open faucet, its speed increases; and if it descends steadily inside a sealed pipe, its pressure increases. The reverse happens as water flows uphill. Water rising from a fountain loses speed as it ascends while water rising steadily in a pipe loses pressure on its way up.

This interchangeability of height, pressure, and speed makes it possible to pressurize plumbing by connecting a tall column of water to the pipes. That's why cities, communities, and even individual buildings have water towers (Fig. 4.2.6). A water tower is built at a relatively high site within the region it serves. A pump fills the water tower with water, and then gravity maintains a constant high pressure throughout the plumbing that connects to it (Fig. 4.2.7). The water is at atmospheric pressure at the top of the water tower, but the pressure is much higher at the bottom; at the base of a 50-m high water tower, for example, the pressure is about 600,000 Pa or six times atmospheric pressure.

In addition to providing a fairly steady pressure in the water mains, a water tower stores energy efficiently and can deliver that energy in a short time. When water is drawn out of the water tower, its gravitational potential energy at the top becomes pressure potential energy at the bottom. The water tower replaces a pump, supplying a steady flow of water at an almost constant high pressure. But unlike a pump, the water tower can supply this high-pressure water at an enormous rate. As long as the water level doesn't drop too far, high-pressure water keeps flowing.

A hydroelectric power plant extracts energy from water that has descended from an elevated reservoir in a pipe. In the reservoir, this energy takes the form of gravitational potential energy. Just before the power plant, what form does the energy take?

If the water pressure at the entrance to a building is 1,000,000 Pa, how high can the water rise up inside the building and how fast will it flow out of a faucet right at the entrance? (A liter of water inside a pipe has a mass of 1 kg.)

straws, siphons, water towers

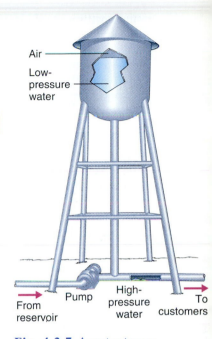

Fig. 4.2.7 A water tower uses the weight of water to create a large water pressure near the ground. The higher the tower, the greater the pressure near the ground. The water tower is able to maintain the water pressure passively and doesn't require constant pumping. Even during periods of peak water consumption, it maintains a fairly steady pressure. When the water level in the water tower drops below a certain set point, a pump refills the tower.

Epilogue for Chapter 4

In this chapter we investigated some of the basic concepts associated with fluids. In *balloons,* we explored the concept of pressure and the way in which air pressure structures and supports the earth's atmosphere. We saw that increased air pressure beneath an object produces an upward buoyant force on that object, and that this buoyant force can float objects, such as hot-air and helium balloons, that are less dense than the surrounding air.

In *water distribution,* we examined how water pressure causes water to accelerate through plumbing, from high pressure to low pressure. We then focused on ways to produce water pressure, either with pumps or with gravity. By studying the forms of energy in water, we were led to Bernoulli's equation, which describes the conversion of a fluid's energy between pressure potential energy, kinetic energy, and gravitational potential energy. Although the most dramatic applications of Bernoulli's equation are ahead of us, we've already used it to understand the changes in pressure and speed that accompany water's movement up and down in pipes and fountains.

Explanation: A Cartesian Diver

The diver floats because its average density—that is, the mass of the vial and its contents divided by the volume of space those two components occupy—is less than the density of water. Since the upward buoyant force on the diver exceeds its downward weight, the diver floats upward toward the top of the bottle. When the diver begins to stick out of the water, it displaces less water and more air and the buoyant force it experiences decreases. Eventually it experiences zero net force and floats without accelerating at the water's surface.

When you squeeze the soda bottle, you increase the pressure inside it. Because water is incompressible, its density doesn't change as the pressure goes up. However, the air bubble inside the vial is compressed and takes up less space inside the vial. Water flows into the vial and increases the average density of the vial and its contents. When the average density of the diver finally exceeds the density of water, the diver sinks.

To keep the diver hovering in the water, you must adjust the water pressure until the diver's average density is exactly that of water. This adjustment is impossible to make without looking at the diver. Even the slightest overpressure or underpressure will cause the diver to drift slowly down or up.

Chapter Summary

How Balloons Work: A hot-air balloon floats because its total weight (basket, envelope, and hot air) is less than that of the cooler air it displaces. By heating the air in the envelope with a flame, the pilot reduces its density. As the air warms up, fewer molecules are needed to fill the envelope, the extra molecules flow out of the envelope through the opening at its bottom, and the balloon becomes lighter.

A helium balloon also weighs less than the air it displaces. But its lower weight is due to the lightness of the individual helium atoms, each of which weighs much less than the air molecule it replaces. By filling a balloon with helium, the balloon's weight is dramatically reduced. Since the buoyant force exerted on the balloon by the surrounding air exceeds the balloon's weight, the balloon accelerates upward.

How Water Distribution Works: Water distribution begins when a pump transfers low-pressure water from a reservoir to high-pressure plumbing. Along a level path, water accelerates toward regions of lower pressure, such as open hoses or shower heads. Pressure imbalances allow the water to negotiate bends in the pipes on route to its destination. During its travels, the water may rise or fall in height; as it does, its pressure changes, the pressure decreasing as the water moves upward and increasing as the water moves downward. In low-lying regions, where the water pressure may be too high to use directly, a pressure regulator may have to be added to the plumbing. In high-lying regions, where the water pressure may be too low to be practical, an additional pump may have to be employed to boost its pressure.

Important Laws and Equations

1. Archimedes' Principle: An object partially or wholly immersed in a fluid is acted upon by an upward buoyant force equal to the weight of the fluid it displaces.

2. The Ideal Gas Law: The pressure of a gas is equal to the product of the Boltzmann constant times the particle density times the absolute temperature, or

$$\text{pressure} = \text{Boltzmann constant} \cdot \text{particle density} \\ \cdot \text{absolute temperature.} \quad (4.1.3)$$

3. Bernoulli's Equation: For an incompressible fluid in steady-state flow, the sum of its pressure potential energy, its kinetic energy, and its gravitational potential energy is constant along a streamline, or

$$\frac{\text{energy}}{\text{volume}} = \frac{\text{pressure potential energy}}{\text{volume}} + \frac{\text{kinetic energy}}{\text{volume}} \\ + \frac{\text{gravitational potential energy}}{\text{volume}} \\ = \text{pressure} + \tfrac{1}{2} \cdot \text{density} \cdot \text{speed}^2 \\ + \text{density} \cdot \text{acceleration due to gravity} \cdot \text{height} \\ = \text{constant (\textit{along a streamline}).} \quad (4.2.4)$$

Check Your Understanding—Answers

Section 4.1 BALLOONS

1. Air pressure.

Why: Because the space between the suction cup and the wall is empty, the pressure there is zero. Pressure of the surrounding air exerts large inward forces on the outsides of both the cup and the wall, squeezing them together. As long as there is no air between them to push outward, the cup and wall remain tightly attached. Once air leaks into the suction cup, it's easily detached from the wall.

2. The pressure of the air trapped in the container increases as its temperature increases, causing the lid to bulge outward.

Why: Whenever a trapped quantity of gas changes temperature, it also changes volume or pressure or both. In this case, warming the air trapped in the container causes its pressure to increase. The unbalanced pressures inside and outside the container cause the lid to bow outward or even to pop off.

3. As you change altitude, the atmospheric pressure changes.

Why: The air inside your ears is trapped, so that its temperature, density, and pressure are normally constant. As your altitude changes, the pressure outside your ear changes and your eardrum experiences a net force. It bows inward or outward, muting the sounds you hear and causing some discomfort. The pressure imbalance is relieved during swallowing, when air can flow into or out of your eardrum.

4. About 1 N.

Why: Air exerts a buoyant force on him equal to the weight of the air he displaces. The density of air near sea level is about 1.25 kg/m³, so 0.08 m³ of air has a mass of about 0.1 kg (1.25 kg/m³ times 0.08 m³) and a weight of about 1 N. So the upward buoyant force on him is about 1 N. This buoyant force due to the air is real and reduces the weight you read when you stand on a scale by about 0.125%.

5. It can lift more on a cold day.

Why: On a cold day, the outside air is relatively dense and the buoyant force on a balloon is larger than it would be on a hot day. The hot air in the balloon will cool off more quickly on a cold day, but the balloon will be able to carry a heavier load. Airplanes also fly better on cold days.

6. It flows down.

Why: Carbon dioxide, found in carbonated beverages, dry ice, and fire extinguishers, is heavier than air because its molecules are heavier than air molecules. The carbon dioxide you pour from the cup has the same pressure and temperature as the air around it and thus the same particle density. But each carbon dioxide molecule weighs more, so the carbon dioxide is the denser gas (mass density) and flows down in air. This tendency to flow along the floor makes carbon dioxide very good at extinguishing low-lying flames by depriving them of oxygen.

Section 4.2 WATER DISTRIBUTION

1. The water pressure (force on a unit of surface area) inside the hose is uniform throughout, so the water experiences no net force and doesn't accelerate.

Why: In the absence of gravity, fluids accelerate only when they experience pressure imbalances. Since water throughout the hose is at the same pressure, there is no pressure imbalance and no acceleration. When you open the nozzle, the pressure at that end of the hose drops and the water accelerates toward it.

2. Pushing it back in usually requires more work.

Why: When you pull the piston out of the cylinder, you are moving air out of the way and creating a partial vacuum inside the cylinder. This action requires a modest amount of work because the air doesn't push terribly hard on the back of the piston and pressure from the water flowing into the cylinder assists you. But as you push the piston back into the cylinder, you are pressurizing the water. Depending on the pressure of water in the outlet hose, the water in the cylinder may exert a very large force on the piston. In that case, you must do a great deal of work on the water as you push the piston inward and drive the water out of the cylinder.

3. Atmospheric pressure.

Why: As the water accelerates out of the hose, its pressure drops. It is converting pressure potential energy into kinetic energy. The water pressure drops until it reaches the pressure of the surrounding air, atmospheric pressure.

4. About 4,000,000 Pa (40 atmospheres) higher.

Why: The weight of water inside the pipe would create an enormous excess pressure near the bottom of the building. Water spraying from an open faucet on the first floor at this enormous pressure could accelerate to 319 km/h (200 mph), as it does in some high-pressure jet washing and cutting machines.

5. Pressure potential energy (and some kinetic energy).

Why: As the water descends inside the pipe, its gravitational potential energy is converted into pressure potential energy. The water reaching the power plant is under enormous pressure, and it's this pressure that exerts the forces needed to turn the turbines that run the generators. Work is required to turn the turbines, so the water gives up much of its energy in the power plant. This energy leaves the power plant via the electric power lines.

Check Your Figures—Answers

Section 4.1 BALLOONS

1. It increases by 8.4%.

Why: To use Eq. 4.1.3 to determine the pressure change, we need temperatures measured on an absolute scale, such as the Kelvin scale. Since 0 °C is about 273 K, 2 °C is about 275 K and 25 °C is about 298 K. We can write Eq. 4.1.3 twice, once for each temperature:

pressure$_{298 K}$ = Boltzmann constant · particle density · 298 K, and:

pressure$_{275 K}$ = Boltzmann constant · particle density · 275 K.

The particle density of the air in the container can't change as it warms up because its volume is fixed. Therefore we can divide the upper equation by the lower one and cancel the Boltzmann constant and particle density on the right-hand side:

$$\frac{\text{pressure}_{298 K}}{\text{pressure}_{275 K}} = \frac{298 \text{ K}}{275 \text{ K}}$$

$$= 1.084.$$

The pressure in the container thus increases by a factor of almost 1.084, or about 8.4%. This elevated pressure will cause the container to emit a "pop" sound when you open it.

Section 4.2 WATER DISTRIBUTION

1. It can rise up about 100 m or emerge from the faucet at about 45 m/s (101 mph).

Why: As the water flows through the pipe (a streamline), its pressure potential energy can become gravitational potential energy or kinetic energy. At the start, the water's energy is all pressure potential energy so, from Eq. 4.2.4, the water's energy-per-volume is 1,000,000 Pa. If the water flows up the pipe, that energy will become gravitational potential energy. We can rearrange Eq. 4.2.4 to find the height it can reach:

$$\text{height} = \frac{\text{energy}}{\text{volume}} \cdot \frac{1}{\text{density} \cdot \text{acceleration due to gravity}}$$

$$\text{height} = 1,000,000 \text{ Pa} \cdot \frac{1}{1000 \text{ kg/m}^3 \cdot 9.8 \text{ m/s}^2} = 102 \text{ m}.$$

If the water flows out the faucet, that energy will become kinetic energy. We can also rearrange Eq. 4.2.4 to find the speed it will obtain:

$$\text{speed} = \sqrt{\frac{\text{energy}}{\text{volume}} \cdot \frac{2}{\text{density}}}$$

$$= \sqrt{\frac{2,000,000 \text{ Pa}}{1000 \text{ kg/m}^3}} = 45 \text{ m/s}.$$

Exercises

1. A helium-filled balloon floats in air. What will happen to an air-filled balloon in helium? Why?

2. A log is much heavier than a stick, yet both of them float in water. Why doesn't the log's greater weight cause it to sink?

3. An automobile will float on water as long as it doesn't allow water to leak inside. In terms of density, why does admitting water cause the automobile to sink?

4. Many grocery stores display frozen foods in bins that are open at the top. Why doesn't the warm room air enter the bins and melt the food?

5. Some clear toys contain two colored liquids. No matter how you tilt one of those toys, one liquid remains above the other. What keeps the upper liquid above the lower liquid?

6. When the car you are riding in stops suddenly, heavy objects move toward the front of the car. Explain why a helium-filled balloon will move toward the rear of the car.

7. Water settles to the bottom of a tank of gasoline. Which takes up more space: 1 kg of water or 1 kg of gasoline?

8. Oil and vinegar salad dressing settles with the oil floating on top of the vinegar. Explain this phenomenon in terms of density.

9. When a fish is floating motionless below the surface of a lake, what is the amount and direction of the force the water is exerting on it?

10. Some fish move extremely slowly, and it's hard to tell whether they are even alive. However, if a fish is floating at a middle height in your aquarium and not at the top or bot-

tom of the water, you can be pretty certain that it's alive. Why?

11. A barometer, which is often used to monitor the weather, is a device that measures air pressure. How could you use a barometer to measure your altitude as you climbed in the mountains?

12. If you seal a soft plastic bottle or juice container while hiking high in the mountains and then return to the valley, the container will be dented inward. What causes this compression?

13. Many jars have dimples in their lids that pop up when you open the jar. What holds the dimple down while the jar is sealed, and why does it pop up when the jar is opened?

14. If you place a hot, wet cup upside down on a smooth counter for a few seconds, you may find it difficult to lift up again. What is holding that cup down on the counter?

15. You seal a container that is half full of hot food and put it in the refrigerator. Why is the container's lid bowed inward when you look at it later?

16. Why aren't there any thermometers that read temperatures down to −300 °C?

17. You use your breath to inflate a large rubber tube and then ride down a snowy hill on it. After a few minutes in the snow the tube is underinflated. What happened to the air?

18. A marshmallow is filled with air bubbles. Why does a marshmallow puff up when you toast it?

19. Wasp and hornet sprays proudly advertise just how far they can send insecticide. How does the pressure inside the spray can affect that distance, and why is the direction of the spray important (vertical vs. horizontal)?

20. Ice tea is often dispensed from a large jug with a faucet near the bottom. Why does the speed of tea flowing out of the faucet decrease as the jug empties?

21. Why must tall dams be so much thicker at their bases than at their tops?

22. Waterproof watches have a maximum depth to which they can safely be taken while swimming. Why?

23. Some small animals, such as bats and opossums, hang upside down when they sleep. If you did that, the increased blood pressure in your head would give you a headache. Why don't small animals experience a similar large increase in blood pressure in their heads?

24. When you turn on a drinking fountain, a thin stream of water rises from its nozzle in a low arc so that you can drink. The drinking fountain contains a pressure-regulating device that limits the water pressure in its chilled water storage tank to just a few thousand pascals above atmospheric pressure. Why is that low pressure important, and what would happen if the pressure inside the tank were much higher?

25. When you stand in a pool with water up to your neck, you find that it's somewhat more difficult to breathe than when you're out of the water. Why?

26. How does pushing on the plunger of a syringe cause medicine to flow into a patient through a hollow hypodermic needle?

27. Why must the pressure inside a whistle teakettle exceed atmospheric pressure before the whistle can begin to make noise?

28. Each time you breathe in, air accelerates toward your nose and lungs. How does the pressure in your lungs compare with that in the surrounding air as you breathe in?

29. You can inflate a plastic bag by holding it up so that it catches the wind. Use Bernoulli's equation to explain this effect.

30. When someone pulls a fire alarm in a skyscraper, pumps increase the water pressure in the section of the building nearest that alarm box. How does this pressure change assist firefighters who must battle the blaze?

31. If you fill a balloon full of water, the water has pressure potential energy. This pressure potential energy becomes kinetic energy when you let the water spray out of the opening. But pressure potential energy isn't actually stored in water, so where is it contained in the water balloon?

32. When you breathe out, air flows rapidly out of your nose. From where is that air's kinetic energy coming?

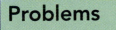

Problems

1. The particle density of standard atmospheric air at 273.15 K (0 °C) is 2.687×10^{25} particles/m^3. Using the ideal gas law, calculate the pressure of this air.

2. How much force is the air exerting on the front surface of this book?

3. If you fill a container with air at room temperature (300 K), seal the container, and then heat the container to 900 K, what will the pressure be inside the container?

4. An air compressor is a device that pumps air molecules into a tank. A particular air compressor adds air molecules to its tank until the particle density of the inside air is 30 times that of the outside air. If the temperature inside the tank is the same as that outside, how does the pressure inside the tank compare to the pressure outside?

5. If you seal a container of air at room temperature (20 °C) and then put it in the refrigerator (2 °C), how much will the pressure of the air in the container change?

6. If you submerge an 8-kg log in water and it displaces 10 kg of water, what will the net force on the log be the moment you let go of the log?

7. If your boat weighs 1200 N, how much water will it displace when it's floating motionless at the surface of a lake?

8. The density of gold is 19 times that of water. If you take a gold crown weighing 30 N and submerge it in water, what will the buoyant force on the crown be?

9. How much upward force must you exert on the submerged crown in Problem 8 to keep it from accelerating?

10. How could you use your results from Problems 8 and 9 to determine whether the crown was actually gold rather than gold-plated copper? (The density of copper is nine times that of water.)

11. Your town is installing a fountain in the main square. If the water is to rise 25 m (82 feet) above the fountain, how much pressure must the water have as it moves slowly toward the nozzle that sprays it up into the air?

12. Rather than putting a pump in the fountain (see Problem 11), the town engineer puts a water storage tank in one of the nearby high-rise office buildings. How high up in that building should the tank be for its water to rise to 25 m when spraying out of the fountain? (Neglect friction.)

13. To clean the outside of your house you rent a small high-pressure water sprayer. The sprayer's pump delivers slow-moving water at a pressure of 10,000,000 Pa (about 150 atmospheres). How fast can this water move if all of its pressure potential energy becomes kinetic energy as it flows through the nozzle of the sprayer?

14. When the water from the sprayer in Problem 13 hits the side of your house, it slows to a stop. If it hasn't lost any energy since leaving the sprayer, what is the pressure of the water at the moment it stops completely?

15. To dive far below the surface of the water, a submarine must be able to withstand enormous pressures. At a depth of 300 m, what pressure does water exert on the submarine's hull?

Cases

1. A large ship floats motionless at the surface of a still sea, supported by the buoyant force. Its average density, including the air it contains, is less than that of the seawater.
  a. What is the net force on the ship?
  b. The ship is displacing both water and air. If the ship were to move upward a few centimeters, what would happen to the average density of the fluid the ship displaces?
  c. If the ship were to move up a few centimeters, what would happen to the buoyant force on the ship?
  d. If the ship were to move up a few centimeters, what would happen to the net force on the ship?
  e. Show that the floating ship is in a stable equilibrium with respect to up and down motion.
  f. Why does the ship settle deeper into the water when another passenger climbs aboard?
  g. The ship's captain can estimate how much cargo it's carrying by how deep the ship rides in the water. How does that technique work?
  h. Does the air exert a buoyant force on the ship? If so, why doesn't the ship float in air?

2. Like all bony fish, a bass has a gas-filled sac or air bladder that allows it to float motionless below the surface of the water.

  a. What forces act on the motionless bass, and what is the net force it experiences?
  b. What is the bass's average density?
  c. Salt water is more dense than fresh water. What are the relative sizes of the air bladders in saltwater and freshwater bass?
  d. A shark is a cartilaginous fish that lacks an air bladder. If it can't move, it will sink. Why?

3. A scuba diver swims below the surface of the water, breathing air from the steel tanks of an aqualung.
  a. It's easiest for her to maintain a constant depth if the buoyant force on her and her equipment exactly balances their weight. Her wet suit floats so she wears a heavy weight belt to compensate. How should her average density compare with that of the water around her?
  ***b.** When she is 20 m below the surface of the water, what is the pressure pushing inward on her chest?
  c. When she breathes, she expects air to flow into her lungs. She can change the pressure in her lungs slightly, using the muscles in her chest and diaphragm. How must the pressure in her mouth compare with that in her lungs in order for air to begin flowing into her lungs?

d. If instead of taking compressed air with her, she were to try to breathe surface air through a straw, the air entering her mouth would be at atmospheric pressure. What would happen when she tried to breathe?

e. The special pressure regulator of an aqualung delivers air to her mouth at exactly the same pressure as that of the surrounding water. She can breathe this air easily. However, the deeper she dives, the faster she consumes air molecules from the tanks and the sooner she must return to the surface. Why does she exhaust her tanks faster by going deeper?

4. Blimps are compact versions of the great airships of the early 20th century. They are filled with helium and are propelled forward by fans. But a blimp has more to it than meets the eye. Inside the cigar-shaped exterior skin are several flexible containers called ballonets. The blimp's buoyancy and orientation are controlled by pumping air into or out of these ballonets.

a. The exterior skin is rigid, so the blimp's total volume doesn't change. Why does pumping air into the ballonets cause the blimp to become less buoyant?

b. Why does pumping air from the forward ballonet to the rearward ballonet cause the blimp to tip its nose upward?

c. When the blimp flies into warmer air, what should it do with its ballonets in order to maintain a constant altitude?

d. The blimp is propelled forward by huge fans, which push the air backward. Why does this procedure help the blimp to move forward?

5. You are washing a car in front of your apartment. You have a garden hose attached to a water faucet and a bucket with a sponge in it.

a. When the water faucet is on, water flows through the hose and out the nozzle. What causes water to accelerate to high speed as it leaves the nozzle?

b. As it arcs through the air, the stream of water is at atmospheric pressure. But when you accidentally hit the bucket with the stream of water, the bucket accelerates away from you and falls off the car. Evidently, the water pressure increases when it touches the bucket. Why does the water pressure increase?

c. You decide that perhaps the water would clean more effectively if it were moving faster, so you let the water spray upward and fall back down on the car. But the falling water is traveling no faster than it was when it first left the nozzle. Why not?

d. The sponge sinks to the bottom of the bucket of water. What can you say about the sponge's average density?

6. A traditional water cooler has a large bottle of water turned upside down so that its neck is submerged in a small chilled water reservoir at the top of the water cooler. The water level in the small reservoir remains just above the neck of the bottle. If you open the valve to let water out of the reservoir, bubbles of air rise up into the water bottle and the level of water in the bottle goes down.

a. Since there is no true seal between the neck of the bottle and the reservoir, what holds the water up inside the water bottle?

b. When you open the valve at the bottom of the water cooler, the water flows out into your glass. What provides the force needed to make the water accelerate?

c. Whenever the water level in the reservoir drops below the lip of the inverted water bottle, air bubbles enter the bottle and some water flows out. What force lifts the air bubbles upward inside the water bottle?

d. Once in a while the delivery person drops a water bottle and makes a tiny crack in its bottom (which is on top when the bottle is upside down on the water cooler). Although water can't pass through the crack, air can. What will happen if you put the cracked bottle on the water cooler, and why will this happen?

7. Firefighters are battling a fire in a tall apartment building. The water pressure in the adjacent fire hydrant is about 500,000 Pa above atmospheric pressure.

*a. Some firefighters take their hose up the stairwell inside the building. What is the highest level at which they can expect water to flow out of their hose without additional pressure?

*b. Firefighters on the ground begin to spray water upward from their hoses. The water enters the hose traveling slowly at 500,000 Pa above atmospheric pressure. How fast will the water be traveling when it leaves the nozzle at atmospheric pressure?

*c. How high will the water in part b rise if the firefighters send it straight up?

*d. To boost the water pressure, the firefighters send it sequentially through pumps in two fire engines. Each pump boosts the pressure by 500,000 Pa, for a total of 1,500,000 Pa above atmospheric pressure. How high will this water rise in hoses carried up the stairs inside the building?

Fluids and Motion

Fluids are fascinating when they move. Stationary water and air may be essential to life, but they're also fairly simple; only their pressures vary from place to place and even these are determined primarily by gravity. But rushing rivers or gusts of wind, with their wonderful variety of simple and complicated behaviors, are much more interesting. And the motion of fluids isn't just interesting; it's also important, since our world is filled with objects and machines that work in whole or part because of the behaviors of moving fluids. In this chapter, we will look at several situations in which fluid motion contributes to the way things work.

EXPERIMENT: A Vortex Cannon

Fluids are real entities, with an existence that doesn't depend on the solids that move through them. This idea is easy to grasp in reference to water, since we can see that water doesn't wait for a boat to sail by before it moves in interesting ways. But

air is harder to visualize in this way, since we seldom see it by itself, apart from its effects on buildings or airplanes or our skin.

To begin seeing air as something tangible and to demonstrate the rich possibilities of motion available to air itself, build a vortex cannon—a device that sends rings of air sailing across a room. While a serious vortex cannon is best constructed from a large drum or crate, you can make a reasonably effective one from an empty cardboard cereal box.

Seal the rectangular box on all edges with tape, and then cut a circular hole 5 centimeters (2 inches) in diameter in the center of one face. To use your cannon, just tap hard on the other face of the box. Rings of air will leap out of the hole and sail across the room at about 5 m/s.

You can watch these rings move by looking for their effects on the objects they meet. They can easily blow out candles or rustle light window drapes across a small room. If you have a friend blow some rings at you, you'll feel exactly where they hit on your face or shirt.

To actually see these rings, fill the cereal box with smoke, mist, or dust. Try to *predict* what the rings will look like when they emerge from the hole. Will the smoke in each ring be stationary, or will it be swirling about the ring in some manner? How will the ring's size and speed of travel depend on the size of the hole from which it emerges? Will it depend on how hard you tap the cereal box? Will the ring's size and speed change in flight?

Now tap the box and *observe* the smoke rings. What motions do you see? *Measure* the size and speed of the rings as they fly. Did you *verify* your predictions? Change the hole size and see how this change affects the rings and their motion. As you can see, the air can execute some very complicated movements all by itself. In this chapter, we will examine how moving air, water, and other fluids affect our everyday lives.

Chapter Itinerary

In particular, we'll explore (1) *garden watering*, (2) *balls and Frisbees*, and (3) *airplanes and rockets*. In *garden watering*, we'll look at how water's pressure and speed vary as it flows through a faucet, a hose, and a nozzle. In *balls and Frisbees*, we'll investigate the effects of air on the motions of balls and other flying toys. Lastly, in *airplanes and rockets*, we'll study the ways in which moving air supports and propels airplanes and rockets in flight.

This chapter continues to develop the concept of energy conservation along a streamline that was introduced in Chapter 4. It also brings up several new types of forces that are present when fluids move past one another or past solid objects. These ideas are present not only in the topics of this chapter, but also in many other commonplace activities, from washing windows with a hose to pumping water with a windmill.

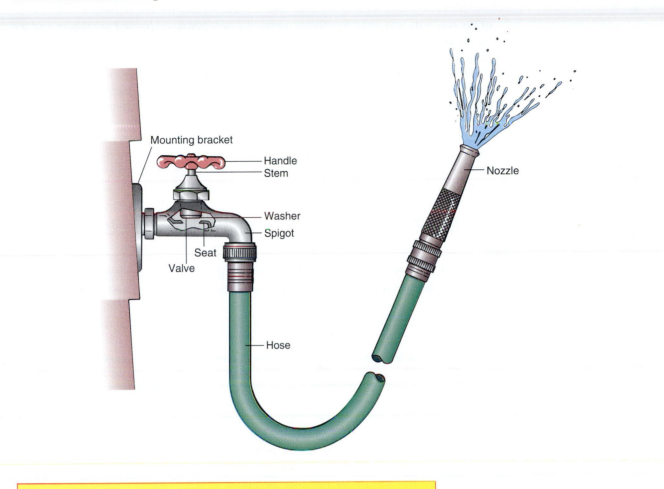

Mounting bracket

Handle
Stem

Washer
Spigot

Seat
Valve

Hose

Nozzle

Garden Watering

Tending a flower garden often involves watering. While this once meant walking the garden's paths with a watering can, modern plumbing has made such effort unnecessary. With a hose and nozzle attached to the faucet, you can do your job without leaving the lawn chair. But while the tools involved—faucets, hoses, and nozzles—are simple and unsophisticated, the principles behind them are not. All three make elegant use of the laws of fluid flow, letting openings and channels control the delivery rate and speed of the water so that it arcs gracefully through the air to the farthest reaches of your garden.

Questions to Think About: *Where is the water when the faucet is closed, and why does it begin flowing through the hose when you open the faucet? What determines the rate at which water flows through the faucet or the hose? If honey ran through your pipes instead of water, how would that affect the flow? Why does water make noise as you sprinkle your garden? Why does a nozzle make the water spray so fast and so far? Why do the pipes sometimes clank when you abruptly close the faucet?*

Experiments to Do: *Open a faucet gradually and watch the water begin to flow. What's pushing the water out of the faucet? What happens to the flow rate and speed* *of the water as you open the faucet further? Look below or behind the faucet and try to determine how the water enters and exits the faucet. When do you hear the water flowing?*

Attach a hose to the faucet. Does the water flow as quickly from the open end of the hose as it did from the faucet alone? Cover most of the hose end with your thumb and watch the water spray out into the air. Why does the water travel so much farther when you almost stop its flow? Do you feel the water pressing against your thumb? Attach a nozzle to the hose and see how the flow rate affects the strength of the spray. Fill a bucket with the open hose and then again while using the nozzle. Which fills the bucket fastest?

Flow from a Faucet: Speed and Viscosity

Having brought water to your home in the previous chapter, we're already well on our way to watering your garden. But first we should learn how to shut the water off. Before your furniture starts floating around the room, it's time to think about faucets.

The operation of a faucet is simple: pressurized water is waiting just inside the faucet and as you move its handle, you open a hole through which the water can accelerate toward low pressure outside. The farther you turn the handle, the larger the hole gets and the more water flows out of the faucet. But a hole is just a hole, so why should its size affect the flow of water? The answer to this question has two parts, anticipating the two paths we'll pursue in this section. The first part deals with exchanges of pressure for speed in an ideal fluid and the second with the wasting of energy in a real fluid as it rubs against its container and itself.

We'll look more carefully at how pressure is exchanged for speed later on when we study nozzles. For the moment, we can understand the first part of the answer by noting that the speed with which water flows through a hole is limited by the water's total energy: the sum of its pressure potential energy, kinetic energy, and gravitational potential energy. Since water is incompressible and the flow here is steady state, Bernoulli's equation applies. As it passes through the hole, the water converts most of its pressure potential energy into kinetic energy. It moves through the hole as quickly as it can, but its energy and peak speed are limited by the water pressure in the plumbing.

Because of this limited speed, only so much water can pass through the hole each second. Think of a two-lane road with a fixed speed limit: it can only carry so much traffic. Widening the hole, however, also widens the flow of water and allows more water to pass—in the same way that widening a two-lane road to four lanes increases its traffic capacity.

The second part of the answer, and the one we'll concentrate on for now, is that the flow of water through a faucet is constrained by the **viscous forces** that appear whenever one layer of fluid tries to slide across another layer of fluid. These forces oppose such relative motion and produce frictionlike effects within the fluid. You observe these effects when you pour honey out of a jar. The honey at the jar's surface is stuck there and remains stationary. But even honey that's far from the walls can't move easily; it experiences viscous forces as it tries to move relative to nearby honey. Since honey is a "thick" or *viscous fluid,* viscous forces act quite effectively to keep all the honey moving with nearly the same velocity. Since the honey at the walls can't move, viscous forces tend to prevent any of the honey from moving.

Water isn't as thick as honey (Table 5.1.1), so it's less resistant to relative motion. The measure of this resistance to relative motion within a fluid is called **vis-**

Table 5.1.1 **Approximate Viscosities of a Variety of Fluids**

FLUID	VISCOSITY[a]
Helium (2 K)	0 Pa·s
Air (20 °C)	0.0000183 Pa·s
Water (20 °C)	0.00100 Pa·s
Olive oil (20 °C)	0.084 Pa·s
Shampoo (20 °C)	100 Pa·s
Honey (20 °C)	1000 Pa·s
Glass (540 °C)	10^{12} Pa·s

[a]The pascal-second (abbreviated Pa·s and synonymous with kg/m·s) is the SI unit of viscosity. Only the superfluid portion of ultracold liquid helium exhibits zero viscosity.

cosity, and water's viscosity is less than that of honey. Actually, hot water is even less viscous than cold water and thus flows more easily. Typical of most liquids, this decrease in viscosity with temperature reflects the molecular origins of viscous forces: the molecules in a liquid stick to one another, forming weak chemical bonds that require energy to break. In a hot liquid, the molecules have more thermal energy, so they break these bonds more easily in order to move past one another (see ❏).

When the faucet is open, water near the center of the opening accelerates toward low pressure and leaves the faucet at high speed. But water at the edges of the opening experiences friction with the walls and is essentially motionless. Since different portions of the water are moving at different velocities, viscous forces appear. These forces act as an internal sliding friction within the water, wasting energy and raising the water's temperature. Because of this wasted energy, the water flowing out of a faucet doesn't reach the speed that Bernoulli's equation predicts, and the faucet delivers less water per second than it would in the absence of viscosity.

➤ Check Your Understanding #1: Keeping Warm on a Windy Day
A loosely woven wool sweater has many tiny air passages between the wool fibers, yet it dramatically reduces the rate at which air flows through to your skin when you stand in a breeze. Why doesn't air flow easily through the gaps between the fibers?

Flow Through a Hose: More Viscous Effects

If viscosity slows the flow of water through a faucet, it should also slow the flow of water through a hose. And so it does. The hose holds the outer layer of water stationary, and this motionless layer exerts viscous forces on the layer of moving water inside it. As this second layer slows, it exerts viscous forces on yet another layer. Layer by layer, viscous forces hold back the moving water until even water at the center of the hose feels viscosity's slowing effects (Fig. 5.1.1).

These viscous forces impede water delivery. Instead of coasting effortlessly through a level hose, real water needs a pressure gradient to keep it moving steadily forward. Like the file cabinet sliding on the sidewalk in Section 2.2, water must be pushed through the hose if it's to maintain a continuous flow.

However, unlike the forces of sliding friction—which don't depend on relative velocities—viscous forces become larger as the relative velocities within a fluid increase. That's because as two layers of water slide past one another faster, their mol-

❏ Your car's engine is protected by motor oil with a carefully chosen viscosity. If that oil were too thin, it would flow out from between surfaces and wouldn't keep them from rubbing against one another. If that oil were too thick, the engine would waste power moving its parts through the oil. Years ago, you had to change your motor oil for the season. Thick 40 Weight motor oil was used in summer because hot weather made it thinner; thin 10 Weight oil was used in winter because cold weather made it thicker. But a modern, multigrade oil maintains a nearly constant viscosity over a wide range of temperatures and need not be changed with the seasons. This oil contains tiny molecular chains that ball up when cold but straighten out when hot. These chains thicken hot oil so that 10W-40 oil resembles 10 Weight oil in winter and 40 Weight oil in summer.

unusual liquids

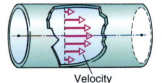

Velocity

Fig. 5.1.1 The speed of water flowing through a pipe is not constant across the pipe. The water near the walls is stationary, while the water at the center of the pipe moves the fastest. The differences in velocity are the results of viscous forces.

ecules collide harder and more frequently. Since it experiences stronger viscous forces, fast-moving water wastes more energy per meter and needs a larger pressure gradient to move steadily through a hose than does slow-moving water.

Because of viscosity, the amount of water flowing steadily through a hose depends on four factors:

1. It's inversely proportional to the water's viscosity. The more viscous the water, the more difficulty it has flowing through the hose.
2. It's inversely proportional to the length of the hose. The longer the hose, the more opportunity viscous forces have to slow the water down.
3. It's proportional to the pressure difference between the hose's inlet and its outlet. This pressure difference determines the water's pressure gradient and thus how hard the water is pushed forward through the hose.
4. It's proportional to the fourth power of the diameter of the hose. Tripling the hose's diameter provides the water with nine times as much room and also allows water near the hose's center to move nine times faster.

We can turn all these proportional relationships into an equation by adding the correct numerical constant ($\pi/128$). The final relationship is called **Poiseuille's law** and can be written as a word equation:

$$\text{volume flow rate} = \frac{\pi \cdot \text{pressure difference} \cdot \text{pipe diameter}^4}{128 \cdot \text{pipe length} \cdot \text{fluid viscosity}}, \qquad \textbf{(5.1.1)}$$

in symbols:

$$\frac{\Delta V}{\Delta t} = \frac{\pi \cdot \Delta p \cdot D^4}{128 \cdot L \cdot \eta},$$

and in everyday language:

It's hard to squeeze honey through a long, thin tube.

> **Poiseuille's Law**
> The volume of fluid flowing through a cylindrical pipe each second is equal to ($\pi/128$) times the pressure difference (Δp) across that pipe times the pipe's diameter to the fourth power, divided by the pipe's length times the fluid's viscosity (η).

❏ To deliver large amounts of water at high pressure or velocity, fire hoses must have large diameters. When filled with high-pressure water, these wide hoses become stiff and heavy, making them difficult to handle. Chemical additives that decrease water's viscosity allow firefighters to use narrower, lighter, and more flexible hoses.

It's hardly surprising that flow rate depends in this manner on the pressure difference, pipe length, and viscosity; we've all observed that low water pressure or a long hose lengthens the time needed to fill a bucket with water and that viscous syrup pours slowly from a bottle. But the dependence of flow rate on the fourth power of diameter may come as a surprise. Even a small change in the diameter of a hose significantly changes the amount of water that hose delivers each second (see ❏s).

We can also look at viscous forces in terms of total energy. By opposing the flow of water through a hose, viscous forces do negative work on it and reduce its total energy—the energy considered in Bernoulli's equation, which doesn't include thermal energy. Just how much total energy the water retains depends on how fast it moves inside the hose. If you allow lots of water to leave the hose, water will move through it quickly and encounter large viscous forces. In the process, most of the water's total energy will be converted into thermal energy and the water will pour gently out of the end of the hose.

But if you partially block the hose's opening with your thumb and reduce the flow, water will travel slowly through the hose and encounter smaller viscous forces. As a result, the water will retain most of its total energy and will still be at high pressure when it reaches your thumb. This high-pressure water will then accelerate to enormous speed as it passes through the narrow opening and sprays out into the air.

❑ Very large diameter pipes are required to transport crude oil across the Alaskan wilderness. The distances are long and the fluid is viscous, particularly during the winter.

The long air ducts used to ventilate homes and businesses usually have very large diameters. These ducts are often visible near the ceilings of modern warehouse-style stores and restaurants as pipes roughly 0.5 m across. Why must the ducts be so large in diameter?

When your friend's house was new, the kitchen faucet could deliver 0.50 liter per second (0.50 l/s). But mineral deposits have built up in the pipes over the years and reduced their effective diameters by 20%. How much water can the faucet deliver now?

narrowing pipes and arteries

Flow from a Nozzle: Pressure and Speed

When you partly cover the hose end with your thumb, you create a simple nozzle—a constriction that causes the water to speed up and exchange its pressure potential energy for kinetic energy. While such exchanges are an important consequence of Bernoulli's equation, they don't happen willy-nilly. They only occur when an obstacle deflects the flowing water, something that doesn't happen in a straight hose. As it redirects the flow, your thumb initiates the pressure drop and speed increase that send the water spraying out into the garden.

To understand why deflection is so important, consider what would happen if water flowing steadily through a straight hose tried to change speed partway along its trip. If it tried to speed up, it would leave an empty space behind it; if it tried to slow down, it would cause a "traffic jam." Since neither outcome can occur in steady-state flow, water passing continuously through a straight hose must maintain constant velocity. This constancy has nothing to do with viscosity, a fact that will permit us to neglect viscosity in the present discussion.

But when flowing water is deflected by an obstacle, its velocity stops being constant. The deflected water accelerates, and its pressure and speed change in interesting ways. Water deflecting away from a surface experiences a rise in pressure and a drop in speed, while water deflecting toward a surface experiences a drop in pressure and a rise in speed. To explore these effects, consider what happens when the straight hose has a steplike flaw in its wall. As shown by the black streamlines in Fig. 5.1.2, p. 152, the water flows up and around the step's surface before continuing through the remaining length of hose.

When it turns away from the surface at the base of the step, the water slows down and its pressure rises. High pressure near the surface is what pushes the water stream away from the step so that it bends upward. In the figure, the slowing appears

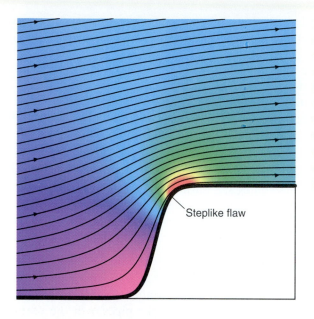

Steplike flaw

Fig. 5.1.2 Water flowing up and over a steplike flaw in a hose's wall experiences changes in speed and pressure. The black streamlines show the paths the water takes as it flows. The spacing between streamlines indicates flow speed (wider space is slower flow), and the background color indicates pressure (violet is higher pressure; red is lower pressure).

as a widening in the separation between streamlines and the pressure rise is indicated by a color shift toward the violet end of the spectrum.

When it turns toward the surface at the top of the step, the water speeds up and its pressure drops. Low pressure near the surface is what allows the surrounding water to push the water stream toward the step so that it bends downward. The speeding up appears as a narrowing in the separation between streamlines, and the pressure drop is indicated by a color shift toward the red end of the spectrum.

It's natural to wonder which happens first: the change in the water's pressure or the change in its speed. The answer is that they occur at the same time. Once the steady-state flow pattern has formed around an obstacle, water following a particular streamline experiences rises and falls in pressure at the same time that it experiences decreases and increases in speed. The two simply go hand in hand.

What happens when water passes through a nozzle? The answer is shown in Fig. 5.1.3. When it first encounters the nozzle's shrinking orifice, water near the

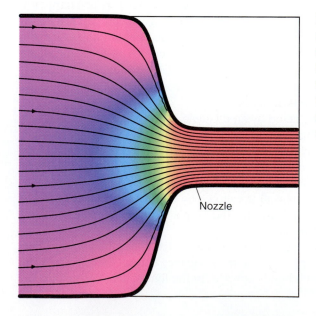

Nozzle

Fig. 5.1.3 Water flowing through a nozzle speeds up, and its pressure drops. The narrowing spacing between streamlines indicates that the flow speed is increasing while the color shift from violet toward red indicates that the pressure is dropping.

edges of the hose slows down and its pressure rises. It bends inward, away from the nozzle's walls. But as this water approaches the center of the nozzle, it bends again in order to pass through the nozzle's narrow neck. This time it bends toward the surface of the nozzle, so its pressure drops and its speed increases. Remarkably enough, this pressure drop at the nozzle's neck affects the entire flow. All the water approaching the nozzle accelerates toward the neck, increasing its speed and lowering its pressure. The flow emerges from the nozzle as a narrow stream of fast moving water. The water is now at atmospheric pressure and flies through the air in a graceful arc.

While viscosity slows the water slightly as it passes quickly through the nozzle's neck, that neck is so short that viscosity has little impact on the water's final speed. Instead, the water retains most of its total energy as it goes suddenly from high pressure and low speed before the nozzle to low pressure and high speed after the nozzle. No wonder you can reach the farthest parts of your garden with water when you use a nozzle.

Check Your Understanding #3: Cleaning House

As air rushes steadily into the narrow opening of a vacuum cleaner attachment, it accelerates to high speed and its pressure drops well below atmospheric pressure. From where does the air's newfound kinetic energy come?

Turbulence and Noise

Water often makes noise as it flows through faucets and nozzles. The sound is a high-pitched hiss that's loudest near the narrowest openings. It's created when water swirls about erratically, a behavior known as **turbulent flow.**

Up until now, we've discussed only **laminar flow**—smooth, silent flow that's characterized by simple streamlines. In laminar flow, adjacent regions of a fluid always remain nearby. For example, if you place two drops of dye near one another in a smoothly flowing stream, they will remain close together indefinitely as they follow streamlines in the laminar flow (Fig. 5.1.4).

But as the stream flows past rocks and obstacles, its streamlines break up into the eddies and churning "white water" that make rafting exciting. The dye is quickly dispersed. In this turbulent flow, adjacent regions of a fluid soon become separated from one another as they move independently in unpredictable directions.

Fig. 5.1.4 Water flows slowly past rocks in the stream on the left, and its viscosity keeps it smooth and laminar. Water flows quickly past rocks in the stream on the right, and its inertia separates it into swirling, splashing pockets of turbulence.

Fig. 5.1.5 Honey's large viscosity keeps it flowing smoothly (laminar flow) when you pour it. Colored water's small viscosity allows it to splash about (turbulent flow).

Whether a flow is laminar or turbulent depends on several characteristics of the fluid and its environment:

1. The fluid's viscosity. Viscous forces tend to keep nearby regions of fluid moving together, so high viscosity favors laminar flow (Fig. 5.1.5).
2. The fluid's speed past a stationary obstacle. The faster the fluid is moving, the more quickly two nearby regions of fluid can become separated and the harder it is for viscous forces to keep them together.
3. The size of the obstacle the fluid encounters. The larger the obstacle, the more likely that it will cause turbulence because viscous forces will be unable to keep the fluid ordered over such a long distance.
4. The fluid's density. The denser the fluid, the less it responds to viscous forces and the more likely it is to become turbulent.

Rather than keeping track of all four physical quantities independently, English mathematician and engineer Osborne Reynolds (1842–1912) found that they could be combined into a single number that permits a comparison of seemingly different flows. The **Reynolds number** is defined as

$$\text{Reynolds number} = \frac{\text{density} \cdot \text{obstacle length} \cdot \text{flow speed}}{\text{viscosity}}. \qquad \textbf{(5.1.2)}$$

The units on the right side of Eq. 5.1.2 cancel one another so that the Reynolds number is dimensionless; that is, it's just a simple number, such as 10 or 25,000, with no dimensions or units. As the Reynolds number increases, the flow goes from laminar to turbulent. At a low Reynolds number, viscous forces dominate the flow and keep it smooth and laminar. At a high Reynolds number, inertia dominates the flow and each portion of fluid moves according to its own momentum. These individual portions of fluid separate or collide frequently, and the flow becomes swirling and turbulent.

In his experiments, Reynolds found that turbulence usually appears when the Reynolds number exceeds roughly 2300. You can observe this transition by moving a 1-cm thick stick through still water. If you move the stick slowly, about 10 cm/s, the Reynolds number will be about 1000 and the flow around the stick will be laminar. But if you speed the stick up to about 50 cm/s, the Reynolds number will rise to about 5000 and the flow will become turbulent.

One of the most common features of turbulent flows is the **vortex,** a whirling region of fluid that moves in a circle around a central cavity. A vortex resembles a miniature tornado, with its cavity created by inertia as the fluid spins. Vortices are easily visible behind a canoe paddle or in a mixing bowl. Once an object moves fast enough through a fluid to create turbulence, these vortices begin to form. Each vortex builds up behind the object but is soon whisked away to form a wake of *shed vortices* (Fig. 5.1.6).

While laminar flow is fully predictable, turbulent flow exhibits chaotic behavior or **chaos:** you can no longer predict exactly where any particular drop of water will go. The study of chaos is a relatively new field of science. Because a **chaotic system**—a system exhibiting chaos—is exquisitely sensitive to initial conditions, even the slightest change in those conditions may produce profound changes in its situation later on.

Although you can't see the vortices in a faucet or nozzle, you can usually hear them. Turbulence appears when fast-moving water changes directions too abruptly and it converts some of the water's total energy into thermal energy and sound. This turbulence slightly reduces the water's speed and pressure as it flows toward your garden.

A different sound occurs when you abruptly close the nozzle and stop the flow of water. The moving water has momentum, and stopping it suddenly requires an enormous backward force. Since the slowing flow is not steady state, Bernoulli's equation doesn't apply and the pressure can surge to astronomical values near the front of the moving water. This pressure surge is what accelerates the water backward to slow it down and also what leads to the loud "thump" sound you hear as the water stops. Known as **water hammer,** the surging pressure in front of stopping water jerks the nozzle, swells the hose, and may even rattle the pipes in your home.

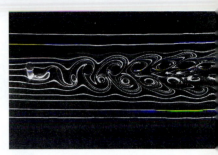

Fig. 5.1.6 When water flows rapidly around a cylinder, its flow becomes turbulent. A pattern of swirling vortices forms to the right of this cylinder.

chaos

water hammer

Check Your Understanding #4: Urban Windstorms

On a windy day in a city with many tall buildings, leaves and papers can be seen swirling about in the air or on the sidewalks. What causes these whirling air currents?

Check Your Figures #2: Wind on the Open Road

Is the flow of air around a convertible laminar or turbulent as the convertible cruises down the highway? (Air's viscosity is given in Table 5.1.1.)

screws, seals

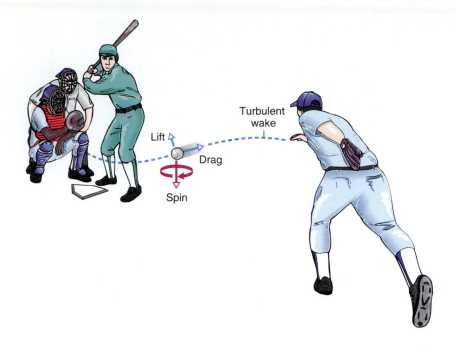

SECTION 5.2

Balls and Frisbees

Much of the subtlety and nuance in games such as baseball and golf come from the way balls interact with air. If baseball were played on the moon, which has no air, the sinking fastball would be the only interesting pitch; other thrown objects, such as fly-ing disks, wouldn't work at all. In this section we will investigate how air affects the flight of balls and other related objects.

Questions to Think About: *Why can you throw a real baseball so much farther than a hollow plastic one? Why does a long fly ball appear to drop straight down when you try to catch it in deep center field? What kind of force could make a curve ball curve? What makes a well-hit golf ball hang in the air before falling to the green? How can a knuckle ball or spitball jitter about in flight? What keeps a Frisbee in the air?*

Experiments to Do: *To make air's effects most apparent, you need a ball that weighs little but has lots of surface area. A beach ball is ideal, but a whiffle ball or hollow plastic ball will also do nicely. See how far you can throw it. How does it stop? Does it slow down and lose height gradually, or does it stop rapidly and fall to the ground? Now make the ball spin as you throw it. Why does the ball curve in flight? Does a fast spin make the ball curve more or less? Which way is the ball spinning, and how is the spin related to the direction of its curve? Change the direction of spin. Which way does the ball curve now?*

When a Ball Moves Slowly: Laminar Airflow

One of the first things you might notice if you joined a new baseball franchise on the moon would be that you could throw the ball farther than back at home. Part of this increased distance would be due to the weaker gravity, since the ball would take longer to fall, but part would stem from the absence of air resistance. One of air's many effects on moving objects is to slow them down. In the previous section, we saw how objects affect moving fluids. Now as we study **aerodynamics,** the science of air's dynamic interactions, we'll see how fluids affect moving objects.

A moving ball experiences **aerodynamic forces**—that is, forces exerted on it by the passing air. These consist of **drag forces** that oppose relative motion and **lift forces** that push the ball to one side or the other. Grouped together, the drag forces are commonly referred to as "air resistance."

Since many of these aerodynamic forces are caused by turbulent airflow, they're present only when the air's motion is dominated by inertia. And since viscosity dominates air's motion around small and slow-moving particles, we'll need a fairly large, fast-moving object to explore turbulent airflow. That's why a ball's aerodynamics are so interesting; except when it moves extremely slowly, the airflow around a ball is turbulent.

Laminar airflow: viscosity dominates the airflow, keeping it smooth and orderly. Laminar airflow occurs around smaller objects that move slowly through the air.

Turbulent airflow: inertia dominates the airflow, ripping it apart into swirling eddies. Turbulent airflow occurs around larger objects that move quickly through the air.

Before we explore turbulent airflow, however, we should look at the pattern of laminar airflow around a slow-moving ball (Fig. 5.2.1). Actually, this pattern will be the same whether the ball moves slowly through the air or the air moves slowly past the ball. For simplicity, we'll study airflow from the reference frame of the ball, so that the ball appears stationary with the air flowing past it.

Fig. 5.2.1 The airflow around a slowly moving ball is laminar. Air slows down in front of and behind the ball (widely spaced streamlines), and its pressure increases (shifts toward the violet end of the spectrum). Air speeds up at the sides of the ball (narrowly spaced streamlines), and its pressure decreases (shifts toward the red end of the spectrum). However, the pressure forces on the ball balance one another perfectly, and it experiences no pressure drag. Only viscous drag is present to affect the ball.

❑ When the airflow around an object is laminar, the pressure forces on it cancel perfectly and it experiences no drag due to pressure imbalances—no *pressure drag.* The absence of pressure drag was a great puzzlement to early aerodynamicists, who knew that the airflow around dust is laminar and that it experiences a drag force. This mystery was named d'Alembert's paradox, after Jean le Rond d'Alembert (1717–1783), the French mathematician who first recognized it. D'Alembert and his contemporaries didn't know about the viscous drag force, which is what really slows dust's motion through the air.

When the air moves slowly enough, it separates neatly around the front of the ball and comes back together behind it, producing a **wake,** an air trail left behind the ball, that's smooth and free of turbulence. But the air's speed and pressure aren't uniform all the way around the ball. At the ball's front, the air bends away from the ball's surface, so the pressure there must be higher than atmospheric. As we saw with nozzles, an outward bend (accelerating away from the surface) causes the air's speed to decrease and its pressure to increase. The slowing is indicated by the widening separation of the streamlines, and the pressure rise is indicated by the color shift toward the violet end of the spectrum.

Near the sides of the ball, the air bends toward the ball's surface, so the pressure there must be below atmospheric. An inward bend (accelerating toward the surface) causes the air's speed to increase and its pressure to decrease. The speed-up is indicated by the narrowing separation of the streamlines, and the pressure drop is indicated by the color shift toward the red end of the spectrum.

The laminar airflow continues around to the back of the ball, where the air again bends outward, so that it slows down and its pressure rises. Just as at the front of the ball, the air pressure behind the ball is actually greater than atmospheric pressure.

It seems strange that the air pressure can be different at different points on the ball, but that is what happens in a flowing stream of air. It's particularly remarkable that low-pressure air at the sides of the ball is able to flow around to the back of the ball, where the pressure is higher! Your intuition may tell you that air should flow from high pressure to low pressure, not the reverse. But when you think that, you forget inertia. The low-pressure air flowing past the sides of the ball has enough forward momentum to carry it around to the back of the ball. When the airflow moves from high pressure to low pressure, it speeds up. When the airflow moves from low pressure to high pressure, it slows down. Everything makes sense.

The airflow around the ball is symmetric, and the forces exerted by air pressure on the ball are also symmetric. These pressure forces cancel one another perfectly so that the ball experiences no net pressure force. In particular, the high pressure in front of the ball is balanced by the high pressure behind it. As a result of this symmetric arrangement, the only aerodynamic force acting on the ball is **viscous drag**—the frictionlike downstream force caused by layers of viscous air sliding across the ball's surface (see ❑).

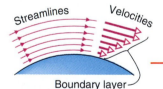

Fig. 5.2.2 Airflow past a surface is laminar until the Reynolds number exceeds about 2000. The freely flowing airstream then becomes turbulent, leaving only the thin layer of laminar flow near the surface that is shown here. This boundary layer is slowed by viscous drag and doesn't become turbulent until the Reynolds number exceeds about 100,000.

Check Your Understanding #1: Smooth Flow in a Stream

When water in a stream flows slowly past a small rock, the water in front of the rock slows down and its increased pressure lifts the water level slightly. The water level behind the rock also rises slightly. Explain.

When a Ball Moves Fast: Turbulent Airflow

Not all objects experience laminar airflow. Turbulence is everywhere, particularly in sports, bringing with it a new type of drag force. When the air flowing around a ball is turbulent, the air pressure distribution is no longer symmetric and the ball experiences **pressure drag**—the downstream force exerted by unbalanced pressures in the moving air. The pressure imbalance produces a net force on the ball that slows its motion through the air.

A ball can experience turbulent airflow and pressure drag when its Reynolds number exceeds about 2000. The Reynolds number, introduced in the previous section, combines the ball's size and speed with the air's density and viscosity to give an indication of whether the airflow is dominated by viscosity or inertia. At low Reynolds numbers, the air's viscosity supports laminar flow over the ball's surface. But at high Reynolds numbers, the air's inertia prevents it from following the ball's curves and it begins to swirl about. This turbulence, however, won't start until something triggers it. Something must cause the first swirling vortex in the air, and that something is viscosity.

To understand viscosity's role, we must look at the air near the ball's surface. Even in a strong wind, air touching the surface is held stationary by viscous forces, which also slow down a thin **boundary layer** of nearby air (Fig. 5.2.2). Discovered by Ludwig Prandtl ❏ with help from Gustave Eiffel (Fig. 5.2.3), this boundary layer moves more slowly and has less total energy than the freely flowing air farther from the surface.

As air flows toward the back of the ball, it travels through an **adverse pressure gradient**—a region of rising air pressure that causes the air to decelerate. While air outside the boundary layer has enough total energy to make it to the back of the ball, air in the boundary layer does not. At low Reynolds numbers, the entire airstream helps to push the boundary layer around the ball. But at high Reynolds numbers, viscous forces between the freely flowing airstream and the boundary layer are too weak to keep the boundary layer moving forward. The boundary layer eventually comes to a stop and reverses directions. This boundary layer reversal disrupts the flow so completely that the freely flowing air separates from the surface and leaves a large turbulent wake or air pocket behind the ball (Fig. 5.2.4).

Because of this turbulent wake, the air behind the ball no longer slows down and its pressure no longer rises. The swirling air upsets the perfect cancellation of pressure forces. Since the air in front of the ball exerts more pressure and force on the ball than the air behind it, the ball experiences pressure drag. In effect, the ball drags the air in the turbulent wake along with it.

Pressure drag slows the flight of almost any ball moving faster than a snail's pace. The pressure drag force is roughly proportional to the cross-sectional area of the turbulent air pocket and to the square of the ball's speed through the air. For a ball moving at a moderate speed, the air pocket is about as large around as the ball and the ball experiences a large pressure drag force.

Fig. 5.2.3 Early experiments in aerodynamics were performed by Gustave Eiffel (French engineer, 1832–1923), who designed the tower that bears his name. In the 1890s, Eiffel dropped objects of various sizes and shapes from his tower and measured the drag that they experienced. His work was used by Prandtl to explain the reduction in drag that accompanies the appearance of turbulent boundary layers.

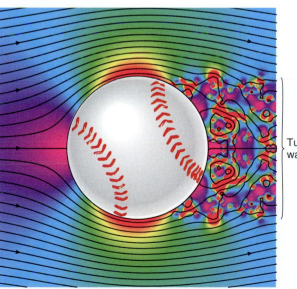

Fig. 5.2.4 When a ball's speed gives it a Reynolds number between about 2000 and 100,000, flow problems in the laminar boundary layer cause the main airflow to separate from the ball's surface, leaving a large turbulent wake. The average pressure behind the ball remains low, and the ball experiences a large pressure drag.

Turbulent wake

➤ **Check Your Understanding #2: Leaving No Trace**

When your canoe coasts extremely slowly across the water of a still lake, it leaves almost no trail in the water behind it. However, when you paddle it swiftly through the water, the canoe leaves a rippling wake. Explain this difference.

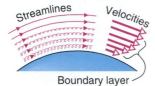

Fig. 5.2.5 When the Reynolds number exceeds about 100,000, the boundary layer of air flowing past a surface becomes turbulent. This whirling gas brings in extra energy from the freely flowing airstream and can travel deep into a region of increasing pressure.

The Dimples on a Golf Ball

If this were the whole story, you would never hit a home run at a baseball game or a 250-yard drive on the golf course. But inertia has yet another card to play. At very high Reynolds numbers the boundary layer itself becomes turbulent (Fig. 5.2.5). Because it's whirling about and exchanging energy with the freely flowing air nearby, air in a turbulent boundary layer has more total energy than air in a laminar boundary layer. This extra energy allows the turbulent boundary layer to flow farther around the back of the ball before the rising pressure there stops its forward motion. The freely flowing air follows this turbulent boundary layer and while the two eventually separate from the back of the ball, the turbulent wake they create is relatively small (Fig. 5.2.6).

As a result of this smaller air pocket, the pressure drag is reduced from what it would be without the turbulent boundary layer. The effect of replacing the laminar boundary layer with a turbulent one is enormous; it's the difference between a golf drive of 70 yards and one of 250 yards! The effects of Reynolds number on the airflow around a ball are summarized in Table 5.2.1.

Delaying the airflow separation behind the back of the ball is so important to distance and speed that the balls of many sports are designed to encourage a turbulent boundary layer (Fig. 5.2.7). Rather than waiting for the Reynolds number to exceed 100,000, the point near which the boundary layer spontaneously becomes turbulent, these balls "trip" the boundary layer deliberately (Fig. 5.2.8). They introduce some impediment to laminar flow, such as hair or surface irregularities, which causes the air near the ball's surface to tumble about and become turbulent. The drop in pressure drag more than makes up for the small increase in viscous drag. That's why a tennis ball has fuzz and a golf ball has dimples.

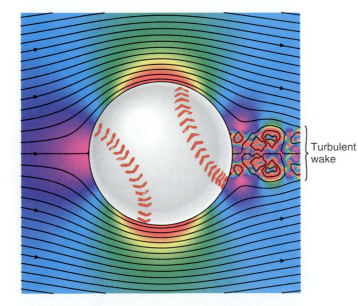

Fig. 5.2.6 When a ball travels fast enough that its Reynolds number exceeds 100,000, its boundary layer becomes turbulent. This turbulent layer travels much of the way around the back of the ball before it separates from the surface. The freely flowing air follows it, and the two leave a relatively small turbulent wake. The ball experiences only a modest pressure drag.

Table 5.2.1	The Effects of Reynolds Number on the Airflow Around a Ball or Other Object		
REYNOLDS NUMBER	BOUNDARY LAYER	TYPE OF WAKE	MAIN DRAG FORCE
<2000	Laminar	Small laminar	Viscous
2000–100,000	Laminar	Large turbulent	Pressure
>100,000	Turbulent	Small turbulent	Pressure

So how much does drag affect balls in various sports? For those that involve rapid movements through air or water, the answer is quite a bit. Drag forces increase dramatically with speed; as soon as a turbulent wake and pressure drag appear, the drag force increases as the square of a ball's speed. As a result, baseball pitches slow significantly during their flights to home plate, and the faster they're thrown, the more speed they lose. A 90-mph fastball loses about 8 mph *en route,* while a 70-mph curve ball loses only about 6 mph.

A batted ball fares slightly better because it travels fast enough for the boundary layer around it to become turbulent, an effect that appears at around 160 km/h (100 mph). While the resulting reduction in drag explains why it's possible to hit a home run, the presence of air drag still shortens the distance the ball travels by as much as 50%. Without air drag, a routine fly ball would become an out-of-the park home run. To compensate for air drag, the angle at which the ball should be hit for maximum distance isn't the theoretical 45° above horizontal discussed in Section 1.2. Because of the ball's tendency to lose downfield velocity, it should be hit at a little lower angle, about 35° above horizontal (Fig. 5.2.9).

Since the ball loses much of its horizontal component of velocity during its trip to the outfield, a long fly ball tends to drop almost straight down as you catch it. Gravity causes it to move downward, but drag almost stops its horizontal motion away from home plate. Drag also limits the speed of a falling ball to about 160 km/h (100 mph). That's the baseball's **terminal velocity,** at which the upward drag force exactly balances its downward weight. Even if you drop a baseball from an airplane, its speed will not exceed this value.

Check Your Understanding #3: Designing a Great Sports Car

As an automobile designer, your job is to minimize the aerodynamic drag experienced by the car on which you are working. Where should you try to locate the point at which the airflow separates from the car?

Curve Balls and Knuckle Balls

The drag forces on a ball push it downstream, parallel to the onrushing air. But in some cases, the ball may also experience lift forces—forces that are exerted per-

Fig. 5.2.7 Early golf balls (*left*) were handmade of leather and stuffed with feathers. Golf became popular when cheap balls made of a hard rubber called gutta-percha became available. But new, smooth "gutties" didn't travel very far; they flew better when they were nicked and worn. Manufacturers soon began to produce balls with various patterns of grooves on them (*bottom* and *right*), and these balls traveled dramatically farther than smooth ones. Modern golf balls (*top*) have dimples instead of grooves.

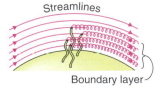

Fig. 5.2.8 The boundary layer can be made turbulent at Reynolds numbers below 100,000 by "tripping" it with obstacles such as fuzz or dimples.

Fig. 5.2.9 Air drag slows the flight of a batted ball so that the ideal angle at which to hit it isn't the theoretical 45° of Fig. 1.2.7. An angle of roughly 35° above horizontal will achieve the maximum distance.

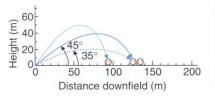

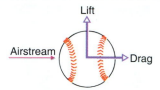

Fig. 5.2.10 The two types of aerodynamic forces exerted on objects by air are drag and lift. Drag is exerted parallel to the onrushing airstream and slows the object's motion through the air. Lift is exerted perpendicular to that airstream so that it pushes the object to one side or the other. Lift is not necessarily in the upward direction.

pendicular to the airstream (Fig. 5.2.10). To experience drag, the ball only has to slow the airstream down; to experience lift, the ball must deflect the airstream to one side or the other. Although its name implies an upward force, lift can also push the ball toward the side or even downward.

Curve balls and knuckle balls both use lift forces. In each of these famous baseball pitches, the ball deflects the airstream toward one side and the ball accelerates toward the other. Action and reaction—the air and the ball push off one another. Getting the air to push the ball sideways is no small trick. Explaining it isn't easy either, but here we go.

A curve ball is thrown by making the ball spin rapidly about an axis perpendicular to its direction of motion. The choice of this axis determines which way the ball curves. In Fig. 5.2.11, the ball is spinning clockwise, as viewed from above. With this choice of rotation axis, the ball curves to the pitcher's right because the ball experiences two lift forces to the right. One is the Magnus force, named after the German physicist H. G. Magnus (1802–1870) who discovered it. The other is a force we will call the wake deflection force.

The **Magnus force** occurs because the spinning ball carries some of the viscous air around with it (Fig. 5.2.11a). The steady-state flow pattern that eventually forms around this ball is asymmetric: the airstream that moves with the turning surface is much longer than the airstream that moves opposite that surface. The longer airstream has a prolonged inward bend and thus a long region of low pressure, while the shorter airstream has mostly an outward bend with its associated high pressure. Because the pressure forces on the ball's sides don't balance one another, the ball experiences a sideways force. This Magnus force points toward the low-pressure side and deflects the ball in that direction.

In laminar flow, the Magnus force is the only lift force acting on a spinning object. But a pitched baseball has a turbulent wake behind it and is also acted on by the **wake deflection force.** The ball's rotation pulls air around the ball on the side turning toward the pitcher and delays separation of the airstream on that side (Fig. 5.2.11b). But on the side of the ball turning toward the batter, the ball's rotation

Fig. 5.2.11 A rapidly rotating baseball experiences two lift forces that cause it to curve in flight. (a) The Magnus force occurs because air flowing around the ball in the direction of its rotation makes a long inward bend and its pressure there drops. (b) The wake deflection force occurs because air flowing around the ball in the direction of its rotation remains attached to the ball longer and the ball's wake is deflected.

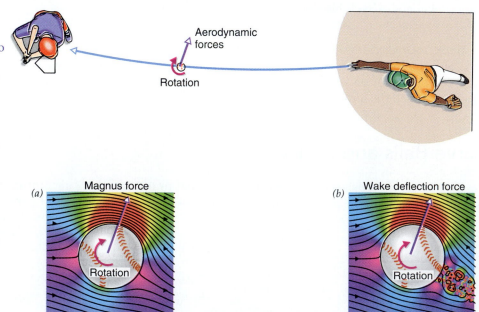

slows air in the boundary layer and hastens separation of the airstream. The overall wake of air behind the ball is thus deflected toward the side of the ball turning toward the batter. Since the ball has deflected the airstream one way, the airstream must push it the other way. The wake deflection force and the Magnus force both deflect the ball in the same direction.

Of these two forces, the wake deflection force is probably the more important for a curve ball, although the Magnus force is usually given all the credit. A skillful pitcher can make a baseball curve about 0.3 m (12 inches) during its flight from the mound to home plate—the more spin, the more curve. The pitcher counts on this change in direction to confuse the batter. The pitcher can also choose the *direction* of the curve by selecting the axis of the ball's rotation. The ball will always curve toward the side of the ball that is turning toward the pitcher. When thrown by a right-handed pitcher, a proper curve ball curves down and to the left, a slider curves horizontally to the left, and a screwball curves down and to the right.

When the pitcher throws a ball with backspin, so that the top of the ball turns toward the pitcher, the ball experiences an upward lift force. In baseball this force isn't strong enough to overcome gravity, but it does make the pitch hang in the air longer than normal. And in golf, where the club can give the ball enormous backspin, the ball really does lift itself upward so that it flies down the fairway like a glider.

However, there are some cases when a ball's behavior stems from its *lack* of spin. In baseball, for example, a knuckle ball is thrown by giving the ball almost no rotation. Its seams are then very important. As air passes over a seam, the flow is disturbed so that the ball experiences a sideways aerodynamic force: a lift force. The ball flutters about in a remarkably erratic manner. Releasing the ball without making it spin is difficult and requires great skill. Pitchers who are unable to throw a knuckle ball legally sometimes resort to lubricating their fingers so that the ball slips out of their hands without spinning. Like its legal relative, the so-called "spitball" dithers about and is hard to hit. The same is true for a scuffed ball.

more spinning balls

knuckle balls

rifle bullets

Check Your Understanding #4: Center Court

One of the most difficult and effective strokes in tennis is the topspin lob, in which the top of the ball spins away from the player who hit it. Which way is the lift force on this ball directed?

Frisbees

Even if there were no such thing as lift, baseball and golf would still look familiar. But sports involving flying disks would not be the same at all. Frisbees and Aerobies are held aloft by aerodynamic lift and would fall like stones without it.

A Frisbee's lift arises from its shape and orientation as it flies through the air (Fig. 5.2.12). With its top surface bowed upward and its leading edge higher than its trailing edge, the Frisbee develops an asymmetric pattern of airflow. The airstream passing over the top of the Frisbee is particularly long and involves an extended inward bend. This bend results in low pressure, fast-moving air above the Frisbee. In contrast, the airstream passing under the Frisbee completes a short outward bend that leads to high pressure, slow-moving air. The Frisbee experiences an overall pressure force that pushes it upward and slightly toward the rear—a strong upward lift force and a weak rearward drag force. The lift force keeps the Frisbee aloft while the drag force gradually slows its forward motion.

Fig. 5.2.12 As this Frisbee flies forward, its curved shape and orientation allow it to deflect the passing air downward. The air reacts by pushing upward on the Frisbee, supporting it in flight.

Like an airplane wing, the Frisbee is an **airfoil,** an aerodynamically engineered surface that's designed to obtain particular lift and drag forces from the air flowing around it. But lift doesn't appear instantly when the Frisbee is thrown; first the airflow around the Frisbee must lose its symmetry. Let's look at how that happens.

The airstream encounters the airfoil near its *leading edge* and leaves the airfoil near its *trailing edge*. At the start of the Frisbee's flight, the airstream splits apart near the bottom of the Frisbee's leading edge and flows over and under the Frisbee (Fig. 5.2.13a). Initially, these two airflows travel at the same average speed and follow equivalent paths around the Frisbee before rejoining near the top of the Frisbee's trailing edge. Each airflow experiences one inward bend with its low pressure and one outward bend with its high pressure, and this symmetry leads to an overall cancellation of pressure forces on the Frisbee. It experiences neither lift nor pressure drag. The Frisbee doesn't deflect the airstream downward and the airstream doesn't push the Frisbee upward.

However, air flowing upward from beneath the Frisbee must turn abruptly around the trailing edge. This portion of the flow pattern is unstable and blows away from the back of the Frisbee partway through the throw. The air here is rotating and becomes a vortex as it's "shed" from the trailing edge of the Frisbee (Fig. 5.2.13b). The Frisbee is left with a new, stable pattern of airflow that gives it upward lift, perpendicular to the horizontal airstream (Fig. 5.2.13c).

frisbee stability

When the Frisbee sheds its vortex, the air flowing over it acquires a higher average velocity than the air flowing under it. This difference in speeds reflects the fact that the airflow around the Frisbee now has angular momentum. The whirling vortex took away angular momentum with it, so the air passing around the Frisbee must be left with an equal amount of angular momentum in the opposite direction. Although the air flowing over the Frisbee is always heading toward its trailing edge, the fact that the top flow is faster than the bottom flow gives it angular momentum. In Fig. 5.2.13, the shed vortex takes with it counterclockwise angular momentum, leaving the airflow around the Frisbee with clockwise angular momentum. Overall, angular momentum is conserved.

drag and sports

> **Check Your Understanding #5: Going Long**
An Aerobie is a thin, ring-shaped flying disk that creates a much smaller wake than a Frisbee. Why does an Aerobie travel so much farther than a Frisbee when the two are thrown with the same initial speed?

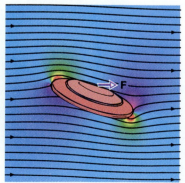
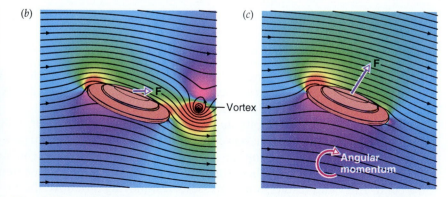

Fig. 5.2.13 (*a*) When a Frisbee starts its flight, the airflow around it is symmetric and it experiences no lift, only drag. (*b*) The symmetric airflow is unstable, and a vortex is soon shed from the Frisbee's trailing edge. (*c*) The Frisbee's wake is then deflected downward, and the Frisbee experiences an upward lift force. The aerodynamic forces on the Frisbee are indicated by the arrow **F.**

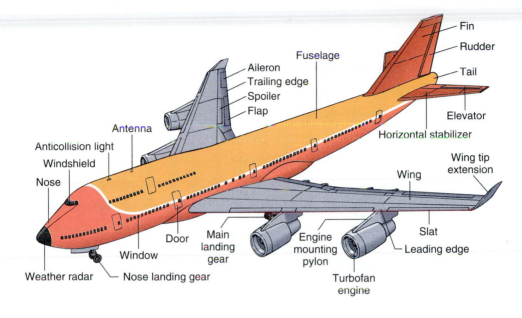

Fin
Rudder
Fuselage
Tail
Aileron
Trailing edge
Spoiler
Flap
Elevator
Antenna
Horizontal stabilizer
Anticollision light
Wing tip extension
Windshield
Wing
Nose
Slat
Door
Leading edge
Main landing gear
Engine mounting pylon
Window
Weather radar
Nose landing gear
Turbofan engine

Airplanes and Rockets

We have now set the stage for the ultimate aerodynamic machines, airplanes and rockets. Freed from contact with the ground, these objects are affected only by aerodynamic forces and gravity, hopefully in that order. Despite their complex appearances, airplanes and rockets employ physical principles that we have already successfully examined. But while this section revisits many familiar concepts, it also explores new territory. For example, you may have already figured out what type of aerodynamic force holds an airplane up, but what type of aerodynamic force keeps it moving forward?

Questions to Think About: *Why are the wings of small, propeller-driven aircraft relatively large and bowed in comparison to those of jets? Why does a commercial airplane extend slats and flaps from its wings during takeoffs and landings? What pushes airplanes and rockets forward in flight? How can some airplanes fly upside down? Why do most fast commercial aircraft employ jet engines and not propellers? Why does a rocket still work in empty space? How fast can a rocket go?*

Experiments to Do: *The best experiment for this section is to take a plane flight or at least to visit the airport and watch the planes. Of course, a rocket flight would be nice, too, but that's a bit much to expect.*

As you sit in the plane during takeoff, feel the plane accelerating forward. If you're on a commercial jet, notice that the slats and flaps on the airplane's wings are extended during takeoff, making the wings wider and more curved. How could this increased width and curvature help the plane take off? The pilot holds the plane on the ground until it reaches the proper speed, then quickly tips it upward into the air. An invisible vortex of air peels away from the trailing edge of the wing and the plane lifts off the ground.

Once airborne, the airplane retracts its landing gear, slats, and flaps. Watch the trailing edge of each wing as the plane turns or changes altitudes; you'll see various surfaces there move up or down. Similar motions occur on the tail. How do these surfaces control the plane's orientation?

Near its destination, the airplane prepares for landing. Again the slats and flaps are extended. Watch as spoilers on the tops of the wings pop up and down noisily. How do these surfaces affect the drag force on the plane? The plane's landing gear extends, and it touches down on the runway. The propellers or jet engines abruptly begin to slow the airplane, assisted by the spoilers on the wings. Feel the plane accelerating backward. Another invisible vortex of air peels away from the trailing edge of the wing, rotating in the opposite direction from the first vortex, and the flight is over.

Airplane Wings

By now you've probably realized that the lift on an airplane wing is the same as that on a Frisbee. The wing is an airfoil, shaped and oriented so that, during flight, air follows a long inward bend over the top surface and a short outward bend under the bottom surface (Fig. 5.3.1). These bends are associated with pressure changes and are responsible for the upward lift force that suspends the airplane in the sky.

To understand how the wing develops this lift, and to review concepts we first encountered with Frisbees, imagine yourself in an airplane that has just begun rolling down the runway. From your perspective, air is beginning to flow past the airplane's wing. This moving air separates into two streams, one traveling over the wing and the other under it. Initially both airstreams move at equal average speeds, bending inward and outward in similar amounts and providing the wing with no overall lift (Fig. 5.3.2).

But the lower airstream must make a sharp upward bend around the wing's trailing edge. Because of the air's limited total energy and speed, this sudden bend is unstable. The kink in the airflow drifts backward along the wing's trailing edge until a vortex of air finally peels away from the wing. A new, stable flow pattern forms in which both airstreams pass smoothly away from the wing's trailing edge (Fig. 5.3.1a,b), a situation named the Kutta condition after the German mathematician M. Wilhelm Kutta (1867–1944).

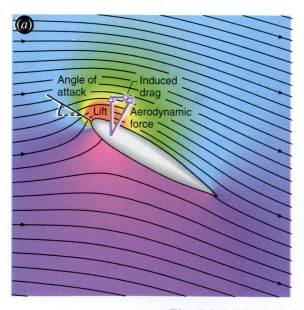

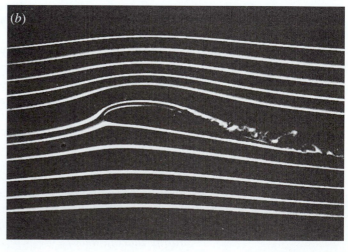

Fig. 5.3.1 (*a*) An airplane wing is an airfoil, shaped and oriented so that air follows a long inward bend over its top surface and a short outward bend under its bottom surface. The wing experiences a large aerodynamic force that points upward and slightly downstream. The upward component of this force is lift. The downstream component is induced drag. (*b*) Smoke trails in a wind tunnel show the airflow past a wing.

Fig. 5.3.2 The initial airflow around a wing is symmetric and produces no lift. Lift doesn't appear until the wing sheds a vortex and the airflow around the wing acquires angular momentum.

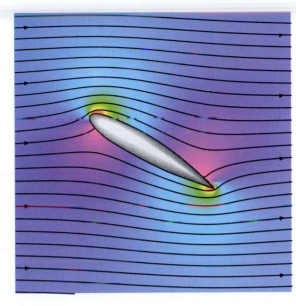

The airstream flowing over the wing is now longer than the airstream flowing under it. The upper airstream bends inward, toward the wing, and experiences a drop in pressure (a shift toward red) and an increase in speed (narrowly spaced streamlines). In contrast, the lower airstream bends outward, away from the wing, and experiences a rise in pressure (a shift toward violet) and a decrease in speed (widely spaced streamlines). Since the upper airstream is flowing faster on average than the lower airstream, the airflow around the wing has angular momentum in the clockwise direction. And because the pressure is now higher under the wing than over it, this new flow pattern produces upward lift. The air now supports your plane and up you go.

Another way to think about this lift is as a deflection of the airflow. Air approaches the wing horizontally but leaves heading somewhat downward. To cause this deflection, the wing must push the airstream downward. In reaction, the airstream pushes the wing upward and produces lift.

But the force on the wing is not directly upward; as was the case with the Frisbee, it tilts slightly downstream. The vertical component of this pressure force is lift, but the downstream horizontal component is a new type of drag force—induced drag. **Induced drag** is a consequence of energy conservation: the deflected air carries kinetic energy away from the plane, and so the plane must slow down. To minimize induced drag, a plane should move as much air as possible as slowly as possible. That way it obtains its lift while giving the deflected air very little kinetic energy. Larger wings move more air and thus produce less induced drag.

But bigger isn't always better, and large wings also experience substantial viscous drag. Therefore, choosing the right size for an airplane's wings is complicated. Moreover, the amount of lift experienced by a wing increases with the curvature of that wing's top surface and its *airspeed*—its speed relative to the air it encounters. Small propeller airplanes that move slowly through the air need relatively large, highly curved wings to support them, while commercial and military jets fly faster and can get by with relatively small, moderately curved wings.

But even at constant airspeed, a wing's lift can be adjusted by varying its angle of attack—the angle at which it approaches the airstream. The larger the angle of attack, the more the two airstreams bend and the greater the lift. Because the wings are rigidly attached to the plane, the pilot tips the nose of the plane upward to in-

tilting the passengers

crease the lift and downward to reduce the lift. That's why raising the plane's nose during takeoff is what finally makes the plane leap into the air.

Since lift depends so strongly on a wing's angle of attack, some planes can be flown upside down. As long as the inverted wing is tilted properly, it obtains upward lift and supports the plane. But because its curvature is inverted, it must have a very large angle of attack in order to obtain lift. Nonetheless, stunt fliers regularly fly upside down at air shows.

Check Your Understanding #1: Blowing in the Wind

The sail of a small sailboat bows forward and outward so that wind traveling around the sail's outside surface completes an inward bend (toward the sail), while wind traveling across its inside surface completes an outward bend (away from the sail). How does this arrangement propel the sailboat across the water?

Lift Has Its Limits: Stalling a Wing

However, there's a limit to how much lift the pilot can obtain by increasing the wing's angle of attack. Beyond a certain angle, the airflow over the top of the wing separates from its surface and the wing experiences **stalling.** This separation occurs when air in the upper boundary layer is unable to flow into the rising pressure beyond the wing's thickest point. As we saw with balls, when the boundary layer stops moving forward, it causes the entire airflow to separate from the surface.

During a stall, the separated airstream over the top of the wing leaves a billowing storm of turbulence beneath it (Fig. 5.3.3). This airflow separation is an aerodynamic catastrophe for the airplane. The wing loses much of its lift because the

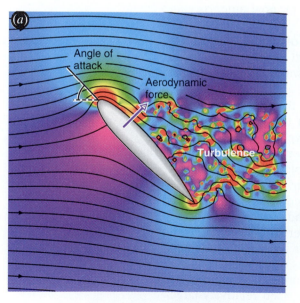
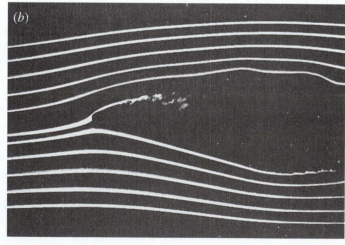

Fig. 5.3.3 (*a*) A wing stalls when the airstream over the top of the wing separates from its surface. A turbulent air pocket forms above the wing, making it much less efficient. The wing's lift decreases because the average pressure above the thickest part of the wing becomes higher, and the drag increases because the average pressure above the trailing edge becomes lower. (*b*) Smoke trails in a wind tunnel show that the air separates from the surface and becomes turbulent as it flows over a stalled wing.

pressure above the thickest part of the wing increases. And the appearance of a turbulent wake heralds the arrival of severe pressure drag. The plane slows dramatically and drops like a rock.

To avoid stalling, pilots keep the angle of attack within a safe range. But the possibility of stalling also limits the minimum speed at which the airplane will fly. As the airplane slows down, the pilot must increase its angle of attack to maintain lift. Below a certain speed, the airplane can't obtain enough lift without tilting its wings until they stall. It can no longer fly.

To avoid stalling, a plane must never fly slower than this minimum speed, particularly during landings and takeoffs. For a small, propeller-driven plane with highly curved wings, the minimum flight speed is so low that it's not a problem. For a commercial jet, however, the minimum airspeed is about 220 km/h (140 mph). Airplanes taking off or landing this fast would require very long runways on which to build up or get rid of speed. Instead, commercial jets have wings that can change shape during flight. Slats move forward and down from the leading edges of the wings, and flaps move back and down from the trailing edges (Fig. 5.3.4). With both slats and flaps extended, the wing becomes larger and more strongly curved, similar to the wings of a small propeller plane, and the minimum safe airspeed drops to a reasonable 150 km/h (95 mph). Vanes near the flaps also emerge during landings to direct high-energy air from beneath the wings onto the flaps. These jets of air keep the boundary layers moving downstream and help prevent stalling. (For another approach to stall prevention, see ❏.)

Once a commercial jet lands, flat panels on the top surfaces of its wings are tilted upward and cause the airflow to separate from the tops of the wings. The resulting turbulence created by these spoilers reduces the lift of the wings and increases their drag. On the ground, the airplane doesn't need lift from the wings, so the spoilers help to keep the plane from flying again. They also slow it down. The spoilers are sometimes used in flight to slow the plane and help it descend rapidly toward an airport.

But what happens to the airflow at the end of a wing? With low pressure above the wing and high pressure below it, air tends to flow around the ends of the wing from bottom to top. This air is soon left behind, but not before it has acquired lots of angular momentum and kinetic energy. A swirling vortex thus emerges from the tip of each wing and trails behind the plane for several kilometers, like an invisible tornado. A wing-tip vortex from a jumbo jet can flip a small aircraft that flies through it or give passengers in a much larger plane an unexpected thrill. For safety, air traffic controllers are careful to keep planes from flying through one another's wakes and schedule them at least 90 s apart on runways. Some modern airplanes have wing-tip extensions that reduce these vortices, both to save energy and to diminish the hazard.

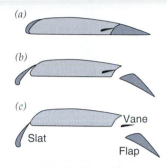

Fig. 5.3.4 At cruising speed, an airplane's wings are moderately curved airfoils (*a*). But during takeoffs (*b*) and landings (*c*), slats are extended from the leading edges and flaps from the trailing edges. The airfoils become much more highly curved, generating more lift at low speeds. During landing, a vane is also extended for boundary layer control to prevent stalling.

❏ Airplane designers can reduce the dangers of stalling by adding special boundary layer control devices to their aircraft. Narrow metal strips called *vortex generators,* which stick up from the surfaces of wings, introduce turbulence into the boundary layers over the wings. This turbulent flow allows higher energy air to mix with the boundary layers so that they can continue forward into rising pressure. This process helps keep the airstreams attached to the surfaces.

Check Your Understanding #2: Stunt Flying

A pilot normally tips the plane's nose upward in order to gain altitude. But if the pilot tries to make the plane rise too quickly, the plane will suddenly begin to drop. What is happening?

Propellers

For a plane to obtain lift, something must push the plane forward so that air flows across its wings. The two most common devices for this purpose are propellers and jet engines.

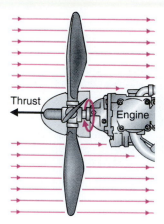

Fig. 5.3.5 A propeller behaves like a rotating wing. As the propeller turns, its blades create lift in the forward direction. This lift pushes the engine and the aircraft forward through the air, so that it's called thrust.

❑ One of the principal sources of noise in submarines is the turbulence created by their propellers. To reduce this turbulence, the propellers of modern nuclear submarines are designed to avoid water flow separation and stalling.

❑ In addition to achieving the first self-propelled flight of an airplane in 1903, **Orville (1871–1948) and Wilbur (1867–1912) Wright (American aviators)** were exceptionally accomplished aerodynamicists. In 1902, Wilbur was the first person to recognize that a propeller is actually a rotating wing. Propellers up until his time were little more than rotating paddles, more effective at stirring the air than propelling the plane. Wilbur's aerodynamically redesigned propeller made flight possible and dominated aircraft design for a decade.

A propeller is an assembly of rotating wings. Extending from its central hub are two or more blades that together form a sophisticated fan (Fig. 5.3.5). These blades have airfoil cross sections and are designed to create a forward force when the propeller turns and the blades move through the air.

As a propeller blade slices through the air, air flows around it just as air flows around the wings of the plane. Because of the shape and orientation of the blade, the air flowing over the front surface of the blade executes an inward bend while air flowing over the rear surface makes an outward bend (Fig. 5.3.6). The pressure in front of the propeller is thus less than behind it, so the propeller experiences a forward lift force. Because it propels the plane forward, that lift force is called **thrust.**

The propeller blades have all the features, good and bad, of airplane wings. Their thrust increases with size, front-edge curvature, angle of attack, and airspeed; in other words, the larger the propeller, the faster it turns, and the more its blades are angled into the wind, the more thrust it produces. And like a wing, a propeller stalls when the airflow separates from the front surfaces of its blades. It suddenly becomes more of an air-mixer than a propeller. This stalled-wing behavior was the standard operating condition for air and marine propellers (see ❑) before the work of Wilbur Wright in 1902 (see ❑). The Wrights were among the first people to study aerodynamics with a wind tunnel (Fig. 5.3.7) and their methodical and scientific approach to aeronautics allowed them to achieve the first powered flight (Fig. 5.3.8).

A propeller also experiences induced drag, which tends to slow its rotation. An engine must continuously do work on the propeller to keep it turning. As the propeller's thrust pushes the plane through the air, the air extracts energy from the propeller with induced drag, so something must do the work of turning the propeller. Propellers are driven by high-performance reciprocating engines, like those found in automobiles, and the turbojet engines that we'll discuss later.

Propellers aren't perfect; they have three serious limitations. First, since a propeller exerts a torque on the air that passes through it, the air exerts a torque on the propeller. This torque can tip over a small plane. That's why many airplanes use pairs of propellers that turn in opposite directions so the net torque on the plane is zero. Planes with a single propeller usually locate it in front, so that the spinning air can return some angular momentum to the plane as it passes over the wings.

A second problem with propellers is that their thrust diminishes as the plane's forward speed increases. When the airplane is stationary, the propeller blade turns past motionless air (Fig. 5.3.9a). But when the airplane is moving very fast, the air ap-

Fig. 5.3.6 As the propeller blade rotates, the flow of air around it creates a low pressure in front of it (left) and a high pressure behind it (right). The blade experiences a lift force that pushes the propeller and plane forward (toward the left). Induced drag tends to slow the rotation of the propeller.

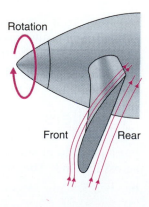

Fig. 5.3.7 The Wright brothers were accomplished aero-dynamicists, using this wind tunnel to study and perfect wings and propellers for their airplanes.

Fig. 5.3.8 The era of powered flight began at 10:35 a.m. on Dec. 17, 1903, when the Wright *Flyer* lifted Orville Wright into the air over Kitty Hawk, NC. His brother Wilbur stands beside him in this unique photograph of that first powered flight.

proaches that same propeller blade from the front of the plane (Fig. 5.3.9*b*). To retain lift or thrust at higher airspeeds, the propeller blade must swivel forward, increasing its pitch to meet this onrushing airstream. The blades themselves have a twisted shape to accommodate the variations in airspeed along their lengths, from hub to tip.

The third and most discouraging problem with propellers, especially in high-speed aircraft, is drag. To keep up with the onrushing air at high airspeeds, the propellers must turn at phenomenal rates. The tips of the blades travel so fast that they exceed the **speed of sound**—the fastest speed at which a fluid such as air can convey forces from one place to another. When the blade tip exceeds this speed, the air near the tip doesn't accelerate until the tip actually hits it. Instead of flowing smoothly around the tip, the air forms a **shock wave**—a narrow region of high pressure and temperature caused by the supersonic impact—and the propeller stalls. That's why propellers aren't useful on high-speed aircraft.

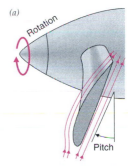

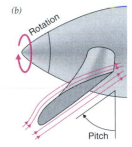

Fig. 5.3.9 At low air-speeds (*a*), the propeller blade approaches nearly sta-tionary air as it rotates. At high airspeeds (*b*), air rushes past the propeller, so the blade must swivel forward to meet it. The blade's angle of attack is called its pitch.

Check Your Understanding #3: Circulating the Air

The only difference between a fan and a propeller is what moves, the air or the object. Which side of each fan blade experiences the lowest air pressure, the in-let or the outlet side?

Jet Engines

Unlike propellers, jet engines work well at high speeds. While a propeller tries to operate directly in the stream of high-speed air approaching the plane, a jet engine first slows this air down to a manageable speed. To achieve this change in speeds, the jet engine makes wonderful use of Bernoulli's equation.

A turbojet engine is depicted in Fig. 5.3.10, p. 172. During flight, air rushes into the engine's inlet at about 800 km/h (500 mph), the speed of the plane. Once inside the inlet, the air slows down and its pressure increases, but its total energy is un-

Fig. 5.3.10 The turbojet operates by compressing incoming air with a series of fanlike blades. Fuel is mixed with the high-pressure air and the mixture is ignited. The high-energy/high-pressure air accelerates out the rear of the jet, does work on the turbines, and leaves at a greater speed than it had when it arrived. The engine has accelerated the air backward and experiences a thrust forward.

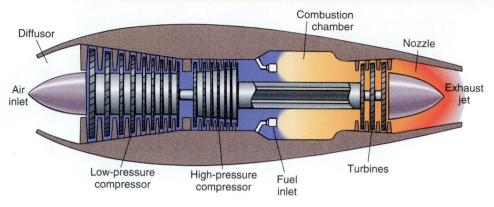

❑ Ramjets are jet engines that have no moving parts. Air that approaches the engine at supersonic speeds interacts with carefully tapered surfaces so that its own forward momentum compresses it to high density. The engine then adds fuel to this pressurized air, ignites the mixture, and allows the hot burned gas to expand out of a nozzle. The engine pushes this exhaust backward, and the exhaust propels the engine and airplane forward. Although the air enters the engine at supersonic speeds, it passes through the combustion chamber much more slowly. In a supersonic combustion ramjet or "scramjet," the fuel and air mixture flows through the combustion chamber at supersonic speeds. This motion makes it extremely difficult to keep the fuel burning because the flame tends to flow downstream and out of the engine. The flame can't advance through the mixture faster than the speed of sound, so it won't spread upstream fast enough to stay in the engine on its own.

changed. The air then passes through a series of fanlike compressor blades that push it deeper into the engine, doing work on it and increasing both its pressure and its total energy. By the time the air arrives at the combustion chamber, its pressure is many times atmospheric.

Now fuel is added to the air and the mixture is ignited. Since hot air is less dense than cold air, the hot exhaust gas takes up more space than it did before combustion. This hot exhaust gas pours out of the combustion chamber, traveling faster than when it entered. Its pressure is still very high and it accelerates out of the back of the engine, exchanging its pressure for speed. On its route out of the engine, the exhaust gas passes through a fanlike turbine. The turbine behaves like a windmill. The gas does work on this turbine and gives up a little of its energy in the process. The turbine drives the compressor for the incoming air.

Overall, the engine has slowed the air down, added energy to it, and then accelerated it back to high speed and its original pressure. Because the engine has added energy to the air, the air leaves the engine traveling faster than when it arrived. The jet engine has exerted a rearward force on the air to accelerate it rearward; the air has pushed back and produced a thrust that propels the airplane forward.

The turbojet isn't quite as efficient as it might be; it moves too little air and gives that air too much kinetic energy. To make the engine more efficient, it should move more air but eject that air at lower speeds.

The turbofan engine solves this problem by adding a huge fan to the front of a turbojet (Fig. 5.3.11), which provides the power needed to keep the fan turning. Air flowing into the turbofan engine's inlet slows down as it approaches the fan, and its pressure increases. The fan does work on this air so that it leaves the fan at a higher pressure than when it entered. While about 5% of this air then enters the turbojet engine, the vast majority of it accelerates out the back of the fan duct and sails off behind the airplane. It converts its pressure energy into kinetic energy and leaves the engine traveling faster than when it arrived. The fan has pushed on the air and the air has pushed back on the fan, producing forward thrust.

Once again, the engine has slowed the air down, added energy to it, and returned it to high speed and its original pressure. Again, the air leaves the engine traveling faster than when it arrived. But the turbofan engine moves more air than a simple turbojet engine, giving that air less energy and using less fuel. The huge fanlike engines on many jumbo jets are turbofans. (For another type of jet engine, see ❑.)

➤ **Check Your Understanding #4: Energy and a Jet Engine**

A jet engine somehow slows the air down, adds energy to it at low speed, and then returns the air to high speed. Why doesn't slowing the air down waste lots of energy?

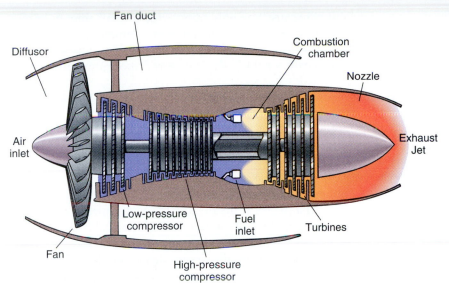

Fig. 5.3.11 The turbofan engine adds a giant fan to the shaft of a normal turbojet engine. Most of the air passing through the fan bypasses the turbojet and returns directly to the airstream around the engine. Because the fan does work on this air, it leaves the engine at a higher speed than it had when it arrived. The air has transferred forward momentum to the engine and the plane.

Rocket Propulsion

A jet engine obtains forward thrust by gathering outside air, adding energy to it, and pushing that air backward. What if you replaced the outside air with air from some internal supply? Would you still obtain thrust?

The answer is yes. You would now have a rocket engine. The only real difference between jets and rockets is that jets use outside gas to obtain thrust while rockets use an internal supply. Carrying its own gas supply allows a rocket to reach enormous speeds and to operate where no air is present, that is, in space. To understand how a rocket obtains thrust from its gas supply and why that thrust depends on the amount of gas ejected and on its speed, let's look at how Newton's third law, the one describing action and reaction, applies to rockets.

Imagine that you're sitting in the middle of a frozen pond with zero velocity and no momentum. It's a warm day and the wet ice is remarkably slippery. Try as you like, you can't seem to get moving at all. How can you get off the ice?

Because of your inertia, the only way you can start moving is if something pushes on you. You could wave your hands like a propeller and try to push the air in one direction, but instead you follow the ideas we discussed on page 61. You remove a shoe and throw it as hard as you can toward the east side of the pond (Fig. 5.3.12). As you throw the shoe, you exert a force on it with your hand. The shoe accelerates and heads off across the ice.

What happens to you? You head off toward the west side of the pond! You're moving because when you pushed the shoe toward the east side of the pond, it pushed you equally hard toward the west side of the pond. In the process, you transferred momentum to the shoe and it transferred momentum in the opposite direction to you. Momentum isn't being created or destroyed, it's only being redistributed. Even after you let go of the shoe, your combined momentum remains at zero. The shoe has as much momentum in one direction as you have in the other.

Of course, you are much more massive than the shoe, so it ends up traveling faster than you do. Momentum is the product of mass times velocity. The more massive the object, the less velocity it needs for the same amount of momentum. The shoe must travel much faster than you do for its momentum to exactly cancel your momentum. Still, you've achieved what you set out to do: you're sliding slowly toward the west side of the pond.

(a)

Man and shoe
(Stationary)

(b)

Man's velocity

Shoe's velocity

Man and shoe
(After throwing)

Fig. 5.3.12 A man who is holding a shoe while standing still on ice has zero momentum. Once he has thrown the shoe to the right, the shoe has a momentum to the right and the man has a momentum to the left. The total momentum of the man and shoe is still zero. Because the man is much more massive than the shoe, the shoe moves much faster than the man.

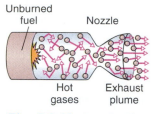

Unburned fuel   Nozzle

Hot   Exhaust
gases   plume

Fig. 5.3.13 A molecular picture of what happens in a chemical rocket engine. The engine burns its fuel in a confined chamber, and the exhaust gas flows out of a nozzle. The nozzle converts the random, thermal motions of the exhaust gas molecules into directed motion away from the rocket engine.

❏ **Swedish inventor and engineer Carl Gustaf de Laval's (1845–1913)** invention of the converging–diverging nozzle predates the modern development of rockets by several decades. He invented this nozzle as a way to make steam turbines more efficient and is credited with laying the foundation for all future turbine technology. De Laval is also known for his invention of the cream separator for milk.

❏ On January 13, 1920, *The New York Times* ran an editorial attacking Robert Goddard for proposing that rockets could be used for travel in space. With modest financial support from the Smithsonian Institution, Goddard was pioneering the development of liquid fuel rockets. The editorial began: "That Professor Goddard, with his 'chair' in Clark College and the countenancing of the Smithsonian Institution, does not know the relation of action to reaction, and of the need to have something better than a vacuum against which to react—to say that would be absurd. Of course he only seems to lack the knowledge ladled out daily in high schools."

Your final speed is limited because you only transferred a small amount of momentum to the shoe and thus received only a small amount of momentum in the opposite direction in return. If you'd been able to throw the shoe faster or if you'd thrown a whole box of shoes, you would have transferred more momentum and would be going faster.

Instead of throwing shoes, you'd have done better to throw very fast-moving gas molecules. Even at room temperature, the molecules in the air are traveling about 1800 km/h. When gas molecules are heated to roughly 2800 °C (5000 °F), as they are in a liquid-fuel rocket engine, they move about three times that fast. If you throw something in one direction at that kind of speed, you receive quite a lot of momentum in the other direction.

That's what a conventional rocket engine does (Fig. 5.3.13). It uses a chemical reaction to create very hot exhaust gas from fuels contained entirely within the rocket itself. What started as potential energy in the stored chemical fuels becomes thermal energy in the hot, burned exhaust gas. This thermal energy is mostly kinetic energy, hidden in the random motion of the tiny molecules themselves. The rocket engine's nozzle steers most of this random motion in one direction, and the engine obtains thrust in the opposite direction.

If you've ever watched the launch of a large rocket, you've probably noticed the huge, bell shaped nozzles through which the exhaust flows (Fig. 5.3.14). Each nozzle allows the rocket to extract as much energy and momentum as possible from its exhaust by releasing the exhaust at a pressure that's only slightly higher than the surrounding air pressure. The shape that makes the most efficient use of the gas's energy is a converging–diverging nozzle, called a de Laval nozzle after its Swedish inventor, Carl Gustaf de Laval ❏.

The first half of a de Laval nozzle converges, narrowing just like the water nozzle we discussed in Section 5.1. This analogy is appropriate because, at low speeds, the exhaust gas is almost incompressible like water. But when the gas reaches the narrowest part of the nozzle, it's traveling at the speed of sound and has become highly compressible. To reduce its pressure further, the gas's density must decrease. The nozzle stops becoming narrower and instead flares outward to allow the gas itself to expand. The gas's density drops and so does its pressure.

By the time the gas reaches the end of the nozzle, its pressure is very low and it has converted most of its thermal energy into kinetic energy directed away from the nozzle. Since the gas actually continues to burn even as it flows through the nozzle, its kinetic energy continues to rise. With the help of the de Laval nozzle, gas molecules leave the rocket at between 10,000 and 16,000 km/h.

As it creates this plume of exhaust, the rocket pushes the gas backward and gives it backward momentum. The gas completes the momentum transfer by pushing the rocket forward. The very act of ejecting the exhaust is all that's required to obtain forward thrust; the rocket doesn't need anything outside to push "against" and will operate perfectly well in empty space (see ❏). When it pushes hard enough on its exhaust, the rocket can not only support its own weight, it can even accelerate upward. The Space Shuttle weighs about 20,000,000 N at launch but its thrust is about 30,000,000 N. That means that the space shuttle can accelerate upward at about half the acceleration due to gravity! As the shuttle consumes its fuel, so that its weight and mass diminish, it can accelerate upward even more rapidly.

Check Your Understanding #5: A Rocket with a Head Start

When a missile is launched from beneath the wing of a fighter aircraft, what does it push against in order to accelerate forward?

The Ultimate Speed of a Rocket

It might seem that the ultimate speed of a rocket is limited by the exhaust velocity of the engine. Remarkably, there is no such limit. As long as the engine keeps pushing material backward, the rocket will continue to accelerate. However, to reach extremely high speed, the rocket must push the vast majority of its initial mass backward as exhaust. For example, if a rocket pushes 90% of its initial mass backward as exhaust, then the remaining portion (with only 10% of the initial mass) might be expected to head forward at about nine times the speed of the exhaust gas. That analysis is oversimplified, since the rocket speeds up as its engine operates. Nonetheless, its implications are correct. Taking into account the rocket's changing speed and the effect of this motion on the exhaust gas, the rocket can actually reach 2.3 times the speed of its exhaust gas. If it can eject more than 90% of its initial mass as exhaust, it can go even faster!

But there's a problem with trying to burn up and eject a huge fraction of the rocket's original mass as exhaust. It's difficult to construct a rocket that is 99.99% fuel. Instead, space-bound rockets use several separate stages, each stage much smaller than the previous stage. Once the first stage has used up all of its fuel, the whole stage is discarded and a new, smaller rocket begins to operate. In this manner, the rocket behaves as though it's ejecting almost all of its mass as rocket exhaust. With the help of stages and lots of fuel, rockets can travel substantially faster than their exhaust velocities and reach earth orbit or the solar system beyond. (For recent developments in single-stage rockets, see ❏ on p. 176.)

Fig. 5.3.14 The space shuttle's nozzles are designed to push its rocket exhaust downward as long glowing plumes. The gas pushes back, lifting the shuttle upward into space.

> **Check Your Understanding #6: Not Everything Is Disposable**
> The Space Shuttle doesn't have the staged look of expendable rockets. How does it manage to eject most of its launch mass as exhaust?

rocket efficiency, model rockets

Orbiting the Earth

Once it reaches a very high horizontal velocity, a spacecraft's engine stops firing and it begins to circle endlessly around the earth. With its engine off and no atmosphere pressing against it, the only significant force acting on the spacecraft is the earth's gravity. The spacecraft's weight causes it to accelerate toward the center of the earth so that, instead of traveling in a straight line at constant velocity, it travels in a huge elliptical loop around the earth.

The spacecraft is in orbit around the earth. An **orbit** is the path an object takes as it falls freely around a celestial object. The spacecraft accelerates directly toward the earth's center at every moment, but its huge horizontal speed prevents it from actually hitting the earth's surface. In effect, the spacecraft perpetually misses the earth as it falls (Fig. 5.3.15). To orbit the earth just above the atmosphere, a spacecraft must travel at the enormous speed of 7.9 km/s (about 17,800 mph) and will circle the earth once every 84 minutes.

However, as the spacecraft moves farther from the earth's surface, its **orbital period,** the time it takes to complete one orbit, increases significantly. First, the spacecraft must travel farther to complete the larger orbit so the trip takes longer. Second, the spacecraft must travel slower in order to follow a circular path around the earth because the earth's gravity becomes weaker with distance.

Fig. 5.3.15 Newton's drawing of a cannonball fired horizontally from the top of a tall mountain. As the cannonball's speed increases, it travels farther from the mountain before hitting the earth. Eventually, the cannonball moves so quickly that the curved earth drops away beneath it and it never hits the earth at all. The cannonball then orbits the earth.

❑ In 1989, the U.S. government began a program to develop a reusable rocket vehicle that could achieve earth orbit with only a single stage. Nothing but fuel would be jettisoned during launch so that the vehicle could travel to and from orbit repeatedly with only refueling and minimal maintenance between flights. The challenges facing this program are formidable. Even with liquid hydrogen and oxygen as its fuels, almost 90% of this vehicle's launch weight must be fuel. Nonetheless, construction and testing of such Single Stage To Orbit (SSTO) vehicles is proceeding rapidly, and a test vehicle, the DC-X1, has already shown the feasibility of the ideas. A full-scale demonstration vehicle, the X-33, is currently under development.

Chapter 1 noted that gravity attracts every object in the universe toward every other object in the universe (see ❑ below). In particular, objects are attracted toward the earth. Near the earth's surface, an object's weight is simply its mass times 9.8 N/kg, the acceleration due to gravity. But as the object's distance from the center of the earth increases, the acceleration due to gravity diminishes. Equation 1.1.2 is only an approximation, valid for objects near the earth's surface.

A more general formula relates the gravitational forces between two objects to their masses and the distance separating them. These forces are equal to the gravitational constant times the product of the two masses, divided by the square of the distance separating them. This relationship, discovered by Newton and called the **law of universal gravitation,** can be written as a word equation:

$$\text{force} = \frac{\text{gravitational constant} \cdot \text{mass}_1 \cdot \text{mass}_2}{(\text{distance between masses})^2}, \qquad \textbf{(5.3.1)}$$

in symbols:

$$F = \frac{G \cdot m_1 \cdot m_2}{r^2},$$

and in common language:

The pull of gravity is strongest between massive objects but diminishes rapidly with distance.

Note that the force on mass_1 is directed toward mass_2 and the force on mass_2 is directed toward mass_1. Those two forces are equal in magnitude but oppositely directed. The **gravitational constant** is a fundamental constant of nature, with a measured value of 6.6720×10^{-11} N·m²/kg².

weightlessness

space shuttle orbit

The Law of Universal Gravitation
Every object in the universe attracts every other object in the universe with a force equal to the gravitational constant times the product of the two masses, divided by the square of the distance separating the two objects.

❑ **English physicist Henry Cavendish (1731–1810)** proved that terrestrial objects do exert gravitational forces on one another. His experiment, performed in 1798, measured the tiny forces that two metal spheres exert on one another, using a very sensitive torsion balance. Comparing the forces between the two spheres with those between the earth and those same spheres (their weights), Cavendish was able to deduce the mass of the earth.

This relationship describes any gravitational attraction, whether it's between two planets or between the earth and you. The effective location of an object's mass is its center of mass, so the distance used in Eq. 5.3.1 is the distance separating the two centers of mass. For a spacecraft orbiting the earth just above its atmosphere, that distance is roughly the earth's radius of 6378 km. But for a spacecraft far above the atmosphere, the distance is larger and the force of gravity is weaker. That spacecraft experiences a smaller acceleration due to gravity. To give it the additional time it needs for its path to bend around in a circle, the high-altitude spacecraft must travel more slowly than the low-altitude spacecraft. This reduced speed explains the long orbital periods of high-altitude spacecraft.

At 35,900 km (22,300 miles) above the earth's surface, the orbital period reaches 24 hours. A satellite in such an orbit is said to be *geosynchronous*. If a geosynchronous satellite orbits the earth eastward around the equator, it's also *geostationary*. Geostationary satellites and the earth turn together in unison. A geostationary satellite always remains over the same spot on the earth's surface. Such a fixed orientation is very useful for communications and weather satellites.

Not all orbits are circular. Many spacecraft follow elliptical orbits so that their altitudes vary up and down once per orbit. At apogee, its greatest distance from the earth's center, a spacecraft travels relatively slowly because it has converted some of its kinetic energy into gravitational potential energy. At perigee, its smallest distance from the earth's center, the spacecraft travels relatively rapidly because it has converted some of its gravitational potential energy into kinetic energy. Of course, the perigee should not bring the spacecraft into the earth's atmosphere or it will crash.

The other type of orbit that a spacecraft can experience is a hyperbolic orbit. If the spacecraft is traveling too fast, the earth will be unable to bend its path into a closed loop and the spacecraft will coast off into interplanetary space. The spacecraft's path near the earth is then a hyperbola. The spacecraft only follows this hyperbolic path once and then drifts away from the earth forever.

A spacecraft usually enters a hyperbolic orbit by firing its rocket engine. It starts in an elliptical orbit around the earth and uses the rocket engine to increase its kinetic energy. This kinetic energy becomes gravitational potential energy as the spacecraft arcs away from the earth. But the earth's gravity becomes weaker with distance and the spacecraft's gravitational potential energy slowly approaches a maximum value even as its distance from the earth becomes infinite. If the spacecraft has more than enough kinetic energy to reach this maximum gravitational potential energy, it will be able to escape completely from the earth's gravity. The speed that a spacecraft needs in order to escape from the earth's gravity is called the **escape velocity.** This escape velocity depends on the spacecraft's altitude and is about 11.2 km/s near the earth's surface. A spacecraft traveling at more than the escape velocity follows a hyperbolic orbital path and heads off toward the other planets or beyond.

Check Your Understanding #7: Speeding Up the Lunar Month

The moon orbits the earth every 27.3 days, at a distance of 384,400 km from the earth's center of mass. For the moon to orbit in less time, how would its distance from the earth have to change?

Check Your Figures #1: Attractive Cars

How much force does a 1000-kg automobile exert on an identical car located 10 m away?

stability and steering, supersonic flight, cabin pressure, navigation, helicopters, rocket history, ion propulsion, solar sails

Epilogue for Chapter 5

In this chapter, we looked at a number of objects that use moving fluids to perform their tasks. In *garden watering,* we saw how water moves through openings and channels. We looked at the effects of viscosity and the importance of Bernoulli's equation in describing the conversion of pressure potential energy into kinetic energy and vice versa. Two types of fluid flow appeared—laminar and turbulent. While laminar flow is smooth and predictable, we saw that turbulent flow involves unpredictability and chaos, a common behavior in our complicated universe.

In the section on *balls and Frisbees,* we examined the ways in which moving air exerts forces on larger objects, both in the downstream direction as drag and in a perpendicular direction as lift. We learned about several different types of drag forces and how these can be controlled or reduced by choosing the shapes or motions of the balls. We also looked at airfoils, aerodynamic objects that are designed to obtain substantial lift as they pass through the air.

Finally, in *airplanes and rockets,* we explored the aerodynamics of those remarkable machines. We saw how they use fluids to support and propel themselves. We also examined the limitations of wings and saw what can go wrong if those limitations are exceeded. We studied propulsion in propeller planes, in jet aircraft, and in rockets and saw that these systems involve Bernoulli's equation and the forward forces that come from pushing gases backward.

Explanation: A Vortex Cannon

The vortex cannon creates ring vortices—tiny tornadoes that are bent into loops so that they have no beginnings or ends. Air swirls forward in the middle of each ring and backward on its outer edge. This circular tornado structure is created when air flows through the hole of the vortex cannon. Air flows forward in the middle of the hole while the hole's edges create the backward flow around the outside of the ring. After it leaves the cannon, each ring vortex crawls forward through the surrounding air until its kinetic energy has been exhausted and it slows to a stop. It finishes its existence swirling in place until viscous forces bring its moving air to rest.

Chapter Summary

How Garden Watering Works: A faucet controls the water's flow rate by opening or closing a hole through which the water must pass. The water's speed through this hole is limited by its total energy and by viscous forces that oppose relative motion in the water. Because of this limited speed, the amount of water flowing through the faucet each second depends on the size of the hole. As this water continues on through a hose, viscous forces in the hose limit its speed, particularly near the walls of the hose. When the water reaches the nozzle, it makes a series of bends that together cause its pressure to drop and its speed to increase. When the water emerges from the nozzle at atmospheric pressure, it's moving rapidly. Turbulence in the flow produces a soft hissing noise as the water arcs gracefully into the garden.

How Balls and Frisbees Work: A ball traveling through the air experiences two major types of aerodynamic force—drag and lift. For a nonspinning ball traveling very slowly, the only drag force is viscous drag; it experiences no lift force. As the ball's speed through the air increases, a large turbulent wake appears behind the ball and the ball experiences pressure drag. At a still higher speed, the boundary layer of air near the ball's surface becomes turbulent and the size of the wake behind the ball shrinks. A ball with a turbulent boundary layer experiences less drag than one with a laminar boundary layer, which is why balls are designed to encourage turbulent boundary layers.

Rotating balls and flying disks experience lift forces. These forces occur because bending and deflecting airstreams push asymmetrically on these objects. Lift forces can cause a ball to curve in flight or keep a flying disk from falling.

How Airplanes and Rockets Work: An airplane is supported in flight by air passing across its wings. Air bends inward (toward the surface) as it passes over the wings and outward (away from the surface) as it passes under the wings. As a result, the air pressure is lower above the wings and the wings experience an upward lift force. The airplane also experiences drag, which tends to slow it down. To keep the airplane moving forward, the plane employs propellers or jet engines. These devices push the air backward and the air reacts by pushing them forward. A propeller works directly in the oncoming air, increasing the air's energy and pushing it backward with rotating blades. A jet engine first slows the air, then increases its energy by burning fuel in it, and finally pushes it backward at high speed.

A rocket obtains its thrust by pushing gas out of an engine in its tail. This gas usually comes from burning chemical fuels contained entirely inside the rocket itself. A carefully designed engine nozzle permits the rocket to make efficient use of the energy stored in the fuel by ensuring that gas leaves the rocket at a very low pressure. The rocket's thrust is used to lift the rocket against the force of gravity and to accelerate it upward. Once in space, with its engine inactive, the rocket orbits the earth.

Important Laws and Equations

1. Poiseuille's Law: The volume of fluid flowing through a pipe each second is equal to $(\pi/128)$ times the pressure difference across that pipe times the pipe's diameter to the fourth power, divided by the pipe's length times the fluid's viscosity, or

$$\text{volume flow rate} = \frac{\pi \cdot \text{pressure difference} \cdot \text{pipe diameter}^4}{128 \cdot \text{pipe length} \cdot \text{fluid viscosity}}. \qquad (5.1.1)$$

2. The Law of Universal Gravitation: Every object in the universe attracts every other object in the universe with a force equal to the gravitational constant times the product of the two masses, divided by the square of the distance separating the two objects, or

$$\text{force} = \frac{\text{gravitational constant} \cdot \text{mass}_1 \cdot \text{mass}_2}{(\text{distance between masses})^2}. \qquad (5.3.1)$$

Check Your Understanding—Answer

Section 5.1 GARDEN WATERING

1. The air at the surfaces of the fibers is stationary, and the air's viscosity slows the motion of air in the vicinity of the fibers.

Why: Although air trying to pass through the gaps may not directly touch the fibers, viscosity tries to keep all of the air moving together at the same velocity. As soon as some air is held up by a fiber, the air nearby is slowed by viscous forces. The fibers of a sweater are close enough together that viscous forces slow all the air trying to pass through the sweater. Imagine trying to pour honey through a sweater.

2. Air's viscosity slows its flow through ductwork. Moving large volumes of air rapidly without a large pressure differ-ence between inlet and outlet requires a very large diameter pipe.

Why: The airflow through long ductwork is dominated by viscous forces. The volume of air moved through ductwork is often enormous, and the pressure difference between the inlet and outlet is normally only a fraction of an atmosphere. To keep the air moving quickly through the ducts, their diameters must be large.

3. Even at atmospheric pressure, air has pressure potential energy. As it flows toward even lower pressure in the attachment, air converts some of that pressure potential energy into kinetic energy.

Why: Vacuum cleaners employ nozzles to get air moving very quickly. Fast-moving air rushing into the vacuum cleaner draws dust with it and cleans your home.

4. The air flowing through the "canyons" created by the buildings becomes turbulent and forms vortices that swirl the leaves and papers.

Why: Whether an object moves through a fluid or a fluid moves past an object, a Reynolds number is associated with the situation. When the Reynolds number becomes high enough that viscosity is unable to keep the fluid flowing in an orderly fashion, turbulence appears. Air is a fluid and when wind blows through in a big city, turbulence is commonplace. Whirling vortices are found everywhere.

Section 5.2 BALLS AND FRISBEES

1. The slow flow of water around the rock is laminar, so its pressure is highest at the front and back of the rock. The increased pressure behind the rock lifts the water level there.

Why: Laminar flow around an obstacle tends to create high-pressure regions at the front and back and low-pressure regions on the sides. In this case, the pressure differences are visible as changes in the water level. At the front and back of the rock, the relatively high pressures push the water level upward while at the sides of the rock, the water level is depressed.

2. The slow-moving canoe experiences laminar flow in the water while the fast-moving canoe experiences turbulent flow.

Why: If the canoe's speed is less than about 1 cm/s, its Reynolds number will be less than 2000 and the water flow around it will be laminar. The water will pass smoothly around the canoe's sides and join back together behind it. But when the canoe is moving fast enough that its Reynolds number exceeds 2000, the water flow becomes turbulent and the canoe leaves a churning wake in the water behind it. This wake produces pressure drag on the canoe and extracts energy from it. Anyone who has paddled a canoe knows that overcoming this pressure drag can be exhausting.

3. As far back on the car as possible.

Why: As for all large, fast-moving objects in air, pressure drag is the main source of air resistance. You want to minimize this drag by keeping the air flowing smoothly over the car until it leaves the rear around a small turbulent wake. The smaller the air pocket behind a car, the better. Aerodynamically designed production cars leave a turbulent wake that is only about a third as large in area as the thickest cross section of the car. While there is still room for improvement, these cars experience far less drag than the boxy cars of earlier times.

4. The lift force is directed downward, so the ball accelerates downward faster than it would by gravity alone.

Why: A ball with topspin falls faster than it would without a spin. In tennis, the topspin strokes appear to dive downward once they cross the net. Their downward curve means

that they can travel very fast and still remain inside the court and are thus very hard to return.

5. Because of its smaller wake, the Aerobie experiences less drag and retains its speed better.

Why: Flying disks aren't perfect and experience drag forces that slow them down. Reducing these drag forces results in longer flights.

Section 5.3 AIRPLANES AND ROCKETS

1. The air traveling around the outside of the sail has a lower pressure than that traveling across the inside of the sail. The sail experiences a lift force that pushes it and the boat across the water.

Why: Sails experience both lift and drag forces. The sail experiences an aerodynamic force that pushes it outward (lift) and slightly downwind (drag) just as an airplane wing experiences an aerodynamic force upward (lift) and slightly downstream (drag). The sailboat's keel, or centerboard, and its rudder provide additional forces so that the net force on the sailboat can be controlled and it can travel in a variety of directions.

2. The plane's wings are stalling.

Why: Tipping the plane's nose upward increases the wings' angle of attack. While this action increases lift up to a point, it can also cause the airflow to separate from the top surface of the wing. The sudden reduction in lift and increase in drag that accompanies stalling can cause the plane to drop. A stall during takeoff or landing is extremely dangerous.

3. The inlet side of each fan blade experiences the lowest air pressure.

Why: Air blowing toward you from a fan is like the air blown back by a propeller. The lowest pressures experienced by a propeller are on its forward surfaces. Similarly, the lowest pressures experienced by a fan are on its inlet side surfaces. This pressure imbalance pushes the fan away from you while the fan pushes the air toward you.

4. As the air is slowed down, its pressure increases. The air's total energy remains constant.

Why: The remarkable result of Bernoulli's effect is that you can slow the air down without squandering its kinetic energy. That energy becomes pressure potential energy, and it passes through the jet engine in that form. As the air leaves the jet engine, its pressure potential energy becomes kinetic energy once again, so that no energy is wasted.

5. It pushes against its own exhaust, as do all rockets.

Why: While it may appear that the plume of exhaust beneath a ground-launched rocket is what lifts that rocket upward, the ground itself doesn't contribute to the thrust. The very action of pushing the gas out the nozzle propels the rocket forward.

6. It's actually staged subtly. Its two solid fuel boosters are effectively the first stage, the external liquid fuel tank is the second stage, and the orbiter itself is the third stage.

Why: Although the Space Shuttle is not stacked one stage on the next like a Saturn or Delta rocket, it doesn't travel from ground to space as a single object. It discards empty fuel containers as it accelerates upward. First to go are the two boosters, followed by the external fuel tank. The final

mass of the orbiter itself is much less than what left the launch pad.

7. The distance between the earth and moon would have to decrease.

Why: The moon behaves just like a spacecraft orbiting the earth at a distance of 384,400 km. Such a spacecraft would also have an orbital period of 27.3 days. To reduce this orbital period, the moon would have to move closer to the earth so that the earth's gravity could bend its path more rapidly.

Check Your Figures—Answer

Section 5.1 GARDEN WATERING

1. About 0.20 L/s.

Why: When the faucet is wide open, water flow is limited by the pipes. The diameters of the old pipes are 20% less than when they were new, or 0.80 times the diameter of the new pipes. We can write Poiseuille's law (Eq. 5.1.1) for the new pipes:

$$0.50 \text{ L/s} = \frac{\pi \cdot \Delta p \cdot D^4}{128 \cdot L \cdot \eta},$$

and the old pipes:

$$\text{volume flow rate} = \frac{\pi \cdot \Delta p \cdot (0.80 \cdot D)^4}{128 \cdot L \cdot \eta},$$

and divide the second equation by the first:

$$\frac{\text{volume flow rate}}{0.50 \text{ L/s}} = \frac{\dfrac{\pi \cdot \Delta p \cdot (0.80 \cdot D)^4}{128 \cdot L \cdot \eta}}{\dfrac{\pi \cdot \Delta p \cdot D^4}{128 \cdot L \cdot \eta}} = 0.80^4 = 0.4096.$$

We don't need to know the pressure difference (Δp), the pipe length (L), or water's precise viscosity (η) because they cancel in the last equation. If we multiply both sides of that equation by 0.5 L/s, we get the volume flow rate:

$$\text{volume flow rate} = 0.4096 \cdot 0.5 \text{ L/s} = 0.2048 \text{ L/s}.$$

Because our measured values are only accurate to two decimal digits (0.50 L/s is not as accurate a measurement as 0.5000 L/s), we can only report a two-digit result: 0.20 L/s for the flow from the old pipes. It clearly doesn't take very

much mineral accumulation to dramatically reduce the flow from a faucet.

2. Turbulent.

Why: To calculate the Reynolds number for the airflow around the car, you need air's viscosity from Table 5.1.1 (0.0000183 Pa·s or 0.0000183 kg/m·s), air's density from Section 4.1 (1.25 kg/m³), the car's size (roughly 3 m), and its speed through the air (roughly 55 mph or 25 m/s). You can then calculate its approximate Reynolds number, using Eq. 5.1.2:

$$\text{Reynolds number} = \frac{1.25 \text{ kg/m}^3 \cdot 3 \text{ m} \cdot 25 \text{ m/s}}{0.0000183 \text{ kg/m·s}}$$
$$= 5.1 \text{ million}.$$

The convertible's Reynolds number is far above the threshold for turbulence (2300), so the air swirls chaotically around the vehicle. That explains why your hair flies around wildly.

Section 5.3 AIRPLANES AND ROCKETS

1. About 6.7×10^{-7} N.

Why: We use Eq. 4.3.1 to obtain the force:

$$\text{force} = \frac{6.6720 \times 10^{-11} \text{ N·m}^2/\text{kg}^2 \cdot 1000 \text{ kg} \cdot 1000 \text{ kg}}{(10 \text{ m})^2}$$
$$\text{force} = 6.6720 \times 10^{-7} \text{ N}.$$

This force is roughly equal to the weight of a grain of sand. No wonder it's hard to feel gravity from anything but the entire earth.

Exercises

1. A favorite college prank involves simultaneously flushing several toilets while someone is in the shower. The cold

water pressure to the shower drops and the shower becomes very hot. Why does the cold water pressure suddenly drop?

2. On hot days in the city people sometimes open up fire hydrants and play in the water. Why does this activity reduce the water pressure in nearby hydrants?

3. Why does a relatively modest narrowing of the coronary arteries, the blood vessels supplying blood to the heart, cause a dramatic drop in the amount of blood flowing through them?

4. Why does hot maple syrup pour more easily than cold maple syrup?

5. Why is "molasses in January" slower than "molasses in July," at least in the northern hemisphere?

6. Why is it so difficult to squeeze ketchup through a very small hole in its packet?

7. A baker is decorating a cake by squeezing frosting out of a sealed paper cone with the tip cut off. If the baker makes the hole at the tip of the cone too small, it's extremely difficult to get any frosting to flow out of it. Why?

8. Why is the wind stronger several meters above a flat field than it is just above the ground?

9. Pedestrians on the surface of a wind-swept bridge don't feel the full intensity of the wind because near the bridge's surface, the air is moving relatively slowly. Explain this effect in terms of a boundary layer.

10. An electric valve controls the water for the lawn sprinklers in your backyard. Why do the pipes in your home shake whenever this valve suddenly stops the water but not when this valve suddenly starts the water?

11. If you drop a full can of applesauce and it strikes a cement floor squarely with its flat bottom, what happens to the pressures at the top and bottom of the can?

12. When you mix milk or sugar into your coffee, you should move the spoon quickly enough to produce turbulent flow around the spoon. Why does this turbulence aid mixing?

13. If you start two identical paper boats from the same point, you can make them follow the same path down a quiet stream. Why can't you do the same on a brook that contains eddies and vortices?

14. When you swing a stick slowly through the air, it's silent. But when you swing it quickly, you hear a "whoosh" sound. What behavior of the air is creating that noise?

15. If you try to fill a bucket by holding it in a waterfall, you will find the bucket pushed downward with enormous force. How does the falling water exert such a huge downward force on the bucket?

16. You sometimes see paper pressed tightly against the front of a moving car. What holds the paper in place?

17. Fish often swim just upstream or downstream from the posts supporting a bridge. How does the water's speed in these regions compare with its speed in the open stream and at the sides of the posts?

18. Why do flyswatters have many holes in them?

19. You have two golf balls that differ only in their surfaces. One has dimples on it while the other is smooth. If you drop these two balls simultaneously from a tall tower, which one will hit the ground first?

20. If you ride your bicycle directly behind a large truck, you will find that you don't have to pedal very hard to keep moving forward. Why?

21. How does running directly behind another runner reduce the wind resistance you experience?

22. When a car is stopped, its flexible radio antenna points straight up. But when the car is moving rapidly down a highway, the antenna arcs toward the rear of the car. What force is bending the antenna?

23. To drive along a level road at constant velocity, your car's engine must be running and friction from the ground must be pushing your car forward. Since the net force on an object at constant velocity is zero, why do you need this forward force from the ground?

24. Racing bicycles often have smooth disk-shaped covers over the spokes of their wheels. Why would these thin wire spokes be a problem for a fast-moving bicycle?

25. A bullet slows very quickly in water but a spear doesn't. What force acts to slow these two objects, and why does the spear take longer to stop?

26. If you hang a tennis ball from a string, it will deflect downwind in a strong breeze. But if you wet the ball so that the fuzz on its surface lies flat, it will deflect even more than before. Why does smoothing the ball increase its deflection?

27. Explain why a parachute slows your descent when you leap out of an airplane.

28. Bicycle racers sometimes wear teardrop-shaped helmets that taper away behind their heads. Why does having this smooth taper behind them reduce the drag forces they experience relative to those they would experience with more ball-shaped helmets?

29. If you want the metal tubing in your bicycle to experience as little drag as possible while you're riding in a race, is cylindrical tubing the best shape? How should it be shaped?

30. In 1971, astronaut Alan Shepard hit a golf ball on the moon. How did the absence of air affect the ball's flight?

31. A water skier skims along the surface of a lake. What types of forces is the water exerting on the skier, and what is the effect of these forces?

32. How would a Frisbee fly on the airless moon?

33. You can buy special golf tees that wrap around behind the ball to prevent you from giving it any spin when you hit it. These tees are guaranteed to prevent hooks and slices (i.e., curved flights). But how do these tees affect the distance the ball travels? Why?

34. A hurricane or gale force wind can lift the roof off a house, even when the roof has no exposed eaves. How can wind blowing across a roof produce an upward force on it?

35. A skillful volleyball player can serve the ball so that it barely spins at all. The ball dithers slightly from side to side as it flies over the net and is hard to return. What causes the ball to accelerate sideways?

36. Why does an airplane have a "flight ceiling," a maximum altitude above which it can't obtain enough lift to balance the downward force of gravity?

37. If you let a stream of water from a faucet flow rapidly over the curved bottom of a spoon, the spoon will be drawn into the stream. Explain this effect.

38. If you put your hand out the window of a moving car, so that your palm is pointing directly forward, the force on your hand is directly backward. Explain why the two halves of the airstream, passing over and under your hand, don't produce an overall up or down force on your hand.

39. If instead of holding your hand palm forward (see Exercise 38), you tip your palm slightly downward, the force on your hand will be both backward and upward. How is the airstream exerting an upward force on your hand?

40. When a hummingbird hovers in front of a flower, what forces are acting on it and what is the net force it experiences?

41. A poorly designed household fan stalls, making it inefficient at moving air. Describe the airflow through the fan when its blades stall.

42. When a plane enters a steep dive, the air rushes toward it from below. If the pilot pulls up suddenly from such a dive, the wings may abruptly stall, even though the plane is oriented horizontally. Explain why the wings stall.

43. Some military planes can swivel their jet engines to draw in air from above and send their exhaust downward. As a result, these planes can take off vertically without using their wings. How do its jet engines provide the upward force needed to support the plane against the force of gravity?

44. If an airplane experienced no drag forces at all, the only aerodynamic forces on it would be lift forces. As a result, the plane could sustain level flight indefinitely, even without engines. Explain why lift forces don't change the energy of a plane in level flight through stationary air.

45. An airboat travels through swamps, propelled by a large fan. How does the fan push the boat forward?

46. When a sharpshooter fires a pistol at a target, the gun recoils backward very suddenly, leaping away from the target. Explain this recoil effect in terms of the transfer of momentum.

47. Which action will give you more momentum toward the north: throwing one shoe southward at 10 m/s or two shoes southward at 5 m/s?

48. Do both of the actions in Exercise 47 take the same amount of energy? If not, which one requires more energy?

49. You are propelling yourself across the surface of a frozen lake by hitting tennis balls toward the southern shore. From your perspective, each ball you hit heads southward at 160 km/h. You have a huge bag of balls with you and are already approaching the northern shore at a speed of 160 km/h (100 mph). When you hit the next ball southward, will you still accelerate northward?

50. Can you use the tennis ball scheme of Exercise 49 to propel yourself to any speed, or are you limited to the speed at which you can hit the tennis balls?

51. You and a friend are each wearing roller skates. You stand facing each other on a smooth, level surface and begin tossing a heavy ball back and forth. Why do you drift apart?

52. The time it takes a relatively small object to orbit a much larger object doesn't depend on the mass of the small object. Use an astronaut who is walking in space near the space shuttle to illustrate that point.

53. As the moon orbits the earth, which way is the moon accelerating?

54. Which object is exerting the stronger gravitational force on the other, the earth or the moon, or are the forces equal in magnitude?

55. To free an Apollo spacecraft from the earth's gravity took the efforts of a gigantic Saturn V rocket. Freeing a lunar module from the moon's gravity took only a small rocket in the lunar module's base. Why was it so much easier to escape from the moon's gravity than from that of the earth?

56. Spacecraft in low earth orbit take about 90 minutes to circle the earth. Why can't they be made to orbit the earth in half that amount of time?

Problems

1. About how fast can a small fish swim before experiencing turbulent flow around its body?

2. How much higher must your blood pressure get to compensate for a 5% narrowing in your blood vessels? (The pressure difference across your blood vessels is essentially equal to your blood pressure.)

3. If someone replaced the water in your home plumbing with olive oil, how much longer would it take you to fill a bathtub?

4. You are trying to paddle a canoe silently across a still lake and know that turbulence makes noise. How quickly can the canoe and the paddle travel through water without causing turbulence?

5. The pipes leading to the showers in your locker room are old and inadequate. While the city water pressure is 700,000 Pa, the pressure in the locker room when one shower is on is only 600,000 Pa. Use Eq. 5.1.1 to calculate the *approximate* pressure if three showers are on.

6. If the plumbing in your dorm carried honey instead of water, filling a cup to brush your teeth could take awhile. If the faucet takes 5 s to fill a cup with water, how long would it take to fill your cup with honey, assuming all the pressures and pipes remain unchanged?

7. How quickly would you have to move a 1-cm diameter stick through olive oil to reach a Reynolds number of 2000, so that you would begin to see turbulence around the stick? (Olive oil has a density of 918 kg/m^3.)

8. The effective obstacle length of a blimp is its width—the distance to which the air is separated as it flows around the blimp. How slowly would a 15-m wide blimp have to move in order to keep the airflow around it laminar? (Air has a density of 1.25 kg/m^3.)

9. If an 80-kg baseball pitcher wearing frictionless roller skates picks up a 0.145-kg baseball and pitches it toward the south at 42 m/s (153 km/h or 95 mph), how fast will he begin moving toward the north?

10. How high above the earth's surface would you have to be before your weight would be only half its current value?

11. As you walked on the moon, the earth's gravity would still pull on you weakly and you would still have an earth weight. How large would that earth weight be, compared to your earth weight on the earth's surface?

12. If you and a friend 10 m away each have masses of 70 kg, how much gravitational force are you exerting on your friend?

13. The gravity of a black hole is so strong that not even light can escape from within its surface or "event horizon." Even outside that surface, enormous energies are needed to escape. Suppose that you are 10 km away from the center of a black hole that has a mass of 10^{31} kg. If your mass is 70 kg, how much do you weigh?

14. In Problem 13, how much work would something have to do on you to lift you 1 m farther away from the black hole?

Cases

1. A glider is a propulsionless airplane that is completely dependent on external energy to remain airborne. It must be pulled forward to obtain the lift it needs to start flying. Once aloft, it can use the air rising off hot spots on the ground to remain in the air almost indefinitely. But for the moment, let's suppose that the air is completely motionless.

    a. One way to launch a glider is to pull it with a car, using a rope that can be detached from the glider when it's in the air. Just after the glider's wheels leave the runway, what four forces are acting on the glider?

    b. When the glider releases the rope, one of the four forces disappears. With only three forces left, what happens to the glider's forward component of velocity?

    c. If the glider begins to descend gradually, the air no longer approaches it horizontally. From the glider's perspective, the air is now blowing upward toward it at a shallow angle. How does this new wind direction affect the directions of the three forces on the glider, as viewed from the ground?

    d. It's possible for the descending glider to maintain a constant forward component of velocity. Explain.

    e. Now suppose that the air is rising quickly, perhaps heated by contact with a hot spot on the ground. In that case, the glider doesn't have to descend in order to maintain a constant forward component of velocity; it can glide straight and level forever. Why?

2. A helicopter suspends itself from a spinning rotor. Each blade in the rotor is actually a wing, with an airfoil shape that creates lift as it slices through the air. To hover, the helicopter's rotor creates just enough upward lift to exactly balance the helicopter's downward weight.

a. The rotor keeps turning at a constant rate. To control the rotor's lift, the helicopter adjusts the angles of attack of the blades as they turn. If a hovering helicopter increases the angles of attack of all of its blades simultaneously, what will happen to the forces on the helicopter and how will it move?

b. The helicopter can also adjust the angle of attack of a single blade. If it increases the angle of attack of whatever blade is currently on the pilot's right, the helicopter will experience a torque about its center of mass. Which way will the helicopter begin to tilt?

c. If the helicopter is tilted nose down and tail high, the helicopter will begin to accelerate forward. Why?

d. To turn the helicopter so that it faces a different direction, something must exert a torque on it about a vertical axis. That's one reason why the helicopter has a small rotor at the end of its tail boom. This rotor pushes air to the right or left, depending on the angles of attack of its blades. Which way should the tail rotor push the air to turn the helicopter to the right?

e. The main rotor blades are actually flexible, but they extend almost straight outward when the rotor is spinning. Why?

3. A parachute slows your descent after you jump out of a plane. In the parachute's frame of reference, wind is blowing upward toward it and its large surface area creates a great deal of upward drag force. This drag opposes your weight and slows your fall.

a. As air rushes past the parachute, the pressures above and below the cloth become different. Why is the pressure below the parachute higher than atmospheric pressure?

b. Why is the pressure above the parachute roughly equal to atmospheric pressure?

c. How does air exert an upward force on the parachute?

d. Why does the amount of upward force exerted on the parachute depend on how fast you're falling?

e. If you open your parachute while you're falling very rapidly, the parachute slows your descent. Which way is the net force on you pointing, and which way do you accelerate?

f. A few seconds after your parachute opens, you reach terminal velocity. You are no longer accelerating. How does the parachute's drag force compare to your weight?

g. How heavy do you feel when you first jump out of the plane? when your opening parachute first begins to slow your descent? when you are descending at terminal velocity?

4. The sail of a sailboat acts like a wing, creating both lift and drag. Because it deflects the wind horizontally, the lift is horizontal, too, and pulls the boat through the water.

a. Wind blows around both sides of the sail. It bends inward, toward the sail, as it travels around the outside. How does the pressure on that side of the sail compare with atmospheric pressure?

b. Air passing across the inside surface of the sail bends outward, away from the sail. How does the pressure on that side of the sail compare with atmospheric pressure?

c. What is the direction of the overall force exerted on the sail by the wind?

d. A sailboat has a flat keel that projects into the water and prevents the boat from tilting or drifting sideways. The keel effectively produces a straight track through the water along which the boat can move. If the wind is blowing toward the boat from north to south and the boat's sail is experiencing an eastward lift force, show that the keel can be turned so that the net force on the boat is toward the northeast. (This is how sailboats manage to sail upwind!)

e. Without a keel, the only horizontal forces pushing the boat through the water come from the wind. These wind forces always push the boat at least partly downwind. Why can't forces from the wind push the boat partly upwind?

f. The wind and keel together can push the boat in an upwind direction. But the boat doesn't accelerate forever; it eventually reaches a maximum forward velocity. What other force acts on the boat to reduce the net force on the boat to zero so that it maintains a constant velocity?

g. As the boat moves forward, it's steered by its rudder. With the rudder turned so that it deflects water to one side as the boat travels forward, the boat begins to rotate. How do the water and rudder exert a torque on the sailboat?

5. Umbrellas are subject to many forces on a windy day.

a. When you hold the umbrella upright, its curved surface is above you like a dome. When the wind blows horizontally, why does the umbrella feel lighter than normal?

b. If you are walking into the wind and tip the umbrella forward so that it's in front of your face, why doesn't your face feel much wind anymore?

c. If you hold your hand just beyond the edge of the umbrella while it's still tipped into the wind, you'll feel that the airspeed there is even higher than that of the wind. Explain.

d. If you accidentally tip the umbrella backward, so that the inner surface of its dome is exposed to the wind, the wind will push the umbrella backward hard. Why is this force stronger than when the umbrella is tipped forward?

e. When the inner surface of the umbrella is exposed to a strong wind, the umbrella may turn inside out. Explain this effect in terms of a pressure imbalance.

6. You are consulting for a screenwriter who is working on a science-fiction movie about the crew of a large, intergalactic spaceship. She wants the script to be realistic from a scientific perspective, so she has asked you to check her work so far. You read through the script, which includes descriptions of special effects, and quickly find several problems with it. Here are some of the scenes that have flaws.

a. In one scene, the spaceship must make an abrupt left turn to avoid hitting an asteroid. The crew members stand anxiously but motionlessly on the command bridge during this maneuver. In reality, what should happen to the crew members?

b. In another scene, a guard fires a projectile weapon at a massive alien creature. The guard remains stationary, but the creature is thrown backward by the weapon's impact, even though the weapon doesn't explode. In reality, what should happen to the guard as he fires the weapon?

c. The main spaceship carries several small fighter spacecraft that resemble high-tech modern military fighter aircraft. In the depths of space, these fighter spacecraft dodge and turn rapidly, even though the rocket exhaust is always sent straight out of the rear of each fighter. In reality, why couldn't such spacecraft turn in space while similar looking aircraft can turn near the ground?

d. One small spacecraft secretly shuttles supplies from one earthlike planet to another, completing eight round trips without replenishing its chemical rocket fuel. You inform the writer that the spacecraft will barely be able to leave a planet even once using chemical rockets, unless that spacecraft is allowed to eject stages. Why are you right?

e. Near the end of the script, a main character falls from a cliff but is rescued only 2 m above the ground when she lands on the wing of a passing spacecraft. You inform the writer that the impact with the wing of the spacecraft would be just as fatal as one with the ground. Why is this the case?

Heat and Thermodynamics

We can't see all of the motion that takes place around us. Some of it is hidden deep inside each object, where thermal energy keeps the individual atoms and molecules jiggling back and forth in an endless flurry of activity. In most situations, we're only aware of this thermal energy because it determines an object's temperature. You can feel with your hand when an object is hot or cold; the more thermal energy it contains, the higher its temperature and the hotter it feels.

What you're feeling when you touch a hot object is its thermal energy flowing into your colder hand and raising the temperature of your skin. When thermal energy flows from a hotter object to a colder one, we call this moving thermal energy *heat.* And when devices move heat around deliberately or use it to do interesting things, they're operating in the domain of thermodynamics. In this chapter, we'll examine temperature, heat, and thermodynamics in order to understand more about our hot and cold world.

EXPERIMENT: A Ruler Thermometer

One effect that a change in temperature has on a typical object is to change its size. You can use this size change to build a crude thermometer. While the size change is so tiny that it's normally hard to detect, you can use mechanical advantage to help you. Here's a way to make a thermometer using only a clear plastic ruler, a pin, a small weight, a piece of stiff paper, and some tape.

Lay the plastic ruler along the edge of a table and tape one end of it securely to the table. Cut a very thin strip of stiff paper, about 3 mm wide (0.1 inch) and 15 cm long (6 inches), and push the pin carefully through one end of the strip. Use a dot of tape to stick the pin's head to the paper. When you're done, the paper strip should be securely attached to the pin so that as the pin turns, the strip turns. This strip is your thermometer's pointer.

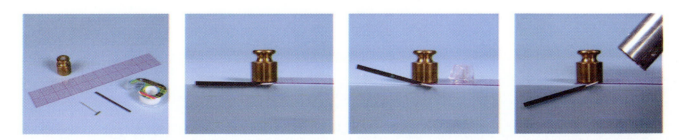

Now slide the pin under the free end of the ruler and place the small weight above it. The weight is there to push the ruler and pin together so that they experience plenty of static friction. That way, as the free end of the ruler moves left or right, the pin will rotate and turn the pointer.

Your thermometer is now complete. If you turn the pin and pointer carefully by hand so that the pointer is horizontal, you can "read" the thermometer by its angle relative to the tabletop. If you heat the plastic ruler by breathing on it, laying your hands on it, or warming it gently with a hair dryer, the ruler will become longer. Its free end will move away from the fixed end and will cause the pin to rotate. You will see a small change in the pointer's orientation as your thermometer reports its new temperature.

Predict what will happen if you don't heat the whole ruler uniformly and then *observe* what happens when you don't. Try to *measure* the effects of nonuniform heating. Did you *verify* your prediction?

You can also make the needle turn the other way by cooling the ruler. Placing a few ice cubes on the ruler will cause the ruler to contract and the needle will turn in the opposite direction.

Chapter Itinerary

In this chapter, we'll examine thermal energy, temperature, and heat in the context of four common types of objects: (1) *woodstoves,* (2) *incandescent light bulbs,* (3) *air conditioners,* and (4) *automobiles.* In *woodstoves,* we'll look at how thermal energy can be produced and at the three principal means by which thermal energy is transferred as heat from hotter objects to colder ones: conduction, convection, and radiation. In *incandescent light bulbs,* we'll study heat transfer by radiation to see that it can include the emission of visible light. In *air conditioners,* we'll look at the rules governing the movement of thermal energy—the laws of thermodynamics—and see how those rules permit an air conditioner to use ordered electric energy to transfer heat from the cold room air to the hot outdoor air. In *automobiles,* we'll investigate the ways in which thermal energy can be used to do work and see how the engine is able to convert thermal energy into work while heat flows from the hot burning gasoline to the cold outside air.

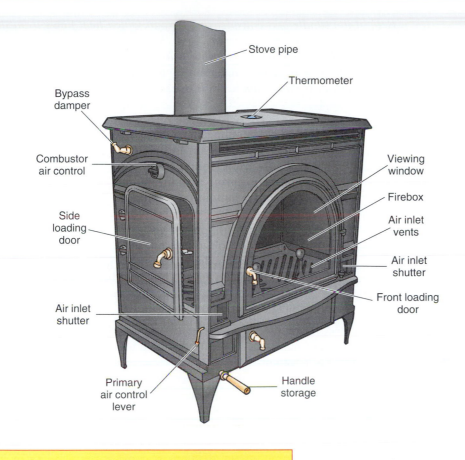

Woodstoves

Winter would be pretty unpleasant for most of us were it not for heating. Heating keeps our rooms warm even when the weather outside is cold. One of the most fashionable types of heating is a woodstove, which burns logs in its firebox and sends thermal energy out into the room. In this section we'll look at how a woodstove produces thermal energy and how this thermal energy flows out of the stove to keep us warm.

Questions to Think About: *What happens to the chemical potential energy in a log when you burn it? Why do you get burned when you touch a hot object? Why does your hand feel hotter when you hold it above a hot surface than next to that hot surface? Why does your skin feel warm when you face a campfire, even if the air around you is cold?*

Experiments to Do: *A burning candle produces thermal energy, providing both light and warmth to a small room. But where does this thermal energy come from and how does it flow into its surroundings?*

Light a short candle and look first at how it releases thermal energy. The flame slowly consumes the wax but it also needs air. Cover the candle with a tall glass, one that won't be touched or damaged by the flame. The glass should prevent room air from reaching the candle. How does the newly sealed environment affect the candle flame? Try to explain this result.

Relight the candle and consider the ways in which heat flows from the flame to you. Carefully pass your hand over the flame, keeping a safe distance above it to avoid being burned. Why does the flame warm your skin so quickly when your hand is directly above it? Now hold your hand beside the flame at a safe distance. You should again feel warmth from the flame. How is heat flowing from the flame to your hand now?

Take a wooden pencil and hold it a few centimeters above the flame for no more than 2 seconds. Then carefully touch the pencil's surface with your fingers. It should warm your fingers. How is heat flowing from the pencil to your hand in this case? Why would it be a painful mistake to try this experiment with a metal pencil?

A Burning Log: Thermal Energy

Fig. 6.1.1 This woodstove transfers heat to the room by conduction through its metal walls, convection of air past its surfaces, and radiation from its black exterior.

heat and irreversibility

A woodstove produces thermal energy and distributes it as heat to the surrounding room (Fig. 6.1.1). We've encountered thermal energy before: in a file cabinet sliding along the sidewalk, in an old ball bouncing inefficiently from a wall, and in honey pouring slowly from a jar. In each case, ordered energy—energy that could easily be used to do work—became disordered thermal energy and the temperatures of the objects increased. But now that we're going to study heating machines themselves, let's reexamine thermal energy and temperature to see how thermal energy moves from one object to another.

When you burn a log in the fireplace or woodstove, you're turning the log's ordered chemical potential energy into thermal energy—a disordered form of energy, contained in the kinetic and potential energies of individual atoms and molecules. The presence of thermal energy in the log is what gives the log a temperature; the more thermal energy the log has, the higher its temperature.

The nature of thermal energy depends somewhat on what it's in. In the hot, burning log, the thermal energy is mostly in the wood's atoms and molecules, which jitter back and forth rapidly relative to one another. When each of these particles moves, it has kinetic energy. When it pushes or pulls on its neighbors, it has potential energy.

But energy associated with the log as a whole isn't part of its thermal energy. Only disordered energy that's internal to the log is included. Thus if you stir the fire about with a poker, the moving log's total kinetic energy increases but not its thermal energy. Since the moving log could do work on the external things it touches, the kinetic energy in its overall motion is neither disordered nor internal to the log.

Similarly, if you lift the burning log with tongs, you increase its gravitational potential energy but not its thermal energy. Since the new potential energy is stored in the gravitational force between the whole log and the earth, it is ordered energy that could do work directly on external objects. The potential energy in thermal energy is stored between the individual atoms and molecules of the log, where it's disordered and can't do work directly.

The air near the burning log also contains thermal energy, but since the atoms and molecules in a gas are essentially free and independent, most of this thermal energy is kinetic energy. The air particles store potential energy only during the brief moments when they collide with one another.

Check Your Understanding #1: A Warm-Up Pitch

If you drop a ball, will its thermal energy increase as it falls?

Forces Between Atoms: Chemical Bonds

To understand how a burning log produces thermal energy, let's take a look at bonds between atoms and the chemical potential energy that's stored in those bonds. Since both result from the forces between atoms, that's where we'll begin.

As you bring atoms close together, they exert attractive forces on one another (Fig. 6.1.2a). These chemical forces are electromagnetic in origin and grow stronger as the two atoms approach. But the attraction diminishes when the atoms start to touch and is eventually replaced by repulsion at very short distances (Fig. 6.1.2b). The separation between atoms at which the attraction ends and the repulsion begins is their *equilibrium separation*—that is, the separation at which the atoms exert no forces on one another (Fig. 6.1.2c). Since atoms are tiny, this equilibrium separation is also tiny, typically only about a ten-billionth of a meter.

Imagine holding two atoms in tweezers and slowly bringing them together. They pull toward one another as they approach, doing work on you and increasing your energy. Since energy is conserved, their energy must be decreasing. They're giving up **chemical potential energy**—energy stored in the chemical forces between atoms.

Once the atoms reach their equilibrium separation, you can let go of them and they won't come apart. Like two balls attached by a spring, the atoms are in a stable equilibrium. Since they've lost some of their chemical potential energy, they can only be separated by the return of that energy. Because it takes work to pull them apart, the atoms are held together by a chemical bond.

The bound atoms have become a molecule. The strength of their bond is equal to the amount of work the atoms did when they drew together. The same amount of work is required to separate them. Bond strengths range from extremely strong in the case of two nitrogen atoms to extremely weak in the case of two neon atoms.

If they have a little extra energy, bound atoms can vibrate back and forth about their equilibrium separation (Fig. 6.1.2d). Whenever the atoms are moving quickly toward or away from one another, most of their energy is kinetic. Whenever the atoms are slowing to turn around, most of their energy is chemical potential. Overall, the molecule's total energy remains constant, and it vibrates back and forth until it transfers its extra energy elsewhere.

But many molecules have more than two atoms. In a large molecule, each pair of adjacent atoms has a chemical bond and an equilibrium separation. If you give this molecule excess energy, it will vibrate in a complicated manner as the energy moves among the various atoms and chemical bonds. The atoms in the molecule will continue to jiggle about until something removes the excess energy from the molecule.

Like all liquids and solids, our burning log is just a huge assembly of atoms and molecules, held together by chemical bonds of various strengths. These particles push and pull on one another as they vibrate about their equilibrium separations. Their motion is *thermal motion,* and the energy involved in this disorderly jiggling is thermal energy. Because thermal energy is fragmented among the atoms and exchanged between them unpredictably, it can't be used directly to do useful work.

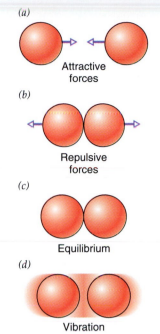

(a) Attractive forces

(b) Repulsive forces

(c) Equilibrium

(d) Vibration

Fig. 6.1.2 (a) Two atoms attract one another at moderate distances but (b) repel when they're too close. (c) In between is their equilibrium separation, at which they neither attract nor repel and are thus in equilibrium. (d) Pairs of atoms with excess energy tend to vibrate about their equilibrium separations.

chemical bond types

Check Your Understanding #2: When Atoms Collide . . .
As two independent nitrogen atoms collide with one another, what forces do they experience?

Heat and Temperature

Everything contains thermal energy, from the hot burning log to the cold metal poker you use to stir the fire. However, that doesn't mean that the thermal energy is equitably distributed. What happens to thermal energy when you push the log with the poker?

When they touch, the poker and the log begin to exchange thermal energy. In effect, the two objects become one larger object, and thermal energy that has been moving among the atoms in each individual object begins to flow across the junction between the two. Since each object starts with some amount of thermal energy, energy moves in both directions across this junction. Nonetheless, there may be some net flow from one object to the other. To allow us to predict the direction of this flow, we define a temperature for each of the objects.

Temperature is the quantity that indicates which way, if any, thermal energy will naturally flow between two objects. If no thermal energy flows when two objects touch, then those objects are in **thermal equilibrium** and their temperatures are equal. But if thermal energy flows from the first object to the second, then the first object is hotter than the second.

A temperature scale classifies objects according to which way thermal energy will flow between any pair. An object with a hotter temperature will always transfer thermal energy to an object with a colder temperature, and two objects with the same temperature will always be in thermal equilibrium. Thus the hot burning log will transfer thermal energy to the cold poker. We say that the burning log is *hot* because it tends to transfer thermal energy to most objects, while the poker is *cold* because most objects tend to transfer thermal energy to it.

Energy that flows from one object to another because of a difference in their temperatures is called **heat.** Heat is thermal energy on the move. Strictly speaking, the burning log doesn't *contain* heat; it contains thermal energy. However, when that log transfers energy to the cold poker because of their temperature difference, it's heat that *flows* from the log to the poker. (For a historical note about the understanding of heat, see ❏.)

Our present definition of temperature can order the objects around us from hottest to coldest, but it doesn't quantify temperature in any unique way. You could make your own temperature scale by comparing every two objects in sight to see which way heat flows between them, but you probably wouldn't enjoy it. You do better to use a standard temperature scale such as Celsius, Fahrenheit, and Kelvin.

Standard temperature scales are based on an object's average thermal kinetic energy per atom. The more kinetic energy each atom has, on average, the more vigorous the object's thermal motion and the more work the object's individual atoms will do on those of a second object. This microscopic work is what actually passes heat between objects. Since an object with more average thermal kinetic energy per atom will pass heat to an object with less, it makes sense to assign temperatures according to average thermal kinetic energies per atom.

The Celsius, Fahrenheit, and Kelvin scales all measure temperature in this manner. In each scale, a 1 degree or unit increase in temperature reflects a specific increase in average thermal kinetic energy per atom. The relationship between average thermal kinetic energy per atom and assigned temperature is based on several standard conditions: absolute zero, water's freezing temperature, and water's boiling temperature. (Recall from Section 4.1 that absolute zero is the temperature at which all thermal energy has been removed from an object.) Once specific temperatures have been assigned to two of these standard conditions, the whole tempera-

❏ Before the time of **Benjamin Thompson, Count Rumford (American-born British physicist and statesman, 1752–1814),** heat was believed to be a fluid called caloric that was contained within objects. Thompson disproved the caloric theory by showing that the boring of cannons produced an inexhaustible supply of heat. Among his scientific and technological contributions, Thompson improved cooking and heating methods. He reshaped fireplaces and developed the damper as ways to reduce smoking and improve heat transfer to the room. Thompson also had a life of sensational escapades and great rises and falls in fortune. He fled New Hampshire in 1775 because he was a British loyalist, he fled London in 1782 under suspicion of being a French spy, and he was, at the time of his studies of heat, among the most powerful people in Bavaria.

converting temperatures

| Table 6.1.1 | Temperatures of Several Standard Conditions, as Measured in Three Temperature Scales: Celsius, Kelvin, and Fahrenheit |

STANDARD CONDITION	CELSIUS (°C)	KELVIN (K)	FAHRENHEIT (°F)
Absolute zero	−273.15	0	−459.67
Freezing water	0	273.15	32
Boiling water	100	373.15	212

ture scale is fixed. For example, the Celsius scale is built around 0 °C being water's freezing temperature and 100 °C being water's boiling temperature. Temperatures for the three standard conditions appear in Table 6.1.1.

Check Your Understanding #3: Frozen Fingers

If you pick up an ice cube, your hand suddenly feels cold. Which way is heat flowing?

Open Fires and Woodstoves

Suppose you needed an easy way to heat your room. The oldest and simplest method would be to start a campfire in the middle of the floor. The burning wood would produce thermal energy, which would flow as heat into the colder room. But how does burning wood produce thermal energy?

This thermal energy is released by a **chemical reaction** between molecules in the wood and oxygen in the air. Recall that atoms do work as they join together in a chemical bond and that the amount of work done depends on which atoms are being joined. While carbon and hydrogen atoms can bond together to form *hydrocarbon* molecules, these atoms form much stronger bonds with oxygen atoms. Thus while it may take work to disassemble a hydrocarbon molecule, the work done by its hydrogen and carbon atoms as they bind to oxygen atoms more than makes up for that investment. As a hydrocarbon molecule burns in oxygen, new, more tightly bound molecules are formed and chemical potential energy is released as thermal energy. The *reaction products* produced by burning hydrocarbons in air are primarily water and carbon dioxide.

Wood is composed mostly of cellulose, a long carbohydrate molecule. *Carbohydrates* contain carbon, hydrogen, and oxygen atoms. Despite the presence of a few oxygen atoms, carbohydrates still burn nicely to form water and carbon dioxide. When you light the wood with a match, you're supplying the energy needed to break the old chemical bonds so that the new bonds can form. This starting energy is called **activation energy**—the energy needed to initiate the chemical reaction. Heat from the match flame gives the wood enough thermal energy to break chemical bonds between various atoms and start the reaction.

Unfortunately, wood isn't pure cellulose. It also contains many complex resins that don't burn well and create smoke. If you plan to breathe the air in which you burn fuel, wood is an awful choice. You'd be better off with kerosene and natural gas, both of which are nearly pure hydrocarbons and burn cleanly. Actually, wood can be converted to a cleaner fuel by baking it in an airless oven to remove all of its volatile resins. This process converts the wood into charcoal, which burns to form nearly pure carbon dioxide, water vapor, and ash.

But even with clean burning fuels, the direct fire-in-the-room heating concept has its disadvantages: it consumes the room's oxygen and presents a safety hazard. Nonetheless, fires have heated dwellings for thousands of years. While fireplaces that burn wood or peat have chimneys that carry away their noxious fumes, the rising smoke takes with it much of the fire's thermal energy and some of the room's air. That's why a room that's heated by a fireplace often feels drafty away from the fireplace itself—cold outside air is seeping in through cracks to replace air drawn up the chimney. Even when clean burning fuels are used without a chimney, there are no simple solutions to the oxygen or safety problems.

Like a fireplace, a woodstove sends fumes from its burning wood up a chimney. But before its thermal energy can follow the fumes outside, a well-designed woodstove transfers most of that energy into the room. A woodstove is an example of a **heat exchanger**—a device that transfers heat without transferring the hot molecules themselves. Its smoke never enters the room but heat from that smoke does. The gas furnace in Fig. 6.1.3 also employs a heat exchanger.

The burning coals and hot gases inside the woodstove contain a great deal of thermal energy and are much hotter than the room air. Because of this temperature difference, heat tends to flow from the fire to the room. What is not so clear yet is how that heat is transferred.

There are three principal mechanisms by which heat moves from the fire to the room: conduction, convection, and radiation. The woodstove makes wonderful use of all three so that most of the thermal energy released by the burning wood is transferred to the room. Let's examine these three mechanisms of heat transport, beginning with conduction.

Fig. 6.1.3 This modern furnace burns natural gas in an S-shaped firebox. A fan at the bottom of the furnace blows fresh air past the hot outer surfaces of the firebox and then circulates the heated air among the rooms.

Check Your Understanding #4: Feeling the Heat

You can make a heat pack by wrapping hot, wet towels in a plastic sheet. This pack will warm an injured muscle but will not get it wet. Is thermal energy moving in this case?

Heat Moving Through Metal: Conduction

Conduction occurs when heat flows through a stationary material. The heat moves from a hot region to a cold region but the atoms and molecules don't. For example, if you place the tip of a metal poker in the fire, the poker's handle will gradually become warm as the metal conducts heat.

Some of this heat is conducted by interactions between adjacent atoms. The vibrating atoms frequently push on one another, doing work in the process and exchanging small amounts of thermal kinetic energy. In this fashion, thermal energy flows randomly from atom to neighboring atom.

But when the poker's tip is hotter than its handle, the flow is no longer completely random. The atoms at the hot tip have more thermal kinetic energy to exchange with their neighbors than atoms at the cold handle. The exchanges statistically favor the flow of thermal energy away from the hot tip and toward the cold handle. This flow of thermal energy from hot to cold through the poker is conduction (Fig. 6.1.4).

However, this atom-by-atom "bucket-brigade" isn't the only way in which materials conduct heat. In a metal, the primary carriers of heat are actually mobile **electrons**—the tiny negatively charged particles that make up the outer portions of atoms. When atoms join together to form a metal, some of the electrons stop belonging to particular atoms and travel almost freely throughout the metal. These mobile electrons can carry electricity (as we'll discuss in Chapter 8) and are also good at transporting heat.

Mobile electrons participate in the bucket-brigade of heat conduction because they, too, can push on vibrating atoms and exchange thermal kinetic energy with them. But while atoms can pass thermal energy only from one neighbor to the next, mobile electrons can travel great distances between exchange partners and can move thermal energy quickly from one place to another.

The ease with which electrons move heat about a metal explains why metals generally have higher thermal conductivities than nonmetals. **Thermal conductivity** is the measure of how rapidly heat flows through a material when it's exposed to a difference in temperatures. The best conductors of electricity—copper, silver, aluminum, and gold—are also the best conductors of heat. Poor conductors of electricity—stainless steel, plastic, and glass—are also poor conductors of heat. There are a few exceptions to this rule. Diamonds, for example, are terrible conductors of electricity but wonderful conductors of heat. Of course, it would be silly to make a woodstove out of diamonds because diamonds burn.

Conduction is what moves thermal energy from the woodstove's inside to its outside. No atoms move through the metal walls of the stove, just heat. So conduction serves as a filter, separating desirable thermal energy from the unwanted smoke and noxious gases that then go up the chimney.

Thus conduction makes the outside surface of the woodstove very hot so that heat should flow from it to the colder room. But what carries heat into the room? If you touch the stove, conduction will immediately transfer a huge amount of heat to your skin and you'll be burned. But even without touching the stove, you're aware of its high temperature. It transfers heat into the room by convection and radiation.

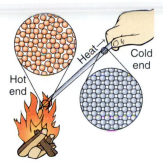

Fig. 6.1.4 When one end of a metal poker is hotter than the other, the atoms at the hot end vibrate more vigorously than those at the cold end. The poker then conducts heat from the hot end to the cold end. Some of this heat is conducted by interactions between adjacent atoms. However, in the metal poker, most of the heat is conducted by mobile electrons, which carry thermal energy long distances from one atom to another.

► Check Your Understanding #5: Too Hot to Handle

Some pot handles remain cool during cooking while others become unpleasantly hot. What determines which handles remain cool and which become hot?

Heat Moving with Air: Convection

Convection occurs when a moving fluid transports heat from a hotter object to a colder object. The heat moves as thermal energy in the fluid so that the two travel together. The fluid usually follows a circular path between the two objects, picking

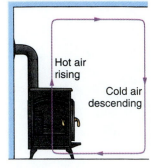

Fig. 6.1.5 When the woodstove is hot, convection carries heat from its surfaces to the ceiling and walls of the room. Warm air rises upward, supported by the buoyant force, and is replaced by cooler air from the floor. The warmed air eventually cools and descends. It then returns toward the stove to repeat the cycle.

up heat from the hotter object, giving it to the colder object, and then returning to the hotter object to begin again.

This circulation often develops naturally. As the fluid warms near the hotter object, its density decreases and it floats upward, supported by the buoyant force. When the fluid cools near the colder object, its density increases and it sinks downward.

Thus air heated by contact with the woodstove rises toward the ceiling and is replaced by colder air from the floor (Fig. 6.1.5). Eventually, this heated air cools and descends. Once it reaches the floor, it's drawn back toward the hot woodstove to start the cycle over. This moving air is a *convection current,* and the looping path that it follows is a *convection cell.* Within the room, convection currents carry heat up and out from the woodstove to the ceiling and walls. When you put your hand over the stove, you feel this convection current rising upward as it transfers heat to your hand.

Natural convection is good at heating the air above the woodstove, but most of that hot air ends up near the ceiling. While some of it will eventually drift downward to where you're standing, convection sometimes needs help. Adding a ceiling fan will help move the hot air around the room and make the woodstove more effective. This forced air circulation still transfers heat from the hot stove to the colder occupants of the room, but it doesn't rely on gravity and the buoyant force to keep the air circulating. The faster the air moves, the more heat it can transport from hot objects to cold objects.

Check Your Understanding #6: Heat and Wind

When sunlight warms the land beside a cool body of water, a breeze begins to blow from the water toward the land. Explain.

Heat Moving as Light: Radiation

There is one more important mechanism of heat transfer: radiation. As the particles inside a material jitter about with thermal energy, they emit and absorb electromagnetic radiation. This radiation consists of electromagnetic waves, which include radio waves, microwaves, and infrared, visible, and ultraviolet light.

We'll study electromagnetic radiation in Chapters 11 and 12. For now, what's most important is that this radiation can carry thermal energy. When heat flows from a hot object to a cold object as electromagnetic radiation, we say that heat is being transferred by thermal radiation or simply **radiation.** Unlike conduction and convection, which depend on atoms, molecules, or electrons to carry the heat, radiation occurs directly through space. Radiative heat transfer happens even when two objects have nothing at all between them.

The types of electromagnetic waves in an object's thermal radiation depend on its temperature. While a colder object emits only radio waves, microwaves, and infrared light, a hotter object can also emit visible or even ultraviolet light. The red glow of a hot coal in the woodstove is that coal's thermal radiation.

Since our eyes are only sensitive to visible light, we can't see all of the thermal radiation emitted by an object, even when it's hot. But whether we see it or not, electromagnetic radiation contains energy and transfers heat to whatever absorbs it. While everything emits thermal radiation, the amount of that emission depends on temperature: the hotter an object gets, the more thermal radiation it emits. Because of this temperature dependence, any exchange of thermal energy via radiation always transfers heat from a hotter object to a colder one.

Radiation transfers a great deal of heat from the woodstove's surface to the surrounding objects. The stove bathes the room in infrared light, which warms everything it reaches. To encourage such radiative heat transfer, the woodstove and its chimney are often painted black. Black not only absorbs light well, but it's also particularly good at emitting light. If you heat a black poker red hot, it will glow much more brightly than one that's white, silvery, or transparent. White, silvery, and transparent surfaces are poor absorbers of light when they're cold and poor emitters of light when they're hot.

Even if air in the room is cold, you can usually feel the invisible infrared light from a woodstove on your face. When you block this light with your hands, your face suddenly feels colder because less heat is reaching your skin. This thermal radiation effect is even more pronounced with a fireplace or campfire, where thermal radiation from the hot coals and flames is the primary mechanism for heat transfer to the surroundings.

Overall, a modern woodstove is an excellent heat exchanger. As convection draws hot smoke up the long black chimney pipe, the smoke heats the stove and the pipe. These metal components conduct heat to their outer surfaces, which then distribute it around the room by convection and radiation. Although the stove consumes some room air, it controls the airflow with dampers so that it draws in only enough air to completely burn the wood. Overall, the stove extracts heat efficiently, cleanly, and safely from the burning wood.

Check Your Understanding #7: Keeping Warm

When you stand under a heat lamp, you feel warm even though the lamp emits very little light. How is heat reaching your skin?

furnaces, heating systems, electric and solar

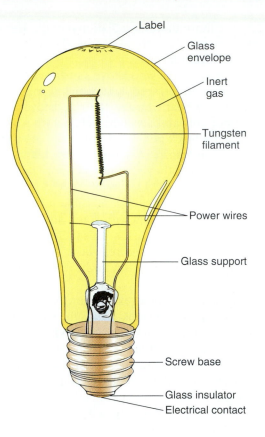

Label
Glass
envelope
Inert
gas
Tungsten
filament
Power wires
Glass support
Screw base
Glass insulator
Electrical contact

SECTION 6.2 — Incandescent Light Bulbs

For more than a century, incandescent light bulbs have provided light at the flip of a switch. Their invention brought to a close the era of candles and gaslights and spurred the development of electric power. While the variety of incandescent bulbs has grown over the years to include everything from heat lamps to halogen head-lights, all incandescent bulbs have at their hearts one simple object: an extremely hot wire filament.

Questions to Think About: *What part of a light bulb emits the light? How is a light bulb similar to a fire or a candle? How is it different? What colors of light can a plain, unpainted light bulb emit? Why does the top of a light bulb darken with age? What happens when a light bulb burns out?*

Experiments to Do: *Take a look at a few incandescent light bulbs. Try turning one on and off. Are the transitions instantaneous? Stand in a darkened room with your eyes closed and open your eyes suddenly, just after you turn the bulb off. Can you see the bulb going dark? How do its brightness and color change with time?*

Compare the color of the light from a conventional bulb with that from an ex-tended life bulb. Which one produces a better simulation of sunlight? Which bulb should you use in your desk lamp? in an inaccessible ceiling lamp?

Now compare both bulbs with a halogen bulb. How do their colors differ? Which bulb do you expect to live the longest in normal use? Is it surprising that halogen bulbs live longer than conventional bulbs?

Light, Temperature, and Color

Light from an incandescent light bulb is part of the thermal radiation emitted by its hot wire filament. While most electromagnetic waves are invisible, our eyes are sensitive to a narrow range of waves that we call visible light. Any object that's hotter than about 400 °C emits enough visible light for us to see it in a dark room. At higher temperatures, that visible light brightens and shifts in color from red to orange to yellow to white. At 500 °C, we see the object glowing a dull red. At 1700 °C, it emits the orange light of a candle. And at 5800 °C, the temperature of the sun's surface, it gives off the brilliant white light of the sun.

To reproduce pure white sunlight, the bulb's filament should be heated to 5800 °C. Unfortunately, nothing is solid at that high temperature. Even tungsten metal, the best filament material known, evaporates quickly at temperatures above 2500 °C. Since incandescent light bulbs must operate at lower temperatures, they can't really reproduce sunlight. Most give off the warm, yellow-white light that's characteristic of tungsten metal at 2500 °C.

As you can see, the filament's brightness and color depend on its temperature (Fig. 6.2.1). We can measure brightness as the number of watts of visible light the filament emits. But how do we characterize color, and what distinguishes visible light from the invisible forms of electromagnetic radiation? Although the full answers to those questions will have to wait until Chapters 11 and 12, we can make a few essential observations about them now.

Visible light is part of a continuous spectrum of electromagnetic radiation that extends from radio waves at one extreme to gamma rays at the other (Fig. 6.2.2). Different types of electromagnetic radiation are distinguished by their **wavelengths**—that is, the distance between their wave crests. Wavelength is easy to see in the waves on a lake or sea, where the crests are visible and you can directly measure the distance from one to the next. But while the wave crests of electromagnetic waves aren't so easy to observe, they do exist and it's possible to measure their spacings.

The electromagnetic radiation produced by a hot filament is mostly infrared, visible, and ultraviolet light. Although this light is just a tiny portion of the overall electromagnetic spectrum, it's particularly important to our everyday world. Figure 6.2.3, p. 200, gives an expanded view of the visible portion of the electromagnetic spectrum. Various colors that we see correspond to specific wavelength ranges. For example, light with a wavelength of 530 nanometers (billionths of a meter, abbreviated nm) appears green to our eyes.

Fig. 6.2.1 As you increase the power to an incandescent light bulb, its filament becomes hotter and emits a brighter and whiter light. The cooler filament on the left is dim and red while the hotter one on the right is bright and yellow-white. Because these bulbs are frosted, you can't see their filaments directly.

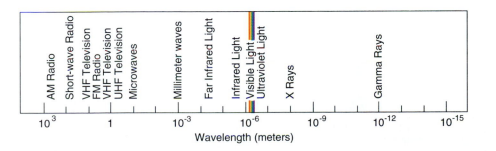

Fig. 6.2.2 The spectrum of electromagnetic radiation, arranged by wavelength. The scale here is logarithmic, meaning that the wavelength decreases by a factor of 10 with each tick mark to the right.

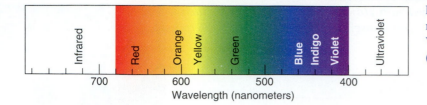

Fig. 6.2.3 A portion of the electromagnetic radiation spectrum around visible light. Wavelengths are measured in nanometers (nm or billionths of a meter).

But the thermal radiation emitted by a filament isn't a single electromagnetic wave with one specific wavelength. Instead, it's many individual waves that cover a broad range of wavelengths. Some of these waves are red light, some green, some blue, and some are invisible.

The distribution of wavelengths emitted by the filament depends on its temperature and surface properties, particularly its **emissivity**—the efficiency with which it emits and absorbs light. Emissivity is measured on a scale from 0 to 1, with 1 being ideal efficiency. A perfectly black object has an emissivity of 1. Although tungsten's emissivity is only 0.43, the filament wire is wound in such a way that it has many dark nooks and crannies. The filament is thus essentially black and its overall emissivity is close to 1.

The distribution of wavelengths emitted by a black object is determined by its temperature alone and is called a **black body spectrum.** As you can see from the examples in Fig. 6.2.4, the spectrum of a black body brightens and shifts toward shorter wavelengths as its temperature increases. An object that isn't black emits somewhat less thermal radiation, but that radiation still brightens and shifts toward shorter wavelengths as the object becomes hotter.

Our eyes make an average assessment of the distribution of wavelengths emitted by a black object, and we observe reddish, orangish, yellowish, whitish, or bluish light,

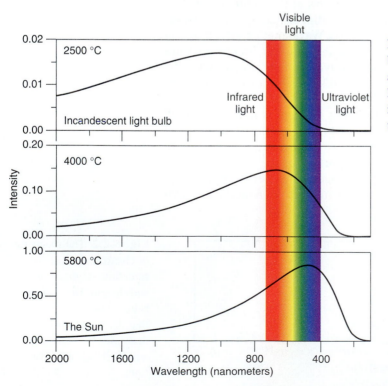

Fig. 6.2.4 The distributions of light emitted by black objects at 2500 °C (top), 4000 °C (middle), and 5800 °C (bottom). In addition to containing a larger fraction of visible light, the 5800 °C object is much brighter than the 2500 °C object (note the different intensity scales).

Table 6.2.1	Temperatures and Colors of Light Emitted by Hot Objects	
OBJECT	TEMPERATURE	COLOR
Heat lamp	500 °C	Dull red
Candle flame	1700 °C	Dim orange
Bulb filament	2500 °C	Bright yellow-white
Sun's surface	5800 °C	Brilliant white
Blue star	>6000 °C	Dazzling blue-white

depending on the object's temperature (Table 6.2.1). The temperature associated with a particular distribution of wavelengths is the **color temperature** of that light.

We can already see two of the principal shortcomings of incandescent light bulbs: their poor efficiency at converting electric energy into visible light and their low color temperature. At 2500 °C, only about 12% of the thermal radiation they emit is visible light. The rest is invisible infrared light. They would have to reach 5000 °C before the infrared fraction of their thermal radiation would drop below 50%. Moreover, their 2500 °C color temperature makes them look yellowish when compared to sunlight because they don't emit enough blue light. Most of the developments in lighting over the past half century have focused on improving energy efficiency and color temperature.

Check Your Understanding #1: Even Plants Can Look Cool

Satellites measure the temperatures of agricultural regions from space, looking for signs of crop distress and disease. How do they make such measurements?

The Filament

The bulb's filament is heated by a current of electrically charged electrons. These particles flow through the filament, where most of their electric energy is converted into thermal energy. The filament heats up until it radiates away thermal energy as quickly as the electricity produces more.

This balance is reached at a specific temperature because the power radiated by the filament is proportional to the fourth power of its absolute temperature. The precise relationship between temperature and emitted power can be written as a word equation:

$$\text{radiated power} = \text{emissivity} \cdot \text{Stefan–Boltzmann constant} \\ \cdot \text{temperature}^4 \cdot \text{surface area}, \tag{6.2.1}$$

in symbols:

$$P = e \cdot \sigma \cdot T^4 \cdot A,$$

and in everyday language:

Warm skin radiates away lots of heat, so cover it up to avoid feeling cold.

This relationship is called the **Stefan–Boltzmann law,** and the **Stefan–Boltzmann constant** (σ) that appears in it has a measured value of 5.67×10^{-8} J/(s·m²·K⁴). Remember that the temperature must be measured in kelvin units.

But while the concept of an incandescent bulb is simple, finding a material that can tolerate extremely high temperatures is not. Early filaments were made of car-

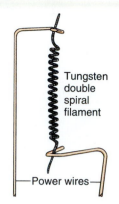

Fig. 6.2.5 The tungsten filament of a modern incandescent light bulb is a double spiral—a coil wound from a smaller coil of extremely fine tungsten wire. The double spiral allows the manufacturers to put a long length of wire in a small space.

special filaments

❑ **Lewis Howard Latimer (African-American scientist and inventor, 1848–1928)** was only eight when the U.S. Supreme Court's Dred Scott decision made his escaped-slave father a fugitive and forced him to disappear. Left behind with his mother, Latimer did well in school and became a skilled draftsman and engineer. While working for Edison's rival, Hiram Maxim, Latimer became an expert in fabricating carbon filaments for incandescent lamps. When Latimer later joined Edison's team of inventors, the "Edison Pioneers," his sturdy carbon filaments quickly replaced Edison's own fragile bamboo ones and provided the crucial ingredient necessary for Edison's lamps to become a commercial success.

bon and platinum. Of these materials, carbon showed the most promise. In 1879, Thomas Edison developed an incandescent lamp with a carbon filament that operated for several hundred hours. His wasn't the first incandescent bulb ever made but rather the first practical one. (For more about the carbon filaments, see ❑.)

However, while carbon has the highest melting temperature of any element (3550 °C), it evaporates atoms directly from the solid at much lower temperatures. This process by which a solid turns directly into a gas is called **sublimation** and occurs because individual atoms can occasionally gather together enough thermal energy to break free from the material. Because of carbon's tendency to sublime, a carbon filament that is heated close to its melting temperature quickly disappears as a gas. When a gap appears in the filament, it stops carrying electricity and "burns out." Furthermore, carbon is flammable, so it must be enclosed in an airtight glass bulb that contains either inert gases or a vacuum.

A better choice for filaments, now used in virtually all incandescent bulbs, is tungsten metal. Tungsten melts at 3410 °C and sublimes only very slowly at temperatures below this melting temperature. Tungsten filaments can be heated to higher temperatures than carbon filaments before the rate of sublimation becomes intolerable. Since a tungsten filament bulb runs hotter than a carbon filament bulb, it produces a richer, whiter light, more like that of the sun. Like carbon, hot tungsten burns in air and must be protected in a glass bulb.

To ensure that most of the electric energy passing through the filament is converted into thermal energy, the filament must be long and thin. A typical 60-W light bulb has about 0.5 m of 25-micron (0.001-inch) tungsten wire, coiled up into a filament only about 2 cm long. To minimize the filament's length, it's wound into a double spiral. First it's wound into a thin springlike coil about 0.25 mm wide. Then this coil is wound into another coil to form the actual filament (Fig. 6.2.5). Fabricating such a complicated tungsten filament is so difficult that it wasn't accomplished until 1937.

Check Your Understanding #2: Here Today, Gone Tomorrow

Snow often disappears from the ground in a period of weeks, even when the temperature remains below freezing. How does the snow disappear?

Check Your Figures #1: The Cold of Deep Space

Suppose that an accident in the depths of space leaves you exposed to an environment near absolute zero. Since your surroundings radiate almost no heat toward you, you are losing heat fast. If your surface area is 2 m^2, your skin temperature is 310 K, and your emissivity is 0.5, how much power will you radiate?

The Glass Bulb

To keep the white-hot filament from burning, it's surrounded by a glass bulb that usually contains oxygen-free inert gas. The gas, typically nitrogen and argon, slows sublimation by bouncing some of the escaping tungsten atoms back onto the filament. Although this gas extends the filament's life, it has at least two drawbacks. First, it allows conduction and convection to carry some heat away from the filament. Second, tiny tungsten particles that form in this gas rise with convection currents to produce a dark spot on the top of the bulb.

The glass bulb presents another challenge: operating a filament inside it involves passing wires right through the glass. This step isn't so easy because the glass and metal must seal to one another perfectly. Complicating this sealing requirement is the fact that materials expand as their temperatures increase. If the glass and metal don't expand equally as the bulb warms up, the bulb may leak.

A material's thermal expansion is caused by atomic vibrations. Because of thermal energy, adjacent atoms vibrate back and forth about their equilibrium separations (Fig. 6.2.6). This vibrational motion isn't symmetric; the repulsive force the atoms experience when they're too close together is stiffer than the attractive force they experience when they're too far apart. As a result of this asymmetry, they push apart more quickly than they draw together and spend most of their time at more than their equilibrium separation. On average, their actual separation is larger than their equilibrium separation, and the material containing them is bigger than it would be without thermal energy.

Increasing an object's temperature moves its atoms even farther apart on average, and the object grows larger in all directions. The extent to which an object expands with increasing temperature is normally described by its **coefficient of volume expansion:** the fractional change in the object's volume caused by a temperature increase of 1 °C. Fractional change in volume is the net change in volume divided by the total volume. Since most materials expand only a small amount when heated 1 °C, coefficients of volume expansion are small, typically about 10^{-5} for metals, about 10^{-6} for special low-expansion glasses, and about 10^{-4} for liquids. In a light bulb, the metal wires and the glass are carefully selected to have similar coefficients of volume expansion. As the bulb warms up, the wires and glass expand together and the seals remain intact.

Fig. 6.2.6 The thermal kinetic energy of a solid increases with temperature, causing its atoms to bounce against one another more vigorously. As they vibrate, the atoms repel more strongly than they attract so their average separation increases slightly.

Average separation

Increasing temperature

krypton bulbs

> ## Check Your Understanding #3: Overcooked
> If you place a pot or saucepan on the stove and fill it to the brim with cold water, it will overflow as you heat it. From where does the extra water come?

Extended Life, Halogen, and Three-Way Bulbs

One way to increase the life of the filament is to make it extra long. This change reduces the amount of thermal power each portion of filament must radiate away so the filament doesn't get quite as hot and doesn't sublime as quickly. The result is an extended life bulb. Unfortunately, extended life bulbs are dimmer and redder than conventional bulbs and also less energy efficient. Because an extended life bulb emits a smaller fraction of its input power as visible light, it must have a higher wattage to give equivalent lighting. As a result, extended life bulbs aren't always a bargain; the money you save on replacement bulbs may well be spent on increased energy costs.

You're much better off buying a halogen bulb, which is both longer-lived and more energy efficient than a conventional bulb. A halogen bulb uses a chemical trick to rebuild its filament continuously during operation. This filament is enclosed in a small tube of quartz or aluminosilicate glass, which can tolerate high temperatures and reactive chemicals (Fig. 6.2.7). The tube contains molecules of the halogen elements bromine and/or iodine. During operation, the tube becomes extremely hot and the halogen reacts with any tungsten atoms on its inside surface. They form a gas of tungsten–halogen molecules that drift about the tube until they encounter the white-hot filament. The molecules then break apart and the tungsten atoms stick to the filament.

The halogens act as scavengers, seeking out tungsten atoms that have sublimed from the filament and using them to rebuild it. Unfortunately, this rebuilding

Fig. 6.2.7 These halogen bulbs operate at higher temperatures than normal incandescent bulbs and produce whiter light. The large glass envelope of the upper bulb protects a smaller lamp inside.

halogen shortcomings

process slowly changes the structure of the filament. The returning tungsten atoms deposit unevenly, so that the filament gradually develops thin spots and eventually burns out. Nonetheless, the filament lives more than 2000 hours, even when it runs several hundred degrees hotter than in a conventional bulb. Its higher filament temperature allows a halogen bulb to produce whiter light than a conventional bulb and increases its energy efficiency.

Bulbs with different power ratings have filaments with different amounts of surface area. The filament of a 100-W bulb has four times as much surface area as a 25-W bulb and thus emits four times as much light. Unlike the elongated filament of a cool-running extended life bulb, the filament of the 100-W bulb is both longer *and thicker* than that of the 25-W bulb. The 100-W filament draws four times as much electric power as the 25-W filament so that they both operate at the same temperature and emit light with the same color temperature.

One way to make an incandescent bulb with a variable light output is to put several independent filaments in it. A "three-way bulb" has two filaments (Figs. 6.2.8 and 6.2.9) that can be turned on and off separately. In a 50-100-150 W bulb, one filament uses 50 W of electric power and the other uses 100 W. If only the low-power filament is on, the bulb appears to be a 50-W bulb. If only the high-power filament is on, it appears to be a 100-W bulb. But when both filaments are on, the bulb appears to be a 150-W bulb. Since one of the filaments burns out before the other, the bulb fails by going from having three light levels to only one.

toasters

Fig. 6.2.8 A three-way bulb has two independent filaments. The filament on the left is shorter and thinner than the one on the right and emits about half as much light. The three different light levels correspond to having the left filament on, the right filament on, and both filaments on.

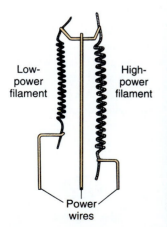

Check Your Understanding #4: Double Wrapping

While conventional light bulbs have thin glass envelopes, halogen bulbs that replace them have surprisingly thick glass shells. If you cut one of these shells open, you'll find a second, much smaller glass bulb inside. Why does the manufacturer go to the trouble of putting two separate bulbs around the filament?

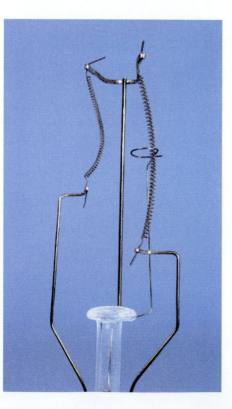

Fig. 6.2.9 The glass envelope of this three-way light bulb has been removed to expose its two filaments. The shorter, thinner filament (left) uses 50 W, while the longer, thicker one (right) uses 100 W.

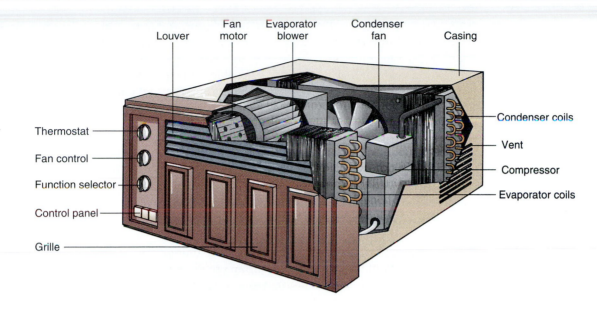

Louver | Fan motor | Evaporator blower | Condenser fan | Casing

Thermostat
Fan control
Function selector
Control panel
Grille

Condenser coils
Vent
Compressor
Evaporator coils

SECTION 6.3 · Air Conditioners

On a summer day, your problem isn't staying warm; it's keeping cool. Instead of looking for something to burn in your woodstove, you turn on your air conditioner. An air conditioner is a device that cools room air by removing some of its thermal energy. But the air conditioner can't make thermal energy disappear. Instead, it transfers thermal energy from the cooler room air to the warmer air outside. Since the air conditioner transfers heat against its natural direction of flow, the air conditioner is a "heat pump." It's also a classic illustration of the laws of thermodynamics in action.

Questions to Think About: Why doesn't heat naturally flow from a colder object to a hotter object? An air conditioner removes thermal energy from room air, so why does the air conditioner require electric energy to operate? Where does this electric energy go? Why does an air conditioner always have an indoor component and an outdoor component? If you put a window air conditioner in the middle of a room and turn it on, what will happen to the temperature of the room?

Experiments to Do: Take a look at a window air conditioner. If you can't find one, examine a refrigerator instead because it's basically a powerful air conditioner cooling a food-storage closet. As the air conditioner (or refrigerator) operates, feel the air leaving the indoor air vent (or inside the refrigerator) and compare its temperature to that of air leaving the outdoor air vent (or near the metal coils on the back of the refrigerator). Which way does the cooling mechanism move heat? If that mechanism were absent, which way would heat flow? Turn off the air conditioner or refrigerator and observe the resulting heat flow. Did you confirm your prediction?

Moving Heat Around: Thermodynamics

On a sweltering summer day, the air in your home becomes unpleasantly hot. Heat enters your home from outdoors and doesn't stop flowing until it's as hot inside as it is outside. You can make your home more comfortable by getting rid of some of its thermal energy. But while we've already looked at ways to *add* thermal energy to room air, we haven't yet learned how to *remove* it. At present, the only cooling method we've discussed is contact with a colder object. Unless you have an icehouse nearby, you need another scheme for eliminating thermal energy. You need an air conditioner.

An air conditioner transfers heat against its natural direction of flow. Heat moves from the colder air in your home to the hotter air outside, so that your home gets colder while the outdoor air gets hotter. There's a cost to transferring heat in this manner. The air conditioner requires ordered energy to operate and typically consumes large amounts of electric energy. It's a type of **heat pump**—a device that uses ordered energy to transfer heat from a colder object to a hotter object, against its natural direction of flow.

Before learning how an air conditioner pumps heat, we should first show that pumping is necessary. There are a number of seemingly reasonable cooling alternatives that we should consider before turning to air conditioning. Three such alternatives are:

1. Letting heat flow from your home to your neighbor's home.
2. Destroying some of your home's thermal energy.
3. Converting some of your home's thermal energy into electric energy.

Unfortunately, these three alternatives can't be done. Still, it will be useful to us to examine them more closely because in doing so, we'll learn about the laws governing the movement of thermal energy, the **laws of thermodynamics.**

The first alternative raises an interesting issue. Your home is in thermal equilibrium with the outside air, meaning that no heat flows from one to the other and they're at the same temperature. Your neighbor's home is also in thermal equilibrium with the outside air. What will happen if you permit heat to flow between your home and your neighbor's home? Nothing. Since both homes are simultaneously in thermal equilibrium with the outside air, they're also in thermal equilibrium with one another. All three are at the same temperature.

This observation is an example of the **zeroth law of thermodynamics,** which says that two objects that are each in thermal equilibrium with a third object are also in thermal equilibrium with one another. This seemingly obvious law is the basis for a meaningful system of temperatures. If you had a roomful of objects at 20 °C and some were in thermal equilibrium with one another while others were not, then "being at 20 °C" wouldn't mean much. Fortunately, every object that has a temperature of 20 °C is in thermal equilibrium with every other object at 20 °C. The zeroth law is observed to be true in nature so that temperature does have meaning. And since your neighbor's home is just as hot as yours, they can relax because you're not going to be sending them any extra heat.

The Zeroth Law of Thermodynamics
Two objects that are each in thermal equilibrium with a third object are also in thermal equilibrium with one another.

The second alternative sounds unlikely from the outset. We've known since the first chapter that energy is special, that it's a conserved quantity. You can't cool your home by destroying thermal energy because energy can't be destroyed. To eliminate thermal energy, you must convert it to another form or transfer it elsewhere.

This concept of energy conservation is the basis for the **first law of thermodynamics,** which states that the change in a stationary object's internal energy is equal to the heat transferred into that object minus the work that object does on its surroundings. Internal energy includes both thermal energy and stored potential energy. This law says that heat added to the object increases its internal energy while work done by the object decreases its internal energy. Since energy is conserved, the only way the object's internal energy can change is by transferring it as heat or work.

The First Law of Thermodynamics
The change in a stationary object's internal energy is equal to the heat transferred into that object minus the work that object does on its surroundings.

Check Your Understanding #1: Stirring Up Trouble
If you put cold water into a blender and mix it rapidly for several minutes, the water will become warm. From where does the additional thermal energy come?

Disorder, Entropy, and the Second Law

The third alternative looks much more promising than the first two. It seems as though you should be able to convert thermal energy into electricity (or some other ordered form of energy). You could then sell it back to the electric company and get credit on your bill. Wouldn't that be great?

But there's a problem. Ordered energy and thermal energy aren't equivalent. You can easily convert ordered energy into thermal energy but the reverse is much harder. For example, you can burn a log to convert its chemical potential energy into thermal energy, but you'll have trouble converting that thermal energy back into chemical potential energy to recreate the log.

The basic laws of motion are silent on this issue. It isn't that the smoke doesn't have the energy to recreate the log. It's that the individual smoke particles must pool their thermal energies together to carry out the reassembly, a remarkably unlikely event. The particles would all have to move in just the right ways to turn the burned gases back into wood and oxygen, an incredible coincidence that simply never happens. Similarly, all of the air particles in your home would have to act together to convert their thermal energy into electricity. Since that coordinated behavior is ridiculously improbable, you're not going to be selling power to the electric company any time soon.

Once energy has been scattered randomly among the individual air particles, you can't collect that energy back together again. Creating disorder out of order is easy, but recovering order from disorder is nearly impossible. As a result, systems that begin with some amount of order gradually become more and more disordered, never the other way around. The best they can do is to stay the same for a while so that their disorder doesn't change. From these observations, we can state that the disorder of an isolated system *never decreases*.

This notion of never decreasing disorder is one of the central concepts of thermal physics. There is even a formal measure of the total disorder in an object: **en-**

tropy. All disorder contributes to an object's entropy, including its thermal energy and its structural defects. Breaking a window or heating it both increase its entropy.

Because disorder never decreases, the third cooling alternative is impossible. Turning your home's thermal energy into electric energy would reduce its disorder and decrease its entropy. But our observations about entropy aren't yet complete. There's one way to decrease your home's entropy: you can export that entropy somewhere else. In fact, you export entropy every time you take out the garbage, though that action also changes the contents of your home. You can also export entropy without modifying your home's contents by transferring heat somewhere else. Heat carries disorder and entropy with it, so getting rid of heat also gets rid of entropy.

Our rule about entropy never decreasing is weakened by the possibility of exchanging heat and entropy between objects. Before asserting that an object or system of objects can't decrease its entropy, we must ensure that it's thermally isolated from its surroundings so that it can't export its entropy. With that in mind, the strongest statement that we can make concerning entropy is that the entropy of a thermally isolated system of objects never decreases. This observation is the **second law of thermodynamics.**

perpetual motion

The Second Law of Thermodynamics
The entropy of a thermally isolated system of objects never decreases.

Because of the second law, the only way to cool your home is to export its thermal energy and entropy elsewhere. Such a transfer would be easy if you had a cold object nearby to receive the heat. But lacking a cold object, you must use an air conditioner. Like all heat pumps, an air conditioner transfers heat and entropy in such a way that the second law of thermodynamics is never violated and the entropy of each thermally isolated system of objects never decreases. As we'll see, the air conditioner lowers the entropy of your home but raises the entropy of the outside air so that, overall, the entropy of the world actually increases.

There's a limit to how much entropy an air conditioner can remove from your home. As it exports thermal energy and entropy, the air conditioner lowers your home's temperature. In principle, your home will eventually approach absolute zero and, as it does, its entropy will approach zero. This relationship between the zero of temperature and the zero of entropy is the **third law of thermodynamics,** which states that as an object's temperature approaches absolute zero, its entropy approaches zero. The third law establishes absolute zero as a destination with no disorder left, but the second law ultimately makes it impossible to extract all the disorder from an object. Because absolute zero is unattainable, the third law refers to *approaching it* rather than to *being there.*

The Third Law of Thermodynamics
As an object's temperature approaches absolute zero, its entropy approaches zero.

Check Your Understanding #2: Something for Nothing
People have tried for centuries to build machines that provide endless outputs of useful, ordered energy without any inputs of ordered energy. Unfortunately, such perpetual motion machines violate the laws of thermodynamics. If such a ma-

chine is thermally isolated, which law does it violate? What about if it's not thermally isolated?

Pumping Heat Against Its Natural Flow

While the second law of thermodynamics doesn't allow the entropy of a thermally isolated system to decrease, it does permit the objects in that system to redistribute their individual entropies. One object's entropy can decrease as long as the entropy of the rest of the system increases by at least as much. This shifting of entropy allows part of the system to become colder if the rest of the system becomes hotter.

For example, suppose that there's a pond of cold water behind your home. You could pump that water through your bathtub and let it draw heat out of the room air. Your home would become colder while the pond would become warmer. This transfer of heat from the hot air in your home to the cold water in the pond satisfies the second law of thermodynamics. The entropy of the combined system—your home and the pool of water—doesn't decrease. In fact, it actually increases!

This entropy increase occurs because heat is more disordering to a cold object than it is to a hot object. Each joule of heat that flows from your home to the pool creates more disorder in the pool than it creates order in your home.

A useful analog for this effect involves two parties taking place simultaneously: the garden society's annual tea party and a four-year old's birthday party. The orderly tea party represents the cold pool while the disorderly birthday party represents your hot home. The analogy to letting heat flow from your hot home to the cold pool is to exchange one lively four-year old from the disorderly birthday party for one quiet octogenarian from the orderly tea party. This transfer will reduce the birthday party's disorder only slightly, but it will dramatically increase the disorder of the tea party. The attendance at each party will be unchanged but their total disorder will increase.

When heat flows from your home to the pool, the overall entropy increases and the second law is more than satisfied. A similar increase in entropy occurs whenever heat flows from a hot object to a cold object, which is why heat normally flows in that direction.

But an air conditioner does the seemingly impossible: it transfers heat from a cold object, your home, to a hot object, the outside air. Heat flows in the wrong direction and creates less disorder as it enters the hot outside air than it creates order by leaving the cold indoor air. The entropy of the combined system decreases, a violation of the second law of thermodynamics!

However, we've omitted an important feature of the air conditioner's operation: the electric energy it consumes. The air conditioner converts this ordered energy into thermal energy and delivers it as heat to the outside air. In doing so, the air conditioner creates enough extra entropy to ensure that the overall entropy of the combined system increases. The second law is satisfied after all.

An air conditioner always consumes ordered energy when it pumps heat from a colder object to a hotter object because it must create the extra entropy required by the second law of thermodynamics (Fig. 6.3.1). The amount of ordered energy it consumes depends on the temperatures of the two objects. If the two objects are close in temperature, the transfer of heat reduces the entropy only slightly so the air conditioner doesn't need to convert much ordered energy into thermal energy. But if the objects are far apart in temperature, the air conditioner must create lots of extra entropy to make up for the entropy lost in the transfer.

This need to keep entropy from decreasing explains why an air conditioner works best when it's cooling your home the least. The greater the temperature dif-

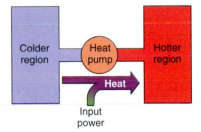

Fig. 6.3.1 A heat pump transfers heat from a colder region to a hotter region. In doing so, it converts some amount of ordered energy into thermal energy in the hotter region. The larger the temperature difference between the two regions, the more ordered energy is required to transfer each joule of heat.

ference between the indoor air and the outdoor air, the more electric energy the air conditioner must consume in order to move each joule of heat. (This energy consumption would become infinite if you tried to cool your home to absolute zero, which is why absolute zero is unattainable!) Moreover, the heat flowing back into your home through its walls is roughly proportional to this temperature difference. Thus the colder you set the thermostat, the larger your electric bills will be.

Check Your Understanding #3: Heat Pumps in Cold Weather

Homes located in mild climates are often heated by heat pumps during the winter. Home heat pumps are essentially air conditioners run backwards. They extract heat from the cold outdoor air and release it to the warm indoor air. Why are heat pumps most effective in mild weather, when the outdoor air isn't too cold?

How an Air Conditioner Cools the Indoor Air

Having determined the air conditioner's goals, it's time for us to look at how a real air conditioner meets them. In most cases, the air conditioner uses a fluid to transfer heat from the colder indoor air to the hotter outdoor air. This *working fluid* absorbs heat from the indoor air and releases that heat to the outdoor air.

The air conditioner has three main components: an evaporator, a condenser, and a compressor (Fig. 6.3.2). The evaporator is located indoors, where it transfers heat from the indoor air to the working fluid (Fig. 6.3.3). The condenser is located outdoors, where it transfers heat from the working fluid to the outdoor air. And the compressor is also located outdoors, where it does work on the working fluid and introduces the additional thermal energy needed to ensure that the total disorder of the system doesn't decrease. To see how these three components pump heat out of your home, let's look at them individually.

We'll begin with the evaporator, a long metal pipe that's decorated with thin metal fins. The evaporator is a heat exchanger, and these fins help heat flow from the warmer room air around it to the cooler working fluid inside it. But the working fluid arrives at the evaporator as a warm, high-pressure liquid. To initiate a drop in its temperature, the fluid passes through a narrowing in its pipe just before it enters the evaporator. This constriction impedes the flow and causes the fluid's pressure to drop dramatically. Thus the working fluid enters the evaporator as a warm, low-pressure liquid.

The chemical bonds that hold the working fluid's particles together are weak and fragile. Near room temperature, the particles need a high pressure to keep them together as a liquid. Once the liquid working fluid passes through the constriction and loses its pressure, its particles begin to separate. One by one, these particles break away from each other so that the working fluid gradually **evaporates** from a liquid of attached particles to a gas of independent particles.

Breaking the weak chemical bonds between particles takes energy— thermal energy. As the working fluid evaporates, some of its thermal energy becomes chemical potential energy in the newly separated gas particles. With its diminished thermal energy, the gaseous working fluid be-

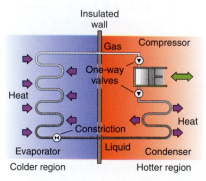

Fig. 6.3.2 A heat pump transfers heat from a colder region to a hotter region. A typical air conditioner does this by condensing a gas to a liquid on the hot side and evaporating the liquid to a gas on the cold side. A compressor provides the necessary input of ordered energy.

Insulated wall

Gas　　Compressor

One-way valves

Heat

Constriction

Evaporator　　Liquid　　Condenser

Colder region　　　　Hotter region

Heat

comes quite cold. This cold gas chills the evaporator and its fins so that heat begins to flow into the working fluid from the room air.

By the time the working fluid emerges from the evaporator, it has evaporated completely and has absorbed considerable thermal energy from the indoor air. It leaves the evaporator as a cool low-pressure gas and travels through a pipe toward the compressor.

Half the air conditioner's job is done: it has removed heat from the indoor air. But the remaining half of its job is more complicated: it must add heat to the outdoor air while ensuring that the total entropy of the combined system doesn't decrease. After all, there's no getting around the second law of thermodynamics.

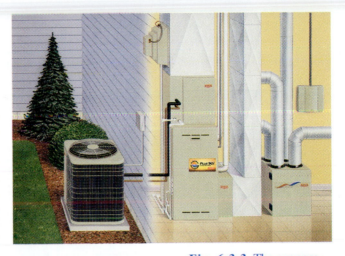

Fig. 6.3.3 The evaporator of this central air conditioning unit extracts heat from the indoor air. This heat is transferred outside, where the compressor and condenser release it into the outdoor air.

> **Check Your Understanding #4: Cooling at the Gas Grill**
> When liquid propane evaporates into a gas in the tank of a propane grill, the tank that contains that liquid propane becomes colder. Why?

How an Air Conditioner Warms the Outdoor Air

Satisfying the second law is the task of the compressor. The compressor receives low-pressure gaseous working fluid from the evaporator, compresses it to much higher density, and delivers it as a gas to the condenser. The compressor may use a piston and one-way valves, like the water pump in Fig. 4.2.2, or it may use a rotary pumping mechanism. In either case, the result is the same: the gaseous working fluid leaves the compressor at much higher density and pressure than it had when it arrived.

Compressing a gas requires work because the compressor must push the gas inward while moving it inward—force times distance. Since work transfers energy, the compressor increases the energy of the gas. The air conditioner usually obtains this energy from the electric company and converts it into mechanical work with an electric motor.

But the only way that the gaseous working fluid can store its new energy is as thermal energy in its individual particles. These particles begin to move about more and more rapidly so that the gaseous working fluid leaves the compressor much hotter than when it arrived.

This hot, high-pressure working fluid then flows into the condenser. Like the evaporator, the condenser is a long metal pipe with fins attached to it. It acts as another heat exchanger, and its metal fins help heat flow from the hotter working fluid inside it to the less hot outside air. The working fluid gradually cools as the outdoor air becomes hotter.

Near room temperature, the working fluid will only remain gaseous if its pressure is low. Otherwise, its particles begin sticking to one another. Once the high-pressure working fluid has cooled enough for this binding to begin, it gradually **condenses** from a gas of independent particles to a liquid of attached particles.

However, when the particles form chemical bonds with one another, they release their chemical potential energy as thermal energy and the condenser must

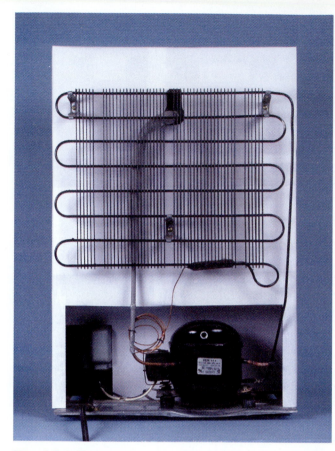

transfer this additional thermal energy to the out-door air. By the time the working fluid emerges from the condenser, it has condensed from a hot high-pressure gas to a warm high-pressure liquid and has released considerable thermal energy to the outdoor air. This heat includes both thermal energy extracted from the indoor air and electric energy converted into thermal energy by the compressor.

The second half of the air conditioner's job is now complete: it has converted ordered energy into thermal energy and it has released heat into the outdoor air. From here, working fluid returns to the evaporator to begin the cycle all over again. The working fluid passes endlessly around the cycle, extracting heat from the indoor air in the evaporator and releasing it to the outdoor air in the condenser. The compressor drives the whole process, thereby satisfying the second law of thermodynamics. The same technique is used to extract heat from the air inside a refrigerator and to release that heat to the room air (Fig. 6.3.4).

Before leaving air conditioners, we should take a moment to look at the working fluid itself. This fluid must become a gas at low pressure and a liquid at high pressure, over most of the temperature range encountered by the air conditioner. For decades, the standard working fluids were *chlorofluorocarbons* such as the various Freons. These compounds replaced ammonia,

Fig. 6.3.4 The compressor (bottom) and condenser coils (top) are visible on the back of this refrigerator. The compressor squeezes the working fluid into a hot, dense gas and delivers it to the condenser. There it gives up heat to the room air and condenses into a liquid. The working fluid evaporates inside the refrigerator, extracting heat from the food.

a toxic and corrosive gas used in early refrigeration.

Chlorofluorocarbons are ideally suited to air conditioners because they transform easily from gas to liquid and back again over a broad range of temperatures. They're also chemically inert and inexpensive. Unfortunately, chlorofluorocarbon molecules contain chlorine atoms and, when released into the air, can carry those chlorine atoms to the upper atmosphere. There they promote the destruction of ozone molecules, essential atmospheric constituents that absorb portions of the sun's ultraviolet radiation. Recently, chlorine-free *hydrofluorocarbons* have replaced chlorofluorocarbons as the working fluids in most air conditioners. Though not as energy efficient and chemically inert as the materials they replace, hydrofluorocarbons appear to be safe for the environment.

Check Your Understanding #5: Air Conditioning or Space Heating?
What would happen to the average air temperature in your room if you placed a window air conditioning unit in the middle of the room and turned it on?

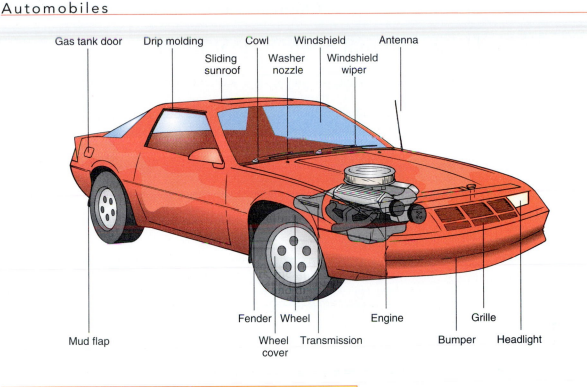

Gas tank door Drip molding Cowl Windshield Antenna
Sliding sunroof Washer nozzle Windshield wiper

Fender | Wheel Engine Grille
Mud flap Wheel cover Transmission Bumper Headlight

SECTION 6.4

Automobiles

Nothing is more symbolic of freedom and personal independence than an automobile. With its keys in your hand, you can go almost anywhere at a moment's notice. The mechanism that makes this instant transportation possible is the internal combustion engine. Though it has been refined over the years, this engine's basic design has changed little since it was invented more than a century ago. It uses thermal energy released by burning fuel to do the work needed to propel the car forward. That thermal energy can do work at all is one of the marvels of thermal physics and the primary focus of this section.

Questions to Think About: What obstacles stand in the way of using burning fuel's thermal energy to propel a car? Why are two objects, one hot and one cold, required in order to convert any thermal energy into useful work? What hot and cold objects does a car have? Why does a car have a cooling system to get rid of waste heat, rather than just converting it into useful work?

Experiments to Do: The recent advances in automobile technology and the increasing demands for pollution control equipment have made automobiles exceedingly complicated. Nonetheless, take a moment to look under the hood of your car or that of a friend. You should be able to identify the engine and its electric support system. You should be able to count four or more spark plug wires heading toward the engine's cylinders. These cylinders convert thermal energy from burning fuel into work to propel the car. Why does the engine need so many cylinders, rather than relying on one larger cylinder?

At the front of the engine compartment, you'll find the radiator. How does this device extract waste heat from the engine? Does heat flow naturally into the radiator and then into the outside air, or is there a heat pump involved?

Using Thermal Energy: Heat Engines

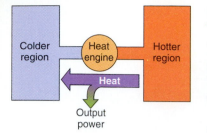

Fig. 6.4.1 A heat engine converts thermal energy into ordered energy as heat flows from a hotter region to a colder region. The larger the temperature difference between the two regions, the larger the fraction of thermal energy that can be converted into ordered energy.

steam engines

The light turns green and you step on the accelerator pedal. The engine of your car roars into action and, in a moment, you're cruising down the road a mile a minute. The engine noise gradually diminishes to a soft purr and vanishes beneath the sound of the radio and the passing wind.

The engine is the heart of the automobile, pushing the car forward at the light and keeping it moving against the forces of gravity, friction, and air resistance. It's not simply a miracle of engineering. It's also a wonder of thermal physics because it performs the seemingly impossible task of converting thermal energy into ordered energy. But the second law of thermodynamics forbids the direct conversion of thermal energy into ordered energy, so how can a car engine use burning fuel to propel the car forward?

The car engine avoids conflict with the second law by being a **heat engine**—a device that converts thermal energy into ordered energy *as heat flows from a hot object to a cold object* (Fig. 6.4.1). While thermal energy in a single object can't be converted into work, that restriction doesn't apply to a system of two objects *at different temperatures*. Because heat flowing from the hot object to the cold object increases the overall entropy of the system, a small amount of thermal energy can be converted into work without decreasing the system's overall entropy and without violating the second law of thermodynamics.

Another way to look at a heat engine is through the contributions of the two objects. The hot object provides the thermal energy that's converted into work. The cold object provides the order needed to carry out that conversion. As the heat engine operates, the hot object loses some of its thermal energy and the cold object loses some of its order. The heat engine has used them to produce ordered energy. Since the heat engine needs both thermal energy and order, it can't operate if either the hot or the cold object is missing.

In a car engine, the hot object is burning fuel and the cold object is outside air. Some of the heat passing from the burning fuel to the outside air is diverted and becomes the ordered energy that propels the car. But what limits the amount of thermal energy the engine can convert into ordered energy?

To answer that question, let's examine a simplified car engine. We'll treat the burning fuel and outside air as a single, thermally isolated system and look at what happens to their total entropy as the engine operates. In accordance with the second law of thermodynamics, this total entropy can't decrease even if the engine produces ordered energy.

When the car is idling at a stop light, its engine is doing no work and heat is simply flowing from the hot burning fuel to the cold outside air. The system's total entropy increases because this heat is more disordering to the cold air it enters than to the hot burning fuel it leaves. In fact, the system's entropy increases dramatically because the burning fuel is extremely hot compared to the cold outside air.

This increase in the system's entropy is unnecessary and wasteful. The second law of thermodynamics only requires that the engine add as much entropy to the cold outside air as it removes from the hot burning fuel. Since a little heat is quite disordering to cold air, the car engine can deliver much less heat to the outside air than it removes from the burning fuel and still not cause the system's total entropy to decrease. As long as the engine delivers enough heat to the outside air to keep the total entropy from decreasing, there's nothing to prevent it from converting the remaining heat into ordered energy!

This conversion starts as soon as you remove your foot from the brake and begin to accelerate forward. Instead of transferring all of the thermal energy in the burning fuel to the outside air, your car then extracts some of it as ordered energy and uses it to power the wheels. The car engine can convert thermal energy into ordered energy, as long as it passes along enough heat from the hot object to the cold object to satisfy the second law of thermodynamics.

Obeying the second law becomes easier as the temperature difference between the two objects increases. When the temperature difference is extremely large, as it is in an automobile engine, a large fraction of the thermal energy leaving the hot object can be converted into ordered energy—at least in theory. Unfortunately, theoretical limits are often hard to realize in actual machines, and the best automobile engines extract only about half the ordered energy that the second law allows. Still, obtaining even that amount is a remarkable feat and a tribute to scientists and engineers who, in recent years, have labored to make automobile engines as energy efficient as possible.

Check Your Understanding #1: Heat Pumps and Heat Engines

An air conditioner uses electric energy to make the air in your home colder than the outside air. Could you use this difference in temperatures to operate a heat engine and generate electric energy?

The Internal Combustion Engine

Invented by the German engineer Nikolaus August Otto in 1867, an internal combustion engine burns fuel directly in the engine itself. Gasoline and air are mixed and ignited in an enclosed chamber. The resulting temperature rise increases the pressure of the gas and allows it to perform work on a movable surface.

To extract work from the fuel, the internal combustion engine must perform four tasks in sequence:

1. It must introduce a fuel–air mixture into an enclosed volume.
2. It must ignite that mixture.
3. It must allow the hot burned gas to do work on the car.
4. It must get rid of the exhaust gas.

In the standard, four-stroke fuel-injected engine found in modern gasoline automobiles, this sequence of events takes place inside a hollow cylinder (Fig. 6.4.2). It's called a "four-stroke" engine because it operates in four distinct steps or strokes: induction, compression, power, and exhaust. "Fuel-injected" refers to the technique used to mix the fuel and air as they're introduced into the cylinder.

Automobile engines usually have four or more of these cylinders. Each cylinder is a separate energy source, closed at one end and equipped with a movable piston, several valves, a fuel injector, and a spark plug. The piston slides up and down in the cylinder, shrinking or enlarging the cavity inside. The valves, located at the closed end of the cylinder, open to introduce fuel and air into the cavity or to permit burned exhaust gas to escape from the cavity. The fuel injector adds fuel to the air as it enters the cylinder. And the spark plug, also located at the closed end of the cylinder, ignites the fuel–air mixture to release its chemical potential energy as thermal energy.

The fuel–air mixture is introduced into each cylinder during its induction stroke. In this stroke, the engine pulls the piston away from the cylinder's closed end

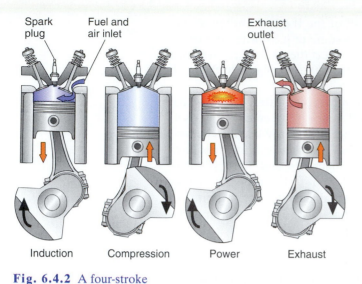

Spark plug   Fuel and air inlet   Exhaust outlet

Induction   Compression   Power   Exhaust

Fig. 6.4.2 A four-stroke engine cylinder. During the induction stroke, fuel and air enter the cylinder. The compression stroke squeezes that mixture into a small volume. The spark plug ignites the mixture and the power stroke allows the hot gas to do work on the automobile. Finally, the exhaust stroke ejects the exhaust gas from the cylinder.

so that its cavity expands to create a partial vacuum. At the same time, the cylinder's inlet valves open so that atmospheric pressure can push fresh air into the cylinder. The cylinder's fuel injector adds a mist of fuel droplets to this air so that the cylinder fills with a flammable fuel–air mixture. Because it takes work to move air out of the way and create a partial vacuum, the engine does work on the cylinder during the induction stroke.

At the end of the induction stroke, the inlet valves close to prevent the fuel–air mixture from flowing back out of the cylinder. Now the compression stroke begins. The engine pushes the piston toward the cylinder's closed end so that its cavity shrinks and the fuel–air mixture becomes denser. Because it takes work to compress a gas, the engine does work on the mixture during the compression stroke. This work appears as thermal energy in the fuel–air mixture, which becomes hotter. Since increases in a gas's density and temperature both increase its pressure, the pressure in the cylinder rises rapidly as the piston approaches the spark plug.

At the end of the compression stroke, the engine applies a high-voltage pulse to the spark plug and ignites the fuel–air mixture. The mixture burns quickly to produce hot, high-pressure burned gas, which then does work on the car during the cylinder's power stroke. In that stroke, the gas pushes the piston away from the cylinder's closed end so that its cavity expands and the burned gas becomes less dense. Since the hot gas exerts a huge pressure force on the piston as it moves outward, it does work on the piston and ultimately propels the car. As it does work, the burned gas gives up thermal energy and cools. Its density and pressure also decrease. At the end of the power stroke, the exhaust gas has cooled significantly and its pressure is only a few times atmospheric pressure. The cylinder has extracted much of the fuel's chemical energy as work.

The cylinder gets rid of the exhaust gas during its exhaust stroke. In this stroke, the engine pushes the piston toward the closed end of the cylinder while the cylinder's outlet valves are open. Because the burned gas trapped inside the cylinder at the end of the power stroke is well above atmospheric pressure, it accelerates out of the cylinder the moment the outlet valves open. These sudden bursts of gas leaving the cylinders create the "poof-poof-poof" sound of a running engine. Without a muffler on its exhaust pipes, the engine would be loud and unpleasant.

Just opening the outlet valves releases most of the exhaust gas, but the rest is squeezed out as the piston moves toward the cylinder's closed end. The engine again does work on the cylinder as it squeezes out the exhaust gas. At the end of the exhaust stroke, the cylinder is empty and the outlet valves close. The cylinder is ready to begin a new induction stroke.

Check Your Understanding #2: Getting Out More Than You Put In

How does the burned gas do more work on the piston during the power stroke than the piston does on the unburned fuel–air mixture during the compression stroke?

Engine Efficiency

The goal of an internal combustion engine is to extract as much work as possible from a given amount of fuel. In principle, all of the fuel's chemical potential energy can be converted into work because both are ordered energies. But it's difficult to convert chemical potential energy directly into work, so the engine burns the fuel instead. This step is unfortunate, for in burning the fuel, the engine converts the fuel's chemical potential energy directly into thermal energy and produces lots of unnecessary entropy.

But all is not lost. Since the burned fuel is extremely hot, a good fraction of its thermal energy can be converted into ordered energy as heat flows from the burned fuel to the outside air. The hotter the burned fuel and the colder the outside air, the more ordered energy the engine can extract. To maximize its fuel efficiency, an internal combustion engine obtains the hottest possible burned gas, lets that gas do as much work as it can on the pistons, and releases the gas at the coldest possible temperature.

It would be ideal if, during the power stroke, the burned gas expanded and cooled until it reached the temperature of the outside air. The exhaust gas would then leave the engine with the same amount of thermal energy it had when it arrived, and the engine would have extracted all of the fuel's chemical potential energy as work. Unfortunately, that would violate the second law of thermodynamics by converting thermal energy completely into ordered energy. Instead, the engine must release the burned gas before it cools to the temperature of the outside air. The engine's exhaust must be hot!

We can see the need for hot exhaust by comparing the normal power stroke with one in which the fuel–air mixture didn't ignite. In the latter case, the power stroke would simply reverse the compression stroke. The unburned mixture would expand, do work on the piston and engine, and cool back down to the temperature of the outside air. It would leave the engine at the same temperature as when it entered and wouldn't carry excess thermal energy away with it.

But the mixture does ignite, and the resulting hot burned gas has far more thermal energy than it can get rid of during the power stroke. Despite expanding and doing work on the piston and engine, this gas doesn't reach atmospheric pressure and temperature by the end of the power stroke. It would have to push the piston much farther outward before its temperature would diminish to that of the outside air. Instead, the burned gas remains substantially hotter than the outside air and its pressure several times atmospheric pressure.

The engine could actually extract a little more energy from the burned gas by allowing the piston to move still farther outward, but the expanding gas would soon reach atmospheric pressure and would still not have cooled to the outside temperature. Ultimately, the engine has no choice but to eject hot exhaust into the outside air because it can't extract any more thermal energy from it. The second law of thermodynamics can't be violated.

But a real internal combustion engine wastes energy and extracts less work than the second law allows. For example, some heat leaks from the burned gas to the cylinder walls and is removed by the car's cooling system. This wasted heat isn't available to produce work. Similarly, sliding friction in the engine wastes mechanical energy and necessitates an oil-filled lubricating system. Overall, a real internal combustion engine converts only about 20% to 30% of the fuel's chemical potential energy into work.

Fuel cells are essentially batteries that convert a fuel's chemical potential energy directly into electric energy, without burning the fuel first. Though harder to build and operate, fuel cells are potentially more energy efficient than combustion engines. Explain.

Improving Engine Efficiency

To obtain the hottest possible burned gas and allow that gas to expand as much as it can before it's released, the burned gas should begin the power stroke in a very small volume and finish the power stroke in a much larger volume. The extent to which the cylinder's volume increases during the power stroke is measured by its compression ratio—its volume at the end of the power stroke divided by its volume at the start of the power stroke. The larger this compression ratio, the higher the initial temperature and pressure of the burned gas and the more energy efficient the engine. While normal compression ratios are between 8:1 and 12:1, those in high-compression engines may be as much as 15:1.

Unfortunately, the compression ratio can't be made arbitrarily large. The more tightly the engine packs the fuel–air mixture during the compression stroke, the greater its pressure, density, and temperature become. If the engine compresses the fuel–air mixture too much, it will ignite all by itself. This spontaneous ignition due to overcompression is called preignition or knocking. When an automobile knocks, the gasoline burns before the engine is ready to extract work from it so much of the energy is wasted.

There are two ways to reduce knocking. First, you can improve the uniformity of mixing between the fuel and air. Fuel injection provides excellent mixing, which is why it has replaced carburetion as the way of introducing fuel in the cylinders of all modern cars. Fuel injection also allows the car's computer to adjust the fuel–air mixture for complete combustion and minimal pollution.

Second, you can use the right fuel. Fuels that are more difficult to ignite resist knocking and are assigned higher "octane numbers." Regular gasoline has an octane number of about 87 while premium has an octane number of about 93. Choosing the right fuel is simply a matter of finding the lowest octane gasoline that your car can use without excessive knocking. Most modern well-tuned automobiles work well on regular gasoline. Only high-performance cars with high-compression engines need premium gasoline so that putting it in a normal car is usually a waste of money.

Since knocking sets the limit for compression ratio, it also sets the limit for efficiency in a gasoline engine. However, diesel engines avoid the knocking problem by separating the fuel and air during the compression stroke (Fig. 6.4.3). Invented by German engineer Rudolph Christian Karl Diesel (1858–1913) in 1896, the diesel engine has no spark plug to ignite the fuel. Instead, it compresses pure air with an extremely high compression ratio of perhaps 20:1 and then injects diesel fuel directly into the cylinder just as the power stroke begins. The fuel ignites spontaneously as it enters the hot, compressed air. Because of its high compression ratio, a diesel engine is more energy efficient than a standard

avoiding knocking

Fig. 6.4.3 A diesel engine cylinder contains pure air during the compression stroke. As the piston does work on it, this air becomes extremely hot. At the start of the power stroke, diesel fuel is injected into the cylinder. The fuel ignites spontaneously, and the hot burned gas does work on the piston and engine during the power stroke.

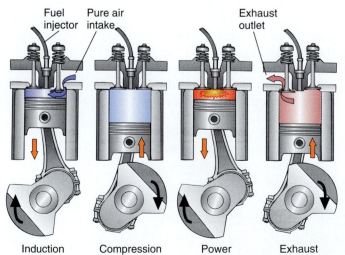

Fuel injector   Pure air intake   Exhaust outlet

Induction   Compression   Power   Exhaust

engine. Unfortunately, it's also harder to start and requires carefully timed fuel injection.

In both gasoline and diesel engines, fuel injection is sometimes combined with a turbocharger or supercharger. These devices are essentially fans that pump outside air into the cylinders during their induction strokes. By squeezing more fuel–air mixture into the cylinders, a turbocharger or supercharger increases the engine's power output. The engine burns more fuel each power stroke and behaves like a larger engine. The fan of a turbocharger is powered by pressure in the engine's exhaust while the engine itself drives the fan of a supercharger.

The downside of turbochargers and superchargers, other than being expensive and wearing out rather quickly, is that they encourage knocking. As they pack air into the cylinders, they also do work on it and heat it up. Since the fuel–air mixture enters the engine hot, it ignites spontaneously during the compression stroke. To avoid knocking in a car equipped with one of these devices, you may need to use premium gasoline. Some cars are equipped with an intercooler, a device that removes heat from the air passing through the turbocharger. By providing cool, high-density air to the cylinders, the intercooler reduces the peak temperature of the compression stroke and avoids knocking.

> ### Check Your Understanding #4: Steam Heat

How can a steam engine be more energy efficient when it operates on 325 °C steam than when it uses 300 °C steam?

Multicylinder Engines

Since the purpose of the engine is to extract work from the fuel–air mixture, it's important that each cylinder do more work than it consumes. Three of the strokes require the engine to do work on various gases, and only one of the strokes extracts work from the burned gas. During the induction stroke, the engine does work to create a partial vacuum. During the compression stroke, the engine does work to compress the fuel–air mixture. During the exhaust stroke, the engine does work to squeeze the exhaust gas from the cylinder. Fortunately, the work done on the engine by the hot burned gas during the power stroke is much greater than the work the engine does on the various gases during the other three strokes.

Still, the engine has to invest a lot of energy into the cylinder before each power stroke. To provide this initial energy, most four-stroke engines have four or more cylinders (Fig. 6.4.4), timed so that there is always one cylinder going through the power stroke. The cylinder that is in the power stroke provides the work needed to carry the other cylinders through the three nonpower strokes, and there is plenty of work left over to propel the car itself.

Fig. 6.4.4 This cut-away view of a V-8 engine shows the eight cylinders arranged in two rows of four, one row in the foreground and the other behind.

While the pistons move back and forth, the engine needs a rotary motion to turn the car's wheels. The engine converts each piston's reciprocating motion into rotary motion by coupling that piston to the crankshaft with a connecting rod. The crankshaft is a thick steel bar, suspended in bearings, that has a series of pedal-like extensions, one for each cylinder. As the piston moves back during the power stroke, it pushes on the connecting rod and the connecting rod pushes on its crankshaft pedal. The connecting rod thus produces a torque on the crankshaft. The crankshaft rotates in its bearings and transmits this torque out of the engine so that it can be used to propel the car. So, while each cylinder initially exerts a force, the crankshaft uses that force to produce a torque.

► Check Your Understanding #5: Hard Starting

Modern cars use an electric motor to start the engine turning but early cars were started with a hand crank. Why was it so hard to turn the crank?

 manual, automatic, pollution control, ignition, starting, differential, suspension, auto brakes, airbags, carburetors, synchromesh

Epilogue for Chapter 6

This chapter examined the roles of thermal energy and heat in a variety of common objects. In *woodstoves,* we saw how combustion converts ordered chemical potential energy into disordered thermal energy and studied the ways in which this thermal energy flows into the woodstove's surroundings: conduction, convection, and radiation. In *incandescent light bulbs,* we found that a sufficiently hot object radiates some of its heat as visible light and saw that the distribution of wavelengths in this light depends on the object's temperature.

In *air conditioners,* we learned how heat pumps use ordered energy to pump heat against its natural flow and observed that the only way to get rid of thermal energy is to transfer it to something else. And in *automobiles,* we saw that heat engines are able to divert some of the heat flowing from a hot object to a cold object and convert it into useful work. We also examined the roles of the two objects in a heat engine, hot and cold, finding that the hot object provides the energy needed to do the work while the cold object provides the order that makes the conversion of thermal energy into ordered energy possible.

Explanation: A Ruler Thermometer

Like most things, the clear plastic ruler expands when you heat it. Its length increases by an amount proportional to its increase in temperature. When you transfer heat to the ruler, by breathing on it, touching it, or exposing it to a hair dryer, its temperature rises, it expands, and its free end turns the needle and pointer. Since the pointer's movement is proportional to the ruler's length change, it's also proportional to the thermometer's change in temperature.

Chapter Summary

How Woodstoves Work: A woodstove burns wood in air to obtain hot gas. The burning process is actually a chemical reaction in which molecules in the wood and air are disassembled into fragments and then reassembled into new, more tightly bound molecules such as water and carbon dioxide. This reassembly process releases more energy than was required to disassemble the original fuel and oxygen molecules. This extra energy appears as thermal energy within the reaction products, so that they're hot. Rather than distribute the hot burned gas directly to the room, a woodstove transfers heat from the burned gas to clean air or water. Heat is conducted through the walls of a woodstove and then flows into the room via convection and radiation.

How Incandescent Light Bulbs Work: An incandescent light bulb produces light as thermal radiation from an extremely hot tungsten filament. The spectrum of light emitted by that filament depends on its temperature—the hotter the filament, the whiter its light. Atoms in the filament sublime during operation so the filament gradually disappears. Eventually it becomes so thin that it breaks. The tungsten filament of a normal bulb operates at 2500 °C because it would burn out too quickly at higher temperatures. Reducing the operating temperature, as is done in an extended life bulb, prolongs the filament's life at the expense of color temperature and energy efficiency. In contrast, adding halogen gas actually increases the filament's life and improves both the color temperature and energy efficiency.

How Air Conditioners Work: An air conditioner moves heat against its natural direction of flow, using a working fluid that travels endlessly through an evaporator, compressor, and condenser. The working fluid flows to the evaporator as a high-pressure liquid before passing through a constriction in the pipe. Its pressure then drops, and it enters the evaporator as a low-pressure liquid. In the evaporator, the working fluid absorbs heat from the indoor air and evaporates into a cold low-pressure gas.

The working fluid then flows to the compressor, which compresses it into a hot, high-pressure gas and sends it to the condenser. As it passes through the condenser, the hot gaseous working fluid releases heat to the outdoor air and condenses into a cool liquid. The compressor uses ordered energy and delivers it as heat to the outdoor air. Without this input of ordered energy, the air conditioner could not move heat against its natural direction of flow.

How Automobiles Work: An automobile engine extracts work from its chemical fuel by burning that fuel inside its cylinders and making the resulting burned gas do work on the engine. Most engines have at least four cylinders, each of which requires four strokes to extract work from the fuel. During the induction stroke, a piston moves out of the cylinder, and fuel and air enter it. During the compression stroke, the piston moves into the cylinder, compressing this fuel–air mixture to high density, pressure, and temperature. An electric spark then ignites the mixture and converts it into extremely hot burned gas. During the power stroke, the piston again moves out of the cylinder while the hot gas does work on it. This work is what powers the car. Finally, during the exhaust stroke, the piston moves into the cylinder and ejects the burned gas. The cylinder then begins again with fresh fuel and air.

Important Laws and Equations

1. The Stefan–Boltzmann Law: The power an object radiates is proportional to the product of its emissivity times the fourth power of its temperature times its surface area, or

$$\text{radiated power} = \text{emissivity} \cdot \text{Stefan–Boltzmann constant} \cdot \text{temperature}^4 \cdot \text{surface area}. \quad (6.2.1)$$

2. The Zeroth Law of Thermodynamics: Two objects that are each in thermal equilibrium with a third object are also in thermal equilibrium with one another.

3. The First Law of Thermodynamics: The change in a stationary object's internal energy is equal to the heat transferred into that object minus the work that object does on its surroundings.

4. The Second Law of Thermodynamics: The entropy of a thermally isolated system of objects never decreases.

5. The Third Law of Thermodynamics: As an object's temperature approaches absolute zero, its entropy approaches zero.

Check Your Understanding—Answers

Section 6.1 WOODSTOVES

1. No, its thermal energy will remain constant.

Why: As the ball falls, its gravitational potential energy is transformed into kinetic energy of the entire ball. However neither energy is internal to the ball so they aren't included in thermal energy. Thus the ball's thermal energy doesn't change.

2. At first, they experience attractive forces. But once they come too close together, the forces will become repulsive. As they separate, the forces will once again be attractive.

Why: As they approach one another, the two nitrogen atoms experience attractive forces and a chemical bond begins to form between them. They accelerate toward one another, converting chemical potential energy into kinetic energy. However, when they are very close together, the forces become repulsive and they bounce off one another. They head apart and the forces again become attractive, but their kinetic energy breaks the bond and they separate forever.

3. From your hotter hand to the colder ice cube.

Why: Heat naturally flows from a hotter object to a colder object. Since the ice cube is colder than your hand, heat flows out of your hand and into the ice cube. Since your hand is losing thermal energy, its temperature drops and you sense cold. While it's tempting to think of cold as something that flows out of an ice cube, the only thing that really moves about is heat. Ice cubes are wonderful absorbers of heat and cool our drinks by reducing their thermal energies.

4. Yes, thermal energy is flowing from the hot towels, through the plastic, to the muscle.

Why: The plastic sheet is acting as a heat exchanger, allowing heat to flow from the hot towels to the cooler muscle but preventing any movement of the hot water itself.

5. The handle's thermal conductivity.

Why: Some handles are made of plastics or stainless steel, which are poor conductors of both electricity and heat. These handles normally remain cool, unless hot gases from the stove directly heat them. Other handles are made from aluminum or copper, which are good conductors of electricity and heat. These handles often become unbearably hot.

6. Convection occurs, with warmed air rising over the land and being replaced by cooler air from above the water. The air moving from over the water to over the land creates the breeze.

Why: Winds are giant convection currents caused by solar heating. Air rises over warm spots on the earth's surface, and surface winds blow toward those warm spots to replace the missing air.

7. Radiation transfers heat from the lamp's filament to your skin.

Why: A heat lamp emits large amounts of invisible infrared radiation. Although you can't see this radiation, you can feel it on your skin.

Section 6.2 INCANDESCENT LIGHT BULBS

1. These satellites can measure the wavelength distributions of thermal radiation emitted by various patches of land and determine their temperatures.

Why: Even objects that are near room temperature emit thermal radiation, although this radiation is entirely in the infrared. While the land is not really black, the infrared light it emits is still an accurate indication of its temperature. A satellite can sense the exact form of the distribution and determine the temperature with great accuracy.

2. The snow sublimes to form water vapor in the air.

Why: Ice sublimes quickly at temperatures near its melting temperature, going from a solid to a gas without ever becoming liquid water.

3. As you heat the pot of water, the water expands more than the pot. Since the water no longer fits in the pot, it overflows.

Why: Both the pot and the water expand with increasing temperature. However, the water has a larger coefficient of volume expansion than the pot, so its volume increases more than that of the pot.

4. The inner bulb aids the tungsten recycling process, while the outer bulb diffuses the light and protects the inner bulb.

Why: The inner bulb must get hot enough for the halogen gas to react with and recycle the tungsten atoms on its surface. The outer bulb is cloudy to soften the glare and ensures that nothing touches the hot inner bulb.

Section 6.3 AIR CONDITIONERS

1. The blender's blade does work on the water, and this work becomes thermal energy.

Why: The first law of thermodynamics states that the change in the water's internal energy is equal to the heat flowing into it minus the work it does on its surroundings. In this case, the water's surroundings are doing work on it by stirring it so its internal energy increases. Since the water can't store this new internal energy as potential energy, the energy becomes thermal energy and the water gets hotter.

2. A thermally isolated perpetual motion machine violates the first law, while one that is not thermally isolated violates the second law.

Why: A thermally isolated perpetual motion machine clearly violates the conservation of energy aspect of the first law of thermodynamics. This isolated machine simply can't export energy forever because it will eventually run out. A perpetual motion machine that is not thermally isolated may not violate conservation of energy because it can absorb heat energy from its surroundings. Instead, it violates the second law of thermodynamics. This machine can't endlessly absorb heat energy and then export it as ordered energy. In doing so, the machine will eventually begin to reduce the entropy of the universe and violate the second law. Sad though it may be, perpetual motion machines can't exist.

3. A heat pump requires more ordered energy to pump heat from a cold object to a hot object when the temperature difference between them is large.

Why: A heat pump becomes less efficient at pumping heat when the temperature of the heat's source becomes much colder than the heat's destination. The colder it is outside, the more ordered energy it takes to move each joule of heat. On bitter cold days, heat pumps aren't able to move enough heat to keep their homes warm, which is why most home heat pumps have built-in electric or gas furnaces to assist them during unusually cold weather.

4. The tank's liquid propane needs heat to evaporate into a gas, and it extracts that heat from its surroundings.

Why: Just as in the evaporator of an air conditioner, the evaporating liquid propane absorbs heat.

5. The room air would become warmer, on average.

Why: The air conditioner would begin to pump heat from its front to its back. The air right in front of the unit would become colder, while the air behind the unit would become hotter. Since the unit would deliver more heat to the hotter air than it would absorb from the colder air, it would increase the total amount of thermal energy in the room. On average, the room would become warmer.

Section 6.4 AUTOMOBILES

1. Yes.

Why: A heat engine is essentially a heat pump operating backward. The air conditioner (a heat pump) uses ordered electric energy to pump heat from the cold air in your home to the hot outdoor air. The heat engine we are considering would use the flow of heat from the hot outdoor air to the cold air in your home to produce ordered electric energy.

2. The pressure is much higher in the burned gas than in the unburned fuel–air mixture.

Why: The amount of work done on the piston by the gas or done on the gas by the piston depends on the pressure inside the cylinder. The higher that pressure, the more outward force the piston experiences and the more work is done on it as it moves. The sudden rise in pressure that occurs when the fuel–air mixture burns explains why the burned gas does so much work on the piston as it moves outward.

3. Because fuel cells don't turn the fuel's ordered energy into thermal energy, they don't have to operate as heat engines. In principle, they can convert all of the fuel's chemical potential energy into electric energy, unlike combustion engines.

Why: Fuel cells remain a promising alternative to combustion engines because they avoid the wasted energy that comes with burning fuel. However, making fuel cells that are efficient and robust is difficult and expensive. At present, fuel cells are only used in special situations, such as in spaceships, where energy efficiency is far more important than cost.

4. Like all heat engines, the steam engine can convert more thermal energy into work when the temperature of its hot object (the steam) increases.

Why: The steam engine converts thermal energy into work as heat flows from the hot steam to the outside air. The greater the temperature difference between those two objects, the more efficient the steam engine can be at turning thermal energy into work. That is why most steam engines use extremely hot steam.

5. The person turning the crank had to do all the work needed to move the engine's pistons through the three non-power strokes.

Why: Before the engine started running on its own, it couldn't provide any of the energy the cylinders needed during the induction, compression, and exhaust strokes. The person turning the crank had to provide this energy. Once the fuel started burning, the power strokes could take over, but up to that point, turning the crank was hard work.

Check Your Figures—Answers

Section 6.2 INCANDESCENT LIGHT BULBS

1. About 262 W.

Why: We can use Eq. 6.2.2 to obtain the radiated power:

$$\text{radiated power} = 0.5 \cdot 5.67 \times 10^{-8} \text{ J/(s·m}^2\text{·K}^4)$$
$$\cdot (310 \text{ K})^4 \cdot 2 \text{ m}^2$$
$$= 262 \text{ J/s} = 262 \text{ W}.$$

This power is much more than the thermal power your body produces while you're standing still. You'll get cold quickly.

Exercises

1. Your body is presently converting chemical potential energy from food into thermal energy at a rate of about 100 J/s, or 100 W. If heat were flowing out of you at a rate of about 200 W, what would happen to your body temperature?

2. You can use a blender to crush ice cubes, but if you leave it churning too long, the ice will melt. What supplies the energy needed to melt the ice?

3. When you knead bread, it becomes warm. From where does this thermal energy come?

4. You throw a ball into a box and close the lid. You hear the ball bouncing around inside as the ball's energy changes from gravitational potential energy to kinetic energy, to elastic potential energy, and so on. If you wait a minute or two, what will have happened to the ball's energy?

5. Why do meats and vegetables cook much more quickly when there are metal skewers sticking through them?

6. Why do aluminum pans heat food much more evenly than stainless steel pans when you cook on a stove?

7. Use the concept of convection to explain why firewood burns better when it's raised above the bottom of a fireplace on a grate.

8. When you heat a pot of water on the stove, the water's temperature is almost uniform throughout the pot. Why?

9. Explain how convection contributes to the shape of a candle flame.

10. Most refrigerators use the freezing compartment to cool the refrigerating compartment as well. Why do most refrigerators place the freezing compartment on top?

11. When you hold a lighted match so that its tip is lower than its stick, the flame travels quickly up the stick. Why?

12. It's often a good idea to wrap food in aluminum foil before baking it near the red-hot heating element of an electric oven. Why does this wrapped food cook more evenly?

13. When you bake a pie in a glass dish over a red-hot electric element, why does the bottom of the pie cook relatively quickly?

14. Why are black steam radiators better at heating a room than radiators that have been painted white or silver?

15. Why does a blackened baking pan heat up more quickly in an oven than one that is shiny?

16. The space shuttle generates thermal energy during its operation in earth orbit. How is it able to get rid of that thermal energy as heat in an airless environment?

17. Mothballs are made from a white solid (naphthalene) that has a strong odor. If you leave a mothball out, it slowly disappears. What happens to the mothball?

18. Frozen vegetables will "freeze dry" if they're left in cold, dry air. How can water molecules leave the frozen vegetables?

19. How would you estimate the temperature of a glowing coal in a fireplace?

20. A light bulb burns out when a small portion of its thinning filament overheats and vaporizes. Why is this event accompanied by a bright flash of blue-white light?

21. The strongest evidence for the Big Bang theory of the origin of the universe is the thermal radiation emitted by that explosion. This radiation has cooled over the years to only 3 K and is now mostly microwaves. Why should 3 K thermal radiation be microwaves?

22. You have a table lamp with a dimmer switch. The dimmer allows you to adjust the temperature and brightness of the lamp's incandescent bulb from very dim red up to brilliant yellow-white. How does the bulb's energy efficiency, the amount of visible light it produces per unit of power consumed, depend on the dimmer's setting?

23. When you operate a 50-100-150 W three-way bulb at its 50-W setting, it emits yellow-white light. When you use a dimmer to operate a regular 150-W bulb on only 50 W of electric power, it emits orangish light. Explain the difference.

24. Astronomers can tell the surface temperature of a distant star without visiting it. How is this done?

25. A doctor can study a patient's circulation by imaging the infrared light emitted by the patient's skin. Tissue with poor blood flow is relatively cool. What changes in the infrared emissions would indicate such a cool spot?

26. The filament of an incandescent bulb is quite small, yet it emits as much as 100 W of thermal radiation. That's almost as much as your whole body emits. What accounts for the strength of the filament's thermal radiation?

27. Why are concrete sidewalks divided into individual squares rather than being left as continuous concrete strips?

28. If you were laying steel track for a railroad, what influence would thermal expansion have on your work?

29. A difficult-to-open jar may open easily after being run under hot water for a moment. Explain.

30. Why do bridges have special gaps at their ends to allow them to change lengths?

31. Drinking fountains that actively chill the water they serve can't work without ventilation. They usually have louvers on their sides so that air can flow through them. Why do they need this airflow?

32. If you open the door of your refrigerator with the hope of cooling your room, you will find that the room's temperature actually increases somewhat. Why doesn't the refrigerator remove heat from the room?

33. The outdoor portion of a central air conditioning unit has a fan that blows air across the condenser coils. If this fan breaks, why won't the air conditioner cool the house properly?

34. If you block the outlet of a hand bicycle pump and push the handle inward to compress the air inside the pump, the pump will become warmer. Why?

35. When the gas that now makes up the sun was compressed together by gravity, what happened to the temperature of that gas? Why?

36. Why is a car more likely to knock on a hot day than on a cold day?

37. A soda siphon carbonates water by injecting carbon dioxide gas into it. The gas comes compressed in a small steel container. As the gas leaves the container and pushes its way into the water, why does the container become cold?

38. A high-flying airplane must compress the cold, rarefied outside air before delivering it to the cabin. Why must this air be air conditioned after the compression?

39. If you drop a glass vase on the floor, it will become fragments. However, if you drop those fragments on the floor, they will not become a glass vase. Why not?

40. When you throw a hot rock into a cold puddle, what happens to the overall entropy of the system?

41. What prevents the bottom half of a glass of water from spontaneously freezing while the top half becomes boiling hot?

42. Suppose someone claimed to have a device that could convert heat from the room into electric power continuously. You would know that this device was a fraud because it would violate the second law of thermodynamics. Explain.

43. Why does snow blanket the ground almost uniformly rather than creating tall piles in certain areas and bare spots in others?

44. If you transfer a glass baking dish from a hot oven to a cold basin of water, that dish will probably shatter. What produces the ordered mechanical energy needed to tear the glass apart?

45. Freezing and thawing cycles tend to damage road pavement during the winter, creating potholes. What provides the mechanical work that breaks up the pavement?

46. The air near a woodstove circulates throughout the room. What provides the energy needed to keep the air moving?

47. Winds are driven by differences in temperature at the earth's surface. Air rises over hot spots and descends over cold spots, forming giant convection cells of circulating air. Near the ground, winds blow from the cold spots toward the hot spots. Explain how the atmosphere is acting as a heat engine.

48. Hurricanes are giant heat engines powered by the thermal energy in warm ocean regions and the order in colder surrounding areas. Why are hurricanes most violent when they form over regions of unusually hot water at the end of summer?

49. On a clear sunny day, the ground is heated uniformly and there is very little wind. Use the second law of thermodynamics to explain this absence of wind.

50. A plant is a heat engine that operates on sunlight flowing from the hot sun to the cold earth. The plant is a highly ordered system with relatively low entropy. Why doesn't the plant's growth violate the second law of thermodynamics?

51. A diesel engine burns its fuel at a higher temperature than a gasoline engine. Why does this difference allow the diesel engine to be more efficient at converting the fuel's energy into work?

52. A chemical rocket is a heat engine, propelled forward by its hot exhaust plume. The hotter the fire inside the chemical rocket, the more efficient the rocket can be. Explain this fact in terms of the second law of thermodynamics.

53. An acquaintance claims to have built a gasoline-burning car that doesn't release any heat to its surroundings. Use the second law of thermodynamics to show that this claim is impossible.

Problems

1. If a burning log is a black object with a surface area of 0.25 m^2 and a temperature of 800 °C, how much power does it emit as thermal radiation?

2. When you blow air on the log in Problem 1, its temperature rises to 900 °C. How much thermal radiation does it emit now? Why did the 100 °C rise make so much difference?

3. If the sun has an emissivity of 1 and a surface temperature of 6000 K, how much power does each 1 m^2 of its surface emit as thermal radiation?

4. A dish of hot food has an emissivity of 0.4 and emits 20 W of thermal radiation. If you wrap it in aluminum foil, which has an emissivity of 0.08, how much power will it radiate?

5. A space vehicle uses a large black surface to radiate away waste heat. How does the amount of heat radiated away depend on the area of that radiating surface? If the surface area were doubled, how much more heat would it radiate, if any?

Cases

1. Baking a potato takes a long time, even in a hot oven, because the inside of a potato warms up very slowly.
  a. What makes it hard for conduction to transfer heat to the center of the potato?
  b. What makes it hard for convection to transfer heat to the center of the potato?
  c. What makes it hard for radiation to transfer heat to the center of the potato?

  d. Why does inserting a metal skewer through the potato help it to cook quickly?

2. An electric hot water heater takes cold water at its inlet pipe and delivers hot water at its outlet pipe. Between these two pipes is a large water container. Many electric hot water heaters have two separate heating elements: one near the top of the water container and one near the bottom.

These elements use a lot of electric power so they don't operate simultaneously. Instead, the top element operates until the water near the top of the hot water heater reaches the desired temperature and then the bottom element operates until the water near the bottom is also at the desired temperature. The water heater then turns itself off and waits until more heating is required.

a. Both heating elements project into the same hot water container inside the water heater. When the top element is heating water at the top of the container, why doesn't convection occur and cause water at the bottom of the container to also become hot?

b. Convection dictates where the water pipes attach to the container. Where should hot water be extracted from the container, and where should cold water be introduced into the container to replace that hot water?

c. Why does the water heater's bottom heating element operate more often and age more quickly than its top heating element?

d. Why is it important to thermally insulate the hot water heater's container?

e. Energy flows into the hot water heater as electricity. Since energy is conserved and can't accumulate forever inside the heater, it must leave somehow. At a time when the hot water is being used, how does most of the energy leave the hot water heater?

3. A typical electric oven has two separate heating elements: one on top and one on the bottom. The bottom element is used for baking while the top element is used to broil foods.

a. When only the bottom element is active and glowing red hot, what heat transfer mechanisms carry most of the heat to the food in the oven?

b. When only the top element is on, what heat transfer mechanism carries most of the heat to the food?

c. How does wrapping the food in shiny aluminum foil affect the rate of heat transfer to the food in these two cases?

d. A convection oven has a fan that circulates air in the oven. How does the fan affect the rate of heat transfer to the food?

4. Sliding friction is useful for starting fires.

a. If you press a wooden peg against another piece of wood and spin the peg rapidly, it can become hot enough to ignite. The peg produces thermal energy no matter how fast it spins, so why is it important that it spin rapidly?

b. If you strike a piece of flint against a piece of steel, you can shave a red-hot spark off the steel. However, moments later the surface of the steel itself barely feels warm. What happens to the thermal energy in that surface?

c. When you rub a match on a matchbook, it suddenly bursts into flames. The principal source of heat here is a chemical reaction in the match head. What purpose does the initial thermal energy serve in starting this reaction?

5. A solar panel is an electronic device that produces electric power when it's exposed to sunlight. It's also a type of heat engine.

a. What object supplies the heat that powers this heat engine?

b. What object receives the waste heat that passes through this heat engine?

c. If you try to run a solar panel by exposing it to light from an incandescent bulb, it will produce relatively little electric power. Explain this result in terms of the temperature of the bulb's filament.

d. Suppose that the solar panel could withstand the sun's temperature. If you were to immerse the panel in the sun, it wouldn't produce any electricity. Why not?

e. A thermoelectric cell produces electricity when you heat one side of it and cool the other. How does this device resemble a solar panel?

6. A refrigerator is a heat pump that cools the contents of a food locker. Like an air conditioner, it has an evaporator, a condenser, and a compressor.

a. Where is the evaporator located in a refrigerator?

b. Where is the condenser?

c. What will happen to the food's temperature if you wrap the refrigerator's condenser and compressor in thermal insulation, so that they can't exchange heat with the room?

d. What will happen to the temperature of the room if you open the door to the refrigerator and let it run for a few hours?

e. Why does it take more electricity to operate a freezer than a refrigerator of similar size and quality?

7. Desalinating salt water is one way arid regions are able to obtain drinking water. However, it's not an easy process.

a. Which has more entropy: a container of salt water or a container of fresh water plus a box of salt?

b. Why does your answer to part a indicate that the desalination process requires either the conversion of ordered energy into thermal energy or the movement of heat from a hotter region to a colder region?

c. One technique for desalination is reverse osmosis, in which salt water is squeezed through a special membrane that permits only pure water to pass. Show that squeezing water through this membrane takes work (ordered energy).

d. Another technique for desalination is distillation, in which salt water is boiled in a hotter region to form water vapor that is then condensed to pure water in a colder region. This desalination process is a heat engine that separates the water from the salt. Explain why it's a heat engine.

e. Still another desalination technique is freezing the salt water into pure ice and salt crystals. In this case the salt water has to be cooled to very low temperatures to freeze it, then warmed to a much higher temperature to melt it. Show that this technique is also using a heat engine to drive the desalination process.

8. It's your first day on the staff of the U.S. Patents Office and you're excited about having some new ideas come across your desk.

a. An inventor comes to you with a small box that is supposed to make batteries obsolete. The inventor claims that the box can produce electricity forever without having to be recharged. Why can you be sure that this claim is nonsense?

b. Another inventor comes in with a small motorlike device that's powered by a burning candle. The inventor claims that this device takes the heat from the candle and turns it entirely into work. The device thus doesn't warm the room at all. Your knowledge of the laws of thermodynamics assures you that this claim, too, is nonsense. Why?

c. As though the entire population of inept inventors was released on you in one day, another character comes in claiming to have a heat pump that can transfer heat out of a box of corn flakes for as long as you like. The cereal just gets colder and colder. Once again, you know that this is impossible. Explain.

d. Just when you thought you were out of the woods, a person comes in with a coffee mug that automatically reassembles itself if you accidentally break it. You're supposed to put the broken mug in a box and wait. In a minute or two, the mug will be as good as new. You agree with the inventor that this reconstruction trick doesn't violate any of the laws of motion. However, you can still be sure that it won't work. Why?

CHAPTER 7

Resonance and Mechanical Waves

Many fascinating motions in the world around us are repetitive ones. Our lives are filled with cycles, from the earth's slow orbit around the sun to a pond's undulating ripples on a rainy day. These cyclic motions are governed by the physical laws and steadily mark our passage through time and space. Some of these cycles structure our lives out of necessity or tradition, while others are simply there to be observed. Still other cycles have become part of our everyday world because they're useful or enjoyable. This chapter is about cyclic motions in three contexts: in clocks, in musical instruments, and at the seashore.

EXPERIMENT: A Singing Wineglass

One simple experiment with cyclic motion involves a crystal wineglass, a little water, and a delicate touch. A crystal wineglass is hard and thin and easily supports

a repetitive mechanical motion called a vibration. The glass's hardness allows it to retain energy in this vibrating motion for a long time so that it rings clearly when you tap it gently with a spoon and can acquire energy slowly in the manner described below.

The experiment itself is an activity discovered by many a bored youth, sitting too long in a fancy restaurant and no toys other than the place settings. If you wet your finger slightly and run it gently around the lip of the wineglass at a slow, steady pace, you should be able to get the wineglass to sing loudly. The walls of the glass will vibrate back and forth and emit a clear tone.

If you can't find a crystal wineglass, you may be out of luck. A normal wineglass doesn't work as well because it converts the energy from your finger into thermal energy rapidly and doesn't emit much sound. In any case, be sure that the wineglass has no sharp edges on its lip so that you don't cut your finger.

What will happen to the sound if you add some water to the glass? Try to *predict* how its pitch will change; how its volume will change. Now *observe* what happens. Can you *measure* the changes? Did you *verify* your predictions? Is anything visible on the surface of the water as you make the glass vibrate?

Chapter Itinerary

The wineglass's vibration is an example of a natural resonance, a cyclic motion of the glass itself. Whether you tap the glass gently with a spoon or rub it with your finger, you're causing it to undergo this characteristic motion. In this chapter we'll examine several other objects that exhibit natural resonances: (1) *clocks,* (2) *violins and pipe organs,* and (3) *the sea and surfing*.

In *clocks,* we'll study pendulums, balance rings, and quartz crystals to see how their natural resonances are used to measure the passage of time. In *violins and pipe organs,* we'll look at the vibrations of strings and air columns to find out how these instruments produce their musical sounds. And in *the sea and surfing,* we'll look at how cyclic motions can move through space as the waves on water.

Clocks

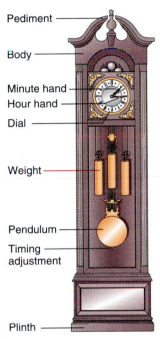

Pediment

Body

Minute hand
Hour hand

Dial

Weight

Pendulum

Timing
adjustment

Plinth

People measure their lives according to the sky, dividing existence into days, months, and years according to the celestial motions of the sun, moon, and stars. But on the less romantic scale of daily life, the sky offers little help. It provides no easy way to measure short periods of time, so people invented clocks.

Early clocks were based on the time it took to complete simple processes—the flow of sand or water, or the burning of candles. However, these clocks weren't very accurate and required constant attention. Better clocks measure time with repetitive motions such as swinging or rocking. In this section, we'll examine the workings of modern clocks based on repetitive motions. As we do, we'll see that repetitive motions are interesting in their own right and appear throughout nature in countless objects besides clocks.

Questions to Think About: What exactly is time? Why do some objects swing or rock back and forth repetitively? How would you use a repetitive motion to measure time? How can you change the rate at which an object swings or rocks? Since repetitive motions don't normally continue forever, how can you keep them going without upsetting their timekeeping ability?

Experiments to Do: You can build the timekeeping portion of a pendulum clock by attaching a small weight to the end of a string and hanging that string from a table or doorway. Push the weight gently so that it swings back and forth. You'll find that it completes this repetitive motion with great regularity. What limits its regularity?

If the string is about 25 cm long, a complete swing (back and forth) will take almost exactly 1 s. Change the length of the string and observe its effect on the swing. Do you think that the weight affects the swing? Change the weight and see if you're right.

Now vary the extent of the swing to see if it affects the time each swing takes. How can it be that a small swing takes the same time as a large swing? Notice that you can get the weight swinging by pushing on it rhythmically, once each swing. Randomly timed pushes won't do; you must push the weight in synchrony with its motion. At what times should you push it to make it swing farther? To make it swing less far? Rhythmic pushes of this sort are what keep the timekeeper in a clock moving, hour after hour.

Time

Before examining clocks, we should take a brief look at time itself. Scientists treat time as a dimension, similar but not identical to the three spatial dimensions that we perceive in the world around us. In total, our universe has four dimensions: three spatial dimensions and one temporal dimension. Thus it takes four numbers to completely specify when and where an event occurs; three numbers identify the event's location, and one identifies the time at which it occurs.

An obvious difference between space and time is that, while we can see space stretched out around us, we can only observe the passage of time. Though we occupy only one location in space at a given moment, we are somehow more aware of the expanse of space around us. It's much harder to sense the whole framework of time stretching off into the past and future; you must use your imagination.

Our perception of space is ultimately based on the need for forces, accelerations, and velocities to travel from one place to another. A city seems far away because we know that traveling there with reasonable forces, accelerations, and velocities would take a long time. Our perception of time is based on the same mechanical principles. If two moments are separated by a long time, then reasonable forces, accelerations, and velocities will permit us to travel large distances between the two moments. In short, our perceptions of space and time are interrelated, and measurements of time and space are connected as well.

We measure space with rulers and time with clocks. But how would you make a ruler? You could construct a rather large ruler by driving a car at constant velocity and marking the pavement with paint once each second. Your ruler wouldn't be very practical, but it would fit the definition of a ruler as having spatial markings at uniform distances. You would be using your movement through time to measure space.

How would you make a clock? You could make a rather strange clock by driving a car at constant velocity down your giant ruler and counting each time you saw one of your marks go by. You would then be using your movement through space to measure time. While most clocks really do use motion to measure time, they use motions that are a bit more compact than a car ride.

ancient clocks

Check Your Understanding #1: The Ultimate Moon-Bounce

To measure the distance from the earth to the moon, scientists bounce light from reflectors placed on the moon by the Apollo astronauts. Light travels at a constant speed. How can a measurement of light's travel time to and from the moon be used to determine the distance from the earth to the moon?

❑ The daughter of a planetarium architect, **Jocelyn Bell (British astronomer 1943–)** acquired an early interest in radio astronomy. Advised to study physics first, she became the only woman in a class of 50 at Glasgow University. While working on her Ph.D. at Cambridge, Bell discovered an extraterrestrial source of radio bursts, occurring precisely 1.33730113 seconds apart. She had discovered the first pulsar, a collapsed star whose angular momentum keeps it turning at an extraordinarily uniform rate. Each burst coincided with one rotation of the star remnant.

Natural Resonances

An ideal timekeeping motion should offer both accuracy and convenience. That rules out some of the obvious choices. The sun, moon, and stars keep excellent time but they fail the convenience test. Sure, conservation of energy, momentum, and angular momentum so dominate their motions that these celestial bodies move steadily and predictably through the heavens, century after century, but what do you do on a cloudy day? And while simple interval timers like sandglasses and burning candles are easy to make and use, they're not very accurate. Besides, who's going to stay up all night lighting fresh candles just to keep the "clock" running? (For an interesting astronomical clock, see ❑.)

Instead, practical clocks are based on a particular type of repetitive motion called a natural resonance. In a **natural resonance,** the energy in an isolated object or system of objects causes it to perform a certain motion over and over again. Many objects in our world exhibit natural resonances, from tipping rocking chairs, to sloshing basins of water, to waving flagpoles. Resonances are very regular motions and can measure the passage of time extremely accurately.

Although a sandglass can be made repetitive by turning it over every time the sand runs out, this manual restarting process introduces timing errors. If a 3-minute sandglass is always turned over within 10 s after the sand runs out, how accurately will it measure time over the course of a day?

Pendulums and Harmonic Oscillators

One of the first natural resonances to find its way into clocks is the swing of a pendulum—a weight hanging from a pivot (Fig. 7.1.1). When the pendulum's center of gravity is directly below its pivot, it's in a stable equilibrium. Its center of gravity is then as low as possible, so displacing it raises its gravitational potential energy and a restoring force begins pushing it back toward that equilibrium position (Fig. 7.1.2). This restoring force is proportional to how far the pendulum is from equilibrium so that the further it's displaced, the stronger the restoring force becomes.

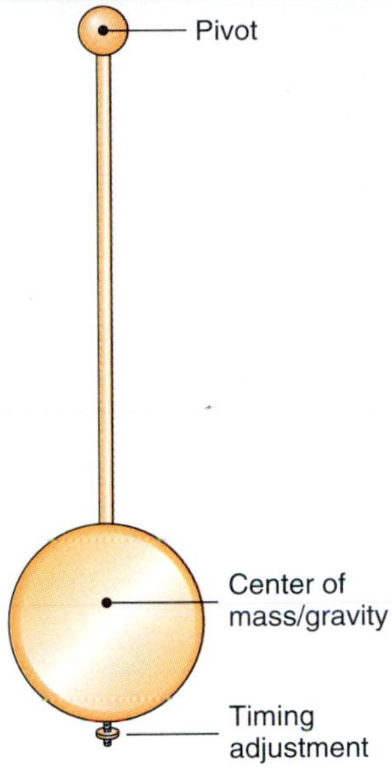

If you displace the pendulum and then let go, it will swing back and forth about its equilibrium position in a repetitive motion called an **oscillation.** As it swings, its energy shifts back and forth between potential and kinetic forms. When the pendulum swings rapidly through its equilibrium position in the middle of each swing, its energy is all kinetic. When it stops momentarily at the end of each swing, its energy is all gravitational potential. This repetitive transformation of energy from one form to another is part of any oscillation and keeps the **oscillator**—the system experiencing the oscillation—moving back and forth until all of its energy is either converted into thermal energy or transferred elsewhere.

But the pendulum isn't just any oscillator. Because its restoring force is proportional to its displacement from equilibrium, the pendulum is a **harmonic oscillator,** one of the simplest and best understood mechanical systems in nature. As a harmonic oscillator, the pendulum undergoes **simple harmonic motion,** a regular and predictable oscillation that makes it a superb timekeeper.

The **period** of any harmonic oscillator—the time it takes for that oscillator to complete one full cycle of its motion—depends only on the stiffness of its restoring force and on the mass of its moving object. **Stiffness** measures how rapidly that restoring force increases as the object moves away from its equilibrium position. The stiffer the restoring force, the harder that force pushes on the displaced object and the faster it oscillates back and forth. But the object's mass opposes acceleration so that the larger that mass, the slower it oscillates.

One of the most remarkable and important features of a harmonic oscillator is that its period doesn't depend on its **amplitude**—its furthest displacement from its equilibrium position. Whether that amplitude is large or small, the harmonic oscillator's period remains the same. This insensitivity to amplitude makes a harmonic oscillator ideal for timekeeping, since accidental changes in amplitude won't make the clock run fast or slow. Virtually all modern clocks are based on harmonic oscillators.

Actually, a pendulum is an unusual harmonic oscillator because its period *doesn't* depend on its mass. That's because increasing the pendulum's mass also increases its weight and stiffens the restoring force. These two changes compensate for one another so that the pendulum's period is unchanged.

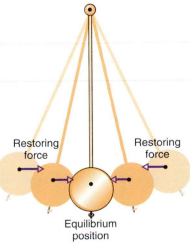

Fig. 7.1.1 A pendulum consists of a weight hanging from a pivot. The pendulum is in a stable equilibrium when its center of gravity is directly below the pivot.

Fig. 7.1.2 If you tilt a pendulum's center of gravity away from its equilibrium position, it experiences a restoring force proportional to its distance from that equilibrium position.

Fig. 7.1.3 The swinging pendulum controls the movement of the clock's hands. The pendulum is 0.248 m long, from pivot to center of mass/gravity, so each cycle takes 1 s to complete and advances the hands by 1 second.

However, a pendulum's period *does* depend on its length and on gravity. When you reduce the pendulum's length—the distance from its pivot to its center of mass—you stiffen the restoring force and shorten its period. Similarly, when you strengthen gravity (perhaps by traveling to Jupiter), you increase the pendulum's weight, stiffen the restoring force, and reduce the pendulum's period. Thus a short pendulum swings back and forth more often than a long one and any pendulum swings faster on the earth than it would on the moon.

On the earth's surface, a 0.248-m pendulum has a period of 1 s (Fig. 7.1.3), making it suitable for a wall clock that advances its second hand by 1 second each time the pendulum completes a cycle. Since a pendulum's period increases as the square root of its length, a 0.996-m pendulum (four times as tall as a 0.248-m pendulum) takes 2 s to complete its cycle and is appropriate for a floor clock that advances its second hand by 2 seconds per cycle.

Since a pendulum's period depends on its length and gravity, a change in either one causes trouble. Materials expand with increasing temperature, so a precision pendulum must be thermally compensated. A compensated pendulum uses several different materials with different coefficients of volume expansion to ensure that its center of mass remains at a fixed distance from its pivot.

While gravity doesn't change with time, it does vary slightly from place to place. To correct for differences in gravity between the factory and a clock's final destination, its pendulum has a threaded adjustment knob. This knob allows you to shift the pendulum's center of mass/gravity slightly to alter its period.

➤ Check Your Understanding #3: Swing Time
A child swinging on a swing set travels back and forth at a steady pace. What determines the period of his motion?

Pendulum Clocks

While a pendulum keeps a steady beat, it's not a complete clock. Something must keep the pendulum swinging and use it to display the current time. A pendulum clock does both. It sustains the pendulum's motion with gentle pushes, and it uses that motion to advance its hands at a steady rate.

The top of the pendulum has a two-pointed *anchor* that controls the rotation of a *toothed wheel* (Fig. 7.1.4). This mechanism is called an *escapement*. A weighted cord wrapped around the toothed wheel's shaft exerts a torque on that wheel, so that the wheel would spin if the anchor weren't holding it in place. Each time the pendulum reaches the end of a swing, one point of the anchor releases the toothed wheel while the other point catches it. The wheel turns slowly as the pendulum rocks back and forth, advancing by one tooth for each full cycle of the pendulum. This wheel turns a series of gears, which slowly advance the clock's hands.

The toothed wheel also keeps the pendulum moving by giving the anchor a tiny forward push each time the pendulum completes a swing. Since the anchor moves in the direction of the push, the wheel does work on the anchor and pendulum, and replaces energy lost to friction and air resistance. This energy comes from the weighted cord, which releases gravitational potential energy as its weight descends. When you wind the clock, you rewind this cord around the shaft, lifting the weight and replenishing its potential energy.

Keeping the pendulum swinging steadily is extremely important. If the pendulum's amplitude varies significantly from cycle to cycle, the toothed wheel will turn erratically and the clock won't be accurate. A clock works best when its pendulum swings freely and easily because any outside force—even the push from the toothed wheel—affects its period. The most accurate timekeepers are those that can oscillate on their own for thousands or millions of cycles. These precision timekeepers need only the slightest pushes to keep them moving and thus have extremely precise periods. That's why a good clock uses an aerodynamic pendulum and low-friction bearings.

But there's a second reason to keep the pendulum's amplitude steady: it's not really a perfect harmonic oscillator. If you displace the pendulum too far, it becomes an **anharmonic oscillator**—its restoring force ceases to be proportional to its displacement from equilibrium, and its period begins to depend on its amplitude. Since a change in period would spoil the clock's accuracy, the pendulum's amplitude must be kept small and steady. That way, the amplitude has almost no effect on the pendulum's period.

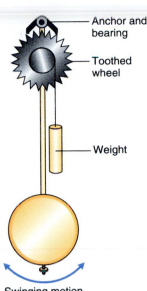

Fig. 7.1.4 A pendulum clock uses a swinging pendulum to determine how quickly a toothed wheel turns and advances a series of gears that control the hands of the clock. The anchor permits the toothed wheel to advance by one tooth each time the pendulum completes a full cycle.

Check Your Understanding #4: Swing High, Swing Low

When pushing a child on a playground swing, you normally push her forward as she moves away from you. What happens if you push her forward each time she moves toward you?

Balance Clocks

Because it relies on gravity for its restoring force, a swinging pendulum mustn't be tilted or moved. A portable clock needs something else as a timekeeper. Since harmonic oscillators are best, a mobile clock needs a gravity-independent restoring force that's proportional to displacement. It needs a spring!

As we saw in Section 3.1, the force a spring exerts is proportional to its distortion. The more you stretch, compress, or bend a spring, the harder it pushes back toward its equilibrium shape. Attach a block of wood to the free end of a spring, stretch it gently, and let go and you'll find you have a harmonic oscillator with a period determined only by the stiffness of the spring and the block's mass (Fig. 7.1.5). Since the period of a harmonic oscillator doesn't depend on the amplitude of its motion, the block oscillates steadily about its equilibrium position and makes an excellent timekeeper.

Fig. 7.1.5 A block attached to a spring is a harmonic oscillator. The oscillator's period is determined only by the stiffness of the spring and the mass of the block.

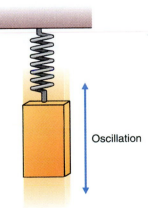

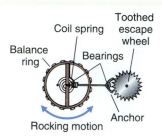

Fig. 7.1.6 A balance clock uses a rocking balance ring to control the rotation of a toothed wheel and the gears that advance the clock's hands. The anchor permits the toothed wheel to advance by one tooth each time the ring completes a full cycle. The clock's energy comes from a main spring (not shown) that exerts a steady torque on the toothed wheel.

jewel bearings

Fig. 7.1.7 The balance ring in this antique Swiss wristwatch is located at the left side of the picture. Below it is the toothed escape wheel. Tiny jewel bearings minimize sliding friction and permit this watch to keep very accurate time.

Unfortunately, gravity complicates this simple system. Although gravity doesn't alter the block's period, it does shift the block's equilibrium position downward. That shift is a problem for a clock that might be tilted sometimes. However, there's another spring-based timekeeper that marks time accurately in any orientation or location. This ingenious device, used in most mechanical clocks and watches, is called a balance ring or simply a balance.

A balance ring resembles a tiny metal bicycle wheel, supported at its center of mass/gravity by an axle and a pair of bearings (Fig. 7.1.6). Any friction in the bearings is exerted so close to the ring's axis of rotation that it produces little torque and the ring turns extremely easily. Moreover, the ring pivots about its own center of gravity so that its weight produces no torque on it.

The only thing exerting a torque on the balance ring is a tiny coil spring. One end of this spring is attached to the ring while the other is fixed to the body of the clock. When the spring is undistorted, it exerts no torque on the ring and the ring is in equilibrium. But if you rotate the ring either way, torque from the distorted spring will act to restore it to its equilibrium orientation. Since this restoring torque is proportional to the ring's rotation away from a stable equilibrium, the balance ring and coil spring form a harmonic oscillator!

Because of the rotational character of this harmonic oscillator, its period depends on the torsional stiffness of the coil spring—how rapidly the spring's torque increases as you twist it—and on the balance ring's moment of inertia. Since the balance ring's period doesn't depend on the amplitude of its motion, it keeps excellent time. And because gravity exerts no torque on the balance ring, this timekeeper works anywhere and in any orientation.

The rest of a balance clock is similar to a pendulum clock (Fig. 7.1.7). As the balance ring rocks back and forth, it tips a lever that controls the rotation of a toothed wheel. An anchor attached to the lever allows the toothed wheel to advance one tooth for each complete cycle of the balance ring's motion. Gears connect the toothed wheel to the clock's hands, which slowly advance as the wheel turns.

Because the balance clock is portable, it can't draw energy from a weighted cord. Instead, it has a main spring that exerts a torque on the toothed wheel. This main spring is a coil of elastic metal that stores energy when you wind the clock. Its energy keeps the balance ring rocking steadily back and forth and also turns the clock's hands. Since the main spring unwinds as the toothed wheel turns, the clock occasionally needs winding. (For an interesting example of a balance clock, see ❏ on p. 237.)

> **Check Your Understanding #5: A Little Light Entertainment**
> When you accidentally strike a chandelier with a broom, this hanging lamp begins to twist back and forth with a regular period. What determines its period of oscillation?

Electronic Clocks

The potential accuracy of pendulum and balance clocks is limited by friction, air resistance, and thermal expansion to about ten seconds per year. To do better, a clock's timekeeper must avoid these mechanical shortcomings. That's why so many modern clocks use quartz oscillators as their timekeepers.

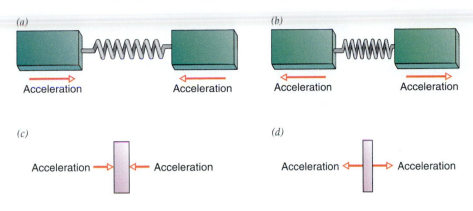

(a) Acceleration   Acceleration   (b) Acceleration   Acceleration

(c) Acceleration →� ← Acceleration   (d) Acceleration ← → Acceleration

Fig. 7.1.8 A quartz crystal acts like a spring with masses at each end. Just as the two masses alternately accelerate (*a*) toward and (*b*) away from one another, so the two halves of the vibrating crystal alternately (*c*) accelerate together and (*d*) apart.

A quartz oscillator is made from a single crystal of quartz, the same mineral found in most white sand. Like many hard and brittle objects, a quartz crystal oscillates strongly after being struck. In fact, it's a harmonic oscillator because it acts like a spring with a block at each end (Fig. 7.1.8*a,b*). The two blocks oscillate in and out symmetrically about their combined center of mass, with a period determined only by the blocks' masses and the spring's stiffness. In a quartz crystal, the spring is the crystal itself and the blocks are its two halves (Fig. 7.1.8*c,d*). Since the forces on the blocks are proportional to their displacements from equilibrium, they're harmonic oscillators.

Because of its exceptional hardness, a quartz crystal's restoring force is extremely stiff. Even a tiny distortion produces a huge restoring force. Since the period of a harmonic oscillator decreases as its spring becomes stiffer, a typical quartz oscillator has an extremely short period. Its motion is usually called a **vibration** rather than an oscillation because vibration implies a fast oscillation in a mechanical system. Oscillation itself is a more general term for any repetitive process and can even apply to such nonmechanical processes as electric or thermal oscillations.

Because of its rapid vibration, a quartz oscillator's period is a small fraction of a second. We normally characterize such a fast oscillator by its **frequency**—the number of cycles it completes in a certain amount of time. The SI unit of frequency is **cycles-per-second,** also called the **hertz** (abbreviated Hz) after German physicist Heinrich Rudolph Hertz. Period and frequency are reciprocals of one another (period equals 1/frequency and vice versa) so that an oscillator with a period of 0.001 s has a frequency of 1000 Hz.

Because the vibrating crystal isn't sliding across anything or moving quickly through the air, it loses energy slowly and vibrates for a long, long time. And because quartz's coefficient of thermal expansion is extremely small, the crystal's period is nearly independent of its temperature. With its exceptionally steady period, a quartz oscillator can serve as the timekeeper for a highly accurate clock, one that loses or gains less than a tenth of a second per year.

Of course, a quartz crystal isn't a complete clock. Like the pendulum and balance, it needs something to keep it vibrating and to display the current time. While these tasks could conceivably be done mechanically, quartz clocks are normally electronic. There are two reasons for this choice. First, the crystal's vibrations are too fast and too small for most mechanical devices to follow. Second, a quartz crystal is intrinsically electronic itself; it responds mechanically to electrical stress and electrically to mechanical stress. Because of this coupling between its mechanical and electrical behaviors, crystalline quartz is known as a piezoelectric material and is ideal for electronic clocks.

quartz crystal vibration

❑ The son and grandson of freed slaves, **Benjamin Banneker (African-American mathematician, astronomer, and writer, 1731–1806)** grew up on a Maryland tobacco farm. He was fascinated by mathematics and science, and supplemented his limited schooling with borrowed books. Though he is best remembered for his work in astronomy and for compiling six almanacs, he also produced one of the first clocks made entirely in America. With only a borrowed pocket watch as a guide, Banneker built his wooden balance clock by hand, using a knife to shape the parts. The clock kept accurate time for half a century and even struck the hours.

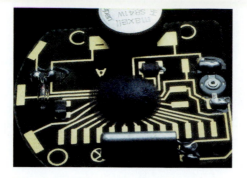

Fig. 7.1.9 This circuit board is part of a quartz wristwatch. The quartz crystal is located in the black cylinder at the bottom. Carefully polished to vibrate at a precise frequency, the crystal keeps the watch accurate to a few seconds a month.

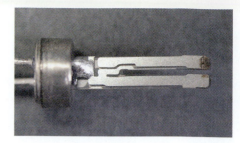

Fig. 7.1.10 Shaped like a tiny tuning fork, this watch crystal vibrates almost exactly 32,768 times per second. Burn marks at its tips were created when its vibrational frequency was tuned by a laser beam.

The clock's circuitry uses electrical stresses to keep the quartz crystal vibrating (Fig. 7.1.9). Just as carefully timed pushes keep a child swinging endlessly on a playground swing, carefully timed electrical stresses keep the quartz crystal vibrating endlessly in its holder. Because the crystal loses so little energy with each vibration, only a tiny amount of work is required each cycle to maintain its vibration.

The clock also detects the crystal's vibrations electrically. Each time its halves move in or out, the crystal experiences mechanical stress and emits a pulse of electricity. These pulses may control an electric motor that advances clock hands or may serve as input to an electronic chip that measures time by counting the pulses.

The quartz crystals used in clocks and watches are carefully cut and polished to vibrate at specific frequencies. The thinner the crystal, the faster it vibrates—less mass and a stiffer restoring force. In effect, these crystals are tuned like musical instruments to match the requirements of their clocks.

While most tiny quartz crystals vibrate millions of times each second, common watch crystals vibrate at 2^{15} Hz or 32,768 Hz. This low frequency prolongs a watch's battery life because counting each pulse consumes some of the battery's energy. To make a small crystal vibrate this slowly, the manufacturer cuts away most of the center of the crystal to weaken its restoring force and slow its oscillations (Fig 7.1.10). The resulting quartz "tuning fork" oscillator is carefully metalized to permit the watch to interact with it electrically, and it's then tuned to exactly 32,768 Hz by burning away some of the metal mass with a laser beam.

Check Your Understanding #6: Heavy Metal Music

If you drop a metal rod on the floor, end first, you hear a high-pitched tone. What's happening?

atomic clocks, electric clocks, tuning fork clocks

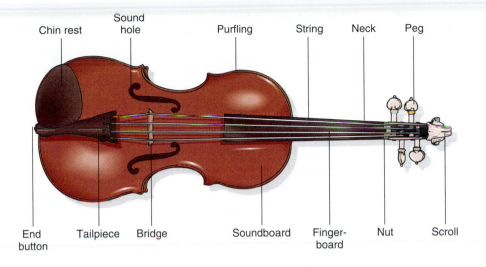

Chin rest · Sound hole · Purfling · String · Neck · Peg

End button · Tailpiece · Bridge · Soundboard · Finger-board · Nut · Scroll

Violins and Pipe Organs

Music is an important part of human expression. While what qualifies as music is a matter of taste, it always involves sound and often involves instruments. In this section, we'll examine sound, music, and two instruments: violins and pipe organs. As examples of the two most common types of instruments, strings and winds, this pair will help us to understand many other instruments as well.

Questions to Think About: *Why are the low-pitch strings on a violin thicker than the high-pitch strings? How does pressing a violin's string against the fingerboard change its pitch? Why does a violin sound different when it's plucked rather than bowed? What purpose does the violin's body serve? What is vibrating inside a pipe organ? Why are some organ pipes longer than others?*

Experiments to Do: *Find a violin or guitar, or stretch a strong string between two rigid supports. Pluck the string with your finger and listen to the tone it makes. The string vibrates back and forth at a particular pitch or frequency even as its amplitude of motion gradually decreases. What kind of oscillator has that behavior?*

Change the string's frequency of vibration by changing its tension or length. What happens when you shorten the string or prevent part of it from moving? What happens if you increase its tension by pulling it tighter? You can also increase the string's mass by wrapping it with tape. How does that affect its pitch?

You can imitate a pipe organ by blowing gently across the mouth of a bottle or soda straw. If done properly, you'll get the air inside the bottle vibrating up and down rhythmically and you'll hear a tone. What happens to its pitch when you add water to the bottle or pinch off the straw at various points? Why does this tone sound different from that of the string, even when the two have the same pitch?

Sound and Music

To understand how violins and pipe organs work, we'll need to know a bit more about sound and music. In air, **sound** consists of pressure waves—patterns of compressions and rarefactions that travel outward from their source at the speed of sound. When a sound passes by, the air pressure in your ear fluctuates up and down about normal atmospheric pressure. Even when these fluctuations have amplitudes less than a millionth of atmospheric pressure, you hear them as sound.

When the fluctuation is repetitive, you hear a *tone* with a *pitch* defined by the fluctuation's frequency. A bass singer's pitch range extends from 80 Hz to 300 Hz, while that of a soprano singer extends from 300 Hz to 1100 Hz. Musical instruments can produce tones over a much wider range of pitches, but we can only hear those between about 30 Hz and 20,000 Hz, and less as we get older.

Most music is constructed around *intervals*—the frequency ratio between two different tones. This ratio is found by dividing one tone's frequency by that of the other. Our hearing is particularly sensitive to intervals, with pairs of tones at equal intervals sounding quite similar to one another. For example, a pair of tones at 440 Hz and 660 Hz sounds similar to a pair at 330 Hz and 495 Hz because they both have the interval 3/2.

The interval 3/2 is pleasing to most ears and is common in Western music, where it's called a *fifth*. A fifth is the interval between the two "twinkles" at the beginning of *Twinkle, twinkle, little star*. If your ear is good, you can start with any tone for the first twinkle and will easily find the second tone, located at 3/2 the frequency of the first. Your ear hears that factor of 3/2 between the two frequencies.

The most important interval in virtually all music is 2/1 or an *octave*. Tones that differ by a factor of 2 in frequency sound so similar to our ears that we often think of them as being the same. When men and women sing together "in unison," they often sing an octave or two apart and the differences in the tones, always factors of 2 or 4 in frequency, are only barely noticeable.

The octave is so important that it structures the entire range of audible pitches. Most of the subtle interplay of tones in music occurs in intervals of less than an octave, less than a factor of 2 in frequency. Thus most traditions build their music around the intervals that lie within a single octave, such as 5/4 and 3/2. They pick a particular standard pitch and then assign *notes* at specific intervals from this standard pitch. This arrangement repeats at octaves above and below the standard pitch to create a complete *scale* of notes. (For a history of scales, see ❑.)

The scale used in Western music is constructed around a note called A_4, which has a standard pitch of 440 Hz. At intervals of 9/8, 5/4, 4/3, 3/2, 5/3, and 15/8 above A_4 lie the six notes B_4, $C^{\#}_5$, D_5, E_5, $F^{\#}_5$, and $G^{\#}_5$. Similar collections of six notes are built above A_5 (880 Hz), which has a frequency twice that of A_4, and above A_3 (220 Hz), which has a frequency half that of A_4. In fact, this pattern repeats above A_1 (55 Hz) through A_8 (7040 Hz).

Actually, Western music is built around 12 notes and 11 intervals that lie within a single octave. Five more intervals account for five additional notes, $B^{\flat}_4$, C_5, $D^{\#}_5$, F_5, and G_5. It's also not quite true that every note is based exclusively on its interval from A_4. While A_4 remains at 440 Hz, the pitches of the other 11 notes have been modified slightly so that they're at interesting and pleasing intervals from one another as well as from A_4. This adjustment of the pitches led to the well-tempered scale that has been the basis for Western music for the last several centuries.

❑ In addition to his contributions to mathematics, geometry, and astronomy, the **Greek mathematician Pythagoras (ca 580–500 BC)** was perhaps the first person to use mathematics to relate intervals, pitches, and the lengths of vibrating strings. He and his followers laid the groundwork for the scale used in most Western music.

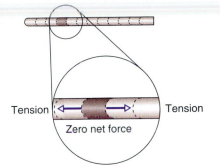

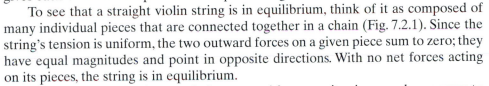
A typical singing voice can cover a range of about 2 octaves, for example, from C_4 to C_6. How broad is this range of frequencies?

How a Violin String Works

The tones produced by a violin begin as vibrations in its strings. But these strings are limp and shapeless on their own and rely on the violin's rigid body and neck for structure. The violin creates tension in its strings and gives each of them an equilibrium shape: a straight line.

To see that a straight violin string is in equilibrium, think of it as composed of many individual pieces that are connected together in a chain (Fig. 7.2.1). Since the string's tension is uniform, the two outward forces on a given piece sum to zero; they have equal magnitudes and point in opposite directions. With no net forces acting on its pieces, the string is in equilibrium.

But when the string is curved, the outward forces on its pieces no longer sum to zero (Fig. 7.2.2). Although the string's uniform tension still gives those outward forces equal magnitudes, they now point in slightly different directions. As a result, each piece experiences a small net force.

The net forces on its pieces are restoring forces because they act to straighten the string. If you distort the string and release it, these restoring forces will cause the string to vibrate about its straight equilibrium shape in a natural resonance. But the string's restoring forces are special; the more you curve the string, the stronger the restoring forces on its pieces become. In fact, the restoring forces increase in proportion to the string's distortion, so that the string is a form of harmonic oscillator!

Actually, the string is much more complicated than a pendulum or a balance ring. It can bend and vibrate in many distinct **modes**—basic patterns of distortion—and each mode has its own period. Nonetheless, the string retains the most important feature of a harmonic oscillator: the period of each vibrational mode is independent of its amplitude. Thus a violin string's pitch doesn't depend on how hard it's vibrating. Think how tricky it would be to play a violin if its pitch depended on its volume!

A violin string has a simplest vibration: its **fundamental vibrational mode.** In this mode, the entire string vibrates in the same direction; its midpoint moves the farthest (the **vibrational antinode**) while its ends remain fixed (the vibrational nodes). The string's shape is a gradual curve based on the trigonometric sine function (Fig. 7.2.3).

In this fundamental mode, the violin string behaves as a single harmonic oscillator. Its kinetic energy peaks as it rushes through its straight equilibrium shape and its potential energy (elastic potential energy in the string) peaks as it stops to turn around. As with any harmonic oscillator, the string's vibrational period depends only on its mass and on the stiffness of its restoring forces. Increasing the string's mass slows its vibration while stiffening the string speeds that vibration up. The frequency of this vibration establishes the violin string's *fundamental pitch.*

Each of a violin's four strings has a different mass and a different fundamental pitch. In a tuned violin, the notes produced by these strings are G_3 (196 Hz), D_4 (293 Hz), A_4 (440 Hz), and E_5 (660 Hz). The G_3 string, which vibrates quite slowly, is the most massive. It's usually made of gut, wrapped in a coil of heavy metal wire.

Tension | Tension
Zero net force

Fig. 7.2.1 A taut violin string can be viewed as composed of many individual pieces. When the string is straight, the two forces exerted on a given piece by its neighbors cancel perfectly.

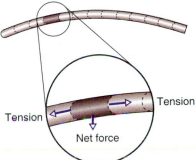

Tension   Tension
Net force

Fig. 7.2.2 When a violin string is curved, the two forces exerted on a given piece by its neighbors don't point in exactly opposite directions and don't balance one another. The piece experiences a net force.

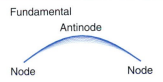

Fundamental
Antinode
Node   Node

Fig. 7.2.3 A string vibrating between two fixed points in its fundamental vibrational mode. The whole string moves together, traveling up and down as a single harmonic oscillator.

The E_5 string, on the other hand, must vibrate quite rapidly and needs to have a low mass. It's usually a thin steel wire.

You can adjust a string's fundamental pitch by changing its stiffness, which itself depends on tension and length. Tightening a string stiffens it by increasing both the outward forces on its pieces and the net forces they experience during a distortion. You tune a violin by adjusting the tension in its strings, using pegs in its neck and tension adjusters on the tailpiece. Since temperature and time can alter a string's tension, you should always tune your violin just before a concert.

A string's length also affects its fundamental pitch. Shortening a string stiffens it by increasing its curvature during a distortion and subjecting its pieces to larger net forces. Shortening the string also reduces its mass. Since a stiffer string with a smaller mass vibrates faster, shortening a string increases its pitch. This effect allows you to raise the pitch of a string by pressing it against the fingerboard in the violin's neck and effectively shortening it. Part of a violinist's skill involves knowing exactly where on the string to press it against the fingerboard in order to produce a particular note.

➤ Check Your Understanding #2: Feeling Tense?
A common way to determine the tension in a cord is to pluck it and listen for how fast it vibrates. Why does this technique measure tension?

The Violin String's Harmonics

The fundamental vibrational mode isn't the only way in which a violin string can vibrate. The string also has **higher-order vibrational modes,** in which the string vibrates as a chain of shorter strings that move in alternating directions (Fig. 7.2.4).

For example, the string can vibrate as two half-strings moving in opposite directions and separated by a motionless vibrational node. In this mode, the violin string not only vibrates as two half-strings, it has the pitch of half-strings as well! Remarkably enough, halving the length of a violin string exactly doubles its vibrational frequency. Frequencies that are integer multiples of the fundamental pitch are called **harmonics,** so this half-string vibration occurs at the *second harmonic pitch* and is called the *second harmonic mode*.

A violin string can also vibrate as three third-strings, with a frequency that's three times the fundamental. The interval between this *third harmonic pitch* and the fundamental pitch is an octave and a fifth (2/1 times 3/2). Overall, the fundamental and its second and third harmonics sound very pleasant together.

While the violin string can vibrate in even higher harmonics, a more important fact is that the string often vibrates in more than one mode at the same time. For example, a violin string vibrating in its fundamental mode can also vibrate in its second harmonic and emit two tones at once. While this multiple vibration complicates the string's shape and motion, the tones it emits are still integer multiples of its fundamental pitch.

Harmonics are important because bowing a violin excites many of its vibrational modes. The violin's sound is thus a rich mixture of the fundamental tone and its harmonics. This mixture of tones is characteristic of a violin, which is why an instrument producing a different mixture doesn't sound like a violin.

Second harmonic

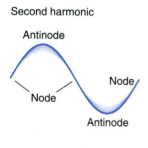

Third harmonic

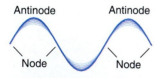

Fig. 7.2.4 A string vibrating between two fixed points in its second and third harmonic modes. The string vibrates as two or three segments, completing cycles at two or three times the fundamental frequency, respectively.

► **Check Your Understanding #3: Swinging High and Low Together**
When two people swing a long jump rope, they can make it swing as a single arc or as two half-ropes arcing in opposite directions. To make the rope swing as two half-ropes, it must be turned faster or with a higher tension. Why?

Bowing and Plucking the Violin String

You play a violin by drawing a bow across its strings. The bow consists of horsehair, pulled taut by a wooden stick. Horsehair is rough and exerts frictional forces on the strings as it moves across them. But most importantly, horsehair exerts much larger static frictional forces than sliding ones.

As the bow hairs rub across a string, they grab the string and push it forward with static friction. Eventually the string's restoring force overpowers static friction, and the string suddenly starts sliding backward across the hairs. Because the hairs exert little sliding friction, the string completes half a vibrational cycle with ease. But as it stops to reverse direction, the hairs grab the string again and begin pushing it forward. This process repeats over and over.

Each time the bow pushes the string forward, it does work on the string and adds energy to the string's vibrational modes. This process is an example of **resonant energy transfer,** in which a modest force doing work in synchrony with a natural resonance can transfer a large amount of energy to that resonance. Just as gentle, carefully timed pushes can get a child swinging high on a playground swing, so gentle, carefully timed pushes from a bow can get a string vibrating vigorously on a violin. Similar pushes can cause other objects to vibrate strongly, notably a crystal wineglass (Fig. 7.2.5) and the Tacoma Narrows Bridge near Seattle, Washington (Fig. 7.2.6). The wineglass's response to a certain tone is also an example of **sympathetic vibration**—the transfer of vibrational energy between two systems that share a common vibrational frequency.

Fig. 7.2.5 Resonant energy transfer makes it possible for sound to shatter a crystal wine glass. When the sound pushes on the glass rhythmically, the sound slowly transfers energy to the glass, until it finally shatters. Because the sound must be extremely loud and at exactly the resonant frequency of the glass, only the most extraordinary opera singers can break it.

Fig. 7.2.6 The Tacoma Narrows Bridge collapsed in November 1940, as the result of resonant energy transfer between the wind and the bridge surface. Shortly after construction, the automobile bridge began to exhibit an unusual natural resonance in which its surface twisted slowly back and forth so that one lane rose as the other fell. During a storm, the wind slowly added energy to this resonance until the bridge ripped itself apart.

Fig. 7.2.7 A violin's bridge transfers energy from its vibrating strings to its belly. The belly moves in and out, emitting sound. Some of this sound leaves the violin through the f-holes in its body.

The amount of energy the bow adds to each vibrational mode depends on where it crosses the violin string. When you bow the string at the usual position, you produce a strong fundamental vibration and a moderate amount of each harmonic. Bowing the string nearer its middle reduces the string's curvature, weakening its harmonic vibrations and giving it a mellower sound. Bowing the string nearer its end increases the string's curvature, strengthening its harmonic vibrations and giving it a brighter sound.

The sound of a plucked violin string also depends on harmonic content and thus on where that string is plucked. But this sound is quite different from that of a bowed string. The difference lies in the sound's *envelope*—the way the sound evolves with time. This envelope can be viewed as having three time periods: an initial *attack*, a middle *sustain*, and a final *decay*. The envelope of a plucked string is an abrupt attack followed immediately by a gradual decay. In contrast, the envelope of a bowed string is a gradual attack, a steady sustain, and then a gradual decay. We learn to recognize individual instruments not only by their harmonic content but also by their sound envelopes.

> **Check Your Understanding #4: What's the Buzz?**
> Sometimes a tone from an instrument or an audio device will cause some object in the room to begin vibrating loudly. Why does this happen?

Turning the String's Vibration into Sound

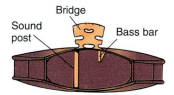

Fig. 7.2.8 The bridge is supported by the bass bar on one side and the sound post on the other. As the strings vibrate back and forth, the bridge experiences a torque that causes the belly of the violin to move in and out and emit sound.

The violin's sound doesn't come directly from its strings because they're much too narrow to push effectively on the air. Instead, the violin creates sound with its top plate or *belly* (Fig. 7.2.7). The strings transfer their vibrational motions to the belly and the belly pushes on the air to create sound.

Most of this vibrational energy flows into the belly through the violin's *bridge*, which holds the strings away from the violin's body (Fig. 7.2.8). Beneath the G_3 string side of the bridge is the *bass bar*, a long wooden strip that stiffens the belly. Beneath the E_5 side of the bridge is the *sound post*, a shaft that extends from the violin's belly to its back.

As a bowed string vibrates across the violin's belly, it exerts a torque on the bridge about the sound post. The bridge rotates back and forth, causing the bass bar and belly to move in and out. The belly's motion produces most of the violin's sound. Some of this sound comes directly from the belly's outer surface, and the rest comes from its inner surface and must emerge through its f-shaped holes.

projecting sound

> **Check Your Understanding #5: Air Guitar**
> Why does an acoustic guitar have a sound box?

Sound from an Organ Pipe

Like a violin, a pipe organ uses vibrations to create sound. However, its vibrations take place in the air itself. An organ pipe is essentially a hollow cylinder, open at each end and filled with air. Because that air is isolated, its pressure can fluctuate up and down relative to atmospheric pressure, and it can exhibit natural resonances.

In its fundamental vibrational mode, air moves alternately toward and away from the pipe's center (Fig. 7.2.9), like two blocks on a spring. As air moves toward the pipe's center, the density there rises and a pressure imbalance occurs. Since the pressure at the pipe's center is higher than at its ends, air accelerates *away from* the center. It eventually stops moving inward and begins to move outward. As air moves away from the pipe's center, the density there drops and a reversed pressure imbalance occurs. Since the pressure at the pipe's center is lower than at its ends, air now accelerates *toward* the center. It eventually stops moving outward and begins to move inward, and the cycle repeats.

The air in an organ pipe behaves as a harmonic oscillator, vibrating about a stable equilibrium of uniform atmospheric density and pressure. The vibrating air's kinetic energy peaks as it rushes through this equilibrium arrangement, and its potential energy (pressure potential energy in the air) peaks as it stops to turn around. The period and frequency of this vibration depend only on the mass of the vibrating air and on the stiffness of the restoring forces the air experiences.

Both the mass of the vibrating air and the stiffness of the restoring forces depend on the length of the organ pipe. A shorter pipe not only holds less air than a longer pipe, it also offers stiffer opposition to any movements of air in and out of that pipe—with less room in the shorter pipe, the pressure inside it rises and falls more abruptly, leading to stiffer forces on the moving air. Together, these effects make the air in a short pipe vibrate faster than that in a long pipe. In general, an organ pipe's vibrational frequency is inversely proportional to its length.

Unfortunately, the mass of vibrating air in a pipe also increases with the air's density, so that even a modest change in temperature or weather will alter the pipe's pitch. Fortunately, all of the pipes shift together so that an organ continues to sound in tune. Nonetheless, this shift may be noticeable when the organ is part of an orchestra.

The organ uses resonant energy transfer to make the air in a pipe vibrate. It starts this transfer by blowing air across the pipe's lower opening (Fig. 7.2.10a), although for practical reasons that lower opening is usually found on the pipe's side (Fig. 7.2.10b). As the air flows across the opening, it's easily deflected and tends to follow any air that's already moving into or out of the pipe. If the air inside the pipe is vibrating, the new air will follow it in perfect synchrony and strengthen the vibration.

This following effect is so effective at enhancing vibrations that it can even initiate a vibration from the random noise that's always present in a pipe. That's how the sound starts when the organ's pump first blows air across the pipe. Once the vibration has started, it grows quickly in amplitude until energy leaves the pipe as sound and heat as quickly as it arrives via compressed air. The more air the organ blows across the pipe, the more power it delivers to the pipe and the louder the vibration.

Like a violin string, an organ pipe can support more than one mode of vibration. In its fundamental vibrational mode, the pipe's entire column of air vibrates together. In the higher-order vibrational modes, this air column vibrates as a chain of shorter air columns moving in alternating directions. If the pipe has a constant width, these vibrations occur at harmonics of the fundamental. When the air column vibrates as two half-columns, its pitch is exactly twice that of the fundamental mode. When it vibrates as three third-columns, its pitch is exactly three times that of the fundamental. And so on.

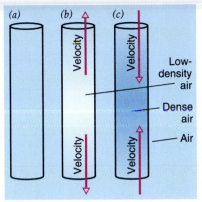

Fig. 7.2.9 In a pipe that's open at both ends (*a*), the air vibrates in and out about the middle of the pipe. (*b*) For half a cycle, the air moves outward and creates a low-pressure region in the middle and (*c*) for half a cycle, the air moves inward and creates a high-pressure region there.

Fig. 7.2.10 (*a*) Air blown across the bottom of an open pipe will follow any other air that's moving into the pipe. If the air in the pipe is vibrating, this effect will add energy to that vibration. (*b*) The lower opening in an organ pipe is cut in its side for practical reasons.

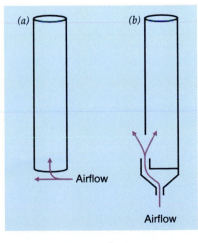

But the air column inside a pipe can vibrate in more than one mode simultaneously. The shape of the organ pipe and the place where air is blown across it determine the pipe's harmonic content and its specific sound. Different pipes can imitate different instruments. To sound like a flute, the pipe should emit mostly the fundamental tone and keep the harmonics fairly quiet. To sound like a clarinet, its harmonics should be much louder. An organ pipe's volume always builds slowly during the attack, so it can't pretend to be a plucked string. However, a clever designer can make the organ imitate a surprising range of instruments.

Check Your Understanding #6: A Pop Organ

If you blow across a soda bottle, it emits a tone. Why does adding water to the bottle raise the pitch of that tone?

complete organ

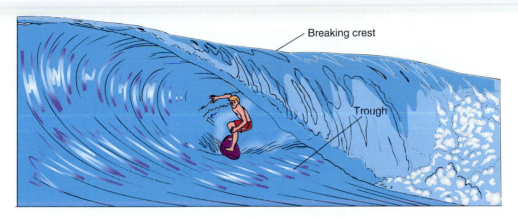

Breaking crest

Trough

SECTION 7.3   # The Sea and Surfing

The sea is never still. If you've visited the seashore, you've probably noticed two of the sea's most important motions: tides and surface waves. In this section, we'll examine the cycle of tides and look at how surface waves travel across water. These water waves can help us to understand other wave phenomena, including the electromagnetic waves that are responsible for light and the pressure waves that are the basis for sound.

Questions to Think About: *Why do the tides vary in height from place to place? Why are there no significant tides in a lake or swimming pool? Why does high tide occur about every 12 hours? What moves in a water wave? Do all waves travel at the same speed? How deep is a wave? Why do waves break near shore?*

Experiments to Do: *While you can make water waves in a basin or tub, they move too quickly to see clearly. You do better to watch waves at the beach. On a calm day, the sea is in its flat equilibrium shape. But when the wind picks up, look out. How does wind affect the sea's surface? If water waves contain energy, where does that energy come from and what forms does it take in the water?*

Watch a floating object move as a wave passes it. Does the object travel with the wave? Does the water itself travel with the wave? Can you think of other cases in which a disturbance moves steadily forward but the material carrying the disturbance remains behind?

Now watch as waves break near the shore. If you look around at different beaches, you'll find waves breaking in two different ways. In some cases a wave will simply collapse into churning froth, while in others its top will dive forward over the water ahead of it. Can you see anything about the beaches or water that might account for this difference?

Fig. 7.3.1 The moon's gravity varies over the surface of the earth. The closer a point is to the moon, the stronger the gravity it experiences. This variation in the moon's gravity produces the tide.

The Tides

If you watch the sea for a few days, you'll notice the *tides*. In a cycle as old as the oceans themselves, the water level rises for $6\frac{1}{4}$ hours to reach *high tide*, drops for $6\frac{1}{4}$ hours to arrive at *low tide*, and then begins rising again. Once a wonderful mystery,

Moon's gravity

Moon

Earth

Fig. 7.3.2 Variations in the moon's gravity produce tidal forces on the earth's surface and cause the earth's oceans to bulge outward in two places. These bulges, which are located nearest and farthest from the moon, move over the earth's surface as it rotates.

variations in moon gravity

Fig. 7.3.3 The tides vary over a lunar month. They're strongest when the sun and the moon are aligned (spring tides) and weakest when they are at 90° from one another (neap tides).

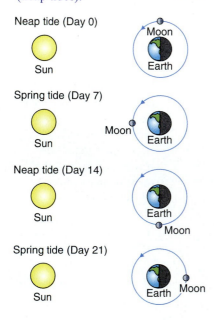

we now know that the tides are caused by the moon's gravity and, to a lesser extent, that of the sun.

On earth, the moon's gravity is so weak that we normally don't notice it. The moon is far away and, as we learned in Section 5.4, gravity diminishes with distance. But this dependence on distance also means that the moon's gravity is stronger on one side of the earth than the other; you experience a stronger pull when you're on the side of the earth nearest the moon than you do when you're on the side opposite it. While you can't feel these variations in moon gravity yourself (Fig. 7.3.1), the earth's oceans respond to them. The oceans are deformed by the moon's gravity (Fig. 7.3.2) and this deformation produces the tides.

The differences between the moon's gravity at particular locations on earth and its average strength for the entire earth give rise to **tidal forces**—residual gravitational forces that act to shift those locations relative to the earth as a whole. The near side of the earth is pulled toward the moon more strongly than average so it experiences a tidal force toward the moon. The far side of the earth is pulled toward the moon less strongly than average so it experiences a tidal force away from the moon.

If the earth weren't so rigid, these tidal forces would stretch it into an egg shape. The near side of the earth would bulge outward toward the moon, while the far side of the earth would bulge outward away from the moon. But while the earth itself is too stiff to deform much, the oceans are not and they bulge outward in response to the tidal forces. Two *tidal bulges* appear: one closest to the moon and one farthest from the moon (Fig. 7.3.2). A beach located in one of these tidal bulges experiences high tide, while one in the ring of ocean between the bulges experiences low tide.

As the earth rotates, the locations of the two tidal bulges move westward around the equator. Since a particular beach experiences high tide whenever it's closest to or farthest from the moon, the full cycle from high to low to high tide occurs about once every 12 hours and 26 minutes. The extra 26 minutes reflects the fact that the moon isn't stationary; it orbits the earth every 27.3 days and thus passes overhead once every 24 hours and 52 minutes, rather than every 24 hours.

Actually, the tides usually lag behind the moon. Because of its inertia, water needs time to accumulate into a bulge, so high tide occurs somewhat after the moon passes directly overhead. This delay increases where shallow water, islands, and channels impede the water flow; local authorities therefore prepare tables of just how many minutes late the tides are at various places along the seashore.

But the moon isn't the only source of tidal forces on the earth's oceans. Although the sun is much farther away than the moon, it's so massive that the tidal forces it exerts are almost half as large as those exerted by the moon. The sun's principal effect is to increase or decrease the strength of the tides caused by the moon (Fig. 7.3.3). When the moon and the sun are aligned with one another, their tidal forces add together and produce extra large tidal bulges. When the moon and the sun are at right angles to one another, their tidal forces partially cancel and produce tidal bulges that are unusually small.

Twice each lunar month, the time it takes for the moon to orbit the earth, the tides are particularly strong. These *spring tides* occur whenever the moon and sun are aligned with one another (full moon and new moon). Twice each lunar month the tides are particularly weak. These *neap tides* occur whenever the moon and sun are at right angles to one another (half moon).

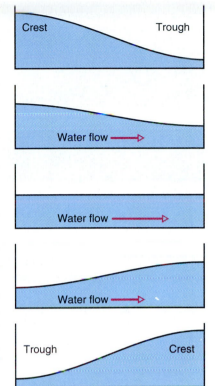

Check Your Understanding #1: Seaside Vacations of the Tidal Rich
The Atlantic and Pacific Oceans come very close to one another in Central America. If the Atlantic side of this isthmus is experiencing high tide, what tide is the Pacific side experiencing?

Tidal Resonances

The sizes of the tides depend on where you are, but high tide is typically a meter or two above low tide. Because the tidal bulges are located near the equator, tides far to the north or south are smaller than that; tides in isolated lakes or seas are smaller still because water can't flow in to create the bulges. However, there are a few special places that have enormous tides. For example, tides in the Bay of Fundy, an estuary between New Brunswick and Nova Scotia, can change the water level by as much as 15 m. How can tides ever get this large?

Giant tides are the results of natural resonances in channels and estuaries. Just as air in an organ pipe can be made to vibrate strongly by a series of carefully timed pushes from a pump, water in a channel or estuary can be made to oscillate strongly by a series of carefully timed pushes from the tides. The giant tides are just another example of resonant energy transfer.

To see how these water oscillations work, consider the behavior of water in a washbasin. In equilibrium, the water's surface is smooth and horizontal. Since gravity produces restoring forces that act to level the water, the water is in a stable equilibrium and will oscillate about that equilibrium if you disturb it. For example, if you shift water from one side of the basin to the other, the water will slosh back and forth until it turns its extra energy into thermal energy or transfers it elsewhere (Fig. 7.3.4).

Because the restoring forces on the water's surface increase in proportion to its displacement from equilibrium, the water-filled basin is yet another harmonic oscillator. Since each mode in which the water can vibrate or slosh has a period that's independent of its amplitude, you can make the water slosh vigorously by pushing on it in sync with one of those periods.

While the period of a washbasin's fundamental mode is only a second or two, that of a large channel may be hours or days. The Bay of Fundy has a fundamental period of roughly 12 hours and 26 minutes. Since this period matches the cycle of the tides, there's a resonant transfer of energy from the moon to the water flowing in the estuary. The tides drive water back and forth in this estuary until, after many cycles, that water is moving so strongly that its height varies dramatically with time (Fig. 7.3.5, on next page).

Fig. 7.3.4 Water sloshing in a basin is an example of oscillation. The water oscillates back and forth, its crest becoming a trough and vice versa, over and over again.

Check Your Understanding #2: Potential Profits
People occasionally propose using giant tides to generate electric power. But there's a problem with this idea. If you extract all of the gravitational potential energy from the water at high tide in the Bay of Fundy, how long will it take for a giant high tide to appear again?

Fig. 7.3.5 The giant tides in the Bay of Fundy can cause its water level to change by as much as 15 m between high and low tide.

Waves on the Surface of Water

As you sit at the seashore, you can't help but notice waves crashing endlessly on the beach. The sea in front of you is covered with ridges that move steadily toward land. These ridges are **surface waves**—disturbances in what would normally be a smooth, level surface. Since distorting the water's surface requires energy, it takes work to make waves. Most ocean waves are driven by the wind.

While it's customary to think of each breaking swell as a separate wave, we can also view the entire pattern of ridges as a single wave. From that perspective, a wave consists of many parallel crests and troughs (Fig. 7.3.6). In a **crest,** the water rises above its equilibrium level. In a **trough,** the water drops below its equilibrium level. As the wave passes across the surface of a deep stretch of open water, the pattern of crests and troughs is evenly spaced and the wave has a *wavelength*—the distance between two adjacent crests.

Crests and troughs can't remain still. As we learned in Section 2.3, a system accelerates in whatever direction reduces its potential energy most rapidly. Since the water in a crest can release gravitational potential energy by flowing into a trough, it accelerates toward the trough. Thus a wave always involves motion.

However, there are two different types of waves: standing waves and traveling waves. In a **standing wave,** the water's surface oscillates up and down vertically, with its crests and troughs always going in opposite directions. Crests become troughs and troughs become crests. The standing wave's pattern of crests and troughs doesn't go anywhere; it simply flips up and down in place at a certain frequency.

Water sloshing in a washbasin is usually a standing wave. The crest on one side of the basin gradually drops to become a trough, while the trough on the other side of the basin gradually rises to become a crest (Fig. 7.3.4). This process reverses and repeats, over and over again. Like all standing waves, the sloshing water's energy goes back and forth between potential and kinetic forms. Its kinetic energy peaks as the water rushes through equilibrium, and its potential energy peaks as the water stops to turn around. For the large *gravity waves* we're considering in this section, the potential energy is gravitational potential energy. However, for the tiny *capillary waves* that occur in a cup of water, it's mostly elastic potential energy in the water's tense and springy surface.

In a **traveling wave,** the pattern of crests and troughs moves smoothly across the surface of the water (Fig. 7.3.7). Because this pattern moves in

Fig. 7.3.6 A surface wave distorts the water's surface away from its equilibrium level. The highest points on the wave are its crests and the lowest are its troughs. The distance between crests is its wavelength. In a traveling wave, the entire pattern of crests and troughs moves across the water's surface at the wave velocity.

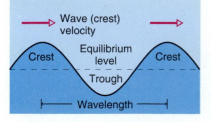

a particular direction with a certain speed, the traveling wave has a velocity—its **wave velocity.** It also has a momentum, and that momentum points in the same direction as the wave velocity. The wave's crests and troughs move steady forward, so the water is never level or motionless. The energy in a traveling wave is always an even mixture of kinetic and potential energies.

Most waves on the open ocean are traveling waves. Despite their steady progress across the water, these traveling waves actually involve oscillation. You can see this oscillation by watching a fixed point on the water's surface. The point fluctuates up and down as the crests and troughs of a traveling wave pass through it, making the wave a form of oscillation and giving it a frequency.

This isn't the first time we've encountered standing and traveling waves. The vibrational modes of a violin string are standing waves; the string bends to create crests and troughs, which interchange over and over again as the string vibrates. Sounds are traveling waves—compressions and rarefactions of the air that travel through space from their source to your ears.

> **Check Your Understanding #3: The Frequency of Seasickness**
> You're sitting in a small boat on the open ocean, rising and falling with each passing wave crest. How could you measure the frequency of the wave passing under your boat?

The Structure of a Water Wave

We've seen that a traveling wave moves across the surface of the water with a certain velocity, wavelength, and frequency. But what is the water itself doing as the wave passes?

You can begin to answer that question by watching a bottle floating on the water's surface (Fig. 7.3.8). As a wave passes it, that bottle rises and falls with the crests and troughs, and ends up traveling in circle. Overall, it doesn't go anywhere. Like the bottle, the water itself doesn't actually move with the wave. Instead, it simply bunches up or spreads out to create each crest or trough. The water moves only short distances as the wave passes and returns to its starting point once the wave has left.

The wave's water moves in a circular pattern (Fig. 7.3.9, p. 252). Water that starts out on top of a crest moves down and forward as the crest passes. It moves down and backward as the trough passes, then up and backward, and finally up and forward as the next crest arrives. Like the bottle in Fig. 7.3.8, the water itself just moves in a circle and returns to its original location each time a crest passes. The wave's direction of travel depends on which way the water is circling. It heads in the same direction as the water on top of a crest.

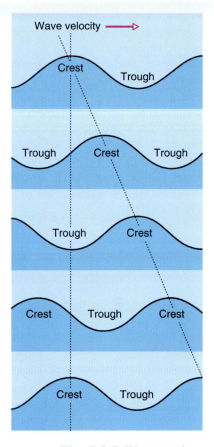

Fig. 7.3.7 Waves on the open ocean are traveling waves. The crest on the left gradually moves to the right before passing out of view. If you watch the water's surface at one point, marked by the dotted line, you will see that it oscillates up and down.

Fig. 7.3.8 You can see that water doesn't move with a wave by watching a bottle floating on the water. The bottle moves in a circle as each crest passes by. Starting at (*a*) a crest, the bottle moves (*b*) down and right, (*c*) down and left, (*d*) up and left, and then (*e*) up and right. It returns to its original position just as the next crest arrives.

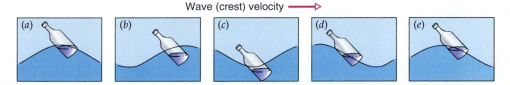

Wave (crest) velocity

Fig. 7.3.9 Surface water moves in a circular motion as a wave passes. The water currently located at each dark dot will follow the circular path outlined around it as time passes. The circles are largest at the surface and become relatively insignificant once you look more than half a wavelength below the surface. The sense of the circular motion (clockwise or counterclockwise) determines the direction in which the wave travels. This wave travels toward the right.

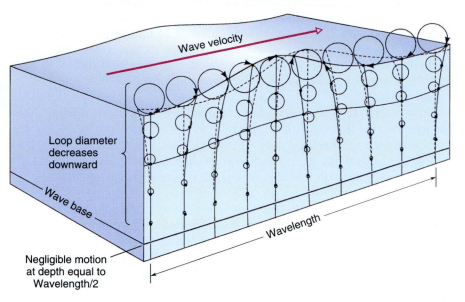

This circular motion occurs only for water that's fairly near the surface. As you look deeper into the water, the circular motion gradually diminishes until the water is barely moving at all. The depth at which the water stops moving with the surface wave is roughly equal to half the wave's wavelength.

Wavelength also affects the wave's velocity. Since any material carrying a mechanical wave must accelerate in various directions as the wave moves through it, the wave moves fastest when the material is stiff and its density is low. That's because its stiffness gives rise to large forces when it's disturbed and its low density allows it to accelerate easily in response to those forces. But in some types of waves, particularly surface waves on water, the stiffness that the material exhibits actually depends on the wave's wavelength. In those cases, waves of different wavelengths actually travel at different speeds!

The longer the wavelength of a surface wave on water, the faster that wave travels. Little ripples have short wavelengths and travel slowly, while large ocean waves have long wavelengths and travel much more rapidly. Giant waves produced by earthquakes and volcanic eruptions—once called "tidal waves" and now called *tsunamis*—have extremely long wavelengths and can travel at hundreds of kilometers-per-hour. Because these giant waves travel so quickly and move water so deep in the ocean, they carry enormous amounts of energy.

> **Check Your Understanding #4: Beneath the Waves**
>
> If you're swimming and want to dive beneath a wave, how deep must you go to avoid any significant motion in the water?

Waves at the Shore

As a wave approaches the shore, it begins to travel through shallow water. Since the water's circular motion extends below the surface, there comes a point at which the wave begins to encounter the bottom. Once the water is shallower than the wavelength of the wave, the bottom distorts the water's circular motion so that it becomes elliptical. The velocity of the wave drops and the crests begin to bunch up. This explains why waves that looked broad and gradual on the open ocean look quite steep and dangerous as they approach the beach. Their crests and troughs have bunched together, and the slopes between the two really have become steeper.

A wave forms each crest by bunching together water from in front of it and behind it. When the crest enters very shallow water, there isn't enough water in front of it to construct its forward side. The crest becomes incomplete and begins to "break."

If the slope of the sea bottom is gradual, the wave breaks slowly to form a smooth, rolling, "boiling" surf (Fig. 7.3.10). If the slope of the sea bottom is steep, the wave breaks quickly by having the top of its crest plunge forward over the trough in front of it (Fig. 7.3.11). The steep slope essentially prevents the forward half of the crest from forming. The rearward half continues through its normal circular motion and dives over the missing half-crest in front of it.

Fig. 7.3.10 When the slope of the sea bottom is very gradual, the wave crests crumble gently into rolling surf.

> ### Check Your Understanding #5: Out Past the Surf
> As you swim away from shore at the ocean, you go through a region where the wave crests are breaking and then reach a more distant region where they don't break. What distinguishes the two regions?

Surfing

Surfing down a wave is a lot like skateboarding down a hill. In either case, the ramp you're on is exerting a downhill force on you, pushing you forward and making you hope your friends don't see you crash.

But the wave is special for two reasons. First, its water exerts a strong uphill drag force on your surfboard so that you don't accelerate downhill forever. Instead, you quickly reach terminal velocity and descend at a steady pace. Second, the wave rises up beneath you so that you don't actually descend at all: you hover!

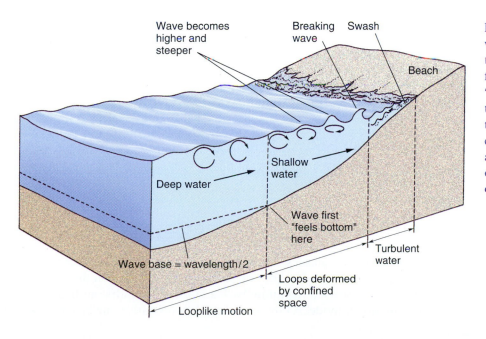

Wave becomes higher and steeper

Breaking wave Swash

Beach

Deep water

Shallow water

Wave first "feels bottom" here

Turbulent water

Wave base = wavelength/2

Loops deformed by confined space

Looplike motion

Fig. 7.3.11 When the water becomes too shallow to form a complete crest for the wave, the wave "breaks." If the slope of the shore is steep enough, the crest will be quite incomplete on its shore side and will plunge forward over the trough in front of it.

Fig. 7.3.12 The surfer is riding this wave like a moving ramp. The wave rises beneath him as he slides forward, so he never reaches the trough. The wave is breaking because the water here is too shallow to form a complete crest.

This hovering (Fig. 7.3.12) reflects the perfect cancellation of forces that occurs at a certain height on a wave's forward edge. When you're at that height, your terminal velocity is just right to keep you moving forward with the wave. If you somehow move up the wave crest, you find it steeper and your terminal velocity increases. You begin to overtake the wave and soon return to the proper height. Similarly, if you somehow move down the wave crest, you find it less steep and your terminal velocity decreases. Now the wave overtakes you and returns you to the proper height. You're in a dynamically stabilized equilibrium!

However, if you ride the wave directly toward shore, you'll hover near the base of the crest. The mild slope there is already steep enough to keep you moving forward at the speed of the wave. For a better ride, you should cut across the forward edge of the crest, so that you approach the shore at an angle. With this angle as a handicap, it's much harder to keep up with the wave so you must travel faster and thus ride up higher and steeper on the crest.

Having the wave break on you is considered poor form and also shortens your life expectancy. Just before the crest breaks, you turn the surfboard sideways to increase its drag. This action slows you down abruptly and the crest passes you by. You then paddle away from shore to catch a ride on another crest.

> **Check Your Understanding #6: Water Works**
> Since drag extracts energy from a surfboard, the wave must constantly replace it. How does the wave transfer energy to the surfer?

Epilogue for Chapter 7

In this chapter, we looked at natural resonances in a variety of objects. In *clocks,* we examined resonances in pendulums, balance rings, and quartz crystals and found that these objects are harmonic oscillators—their periods don't depend on the amplitudes of their motions. In *violins and pipe organs,* we explored the vibrations of strings and air columns to see that they, too, behave as harmonic oscillators, but with many modes of vibration and thus more complicated behaviors.

In *the sea and surfing,* we found that there are two different types of waves: standing waves and traveling waves. We saw that both types of waves are found in water, with standing waves appearing in tidal resonances and traveling waves heading out over the open ocean.

Explanation: A Singing Wineglass

Rubbing the lip of the glass is similar to bowing a violin string; both procedures cause resonant energy transfer. Your finger alternately pushes on the glass and then slips, helping the glass itself to vibrate back and forth beneath your finger in its fundamental vibrational mode. As the glass begins to vibrate, your finger does a little work on it each time the glass is moving in the same direction as your finger. Once the glass is singing loudly, you can keep it singing by circling your finger steadily around the glass. Although the glass continues to emit its energy as sound, you keep giving it more energy by doing work on it with your finger.

Chapter Summary

How Clocks Work: Clocks are usually based on harmonic oscillators because of their extremely steady periods. Most importantly, the period of a harmonic oscillator doesn't depend on the amplitude of its motion. Common harmonic oscillator timekeepers include pendulums, balance rings, and quartz crystals.

In a pendulum clock, a swinging pendulum controls the rotation of a toothed wheel, which in turn controls the rotation of the clock's hands. Energy needed to keep the pendulum swinging and to advance the hands comes from the descent of a weighted cord. In a balance clock, a rocking balance ring controls the toothed wheel. Energy for this clock comes from a wound coil spring. A quartz crystal clock detects the vibration of its crystal electronically and uses that vibration to control a motor that advances its clock hands or an electronic circuit that measures time by counting the crystal's vibrations. Small, carefully timed electric pulses from the clock provide the energy that keeps the crystal vibrating.

How Violins and Pipe Organs Work: A violin string exhibits natural resonances by vibrating back and forth about its equilibrium shape, a straight line. Bowing the string causes it to vibrate by pushing it away from its equilibrium shape and then allowing it to slip back. The bow transfers energy to the vibrating string by pushing it each time it moves in the bow's direction. The string's fundamental pitch is determined by its mass, tension, and length. After selecting a string of the correct mass, you tune that string by adjusting its tension. During play, you change the string's length and pitch by pressing it against the fingerboard. The belly of the violin serves to convert the string's vibration into sound by moving in and out as the string vibrates.

The air inside an organ pipe also exhibits natural resonances. As the organ blows air across the opening at the bottom of a pipe, the air inside that pipe begins to vibrate. The frequency of this vibration is determined principally by the pipe's length, so that a short pipe has a higher pitch than a long one. Pipes with different shapes produce different amounts of harmonics and yield different sounds.

How the Sea and Surfing Work: The sea exhibits two interesting motions: tides and waves. Tides are caused by bulges in the earth's oceans, created by gravitational tidal forces from the moon and sun. The moon dominates the tides so that bulges appear on regions of the earth closest to and farthest from the moon. Since the earth is rotating, these bulges move with respect to land and give the tides their $12\frac{1}{2}$-hour cycle.

Waves occur because the water's surface has a stable equilibrium. When it's disturbed from its flat, level equilibrium, this surface will oscillate in standing waves and/or traveling waves. The familiar ripples on the ocean's surface are traveling waves that head toward shore at speeds that increase with wavelength. As they approach the shore, these waves form incomplete crests, which eventually break. A surfer riding one of these crests is effectively sliding down a hill. What makes surfing so interesting is that the "hill" moves; its surface is always rising upward so that the surfer can continue to slide down the hill almost indefinitely.

Check Your Understanding—Answers

Section 7.1 CLOCKS

1. Since light travels at a constant speed, the distance it travels is equal to the product of its speed times its travel time. If you know the travel time and the speed, you can determine that distance.

Why: Many distance measurements are made by measuring time. Surveyors routinely use light's travel time to measure distances. Some cameras use sound's travel time to measure distances. In general, the motion of an object at constant velocity can be used either to measure the distance traveled if you know the elapsed time or to measure the elapsed time if you know the distance traveled.

2. It may have lost as much as 80 minutes after 24 hours.

Why: If the operator of a sandglass waits 10 s before turning it over each time the sand runs out, the clock will lose 10 s every 3 minutes, 200 s every hour, and 4800 s every day. If you could be sure that the operator would always wait 10 s, you could include it in the clock's design. However, the operator might be faster one time than the next, and this uncertainty makes the clock unreliable and inaccurate. For a repetitive clock to keep accurate time, it must repeat its motion with almost perfect regularity.

3. The strength of the earth's gravity and the length of the swing's chains.

Why: A child swinging on a swing set is a form of pendulum. As with any pendulum, the child's period of motion is determined only by the strength of gravity and the length of the pendulum. In this case, the length of the pendulum is approximately the length of the swing's supporting chains. Thus a tall swing has a longer period than a short swing.

4. The amplitude of her motion will gradually decrease so that she comes to a stop.

Why: To keep her swinging, you must make up for the energy she loses to friction and air resistance. By pushing her forward each time she moves away from you, you do work on her and increase her energy. But when you push her as she moves toward you, she does work on you and you extract some of her energy. You are then slowing her down rather than sustaining her motion.

5. The torsional stiffness of the chandelier's supporting cord and the chandelier's moment of inertia.

Why: The hanging chandelier is a harmonic oscillator. Its supporting cord opposes any twists by exerting a restoring force on the chandelier. Once you twist it away from its equilibrium orientation, the chandelier oscillates back and forth with a period determined only by the cord's torsional stiffness (its stiffness with respect to twists) and the chandelier's moment of inertia. As with any harmonic oscillator, the amplitude of the chandelier's motion doesn't affect its period.

6. The rod is vibrating as a harmonic oscillator, with its two halves first approaching one another and then moving apart.

Why: The metal rod vibrates in the same manner as a quartz crystal. The body of the rod exerts restoring forces on its two halves. After hitting the floor, these halves move toward and away from one another rapidly, emitting the tone that you hear. Because it's a harmonic oscillator, the frequency (and pitch) of the tone doesn't change as the amplitude of motion decreases.

Section 7.2 VIOLINS AND PIPE ORGANS

1. There is a factor of 4 in frequency between the lowest and the highest notes that voice can sing.

Why: Since notes separated by an octave are separated by a factor of 2 in frequency, notes separated by two octaves are separated by a factor of 4.

2. The frequency of a string's fundamental vibrational mode depends on its tension.

Why: Any cord that is drawn taut from its ends will exhibit natural resonances like those in a violin string. The tauter the string, the higher the frequencies of those resonances.

3. A jump rope is essentially a vibrating string. The two half-rope pattern is the second harmonic mode, with a vibrational frequency twice that of the normal fundamental arc.

Why: Although it swings around in a circle, the jump rope is actually vibrating up and down at the same time it's vibrating forward and back. Together, these two vibrations create the circular motion. To make the rope vibrate in its second harmonic mode (as two half-ropes) without changing its tension, it must be swung twice as fast as normal.

4. The object has a natural resonance at the tone's frequency, and sympathetic vibration is transferring energy to the object.

Why: Energy moves easily between two objects that vibrate at the same frequency. A note played on one instrument will cause the same note on another instrument to begin playing. Even everyday objects will exhibit sympathetic vibration when the right tone is present in the air.

5. To transfer the vibrational energy of the strings to the air.

Why: Guitar strings are too narrow to push effectively on the air and emit sound. They do better to transfer their energy to the body of the acoustic guitar so that its flat surfaces can push on the air. An electric guitar avoids the need for a sound box by converting the string's vibrations directly into electric currents and from there into movements of an audio speaker.

6. The water shortens the column of moving air inside the bottle and increases the frequency of its fundamental vibrational mode.

Why: A water bottle is essentially a pipe that is open at only one end. It has a fundamental vibrational mode with a frequency that is half that of an open pipe of equal length. As you add water to the bottle, you shorten the effective length of the pipe and raise its pitch.

Section 7.3 THE SEA AND SURFING

1. High tide, too.

Why: When the Atlantic side of the Central American isthmus is experiencing high tide, there is a tidal bulge over all of Central America. Both sides of the isthmus experience the same (high) tide.

2. Many days.

Why: The sloshing water in the Bay of Fundy acquires its energy via resonant energy transfer. The tides do work on the water, slowly increasing its energy over many cycles of its periodic motion. Since each cycle takes half a day, many days are required to build up the energy needed for a giant

tide. If you were to extract all this stored energy at once, the bay would have to start sloshing all over again.

3. Count the number of up and down cycles you complete in a certain amount of time.

Why: Your boat's up and down motion is caused by the oscillation of the ocean's surface. You are rising and falling at the frequency of the passing wave.

4. You must dive about half a wavelength beneath the water's surface.

Why: A wave causes motion in the water up to a depth of roughly one wavelength. A typical wave has a wavelength of about 5 m, so you must dive 2.5 m under water before the water will remain essentially still.

5. The breaking region is too shallow for complete crests to form. The nonbreaking region is deep enough to form complete crests.

Why: Wave crests break in shallow regions, where complete crests can't be formed. As long as the water is deep enough, the waves travel intact. But if there is a shallow region such as a sand bar, even quite far from shore, the crests may break as the wave passes through it.

6. The water pushes the surfer forward and upward, and the surfer moves forward. Thus the wave does work on the surfer.

Why: Like any ramp, the side of the wave crest exerts a support force on the surfer that is perpendicular to the water's surface. That force points forward and upward. Since the surfer and wave head forward, partly in the direction of the force, the force does work on the surfer.

Exercises

1. The acceleration due to gravity at the moon's surface is only about one-sixth that at the earth's surface. If you took a pendulum clock to the moon, would it run fast, slow, or on time?

2. A clothing rack hangs from the ceiling of a store and swings back and forth. Why doesn't the period of this motion depend on how many dresses the rack is holding?

3. If a child stands up on the seat of a playground swing, how will the swing's period be affected?

4. If you pull a small tree to one side and suddenly let go, it will swing back and forth several times. The period of this motion won't depend on how far you bend the tree. How does the restoring force that returns the tree to its upright position depend on how far you bend the tree?

5. A flagpole is a harmonic oscillator, flexing back and forth with a steady period. If you want to increase the amplitude of the pole's motion by pushing it near its base, when should you push—as it flexes toward you or away from you?

6. Why do thicker quartz crystals vibrate more slowly than thinner quartz crystals?

7. A ball bouncing on a table is poorly suited to use as a timekeeper because its period varies with amplitude. Explain why the period of this anharmonic oscillator varies with amplitude.

8. Depending on how the base of a rocking chair was cut, the period of its motion may or may not depend on how hard you're rocking it. What can you say about the restoring forces acting in these two cases?

9. Some cameras measure the distance to their subjects by bouncing sound off them. Sound travels at approximately 331 m/s. How can a camera use a sound emitter, a sound receiver, and a timer to measure how far away its subject is?

10. If there were no gravity, a pendulum wouldn't have gravitational potential energy. It also wouldn't oscillate. How are these two facts related?

11. If nothing exerted a tension on a violin string, it wouldn't be stretched and wouldn't have any elastic potential energy. It also wouldn't vibrate. How are these two facts related?

12. When you press a key on a piano, a felt-covered hammer strikes one or more taut strings and causes them to vibrate. The hammer bounces off the string so quickly that the string is still moving away from its straight equilibrium shape when the hammer loses contact with it. Why is it important to the piano's volume that the hammer leave the string before the string begins to return toward its equilibrium shape?

13. Why doesn't the rebounding hammer in Exercise 12 take all the collision energy with it, leaving the string motionless?

14. Which of the following clocks would keep accurate time if you took them to the moon: a pendulum clock, a balance clock, and a quartz watch? Why?

15. To modify the pitch of a guitar string you could change its mass, tension, or length. To raise its pitch, how should you change each of these three characteristics?

16. If you stretch a rubber band between your fingers and pluck it, it will vibrate and emit a tone. If you now stretch that rubber band to twice its original length and pluck it again, it will emit almost the same tone. Why

doesn't the rubber band's pitch change very much when you stretch it?

17. If you stretch the rubber band in Exercise 16 to its maximum length, so that it becomes stiff and taut, its pitch will increase significantly. Explain this increase in pitch.

18. Why do changes in temperature affect a violin string's pitch?

19. The strings that play the lowest notes on a piano are made of thick steel wire wrapped with a spiral of heavy copper wire. The copper wire doesn't contribute to the tension in the string, so what is its purpose?

20. Why are the highest pitched strings on most instruments, including guitars, violins, and pianos, the most likely strings to break?

21. Some wind chimes consist of sets of metal rods that emit tones when they're struck by wind-driven clappers. These rods vibrate like the baseball bat in Fig. 3.2.7. Why do the longer rods emit lower pitched tones than the shorter rods?

22. Why would replacing the air in an organ pipe with helium raise its pitch?

23. A flute and a piccolo are both effectively pipes that are open at both ends, with holes in their sides to allow them to produce more tones. The piccolo is very nearly a half-size version of the flute. How does this fact explain why the piccolo's tones are one octave above those of a flute?

24. A bugle is effectively a pipe that's open at its flared "bell" and closed at its mouthpiece. As air vibrates in and out of the bell, the pressure at the mouthpiece fluctuates up and down. The bugler provides power to the vibrations by injecting pulses of air into the bugle through its mouthpiece. Even though there is no way to change the length of the bugle's pipe, a bugler can play several different tones, including those used in familiar pieces such as *reveille* and *taps*. What is happening to the air moving inside the bugle when it produces these different tones?

25. A trumpet resembles a bugle (see Exercise 24), except that it has three valves. These valves let the trumpeter make small changes to the pipe's length. Why do these changes allow the trumpet to produce more tones than a bugle?

26. The most important difference between a trumpet and a tuba is in the lengths of their pipes. The tuba's pipe is much longer than that of the trumpet. How does this difference affect the relative pitches of these two instruments?

27. In the Mediterranean Sea, high tide is only 30 cm above low tide. Why?

28. Why are the tides relatively weak near the north and south poles?

29. The momentum of a water surface wave points in the direction of its wave velocity. Show that ocean waves always move in the same direction as the wind that created them.

30. We are fortunate that a sound wave's speed doesn't depend on its wavelength or frequency, the way a surface wave on water does. If the speed of sound did depend on its wavelength or frequency, what would you hear when you listened to an orchestra?

31. If you're carrying a full cup of coffee and take your steps at just the wrong frequency, the coffee will begin to slosh wildly in the cup. What is causing this energetic motion?

32. When you throw a stone into a pool of still water, small ring-shaped ripples begin to spread outward at a modest pace. Why do these ripples travel so much more slowly than waves on the ocean?

33. When you throw a rock into calm water, ripples head outward as several concentric circles. As the circumferences

of these circles grow larger, their heights grow smaller. Explain this effect in terms of conservation of energy.

34. If you pull downward on the middle of a trampoline and let go, the surface will fluctuate up and down several times. Why is this motion an example of a standing wave?

35. Why is a violin string vibrating up and down in its fundamental vibrational mode an example of a standing wave rather than a traveling wave?

36. When you pluck the end of a kite string, a ripple will head up the string toward the kite. Why is this motion an example of a traveling wave rather than a standing wave?

37. Plucking the end of a kite string sends a ripple up the string toward the kite. The more tension there is in the string, the faster this traveling wave moves. In general terms, why does the increased tension increase the wave's speed?

38. If you replaced the kite string in Exercise 37 with a thinner, lighter string, how would that change affect the speed of traveling waves on the string?

39. As waves pass over a shallow sand bar, the largest ones break. What causes these waves to break and why only the largest ones?

40. Even when waves don't break as they pass over a sand bar (Exercise 39), the sand bar is noticeable because you can see the wave crests move closer together. What is happening to cause this bunching?

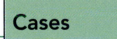

Cases

1. Because sound carries energy, musical instruments need energy to operate. The musician, who does mechanical work on the instrument, normally provides this energy. How does a musician transfer energy to
   a. an accordion?
   b. a piano?
   c. a xylophone?
   d. a violin?

2. When two people swing a jump rope between them, it forms a moving arc in the air. The rope's behavior is similar to that of a violin string, except that the jump rope swings around in a circle rather than vibrating directly back and forth.
   a. What happens to the speed at which the jump rope turns if you increase the length of the rope but keep the tension on that rope the same?

   b. What happens to the speed at which the jump rope turns if you use a thicker, more massive rope but keep the rope's length and tension the same?
   c. What happens to the rope's motion if the people holding its ends pull harder on the rope and increase its tension?
   d. How quickly must the people swing the rope to get it to form an S-shape, with a stationary node in the middle?

3. Imagine a stiff ruler that's lying at the edge of a table so that one end of it is on the table and the other end is free in the air. If you hold the ruler's supported end against the table with one hand and pluck its free end with the other hand, the ruler will vibrate up and down.
   a. Why does the ruler's frequency of oscillation decrease when you extend more of it off the table?

b. Why doesn't the ruler's frequency of oscillation depend on how hard you pluck it?

c. What forms does the ruler's energy take as it vibrates up and down?

d. What eventually happens to the plucked ruler's energy?

e. If the ruler vibrates slowly enough, you can increase the energy of its vibration by pushing down on it rhythmically, once per cycle, just after it completes its upward motion and begins to head downward. Why must you push it downward as it heads downward, rather than pushing it downward as it heads upward?

4. In bungee jumping, a strong elastic cord or "bungee" is used to break the fall of a person who has leapt from a tall platform. This cord runs from the platform to the person's ankles and is short enough that the person doesn't hit the ground. After a few terrifying seconds of freefall, the cord pulls taut and the jumper bounces back upward. On this first bounce, the cord goes slack and the jumper once again experiences freefall. But in subsequent bounces, the cord remains taut and the jumper bounces up and down smoothly.

a. Why must the bungee cord stretch relatively easily to avoid injuring the jumper?

b. As a jumper bounces up and down on the cord, the time it takes to complete each bounce doesn't depend on how high that bounce is. Explain.

c. Why does a heavier person bounce up and down more slowly than a lighter person?

d. Who experiences the more extreme feelings of heaviness while coming to a stop at the end of the cord: a heavier person or a lighter person?

e. Why must the cords used in bungee jumping be able to withstand tensions that exceed the weights of the heaviest jumpers?

f. If the jumper used two bungee cords instead of one—perhaps one attached to each ankle—how would that affect the time it takes to complete each bounce up and down?

5. The body of a car is suspended above its wheels by four strong springs. These springs allow the wheels to respond quickly to bumps in the road without having the entire body respond, too. However, the wheels are also connected to the body by shock absorbers. Unlike the springs, which store energy when work is done on them, the shock absorbers use frictionlike effects to convert work into thermal energy.

a. You remove the shock absorbers from your parked car and then jump into the driver's seat. The body bounces up and down for a long time. Explain this behavior.

b. If the car's first bounce takes 1 s, how long does its second bounce take? its tenth bounce?

c. You then drive the car off a high curb and onto the road. Describe the body's motion for the next ten seconds.

d. You reinstall the shock absorbers. Now every time the car body moves toward or away from the wheels, it does work on the shock absorbers and this work becomes thermal energy. You jump back into the driver's seat of the parked car. How does it respond?

e. You again drive the car off the high curb and onto the road. The shock absorbers make a difference. Describe the body's motion for the next ten seconds.

6. An earthquake generates waves that travel through the earth in all directions. While the earth's internal structure complicates the passages of these waves to distant sites, they follow simple paths to nearby places. However, there are at least two types of waves, and they travel at different speeds.

a. The waves that travel fastest in the earth's crust are those that resemble sound. They're called "P" or "primary" waves because they arrive first. As these waves travel forward, the ground moves alternately forward and backward, along the direction of the wave velocity. For example, if a "P" wave is heading north, the ground oscillates north and south as the wave passes. A *seismometer* can detect this motion by hanging a pendulum above a sheet of paper that's resting on the ground. When the wave passes, the pendulum marks the paper along the direction of the wave velocity. Why does the pendulum shift back and forth relative to the paper?

b. The ground along the path of a "P" wave doesn't all shift north or south at once. These shifts are actually the "P" wave's crests and troughs, and they move northward with the wave. When the ground underfoot shifts first northward and then southward, it's because a crest passed by, followed by a trough. At the moment the crest is under you, the trough is still a long way behind it. In fact, it's exactly half a wavelength behind it. How could you use several seismometers to determine the wavelength of a passing "P" wave?

c. How could you use one seismometer to determine the frequency of the "P" wave?

d. How could you use several seismometers to measure the wave velocity of the "P" wave?

e. Waves that involve sideways or vertical motion of the ground travel more slowly and arrive after the "P" waves. They're called "S" or "secondary" waves. For example, in one type of "S" wave that is heading north, the ground oscillates east and west as the wave passes. How will a hanging pendulum seismometer respond when this "S" wave passes?

f. In another northbound "S" wave, the ground moves up and down. How could you use a weight and a spring to detect this wave as it passes by?

g. The wave from an earthquake spreads out in all directions like a ripple on a pond. Use conservation of energy to explain why the wave's amplitude of motion must decrease as it gets farther from the earthquake's center and why the energy passing under a city near the earthquake must be much greater than the energy passing under a similar-sized city far away from the same earthquake. (This weakening with distance occurs even when no energy is lost as thermal energy. It protects us from all but relatively nearby earthquakes.)

7. When the paper cone in your stereo speaker moves forward and backward rhythmically, it produces a sound wave that travels through the air at about 331 m/s. The crests of this wave consist of slightly compressed air and the troughs consist of slightly rarefied air.

a. Suppose that you have an instrument that can monitor the pressure at one point in space. You place this instrument in front of the speaker and begin measuring. Draw a rough graph of the pressure that the instrument measures versus the time that has elapsed since the measurement started.

b. Suppose that instead of keeping the instrument in one place, you began to move the instrument so that it continues to detect a region of slightly compressed air (one of the crests in the wave). How fast must you move the instrument to keep up with the crest?

c. You spray a mist of water into the air so that you can see if the air itself moves as the sound travels through it. The mist remains still, so you know that the air isn't moving significantly, even though the sound wave is. Explain this surprising result.

d. How could you determine the wavelength of the sound wave emerging from the speaker?

e. What two types of energy does a sound wave contain?

f. The oscillation of air inside an organ pipe is actually a standing sound wave. How is it different from the traveling sound wave that's passing through your room?

8. If astronauts landing on Mars were to find liquid water on the surface, the waves that they would see would behave somewhat differently. That's because the acceleration due to gravity on Mars is only 3.71 m/s^2.

*a. A wave always needs two forms for its energy. In a surface wave on water, one of those forms is gravitational potential energy. How would the gravitational potential energy of a wave on Mars compare with that of an identical looking wave on Earth?

b. The "stiffness" of the water's surface depends on gravity. When gravity is weak, it takes less work to deform the water's surface. Use your answer to part a to show that this is the case.

c. The surface of water on Mars is more easily deformed and thus less "stiff" than the surface of water on Earth. On which planet would surface waves on water travel faster, Mars or Earth?

d. Which wave carries more energy, a surface wave on Mars or an identical looking wave on Earth?

e. On a spaceship that's far away from stars, planets, and other celestial objects, there is almost no gravity. If the astronauts on that spaceship put water in a small basin, would they be able to produce normal surface waves in that water? Why or why not?

CHAPTER 8

Electric and Magnetic Forces

Electric and magnetic forces are all around us. They play important roles in machines we use every day and figure prominently in industry and technology. While these forces are often useful—sticking food wrap to a bowl or a magnet to your refrigerator—they can also make your clothes cling together out of a hot dryer or erase the information on your credit card. Moreover, electric and magnetic forces are what hold atoms, molecules, and materials together.

EXPERIMENT: Moving Water Without Touching It

Unlike gravity, which always pulls objects toward one another, electric and magnetic forces can be either attractive or repulsive. You can experiment with elec-

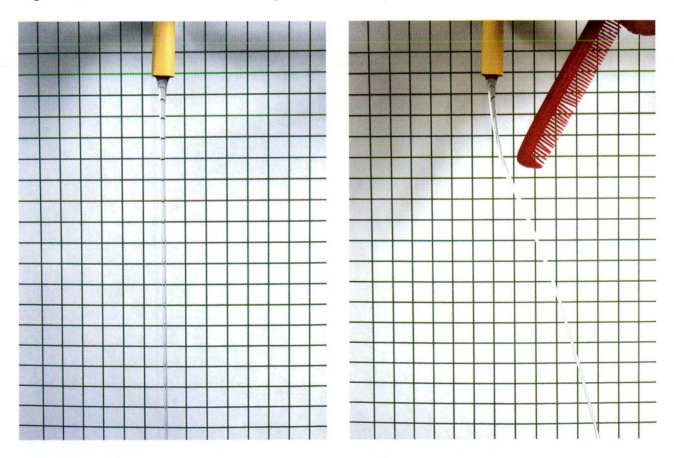

tric forces using a thin stream of water and an electrically charged comb. First, open a water faucet slightly so that the flow of water forms a thin but continuous strand below the mouth of the faucet. Next, give your rubber or plastic comb an electric charge by passing it rapidly through your hair or rubbing it vigorously against a wool sweater. Finally, hold the comb near the stream of water, just below the faucet, and watch what happens to the stream. Is the electric force that you're observing attractive or repulsive? Why does this force change the path of the falling water?

Rubbing the comb through your hair makes it electrically charged. Find other objects that can acquire and hold a charge when you rub them across hair or fabric. Try to *predict* which of these objects will work best. Now rub these objects to charge them and *observe* their behaviors. You can use the water stream to *measure* their charges. Did you *verify* your predictions? Which works better: a metal object or one that's an insulator? Why?

Chapter Itinerary

While many of us have experienced electric and magnetic forces as novelties or nuisances, there are also many important devices that depend on them. In this chapter, we'll examine three of these devices: (1) *electronic air cleaners,* (2) *xerographic copiers,* and (3) *magnetically levitated trains.* In *electronic air cleaners,* we'll look at the problems involved in removing tiny particles from polluted air and how the electric forces between charged particles make such purification possible. In *xerographic copiers,* we'll see how these same electric forces work together with light to control the placement of colored powders, in order to reproduce an image on a sheet of paper. In *magnetically levitated trains,* we'll investigate the origins of magnetic forces and the ways in which these forces can be used to suspend high-speed trains above their tracks.

This chapter introduces the concepts of electric charge and magnetic pole and highlights the similarities and differences between the two. It also begins to describe the relationships between the two types of forces, electric and magnetic, for they're not truly independent of one another. This relationship will become even more apparent in subsequent chapters.

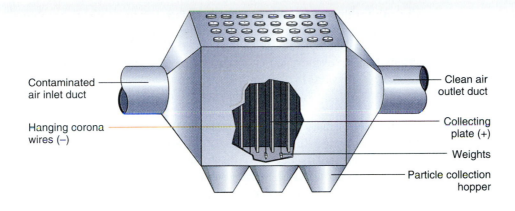

Contaminated air inlet duct

Clean air outlet duct

Hanging corona wires (−)

Collecting plate (+)

Weights

Particle collection hopper

Electronic Air Cleaners

When a ray of sunlight enters a darkened room, it illuminates tiny dust grains drifting in the air. Some of these particles occur naturally, but many others originate in our factories and vehicles. They're a cosmetic nuisance—darkening our air and dirtying our surroundings—and a health hazard. To control particle pollution, many industries filter particles from their smokestacks and many homes remove them from their ventilation systems.

This filtering isn't easy because air resistance makes it hard to pull tiny particles out of the air. That's why many companies and households use electronic air cleaners. These cleaning devices pluck particles from the air with the forces of static electricity—the same forces that cause clothes to cling to one another as you remove them from the dryer. In this section, we'll examine static electricity to see how electronic air cleaners work.

__Questions to Think About:__ What keeps dust in the air and why doesn't gravity remove it? If gravity can't clear the air, how can any other forces succeed? Why don't companies use paper filters to clean dust from smoke? How can it be that some newly cleaned clothes push apart rather than cling to one another?

__Experiments to Do:__ You can study static electricity by rubbing a toy balloon vigorously through your hair or against a wool sweater. Though the balloon won't look any different, it'll begin to attract other things, particularly your hair. What has happened to the balloon? to your hair? Why does the balloon even attract things that weren't rubbed?

Try to get rid of the balloon's attractiveness by letting a thick stream of water flow over its surface. Why does this process return the balloon to normal? What did you "wash" off the balloon? Now rub two identical balloons through your hair and see whether they attract or repel one another. Does the result make sense?

Dust and Electric Charge

Dust doesn't float in air. Like the rocks, dirt, soot, ash, and pollen from which it's made, dust is denser than air and belongs on the ground. But air resistance makes it

why dust won't fall

❑ Although best remembered for his political activities, **American statesman and philosopher Benjamin Franklin (1706–1790)** was also the preeminent scientist in the American Colonies during the mid-1700s. His experiments, both at home and in Europe, contributed significantly to the understanding of electricity and electric charge. In addition to demonstrating that lightning is a form of electric discharge, Franklin invented a number of useful devices, including the Franklin stove, lightning rods, and bifocals.

difficult for dust to fall. With so much surface relative to its tiny volume, a dust grain experiences severe viscous drag forces as it moves through air. Even at millimeter-per-second speeds, the drag force on a dust grain may exceed its weight. Thus dust grains have extremely slow terminal velocities and take minutes or hours to drop through a few meters of still air to the ground.

Moreover, moving air carries dust along with it. Since drag forces oppose relative motion between dust and air, the slightest upward breeze can keep dust aloft indefinitely. Gravity is too weak a force to pluck dust from the air quickly. It needs help from stronger forces, the forces of static electricity.

Static electricity begins with **electric charge,** an intrinsic property of matter that we've encountered before but never discussed in detail. Electric charge is present in many of the **subatomic particles** from which matter is constructed, and these particles incorporate their charges into the objects that contain them. Because electric charge is so intimately connected to the objects that have it, we'll often refer to those objects as **electric charges** or just **charges.**

Electric charges exert forces on one another. Some charges attract while others repel. Although people have known for thousands of years that such forces must involve two different types of charges, it was Benjamin Franklin ❑ who finally named them. He called what appears on glass when it's rubbed with silk *positive* charge and what appears on hard rubber when it's rubbed with animal fur *negative* charge.

Two like charges (both positive or both negative) push apart, each experiencing a repulsive force that pushes it directly away from the other (Fig. 8.1.1a,b). Two opposite charges (one positive and one negative) pull together, each experiencing an attractive force that pulls it directly toward the other (Fig. 8.1.1c). These forces between stationary electric charges are called **electrostatic forces.**

Electronic air cleaners use electrostatic forces to pull dust from the air. A typical air cleaner gives each dust grain a negative charge and collects that grain on a plate coated with positive charges. But what does it mean to give the grain a negative charge?

The answer to that question has three parts. First, giving the dust grain a negative charge doesn't mean that the grain won't contain any positive charges. Like all ordinary matter, the dust grain contains a mixture of positive and negative charges. In fact, the electrostatic forces between those charges are what hold the grain together. But giving the grain a negative charge does mean that its negative charges will outnumber its positive charges and that it will be attracted toward positive charges on the collecting plate.

However, it's awkward having to count up positive and negative charges separately before comparing them, as though the two were separate entities. They're not separate at all; both types of charges carry just one physical quantity: electric charge. Positive charges carry *positive amounts* of electric charge, while negative charges carry *negative amounts.* Therein lies the beauty of the names Franklin chose: to obtain the **net electric charge** for an object such as a dust grain, you simply add up the charges of its constituents, with negative charges appearing as negative amounts in that sum. Giving the grain a negative charge means ensuring that the grain's net electric charge is negative.

Second, for the air cleaner to give the dust grain a negative charge, it must transfer charge between objects. Like momentum, angular momentum, and energy, electric charge is a conserved physical quantity and can't be created or destroyed. That's why the air cleaner transfers charge from the dust grain to its collecting plate. Having lost charge, the grain ends up with a negative net charge and having gained charge, the plate ends up with a positive net charge. Here, as elsewhere in this book, we assume that the amount of charge transferred or moved is positive.

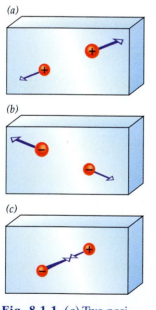

(a)

(b)

(c)

Fig. 8.1.1 (a) Two positive charges experience equal but oppositely directed forces exactly away from one another. (b) The same effect occurs for two negative charges. (c) Two opposite charges experience equal but oppositely directed forces exactly toward one another.

It's like with money: when you say that you gave money to a charity, we assume that you gave a positive amount.

Just as money is measured in standard units such as dollars or drachmas, so charge is measured in standard units. The SI unit of electric charge is the **coulomb** (abbreviated C). One coulomb is a lot of charge. On a dry winter day, the negative charge that you accumulate when you walk across a wool carpet in rubber-soled shoes is only about -1×10^{-6} C (-0.000001 C).

Third, if the air cleaner is going to give the dust grain a negative charge, the grain shouldn't be negative to start with. In fact, the dust grain probably starts with a net charge of zero. That's because a charged dust grain attracts opposite charges, which flock to it, stick, and gradually reduce its net charge to zero. When it reaches zero charge, the grain is said to be electrically **neutral.**

But arriving at *exactly* zero net charge is possible only because of another remarkable feature of electric charge: it's **quantized**—charge always appears in integer multiples of the **elementary unit of electric charge.** This elementary unit of charge is extremely small, only 1.6×10^{-19} C, and is the magnitude of charge found on most subatomic particles. An electron has -1 elementary unit of charge, while a **proton**—one of the subatomic particles found in atomic nuclei—has $+1$ elementary unit of charge. Since the only charged subatomic particles in normal matter are electrons and protons, a dust grain becomes electrically neutral simply by having the same number of electrons as protons.

Returning to the original question, we now know what an electronic air cleaner must do to a neutral dust grain in order to give that grain a negative charge. It must add electrons to and/or remove protons from the grain. Either transfer will upset the charge balance of the neutral grain and give it a negative net charge.

One last word about electric charge: Franklin's charge-naming scheme was brilliant in concept but unlucky in execution. It reduced the calculation of net charge to a simple addition problem, but it required Franklin to choose which type of charge to call "positive" and which to call "negative." Unfortunately, his seemingly arbitrary choice made electrons, the main carriers of electricity, negatively charged. Scientists and engineers have had to deal with negative amounts of charge flowing through wires ever since. Imagine the awkwardness of having to carry out business using currency printed only in negative denominations!

Check Your Understanding #1: A Bolt Out of the Blue

Just before lightning strikes, the clouds overhead have an enormous net charge. From where did this net charge come?

Collecting Dust with Electrostatic Forces

We'll return to dust-charging techniques shortly, but first let's see what happens to the negatively charged dust grains as they flow through the cleaner along with the air. Near the positive charges on a collecting plate, these grains experience strong electrostatic forces. While gravity is too weak to clear the air stream, electrostatic forces easily overwhelm viscous drag and pull the dust grains from the air. They collect on the charged plate, and the air continues on without them.

Because it precipitates clumps of dust on its collecting plates, this type of electronic air cleaner is called an *electrostatic precipitator*. Its chief advantage over a paper or fiber air filter is that huge amounts of dust can accumulate on its plates without blocking the airflow. It's also easy to clean. Just tap or rinse.

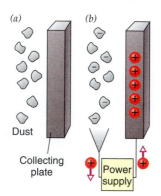

(a)   *(b)*

Dust

Collecting
plate

Power
supply

Fig. 8.1.2 (*a*) By themselves, dust and a collecting plate soon become electrically neutral. (*b*) But the power supply in an electrostatic precipitator transfers charge from the dust to the collecting plate, giving each of them a net electric charge.

But an electrostatic precipitator isn't a passive device. If it were, the negatively charged dust grains arriving on its positively charged collecting plates would quickly neutralize those plates and it would stop cleaning the air. To continue working, the precipitator must steadily transfer electric charge from the incoming dust to the collecting plates, using a device called a power supply.

A *power supply* pumps electric charge against its natural direction of flow. Because opposite charges attract, they tend to join together until everything is electrically neutral. But the precipitator's power supply reverses this process by transferring charge from the dust grains to the plates (Fig. 8.1.2). Recall that we're assuming this transferred charge is a positive amount. The net charge on the dust grains becomes negative, and the net charge on the plates becomes positive. These net charges are equal but opposite, an arrangement called **separated electric charge.** Motionless or slow-moving accumulations of separated electric charge are known as **static electricity.**

The precipitator's power supply must overcome electrostatic forces as it pulls charge away from the negative dust and pushes it onto the positive collecting plates. It therefore does work to produce the separated electric charge. This work is stored in the separated charge's electrostatic forces and is called **electrostatic potential energy.** Electrostatic potential energy is released when the separated charge gets back together again.

Since the power supply does work on each unit of charge it transfers, it's often useful to know how much electrostatic potential energy each unit of charge has at a particular location, a quantity known as **voltage.** When charge moves between two places with the same voltage, its electrostatic potential energy doesn't change. But when charge flows to a place with lower voltage, something it does spontaneously, it releases energy. And when charge is moved to a place with higher voltage, it requires additional energy, which is why the precipitator needs a power supply.

You can think of a location's voltage as the work you'd have to do on a unit of charge to transfer it from a distant, electrically neutral object to that location. Equivalently, you can think of it as the work that a unit of charge at that location would do as it returned to the distant, electrically neutral object. Since the SI unit of energy is the joule and the SI unit of electric charge is the coulomb, the SI unit of voltage is the joule-per-coulomb, also called the **volt** (abbreviated V). Note that a location's voltage can be negative if transferring a unit of charge to that location takes a negative amount of work!

In many situations, including the electrostatic precipitator, charge is transferred between electrically charged locations. In such cases, we're most interested in the voltage *difference* between those locations. This difference is the work you'd have to do on a unit of charge to transfer it from its initial location to its final location. If transferring 1 C of charge takes 100 J of energy, then the voltage of the final location is 100 V higher than that of the initial location. A good example of this concept is a household battery, a simple power supply that uses chemical reactions to pump charge from its negative terminal to its positive terminal. A typical battery does roughly 1.5 J of work on each coulomb of charge it pumps, so that the voltage of its positive terminal is about 1.5 V higher than that of its negative terminal.

The power supply in an electrostatic precipitator is much stronger than a simple battery. It does so much work on the charge it transfers that the voltage difference between the dust and the collecting plates is more than 10,000 V. Because this powerful charge transfer leaves the dust grains negatively charged and the collecting plates positively charged, the dust grains move quickly through the air toward the collecting plates. Although drag turns most of the separated charge's electrostatic potential energy into thermal energy, the grains still move fast enough to stick to the surface when they hit. There are a variety of techniques for producing high voltages, including the one shown in Fig. 8.1.3.

▶ Check Your Understanding #2: Carpet Sparks
When you wipe your feet on a carpet, sliding friction transfers charge from your shoes to the carpet. What happens to your voltage and that of the carpet when this transfer occurs?

Charging the Dust

An electrostatic precipitator uses an effect called a corona discharge to give dust a negative charge. To understand a corona discharge, we must first examine the forces between charged particles in a little more detail.

The forces between a pair of charges depend on two things. First, they're proportional to the amount of each charge. That means that doubling the amount of either charge doubles the forces. Also, replacing a positive amount of charge with a negative amount turns attractive forces into repulsive ones or vice versa.

Second, the forces depend on the distance separating the two charges (Fig. 8.1.4). The forces are inversely proportional to the square of the separation, becoming weaker as the charges move away from one another.

These ideas can be combined to describe the forces acting on two charges and written as a word equation:

$$\text{force} = \frac{\text{Coulomb constant} \cdot \text{charge}_1 \cdot \text{charge}_2}{(\text{distance between charges})^2}, \qquad (8.1.1)$$

in symbols:

$$F = \frac{k \cdot q_1 \cdot q_2}{r^2},$$

and in everyday language:

When two clouds with large opposite charges approach, expect lightning.

The force on charge$_1$ is directed toward or away from charge$_2$, and the force on charge$_2$ is directed toward or away from charge$_1$.

This relationship is called **Coulomb's law,** after its discoverer, French physicist Charles Augustin de Coulomb ❑. The **Coulomb constant** is about 8.988×10^9 N·m^2/C^2 and is one of the physical constants found in nature. Consistent with Newton's third law, the force exerted on charge$_1$ by charge$_2$ is equal in amount but oppositely directed from the force exerted on charge$_2$ by charge$_1$.

> **Coulomb's Law**
> The magnitudes of the electrostatic forces between two objects are equal to the Coulomb constant times the product of their two electric charges divided by the square of the distance separating them. If the charges are like, then the forces are repulsive. If the charges are opposite, then the forces are attractive.

Fig. 8.1.3 Static electricity can be produced by mechanical processes. In this Van de Graaff generator, a moving rubber belt transfers negative charges from the base to the shiny metal sphere. This negative charge creates dramatic sparks as it returns through the air toward the positive charge it left behind.

(a)

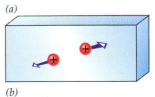

(b)

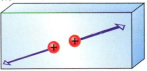

Fig. 8.1.4 The electrostatic forces between two charges increase dramatically as they become closer. As the distance separating two positive charges decreases by a factor of 2 between (a) and (b), the forces those two charges experience increase by a factor of 4.

❏ In 1781, after a career as a military engineer in the West Indies, **French physicist Charles Augustin de Coulomb (1736–1806)** returned to his native Paris in poor health. There he conducted scientific investigations into the nature of forces between electric charges and published a series of memoirs on the subject between 1785 and 1789. His research came to a close in 1789 when he was forced to leave Paris because of the French Revolution.

❏ Contrary to popular belief, lightning rods don't simply attract lightning strikes so as to protect the surrounding roof. Instead, they produce corona discharges that diminish any local build-ups of electric charge. By neutralizing the local electric charge, the lightning rod reduces the chances that lightning will strike the house. Similar devices, called static dissipaters, are found near the tips of airplane wings and protect planes from lightning strikes.

According to Coulomb's law, the repulsion between like charges increases dramatically as they approach one another. Thus when several like charges are bunched tightly together on a surface, their intense repulsions can actually push some of them out of the surface and onto passing gas molecules and dust grains. This flow of charge from a surface to nearby gas or dust is a **corona discharge.**

Corona discharges form easily around sharp metal points and thin metal wires. That's because like charges tend to bunch together on these structures and push one another into the air. While like charges repel and normally spread out fairly evenly on the outside of a metal object, a fraction of these charges naturally flow onto any exterior point or wire the object may have so as to get farther away from the other charges. Although these few charges repel one another ferociously, their numbers are small and the voltage on the point or wire (the energy per unit of charge) is the same as the voltage elsewhere on the metal object.

To initiate a positive corona discharge, the precipitator's power supply could push positive charges onto the point or wire until they began leaping into the air. Such a discharge wouldn't start until the charges were packed extremely tightly and the voltage on that surface was roughly 10,000 V. But the precipitator we've been discussing gives dust a negative charge and has a negative corona discharge. Such a discharge proceeds in exactly the same way as a positive corona except that it involves tightly packed negative charges. Because voltage is defined as the electrostatic potential energy per unit of *positive* charge, the voltage on the negatively charged metal surface emitting the corona is −10,000 V.

Because electrostatic potential energy is released when a charge escapes from the surface, corona discharges are often accompanied by light. This light can be seen around high-tension wires or near the tips of lightning rods (see ❏). On sailing ships, corona discharges were sometimes seen near masts, where they were known as St. Elmo's fire.

Not all electrostatic precipitators put negative charge on the dust. Some put positive charge on the dust and attract it to negatively charged surfaces. While negative corona discharges are less susceptible to accidental sparking and are easier to maintain, they also produce ozone, an irritating and toxic form of oxygen. Most industrial precipitators simply live with ozone production, but household units avoid it by using positive corona discharges.

Check Your Understanding #3: A Safety Pin

You can avoid the shocks of static electricity by carrying a very sharp needle in your hand and holding it out before touching a metal doorknob or wall. How does that needle protect you from static electricity?

Check Your Figures #1: Moving Out

You have two positively charged balls, each of which is experiencing a force of 1 N away from the other. If you separate them by twice as much, what force will each one exert on the other?

Ion Generators

Household ion generators also remove dust and smoke from room air. These machines resemble electrostatic precipitators but have no internal collecting plates. While they still use corona discharges to charge passing dust grains, they don't attempt to remove the dust from the air. Instead, they let those electrically charged particles, or **ions,** drift around the room on their own.

The ions don't drift long because they're attracted to neutral surfaces. A neutral surface contains countless electric charges (Fig. 8.1.5), and a charged dust grain will attract the surface's opposite charges while repelling its like charges. Because of these forces, the surface's positive and negative charges shift slightly in opposite directions and the surface becomes electrically **polarized.** A polarized object is neutral overall but has a positively charged end and a negatively charge end.

The dust grain and the surface it polarizes attract one another and stick. You can see this effect by charging a toy balloon in your hair and then touching it to a wall. The balloon's charge will polarize the wall, and the two will cling to one another. Thus an ion generator removes particles from the air by sticking them to surfaces in the room. This method is cheap and effective, but the dust ends up decorating the walls and furniture, a problem unless you're fond of the color gray.

types of ions

Check Your Understanding #4: Static Cling
When you remove socks from the dryer, they cling to everything. How can a sock cling to the wall or to a piece of lint?

precipitators, filters

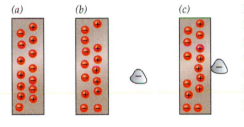

(a)   (b)   (c)

Fig. 8.1.5 (*a*) A neutral wall contains countless positive and negative charges. (*b*) As a negatively charged dust grain approaches the wall, the positive charges move toward it and the negative charges move away. (*c*) The polarized wall continues to attract the grain and holds it in place.

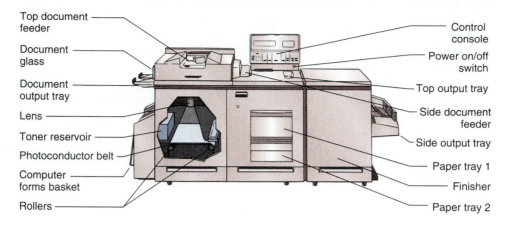

Top document feeder
Document glass
Document output tray
Lens
Toner reservoir
Photoconductor belt
Computer forms basket
Rollers

Control console
Power on/off switch
Top output tray
Side document feeder
Side output tray
Paper tray 1
Finisher
Paper tray 2

Xerographic Copiers

The days of carbon paper and mimeograph machines are long past. What modern office could operate without a xerographic copier? Advertisements for copiers are everywhere, and while manufacturers often claim that their copiers are far better than those of the competition, that's mostly just salesmanship. In reality, all xerographic copiers are based on the very same principles, discovered in 1938 by Chester Carlson. In this section, we'll examine xerographic copiers and the ideas that make them possible.

Questions to Think About: *How could you use static electricity to position black powder on a sheet of paper? How would you put that static electricity on the paper? In order for characters to appear on the sheet, how should its static electricity be distributed? In a copier, what should light do to the static electricity to produce a copy of the original?*

Experiments to Do: *To get a feel for how a copier works, cut a small sheet of paper into tiny squares, about 1 mm on a side. Put the squares on a table and suspend a thin plate of clear plastic above them, a few millimeters away. The top of a clear plastic box will do. Now run a plastic comb through your hair or against a sweater several times and touch it to the top of the plastic plate. Squares of paper will leap off the table and stick to the plastic plate. What's holding the squares against the plastic? If the paper were black, how could you form letters on the surface of the plastic?*

Controlling Static Electricity: Electrons in Solids

The image that a xerographic copier forms on a sheet of paper begins as a pattern of tiny black toner particles. The copier uses static electricity to arrange these particles on a drum or belt and then transfers them to the paper. Since the copier is duplicating an original document, it uses light from that document to control the placement of static electricity, so that the pattern of toner on the copy is identical to the printing on the original.

At the heart of the xerographic copier is a thin layer of material that controls the placement of static electricity. This layer is a **photoconductor**—a solid material through which electric charges can move only when it's exposed to light. In the dark, it's an **electric insulator**: it prevents any net movement of electric charge. In the light, it's an **electric conductor**: it allows charge to move freely.

While a darkened photoconductor can keep positive and negative charges apart, these charges quickly attract one another when light hits the photoconductor (Fig. 8.2.1). This behavior allows light from the original document to determine the pattern of static electricity on the copying drum or belt and hence the placement of toner on a piece of paper. Photoconductors are so central to xerographic copying that we'll spend the next several pages examining photoconductivity and the behaviors of electrons in solids.

Photoconductivity is a consequence of quantum physics, which has enormous influence over the motions of small particles. In particular, quantum physics limits the possible paths that a small particle can take when it travels in an enclosed space. This limitation is particularly important for electrons, which are the most mobile charged particles in matter and thus the main carriers of electricity. An electron in an atom can only orbit the atom's nucleus in one of the paths that quantum physics allows. These allowed paths are called **orbitals.** Similarly, quantum physics limits an electron's motion through a solid object to allowed paths or trajectories that are called **levels.**

One of the most remarkable observations of quantum physics is that every indistinguishable electron must have its own level. This law is called the **Pauli exclusion principle,** after its discoverer, Wolfgang Pauli ❑. The principle applies to a whole class of subatomic particles, the **Fermi particles,** that includes all of the basic constituents of matter: electrons, protons, and **neutrons**—the electrically neutral subatomic particles that, together with protons, make up atomic nuclei. Two indistinguishable Fermi particles can never travel in the same level.

The Pauli Exclusion Principle
No two indistinguishable Fermi particles ever occupy the same quantum level.

However, a peculiar property of electrons allows *two* of them to share an orbital or level. Electrons have two possible internal states, usually called spin-up and spin-down. A spin-up electron is distinguishable from a spin-down electron, so that one spin-up electron and one spin-down electron can share a single orbital or level. However, two is the absolute maximum allowed by quantum physics and the Pauli exclusion principle.

An electron's level also determines its total energy, the sum of its kinetic and potential energies. Since electrons in different levels follow different paths at different speeds, each level corresponds to a particular energy. Electrons in the lowest energy levels generally travel slowly or stay close to positively charged atomic nuclei, while those in higher energy levels usually travel more quickly or avoid nuclei.

Because of frictionlike effects, an object's electrons lose energy until they're traveling in the lowest energy levels. However, since no more than two electrons can share each level, they can't all be in the lowest level. Instead, they fill the levels from lowest energy on up, two to each level. This filling continues up to the **Fermi level**—the highest energy level used by an electron. If we represent these levels by boxes and arrange them vertically according to energy (Fig. 8.2.2), then the levels (boxes)

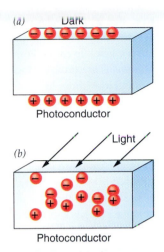

Fig. 8.2.1 (*a*) In the dark, a photoconductor is an electric insulator so that separated electric charges on its surfaces remain there indefinitely. (*b*) But when the photoconductor is exposed to light, it becomes an electric conductor and the opposite electric charges soon join one another.

❑ **Austrian physicist Wolfgang Pauli (1900–1958)** rose to fame at 21 by writing an article on relativity that impressed even Einstein. He went on to discover the exclusion principle, a fundamental part of quantum theory. He was renowned for his intensely critical attitude toward new ideas, considering them all "rubbish" until convinced otherwise. Pauli was also quite interested in psychology, corresponding with Carl Gustav Jung and writing a number of articles on the subject.

Fig. 8.2.2 The levels in an object are normally filled from the lowest energy level up to the Fermi level. The levels are grouped together in bands. Here, those levels are represented by boxes that can hold at most two electrons: one spin-up electron (blue) and one spin-down electron (red). For simplicity, we'll imagine that those levels are distributed in space from the left side of the object to the right side. The positive charges shown are the nuclei of atoms that make up this electrically neutral object.

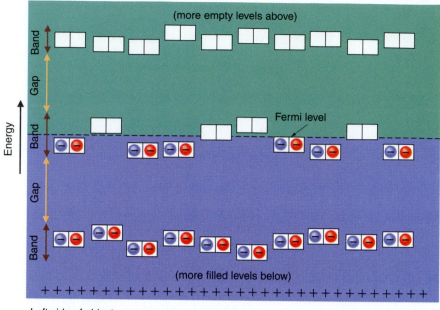

up to the Fermi level each contain two electrons, while those above the Fermi level are empty. Although thermal energy complicates this picture somewhat, we can safely ignore that detail near room temperature or below.

Since levels involve motion and kinetic energy, an electron in a particular level doesn't really have a well-defined location in space. But to facilitate our understanding of photoconductors now and semiconductor electronics in Chapter 10, let's imagine that each level places its electrons near a particular location in the object, as shown in Fig. 8.2.2. While this picture is oversimplified, it's accurate enough to illustrate much of the physics of charge motion in materials.

Of course, electrons aren't the only charged particles in an object. The atoms also have positively charged nuclei. But those nuclei are essentially immobile and rarely participate in the flow of electricity. Instead, they form a uniform background of positive charge, shown schematically in Fig. 8.2.2, so that the object is roughly neutral throughout.

Check Your Understanding #1: Taking It to a Higher Level

If you add one extra electron to an otherwise neutral metal ball, into which level will that electron go?

Metals, Insulators, and Photoconductors

The levels in a solid occur in groups called **bands.** Each band corresponds to a particular type of path through the solid. Since the levels in a band involve similar trajectories, they also involve similar energies. Between these bands of levels there are sometimes *band gaps*—ranges of energy in which there are no levels at all.

Bands and band gaps are what distinguish metals, insulators, and photoconductors. When a band gap is located close to the Fermi level, it can prevent the electrons in a solid from responding to outside forces. To see how that happens, let's examine first a metal and then an insulator.

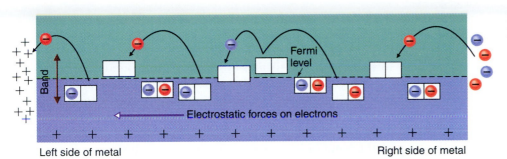

Left side of metal   Right side of metal

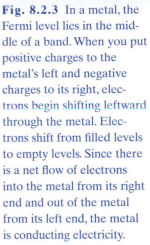

Fig. 8.2.3 In a metal, the Fermi level lies in the middle of a band. When you put positive charges to the metal's left and negative charges to its right, electrons begin shifting leftward through the metal. Electrons shift from filled levels to empty levels. Since there is a net flow of electrons into the metal from its right end and out of the metal from its left end, the metal is conducting electricity.

In a metal, the Fermi level lies in the middle of a band (Fig. 8.2.3). Because the band's empty levels are just above its filled levels, very little energy is required to shift electrons from filled levels to empty levels. This feature allows the metal to conduct electricity. When you put positive charges on the metal's left side and negative charges on its right, its electrons experience leftward electrostatic forces and begin to move left. They move by shifting from filled levels to empty levels (Fig. 8.2.3), obtaining the energy needed to reach those empty levels from the work done on them by the electrostatic forces. Overall, electrons enter the metal from its right and leave from its left, so the metal conducts electricity!

For a useful analogy to a metal, imagine a theater in which only about half the ground-floor chairs are filled. If you ask everyone in this theater to begin shifting left, they can do it easily. Each person finds an empty seat nearby on the left and moves over. New people are then able to enter the theater from the right while others leave the theater from the left. This "metal" theater would be "conducting" people.

Unlike the situation in a metal, an insulator's Fermi level lies at the top of a band, and there's a large band gap just above it (Fig. 8.2.4). With no easily accessible empty levels available, a great deal of energy is required to shift electrons from filled levels to empty levels. When you put positive charges on the insulator's left side and negative charges on its right, its electrons experience leftward electrostatic

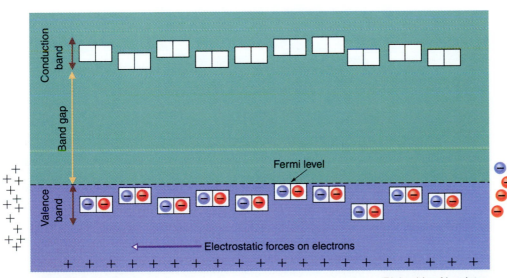

Left side of insulator   Right side of insulator

Fig. 8.2.4 In an insulator, the Fermi level lies at the top of a band. When you put electric charges near the insulator, the electrons in the filled band can't shift to produce a net flow of charge through the insulator. The insulator can't conduct electricity.

forces but are unable to move. To shift into one of the empty levels, an electron near the Fermi level would need more energy than it can get from the electrostatic forces. Since no net charge flows through the insulator, it doesn't conduct electricity!

Our analogy to an insulator is a theater in which all the ground-floor chairs are full and in which the balcony seats are empty. When you ask people in this theater to begin shifting left, they can't do it. All of the ground-floor seats to the left are filled and they can't reach the balcony to make use of its empty chairs. This "insulator" theater would be unable to conduct people.

In a real metal, the band of levels containing the Fermi level is only partially filled, and electrons can easily shift from filled to unfilled levels. In an insulator, the band containing the Fermi level—the **valence band**—is full, and the band gap separating it from the first band of empty levels above it—the **conduction band**—makes such shifts extremely difficult.

But even in an insulator, an electron can shift from a **valence level** (a level in the valence band) to a **conduction level** (a level in the conduction band) if something provides the necessary energy. One such energy source is light. When an insulator is exposed to the light of the right frequency, that light can shift electrons from the material's valence band to its conduction band (Fig. 8.2.5).

Once electrons appear in the normally empty conduction band and empty levels appear in the normally full valence band, electrons can respond to electrostatic forces. They can shift from filled levels to nearby empty levels and thus travel through the material. Electrons can then enter the material from one side and leave from the other, so the material conducts electricity. And because light has made this insulator a conductor, we call the material a *photoconductor*.

Turning again to our analogy, light's role in the insulator theater is performed by a playful gorilla that walks about the ground floor, tossing patrons into the balcony. With some of the ground-floor seats suddenly empty and some of the balcony seats suddenly occupied by dazed theatergoers, the crowd can now respond to your request to move left. The gorilla has made the insulator theater a conductor of people—what you might call a gorillaconductor.

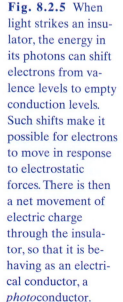

Fig. 8.2.5 When light strikes an insulator, the energy in its photons can shift electrons from valence levels to empty conduction levels. Such shifts make it possible for electrons to move in response to electrostatic forces. There is then a net movement of electric charge through the insulator, so that it is behaving as an electrical conductor, a *photo*conductor.

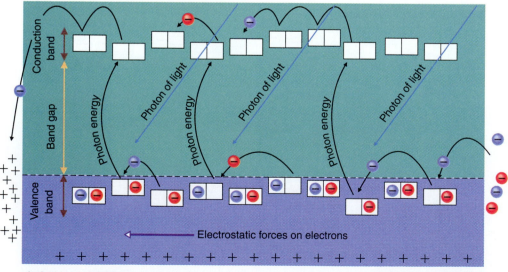

Returning to the real photoconductor, we note that light's frequency is important because light is emitted and absorbed in energy packets or *quanta* called **photons.** This observation is another consequence of quantum physics. The higher the frequency of light, the more energy each of its photons contains. To shift an electron across the large band gap in a typical insulator, high-energy, high-frequency light is needed; the insulator must be exposed to violet or even ultraviolet light.

But nature also provides materials with smaller band gaps that can be crossed with the help of low-energy, low-frequency red or even infrared light. These materials are called **semiconductors** because their properties lie somewhere in between those of conductors and insulators. In our analogy, a semiconductor theater is an insulator theater with a low balcony, so that even a baby gorilla can toss people into it.

The photoconductors found in most xerographic copiers are semiconductors. They have band gaps just above the Fermi level, so they don't normally conduct electricity. But when they're exposed to visible light, electrons in these photoconductors shift from valence levels to conduction levels and allow electric charge to flow through the material. The photoconductor can thus use light to control static electricity so that the copier can reproduce the original.

Check Your Understanding #2: Stop and Go Shopping

In many grocery checkout counters, a conveyor belt carries food to the register but stops when the food reaches the end and blocks a beam of light. How might a photoconductor be used to sense this blockage?

Moving Electric Charge with Electric Fields

The xerographic copier coats its photoconductor with electric charges, and these charges move about when the photoconductor is exposed to light. But before we examine the copier itself, let's think again about what makes those charges move.

The copier puts charges on two sides of a thin sheet of photoconductor, negative charge on one side and positive charge on the other. When the photoconductor is illuminated, electrons flow from the negative side to the positive side and the separated charge disappears. Each electron that completes this journey does it by shifting among the valence and conduction levels so that it can accelerate toward the positive charges in front of it and away from the negative charges behind it.

But instead of thinking about the interaction between the traveling electron and all of the individual charged particles around it, we can view it as an interaction between the traveling electron and something local: an electric field. An **electric field** is an attribute of space that exerts forces on charged particles.

From this new perspective, the traveling electron accelerates forward because it's in an electric field, one that's created by all of the nearby charged particles. In this case, the electric field is simply an intermediary, allowing us to think separately about the traveling electron and the charges surrounding it; the surrounding charges create the field, and the field causes the traveling electron to accelerate. But in later sections, we'll see that electric fields are more than intermediaries, more than seemingly unnecessary fictions. That's because electric fields truly exist in space, independent of the charges that produce them. In fact, electric fields are often created by things other than charges.

Fig. 8.2.6 The electric field produced by two opposite electric charges. At each point in space marked by a red dot, an arrow indicates the electric field's direction and the arrow's length indicates the field's strength. A positive charge located at the dot would experience an electrostatic force in the direction of the arrow and of an amount proportional to the length of the arrow.

An electric field usually varies from place to place, so that the force it exerts on a charge depends on that charge's position. At each point in space, the field has a magnitude and a direction, where the magnitude specifies the amount of force it exerts on each unit of charge and the direction specifies which way it pushes on a positive charge (Fig. 8.2.6). Because an electron has a negative charge, it's pushed in the direction opposite the electric field.

The electric field inside the copier's photoconductor sheet always points from the positive side of the sheet to the negative side (the direction in which a positive charge would accelerate). When light strikes the photoconductor, electrons flow from the negative side to the positive side, pushed along by the electric field.

Check Your Understanding #3: Medical Electrons

A medical linear accelerator uses a strong electric field to accelerate electrons forward and give them enormous kinetic energies. These high-energy electrons enter the patient and kill cancer cells. In which direction does the accelerator's electric field point?

Xerography

The light-sensitive component in a xerographic copier is a metal drum or belt that's covered with a thin layer of photoconductor. The copier coats this photoconductor with electrons, which remain in place as long as the photoconductor is in the dark. But wherever light strikes the photoconductor, it becomes conductive and allows the electrons to escape into the metal. Only the unilluminated portions of the photoconductor retain their static electric charge and eventually attract black toner particles. In that manner, the darkened parts of the photoconductor produce the dark parts of the final copy.

Modern xerographic copiers, such as the one shown schematically in Fig. 8.2.7, use the xerographic process outlined in Fig. 8.2.8. While Chester Carlson invented this process in 1938 ❑, the first commercially successful xerographic copiers weren't produced until 1960.

❑ Impoverished as a youth, **American inventor Chester F. Carlson (1906–1968)** supported his family by washing windows and cleaning offices after school. His work in a print shop as a teenager started him thinking about copying and he began to experiment with electrophotography. After attending Caltech, he worked for Bell Laboratories but was laid off in the Depression. While attending law school, he continued his experiments and invented the xerographic copying process in 1937–1938. Development of commercial copiers was slow and it wasn't until 1960 that Haloid Xerox Corporation produced its first successful copier, Model 914. Carlson became extremely wealthy but gave most of his money away anonymously.

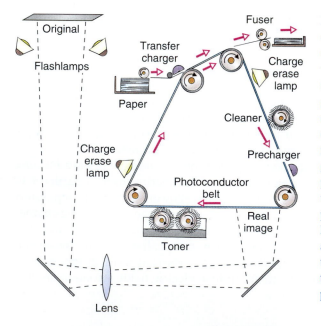

Fig. 8.2.7 This xerographic copying machine uses a photoconductor belt to form black and white images of an original document. The copying process begins with the precharger, which coats the photoconductor with charge. The optical system then forms a real image on a flat region of the photoconductor belt, producing a charge image. After the charge image picks up toner particles, the first charge erase lamp eliminates the charge image and weakens the toner's attachment to the belt. The toner is then transferred and fused to the paper.

The copier starts by applying a uniform charge to the surface of the photoconductor. Although some copiers apply positive charge, for the present discussion, let's assume that it's negative. This charge is applied by a *corotron*, a fine wire centered in a half-cylinder of metal. A power supply pumps electrons onto the fine wire until they're emitted into the air as a corona discharge. When these electrons approach the photoconductor, they polarize it and stick to it, just as negative ions from a simple ion generator stick to the walls of your house. The photoconductor becomes uniformly charged (Fig. 8.2.8a), with about 10^{-7} C of negative charge per cm^2 of surface.

On the other side of the photoconductor is a grounded metal surface. It's grounded in the sense that it's electrically connected to the earth so that charges are free to flow between the two. The negative charge on the free surface of the photoconductor repels electrons so that they leave the metal for the earth. As a result, the metal side of the photoconductor acquires a net positive charge.

After this charging, the copier exposes the photoconductor to light from the original document. It uses a lens to cast an image of the original onto the photoconductor's surface. We'll examine lenses and the formation of images when we look at photographic cameras in Chapter 13. For now, the important point is that light only hits the photoconductor in certain places, corresponding to white parts of the original document.

There are two standard techniques for exposing the photoconductor to light. Some copiers illuminate the original document with the brilliant light of a flash lamp and cast its entire image onto a flattened portion of a photoconductor belt. In an instant, charge flows through any regions of the photoconductor that were exposed to light, leaving these regions electrically neutral (Fig. 8.2.8b). In other copiers, a moving lamp or mirror illuminates the original a little at a time and the image is built piece by piece on a rotating photoconductor drum. The document's image moves with the turning drum, gradually exposing the photoconductor and allowing parts of it to become electrically neutral. When the exposure is over, the photoconductor carries a charge image of the original document (Fig. 8.2.8c).

To develop this charge image into a visible one, the xerographic copier exposes the photoconductor to charged toner particles (Fig. 8.2.8d). The toner is a fine insulating plastic powder that contains a colored pigment. Most often this pigment is black, but it can be any color. The toner is charged opposite to the photoconductor so that the two attract. In our case, the toner is positive.

Applying toner to the photoconductor must be done gently, and it's often accomplished with the help of Teflon-coated iron balls. These tiny balls are held together in long filaments by a rotating magnetic shaft, so that the shaft resembles a spinning brush with extraordinarily soft bristles. These bristles wipe toner particles out of their storage tray and onto the photoconductor. During this transfer, the toner particles become positively charged so that they stick to the negatively charged portions of the photoconductor (Fig. 8.2.8e).

The photoconductor now carries a black image of the original document. But to create a copy, this black image must be transferred to paper. To begin this transfer, the copier illuminates the photoconductor with a charge erase lamp so that the photoconductor's negative charge escapes into the metal (Fig. 8.2.8f). The toner remains in place, but it's very weakly attached (Fig. 8.2.8g).

The copier than transfers the toner to a nearby sheet of paper by applying negative charge to the paper's back (Fig. 8.2.8h). The positively charged toner is attracted to the negatively charged paper, and the two leave the photoconductor together. The copier then heats and presses the copy, permanently fusing the toner into the paper (Fig. 8.2.8i). Sometimes, when a copier jams, you may remove a sheet

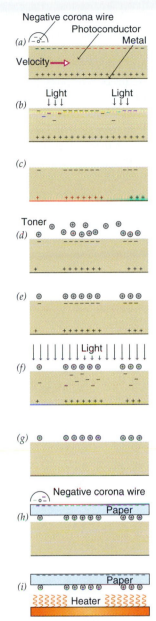

Fig. 8.2.8 The photoconductor is first coated (*a*) with a uniform layer of negative charge. Exposure to light (*b*) erases some charge to form a charge image (*c*). The charge image attracts (*d*) positively charged toner particles (*e*). The charge image is erased (*f*) to release the toner particles (*g*). The toner is transferred to the negatively charged paper (*h*) and fused to the paper with heat (*i*).

Fig. 8.2.9 This xerographic copier places the photoconductor drum, toner supply, and a corona wire inside a disposable cartridge. After the paper passes through the cartridge from right to left, toner is fused to its surface and it leaves the copier.

before it has been fused. The image looks complete and normal but wipes off when you touch it because it's held in place only by electrostatic forces.

Once the image has been transferred to the paper, it's time to get the photoconductor surface ready for the next copy. The charge erasure and transfer processes leave a little charge and toner on the photoconductor so the copier performs an extra cleaning. A second charge erase lamp eliminates the charge, and a brush or squeegee mops up the toner. Once this cleanup is done, the photoconductor is back to its original state and ready to be used again. In a personal copier, the photoconductor and toner supply are contained in a single disposable cartridge (Fig. 8.2.9).

Check Your Understanding #4: Sticky Copies

When the copies emerge from a xerographic copier, they tend to stick to things and attract lint. What causes this effect?

 color copies

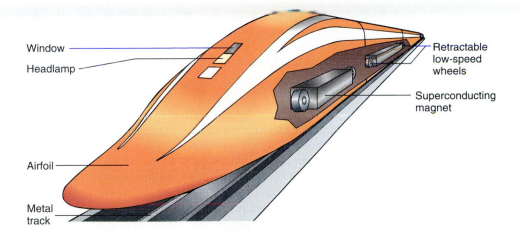

Window

Headlamp

Retractable
low-speed
wheels

Superconducting
magnet

Airfoil

Metal
track

SECTION 8.3

SECTION 8.3   Magnetically Levitated Trains

While jet airplanes still carry most travelers between distant cities, modern trains have eroded a jet's advantage for intermediate distances. Bullet trains are used in several countries, moving their passengers at about 200 km/h (125 mph). However, mechanical interactions between a train and its rails limit its maximum speed. The next generation of ultra-high-speed trains will probably fly above the rails on cushions of magnetic force.

Questions to Think About: What forces do a train's wheels exert on the rails, and why should these forces become larger as the train travels faster? If two properly aligned magnets repel one another, how could they be arranged to support a train? How could you support a train with two magnets that attracted one another? How should the upward magnetic force on the train vary with height to keep the train hovering at a particular level? What would happen to the train if the force varied the wrong way?

Experiments to Do: Permanent magnets are wonderful toys. There is something fascinating about two objects that can attract or repel one another across centimeters of empty space. Play with two strong magnets to get a feel for how the forces between them work. Simple disk-shaped refrigerator magnets are best since some others, such as flexible strip magnets, are rather complicated. You should find that the two disks repel one another in one configuration but attract when you flip one disk around. The closer they are, the stronger the forces. Try to suspend one magnet above the other by repulsion. Did you succeed? Actually, it can't be done. Why not? Now try to suspend one magnet a centimeter or so below the other by attraction. Any luck this time? It can't be done either. Evidently, magnetic suspension isn't so easy, after all.

The Need for Magnetic Levitation

Trains operate more or less as they have since the early 19th century. Cars still roll on wheels, pulled forward by friction between locomotives and the rails beneath them. But while this scheme works well at speeds below about 160 km/h (100 mph), two serious problems appear at higher speeds.

First, a high-speed train has trouble following the rails because it has so little time to move up, down, left, or right. Sudden shifts in direction require rapid accelerations and involve large forces between the rails and the wheels. To minimize the forces and accelerations, the rails must be straight and exceedingly smooth, with only the most gradual curves. But even carefully constructed rails take a beating from the wheels and produce noise, vibration, and mechanical wear. Both the rails and the wheels must be replaced frequently so that, above about 320 km/h (200 mph), maintenance becomes prohibitively expensive.

Second, air resistance becomes so strong at high speeds that traction between the locomotive's wheels and the rails isn't enough to keep the train moving forward. The train can't maintain its cruising speed with friction alone, and it can't stop quickly enough using friction either. Just for safety reasons, conventional rail travel shouldn't exceed about 320 km/h (200 mph).

To overcome these limitations, a whole new technology for ultra-high-speed trains is being developed. Instead of running on metal wheels, these new trains are suspended above their tracks by magnetic forces. Because these magnetically levitated or *maglev* trains don't touch their tracks, those tracks don't have to be very smooth. The trains ride forward on magnetic cushions.

Conventional propulsion won't work for a maglev train because it doesn't have any wheels or friction. Instead, a maglev train uses additional magnets on the train and track to propel the train forward and to stop it during braking.

Check Your Understanding #1: Bumper Cars

An otherwise straight track has a small bump in it. As it rolls over the bump, a train car experiences an upward force that exceeds its downward weight. How does this force affect the car's motion?

Supporting a Train with Magnetic Forces

Instead of rolling forward on wheels, a maglev train rides above its track on a magnetic suspension. Magnets in the train and track exert forces on one another and support the train without any direct contact. But what is magnetism, and what are the forces between magnets?

Magnetism is a phenomenon that closely resembles electricity. Just as there are two types of electric charges exerting electrostatic forces on one another, so there are two types of **magnetic poles** exerting **magnetostatic forces** on one another. The word "pole" was chosen to distinguish magnetism from electricity; **poles** are magnetic while charges are electric. The two types of poles are called north and south, respectively, and in keeping with this geographical naming, they're exact opposites of one another. Both types of poles carry just one physical quantity: **magnetic pole.** North poles carry positive amounts of magnetic pole while south poles carry negative amounts. It should come as no surprise that like poles repel while opposite poles attract. Furthermore, the magnetostatic forces between two poles grow weaker as they move apart and are inversely proportional to the square of the distance between them. The similarities between electricity and magnetism are striking.

Since like poles repel one another, it might seem as though you could levitate a train simply by placing north poles on both it and the track. However, this scheme will fail for two reasons. First, isolated magnetic poles don't seem to exist and second, even if they did, the train would soon fall off its magnetic cushion.

The first problem points out an important difference between electricity and magnetism. While subatomic particles that carry positive or negative electric charges are common, particles that carry north or south magnetic poles have never been found. It's not even clear whether such **magnetic monopoles** exist. But while isolated magnetic poles aren't available in nature, pairs of magnetic poles are. These pairs consist of equal north and south poles, separated from one another in an arrangement called a **magnetic dipole.** Since the two opposite poles have equal magnitudes, they sum to zero and the magnetic dipole has zero **net magnetic pole.**

Since these magnetic dipoles can't be made from isolated magnetic poles, they must be created by something else: electricity. Although stationary electric charges exert no forces on stationary magnetic poles and vice versa, electricity and magnetism are nonetheless intimately related. In addition to resembling one another, they can actually produce one another, but only through *change or motion*. Changing or moving electric phenomena produce magnetic phenomena, and changing or moving magnetic phenomena produce electric phenomena.

The connection between electricity and magnetism involves their fields: electric fields and magnetic fields. Just as an electric field exerts a force on an electric charge, a **magnetic field** exerts a force on a magnetic pole. The magnetic field at a particular point in space indicates the magnitude and direction of the force that an isolated north pole would experience there, if it existed.

There are two electric phenomena that create magnetic fields: moving electric charge and an electric field that changes with time. That moving electric charge produces magnetic fields was discovered by Danish physicist Hans Christian Oersted (1777–1851) in 1820 and studied by French physicist André-Marie Ampère ❑ over the next seven years. In 1865, Scottish physicist James Clerk Maxwell (1831–1879) realized that changing electric fields also produce magnetic fields and included this discovery in his complete electromagnetic theory.

The most important source of magnetic fields for train levitation is moving electric charge, also called **electric current.** Electric currents are magnetic, meaning that whenever an electric current flows through a wire as "electricity" or leaps through the air as a spark, it produces a magnetic field. The structure of this magnetic field can be complicated, but there are a few special arrangements of current that produce simple magnetic fields.

One such arrangement is a loop of current. When current flows around a wire loop, it produces the symmetric magnetic field shown in Fig. 8.3.1a. This field is remarkably similar to the field that two opposite magnetic poles would produce when separated by a small distance, a true magnetic dipole (Fig. 8.3.1b). In fact, the current loop is a magnetic dipole. In effect, it has a north pole at one side of the loop and a south pole at the other. If you take two of these magnets and arrange them so that their like poles are closest together, they repel. If their opposite poles are closest, they attract. And if they're oriented at odd angles, they exert torques on one another and twist each other around.

Even permanent magnets use currents to produce their magnetic fields. We noted in the previous section that electrons have an internal property called spin, and this spin is associated with angular momentum. In effect, electrons are spinning charged objects and behave like tiny loops of electric current. They're dipole magnets! In most materials, their magnetism is hidden because their spins are oriented equally in all directions and their magnetic fields cancel one another. But a few mag-

❑ Self-educated before the French Revolution, during which his father was executed, **French physicist André-Marie Ampère (1775–1836)** became a science teacher in 1796. He served as a professor of physics or mathematics in several cities before settling at the University of Paris system in 1804. In 1820, only a week after learning of Oersted's experiment showing that an electric current causes a compass needle to deflect, Ampère published an extensive treatment of the subject. Evidently, he had been thinking about these ideas for a long time.

(a)

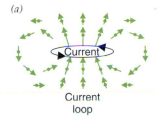

Current loop

(b)

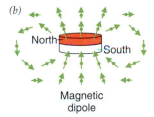

Magnetic dipole

Fig. 8.3.1 (*a*) The magnetic field around a loop of current-carrying wire points up through the loop and down around the outside of the loop. The arrow through each dot indicates the magnitude and direction of the force on a north pole at that location. (*b*) The field produced by two isolated magnetic poles, located one above the other.

Fig. 8.3.2 These two disk magnets repel because their south poles are facing one another. However, the stick is needed to prevent the top disk from falling off the magnetic cushion that supports it.

netic materials avoid this cancellation and can be used to make permanent magnets. A typical permanent magnet has a north pole at one end and a south pole at the other, both created by the spinning electrons inside.

The forces between dipole magnets can be surprisingly strong. If you cover the top of a track and the bottom of a train with dipole magnets, turned so that like poles face each other, the train and track will repel one another. This repulsion becomes stronger as the two approach one another and can exceed the train's weight. When it does, the train stops accelerating downward and can remain suspended over the track on a magnetic cushion. This is magnetic levitation.

Unfortunately, there are at least two problems with this simple levitation scheme. First, a track covered with magnets would attract magnetic junk—old steel cans, nuts and bolts, scraps of iron—making it a nightmare to keep clean. But more importantly, a magnet suspended above another magnet is inherently unstable (Fig. 8.3.2). While it's possible to achieve dynamic stability (see Section 3.4) in a few special cases (Fig. 8.3.3), a train can't do this. Instead, the train will tend to slip sideways onto the ground, just as a marble sitting on top of a dome will tend to roll sideways onto the ground. The train may hover briefly on its magnetic cushion, but it won't stay there for long without help.

Check Your Understanding #2: Two Halves Make a Whole

You have a disk-shaped permanent magnet. The top surface is its north pole and the bottom surface is its south pole. If you crack the magnet into two half-circles, the two halves will push apart. Why?

Fig. 8.3.3 This spinning magnetic top hovers above a repelling magnet, hidden in the wooden base. The top's dynamic stability involves gyroscopic effects that would be impractical for a maglev train.

Stability and Feedback

A train supported by permanent magnets is in an unstable equilibrium. As we discussed in Section 3.4, an object that's in equilibrium experiences zero net force and doesn't accelerate. If the equilibrium is a stable one, the object will naturally return there after being disturbed because a restoring force will push it back (Fig. 8.3.4a). But when the equilibrium is unstable, the disturbed object won't return because it will begin to experience a force that pushes it away from the equilibrium (Fig. 8.3.4b).

A theorem known as Earnshaw's theorem states that no arrangement of electric charges can be in a stable equilibrium as the result of electrostatic forces alone. Similarly, no arrangement of magnetic poles can be in a stable equilibrium as the result of magnetostatic forces alone. Thus no matter how you arrange permanent magnets on the train and track, the train will be in an unstable equilibrium. While the repulsion between magnets can keep the train at a stable height (Fig. 8.3.5), the train's sideways motion is unstable. If the train isn't perfectly centered above the track, the repulsive forces will push it toward the side and it will fall off its magnetic cushion.

Although it might seem that some clever arrangement of permanent magnets could make the train stable, there's absolutely no way around Earnshaw's theorem. The only way to stabilize the train and keep it centered above the track is to use magnets that can be adjusted—turned on and off—to push the train back to center if it starts to fall. Actually, adjustable magnets are easy to make. Since magnetic fields are created by electric currents, you can turn a magnet on or off simply by changing its current.

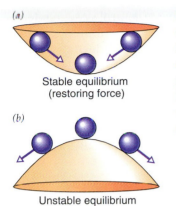

(a)

Stable equilibrium
(restoring force)

(b)

Unstable equilibrium

Fig. 8.3.4 (*a*) A ball disturbed from a stable equilibrium position experiences a restoring force that pushes it back toward equilibrium. (*b*) A ball disturbed from an unstable equilibrium doesn't return.

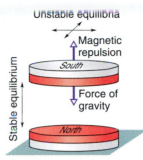

Unstable equilibria

Magnetic repulsion

South

Force of gravity

North

Fig. 8.3.5 If the repulsion between permanent magnets is used to suspend a train above a track, the train's height will remain stable, but its horizontal position will not. If the train isn't perfectly centered, the repulsive forces will push the train sideways until it falls.

Because the currents inside a permanent magnet are hard to stop, you'll need electromagnets. An **electromagnet** is a coil of wire that's magnetic only when an electric current flows through it (Fig. 8.3.6). To make the electromagnet even stronger, its wire is usually wound around iron or another magnetic material. The iron's spinning electrons then align with and reinforce the coil's magnetic field. We'll discuss iron's magnetism in Section 9.3.

To keep the train stable, a control system must monitor the train's position and adjust its electromagnets accordingly. This technique of using information about the current situation to control the way in which the situation changes is called **feedback.** Feedback is used frequently in engineering to stabilize systems that are inherently unstable and to modify the ways in which systems respond to external stimuli (Fig. 8.3.7). Feedback can effectively create restoring forces where none exists naturally.

Some of the maglev train systems currently under development or in trial use feedback stabilized magnetic levitation (Fig. 8.3.8, p. 286). Most support the train's weight with attractive forces between opposite poles and carefully control the electromagnets to maintain a constant separation between those attractive poles.

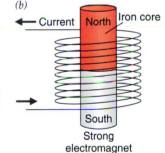

(a)

Current | North

South

Weak
electromagnet

(b)

Current | North | Iron core

South

Strong
electromagnet

Fig. 8.3.6 (*a*) When current flows through a wire coil, the coil becomes a weak dipole magnet. The direction of current flow through the coil determines which end is north and which end is south. (*b*) When iron is inserted into the coil, its spinning electrons align with the coil's magnetic field. The iron becomes a strong dipole magnet.

> **Check Your Understanding #3: A Delicate Balance**
> Why is it much easier to balance a broom on your hand when you can look at the broom than when you don't look at it?

Alternating Current Levitation

Feedback stabilization requires very sophisticated control systems and nearly perfect trains and tracks. There's another magnetic technique that can provide a restor-

Fig. 8.3.7 This magnetic globe is suspended in midair by the electromagnet above it. A control system uses a light beam to measure the height of the globe. It then adjusts the strength of the electromagnet so as to keep the globe floating at a constant height.

Fig. 8.3.8 This maglev train is supported by attractive forces between magnets located on the rail and magnets located in the train arms that reach beneath the rail. Feedback provides stability to this system.

❑ With only a primary education, **English chemist and physicist Michael Faraday (1791–1867)** apprenticed with a bookbinder at 14. At 21, he became a laboratory assistant to Humphry Davy, a renowned chemist. Faraday's experiments with electrochemistry and his knowledge of work by Oersted and Ampère led him to think that, if electricity can cause magnetism, then magnetism should be able to cause electricity. Through careful experimentation, he found just such an effect. Toward the end of his career, Faraday became a popular lecturer on science and made a particular effort to reach children.

ing force in all directions and can levitate a train without any control system or feedback. Instead of putting permanent magnets or electromagnets on both the train and track, some trains use *electromagnetic induction* to make conducting surfaces magnetic temporarily.

To understand induction, recall that there are two ways to produce magnetic fields: moving electric charge and electric fields that change with time. We can make a similar observation about electric fields; there are two ways to produce electric fields: electric charge and magnetic fields that change with time. We already knew that electric charge creates electric fields, but the changing magnetic field part is new. This effect, discovered in 1831 by Michael Faraday ❑, allows magnetism to create electricity. The sources of electric and magnetic fields are summarized in Table 8.3.1.

So what happens when a magnetic field changes inside an electric conductor? The changing magnetic field creates an electric field, which pushes on the conductor's mobile electric charges. These charges begin to move and form an electric current. This process, in which a changing magnetic field *induces* an electric current, is **electromagnetic induction.**

Since moving electric charge produces a magnetic field, this new electric current is itself magnetic. A changing magnetic field thus turns an electric conductor into a magnet! If you hold an electromagnet near an electric conductor and turn it on, the magnetic field in the conductor will change and the conductor will become magnetic. But how are the conductor's magnetic poles oriented?

The conductor becomes magnetic in such a way that it *repels* the electromagnet. For example, if the electromagnet turns on with its south pole nearest the conductor, then the conductor becomes magnetic with its south pole nearest the electromagnet (Fig. 8.3.9). Because the two magnetic fields point in opposite directions, they partially cancel one another. In effect, the induced current is opposing the increase in the electromagnet's magnetic field. In fact, currents generated by electromagnetic induction always produce magnetic fields that oppose the magnetic field change that generated them. This observation is called **Lenz's law,** after Estonian physicist Heinrich Emil Lenz (1804–1865) who discovered it.

Lenz's Law
Current induced by a changing magnetic field always produces a magnetic field that opposes the change.

Because of Lenz's law, a light electromagnet can hover above a sheet of metal. When you turn the electromagnet on, it repels the metal and rises above the surface. Remarkably, this form of magnetic levitation can be stable in all directions. For example, if the electromagnet is in a metal bowl, it will lift itself up and hover above the center of the bowl. If you bump it, it will experience restoring forces that return it to its equilibrium position.

Unfortunately, the metal's current and magnetic field gradually disappear as frictional effects slow the moving charged particles in the metal. To keep the electro-

Table 8.3.1 **Sources of Electric and Magnetic Fields**

SOURCES OF ELECTRIC FIELDS	SOURCES OF MAGNETIC FIELDS
Electric charge	Moving electric charge
Changing magnetic fields	Changing electric fields

magnet hovering above the metal, you must periodically reverse the electromagnet's poles. Each time you do, new currents flow in the metal's surface and these currents repel the electromagnet.

You can reverse the poles of the electromagnet by reversing the direction in which current flows through its wires. An electric current that reverses directions periodically is called an **alternating current,** while a current that flows continuously in one direction is called a **direct current.** Sending an alternating current through the electromagnet, so that its poles reverse periodically, keeps it levitated indefinitely.

Scientists and engineers have used *alternating current magnetic levitation* to suspend trains above U-shaped tracks. In some cases, they have put the electromagnets on the train, and in others they have put the electromagnets on the track. But while both of these schemes work well, currents in the metal surfaces and the electromagnets encounter frictional effects that convert their electric energies into thermal energy. Because that electric energy must be replaced continuously, these maglev trains need too much power to be cost effective.

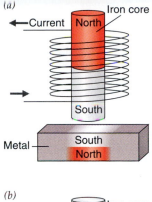

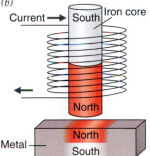

Fig. 8.3.9 (*a*) When an electromagnet turns on with its south pole down, it induces a current in the metal below it. That metal becomes magnetic with its south pole up and repels the electromagnet. (*b*) Reversing the electromagnet's current reverses all poles so the electromagnet is still repelled.

> ➤ Check Your Understanding #4: Field the Heat
> Induction furnaces are common in industry. A piece of metal is placed near or inside a strong electromagnet, and the electric current through that electromagnet is reversed rapidly. The piece of metal becomes very hot. Why?

Electrodynamic Levitation

A more practical induction-based levitation scheme is *electrodynamic levitation,* in which a train containing permanent magnets slides rapidly across a metal track. As the magnets move, the magnetic field in the track changes and an electric field appears inside the track. This electric field propels an electric current through the track, and the track becomes magnetic. The track's magnetic poles repel those of the train so the train rises on a cushion of magnetic force (Figs. 8.3.10 and 8.3.11, p. 288).

Electrodynamic levitation works best when the train moves very fast and its magnets pass quickly over each region of track. That's because the magnetic poles that appear in the track don't just support the train; they also push it backward as it approaches and forward as it leaves. The forward push is particularly important because without it the train would gradually slow down.

The strength of the forward push depends on the train's speed. If the train moves quickly, the track's magnetic poles will push the train forward when it leaves almost as strongly as they pushed it backward when it arrived. But if the train moves too slowly, the currents in the track will have plenty of time to decay and the track will lose much of its magnetism while the train is still nearby. Since the train will receive only a weak forward push as it leaves, it will slow down and some of its kinetic energy will become thermal energy in the track.

Even at high speeds, the train experiences a backward magnetic force. This force, called *magnetic drag,* is the sum of all the forward and backward magnetic forces on the train. Because the poles in the track ahead of the train's magnet are new and fresh, they're relatively strong. The poles in the track behind the train's magnet are weaker because they were created earlier and the currents that produce them have lost energy. With a stronger repulsion from the poles ahead of it than from those behind it, the train experiences a backward force.

Fig. 8.3.10 This maglev train is supported by repulsive forces between superconducting magnets on the train and induced magnetic poles in the metal track.

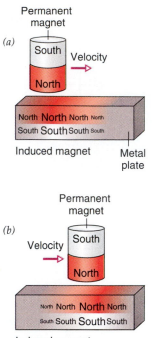

Fig. 8.3.11 When a permanent magnet moves rapidly across a metal surface, it induces currents in that surface. The metal becomes magnetic and repels the moving magnet. The metal's magnetized region moves with the magnet.

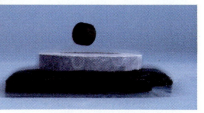

Fig. 8.3.12 A magnetic cylinder floats above the surface of a high-temperature superconductor. Currents flowing in the superconductor make it magnetic and cause it to repel the magnetic cylinder.

permanent magnet levitation

This magnetic drag force is most severe at about 30 km/h (19 mph), the minimum speed at which the train can levitate, and diminishes with increasing speed. The faster the train moves, the less time there is for the poles in the track to lose energy and the more the forward and backward magnetic forces on the train balance one another. Above about 300 km/h (190 mph), magnetic drag is insignificant when compared to air resistance.

Of course, electrodynamic levitation can't support the train when it's stationary, so the train needs retractable "landing gear" to support it while it's starting or stopping. The train leaves the station on wheels, picks up speed, and then "lifts off" onto the magnetic cushion. It retracts its wheels during high-speed movement and lowers them again before "landing" near the next station.

With suitable permanent magnets on board, an electrodynamically levitated train can hover more than 15 cm above an aluminum track. It easily clears small obstacles and provides an extremely smooth ride. Small imperfections in the railway can't be felt at all.

However, ordinary permanent magnets are heavy and expensive, so most electrodynamic levitation schemes use electromagnets instead. But the most energy-efficient electromagnets are ones that use special wires made of **superconductors**—materials that at low temperatures become perfect conductors of electricity. While currents in normal metals experience frictionlike effects and gradually slow down, currents in superconductors flow perfectly freely. With nothing to slow its motion, any current you start in a superconductor will continue to flow until you stop it (Fig. 8.3.12). Since currents are magnetic, a superconducting electromagnet remains magnetic as long as it carries current and behaves like a light, superstrong permanent magnet. A maglev train using superconducting magnets only needs to keep its magnets cold to keep them magnetic, so it can hover easily without requiring much electric power.

Check Your Understanding #5: Magnetic Brakes

If you move a strong permanent magnet rapidly across a sheet of copper or aluminum, a few millimeters above the surface, you will feel a significant force opposing the magnet's motion. What sort of force is this?

Epilogue for Chapter 8

This chapter dealt with three devices that depend on electric and magnetic forces for their operation. In *electronic air cleaners,* we introduced the concept of electric charge and discussed the attractive or repulsive forces that charged particles exert on one another. We learned about the energy stored in separated electric charge and how this energy can be used to pluck dust particles from the air.

In *xerographic copiers,* we saw that quantum physics and the Pauli exclusion principle prevent electrons from moving through a photoconductor in the absence of light. We also examined electric fields as another way to look at the forces that charged particles exert on one another: one charged particle creates an electric field, and the second charged particle accelerates under its influence.

In *magnetically levitated trains,* we looked at the forces between magnetic poles. We discussed the apparent absence of isolated magnetic poles from our universe and explained how currents of moving electric charges can create magnetic fields and dipoles.

Explanation: Moving Water Without Touching It

superconducting magnets,
maglev propulsion

The experiment uses sliding friction to give the comb an electric charge. Electrons rub off your hair and onto the comb, leaving your hair positively charged and the comb negatively charged. Because the comb is an electric insulator, its negative charge remains trapped on its surface for a long time.

As you hold the negatively charged comb close to the water stream, the comb attracts positive charges present in the water and repels negative charges. Because the water conducts electricity somewhat, negative charges flow up the water stream and into the faucet. The positive charges are left behind in the water stream and give that water a net positive charge. Thus the presence of the negatively charged comb makes the water stream positively charged and the water and comb attract one another. The water accelerates toward the comb and arcs sideways as it falls.

Chapter Summary

How Electronic Air Cleaners Work: An electrostatic precipitator removes dust grains from the air by transferring electric charge to them and attracting them to oppositely charged collecting plates. A corona discharge around a thin metal wire charges the dust. A power supply pumps charge from the wire to the collecting plates, leaving the wire with a large negative net charge. As the negative charge increases, so does its electrostatic potential energy. When the voltage on the wire falls below about −10,000 V, negative charges begin to leave the wire and attach themselves to passing molecules and dust grains. The negatively charged grains are then attracted toward the positively charged collecting plates. Although air resistance impedes their motion, the charged particles experience such strong electrostatic forces that they collide with the collecting plates and remain there until they're removed.

How Xerographic Copiers Work: The photoconductor in a xerographic copier allows light to control the distribution of electric charge. An optical image of the original document is projected onto a charged region of the photoconductor. Wherever light hits the photoconductor, charge escapes. The result is a charge image on the surface of the photoconductor. Tiny black toner particles, oppositely charged from the unilluminated portions of the photoconductor, are brought near the charge image. These toner particles stick to the charged portions of the photoconductor, forming a visible image of the document. These toner particles are then transferred to and fused into a sheet of paper to create a finished copy.

How Magnetically Levitated Trains Work: The most promising levitation scheme is electrodynamic levitation. Strong, superconducting magnets on the train create large magnetic fields beneath it. As the train moves along its metal track, these magnets induce currents in the track and the track becomes magnetic. The poles of the track are opposed to those of the train's magnets, and the train and track push against one another. This magnetic repulsion suspends the train above the track.

Important Laws and Equations

1. Coulomb's Law: The magnitudes of the electrostatic forces between two objects are equal to the Coulomb constant times the product of their two electric charges divided by the square of the distance separating them, or

$$\text{force} = \frac{\text{Coulomb constant} \cdot \text{charge}_1 \cdot \text{charge}_2}{(\text{distance between charges})^2}. \quad (8.1.1)$$

If the charges are like, then the forces are repulsive. If the charges are opposite, then the forces are attractive.

2. The Pauli Exclusion Principle: No two indistinguishable Fermi particles ever occupy the same quantum level.

3. Lenz's Law: Current induced by a changing magnetic field produces a magnetic field that opposes the change.

Check Your Understanding—Answers

Section 8.1 ELECTRONIC AIR CLEANERS

1. The ground.

Why: Since charge is a conserved physical quantity, the cloud can't simply manufacture it. Charge must migrate upward to the clouds from the ground to create static electricity. The precise mechanisms by which the clouds become charged are still not well understood.

2. The carpet's voltage increases relative to your voltage.

Why: As charge leaves you and enters the carpet, the carpet becomes positively charged and you become negatively charged. Since positive charge is attracted toward negative charge, a unit of positive charge on the carpet can release electrostatic potential energy by flowing to you. That means that the carpet's voltage is higher than your voltage. On a dry winter day, the carpet's voltage can exceed yours by as much as 50,000 V.

3. The needle emits charge into the air via a corona discharge.

Why: The needle acts as your personal lightning rod. When sliding friction gives you a net electric charge, some of that charge flows onto the needle. As soon as the repulsion between charges on the needle becomes strong enough, a corona discharge begins. This discharge limits your net electric charge and thus the size of any shock you may experience.

4. The sock has a net electric charge and it polarizes nearby objects. Once polarized, these objects attract the sock.

Why: Your electrically charged sock exerts electrostatic forces on all nearby charges. The charges in the wall or the lint shift in response to these forces and these objects become electrically polarized. The end of each object that is closest to the sock has a charge that attracts the sock, so the two cling to one another.

Section 8.2 XEROGRAPHIC COPIERS

1. The electron will go into the lowest energy empty level, the one just above the Fermi level.

Why: Since each electron goes into the lowest energy level that's available, this new electron will fill the level just above the Fermi level. Actually, if the metal ball had an odd number of electrons, then the Fermi level only contained one electron. In that case, the new electron will fill the other opening in the Fermi level.

2. The beam of light shines on the photoconductor, allowing charge to move through it and to operate the conveyer belt's motor. When food blocks the light beam, the photoconductor becomes an electric insulator and the belt stops moving.

Why: Photoconductors are commonly used in light sensors. When light hits the photoconductor, it behaves as an electric conductor and permits electricity to flow. This electricity can be used to operate machinery, trigger a burglar alarm, or turn lights on or off. In the present case, it runs the conveyer belt's motor.

3. It points backward, away from the patient.

Why: Since electrons are negatively charged, they accelerate in the direction opposite to the field. Since the accelerator must push the electrons forward, toward the patient, its field must point away from the patient.

4. The charge that was placed on the paper to attract the toner isn't always removed completely. Moreover, the toner itself is charged.

Why: The final transfer process, lifting the toner particles from the photoconductor to the paper, is done by charging the paper, and some of this charge remains on the paper when it leaves the copier. Copier transparencies are particularly clingy because plastic retains charge so well.

Section 8.3 MAGNETICALLY LEVITATED TRAINS

1. The car accelerates upward.

Why: By giving the car a net upward force, the track lifts the car over the bump. The time during which this lifting can be done depends on the car's speed. Doubling this speed halves the lifting time and actually involves four times the acceleration and four times the upward force from the track. That's because the car must move upward twice as fast as before, with only half the time to acquire the doubled upward speed. These rapidly increasing forces limit conventional trains to about 320 km/h (200 mph).

2. The top surfaces of both halves are still north poles, and the bottom surfaces are still south poles. The two tops repel, as do the two bottoms.

Why: This puzzling phenomenon, where a shattered permanent magnet opposes attempts to reassemble it, is an illustration of the potential energy contained in a permanent magnet. The magnet is a collection of many tinier magnets, all aligned with their north poles together and their south poles together. Like poles repel one another, so the tiny magnets are difficult to hold together. Given a chance, the magnet will push apart into fragments. Very strong permanent magnets release so much potential energy when they break that they practically explode when cracked.

3. A balanced broom is in an unstable equilibrium, and you use visual feedback to keep it balanced.

Why: You need to know where the broom's center of gravity is located in order to keep it balanced. On average, your hand has to be located beneath that center of gravity. Although you can feel the motion of the broom to some extent, it helps to be able to see it.

4. The changing magnetic field of the electromagnet induces currents in the piece of metal. As these currents decay, some of their energy is converted to thermal energy and the metal becomes very hot.

Why: In the presence of changing magnetic fields, a piece of metal has currents flowing through it. These currents can warm the metal just as they warm the filament of a light bulb.

5. Magnetic drag force.

Why: The moving magnet induces currents in the metal sheet and makes it magnetic. The poles of the sheet are opposite to those of the moving magnet, so there is repulsion between the sheet and magnet. However, there is also a slowing force on the magnet. The strongest poles of the sheet are located ahead of the moving magnet and oppose its approach. The sheet pushes backward on the magnet and tries to slow it down.

Check Your Figures—Answers

Section 8.1 ELECTRONIC AIR CLEANERS

1. 0.25 N.

Why: According to Coulomb's law, the force on each charge varies inversely with the square of the separation. By increasing that separation by a factor of 2, you reduce the electrostatic force by a factor of 4.

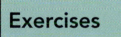

Exercises

1. When you take clothes out of a hot dryer, they often cling to one another. Two identical socks, however, usually repel. Explain their repulsion.

2. It may seem dangerous to be in a car during a thunderstorm, but it's actually relatively safe. Since the car is essentially a metal box, the inside of the car is electrically neutral. Why does any charge on the car move to its outside surface?

3. In industrial settings, neutral metal objects are often coated by spraying them with electrically charged paint or powder particles. How does placing charge on the particles help them to stick to an object's surface?

4. The paint or powder particles discussed in Exercise 3 are all given the same electric charge. Why does this type of charging ensure that the coating will be highly uniform?

5. If two objects repel one another, you know they have like charges on them. But how would you determine whether they were both positive or negative?

6. A bowling ball contains an enormous number of electrically charged particles. Why don't two bowling balls normally exert electrostatic forces on each other?

7. You're holding two oppositely charged balloons in your hands. Compare their voltages.

8. You begin to separate the two balloons in Exercise 7. You do work on the balloons, so your energy decreases. Where does your energy go?

9. When you separate the balloons in Exercise 8, do their voltages change? If so, how do those voltages change?

10. A car battery is labeled as providing 12 V. Compare the electrostatic potential energy of positive charge on the battery's negative terminal with that on its positive terminal.

11. If a very small piece of material contains only 10,000 electrons and those electrons have as little energy as possible, how many different levels do they occupy in that material?

12. If electrons had four different internal states that could be distinguished from one another, how many electrons could occupy the same level without violating the Pauli exclusion principle?

13. If electrons were not Fermi particles, any number of them could occupy a particular level. How would these electrons tend to arrange themselves among the levels in an object?

14. Thermal energy can shift some of the electrons in a hot semiconductor from valence levels to conduction levels. What effect do these shifts have on the semiconductor's ability to conduct electricity?

15. Based on the result of Exercise 14, would heating the photoconductor in a xerographic copier improve or diminish its ability to produce sharp, high-contrast images?

16. Many night-lights and security lamps use a strip of photoconductor to tell when it's dark enough for them to turn on. Should the light turn on when the photoconductor is behaving as a conductor or an insulator?

17. Some photoconductors are insensitive to red light. Xerographic copiers based on such photoconductors turn red parts of an original document into black when preparing a copy because they can't detect red light. What prevents them from responding to red light?

18. Suppose that you had an electrically charged stick. If you divided the stick in half, each half would have half the original charge. If you split each of these halves, each piece would have a quarter of the original charge. Can you keep on dividing the charge in this manner forever? If not, why not?

19. American physicist Robert Millikan (1868–1953) discovered the elementary unit of electric charge by balancing the downward weight of a tiny, electrically charged oil drop against the upward force from an electric field. Without any electric field, the drop would fall under its own weight. However, an electric field could exert an upward force on the charged drop to stop its descent. If the oil drop were negatively charged, in which direction would the electric field have to point in order to stop the oil drop from falling?

20. Millikan (see Exercise 19) observed that different oil drops would stop falling in different electric fields, an observation he attributed to their different electric charges. If the electric field needed to support one oil drop is twice as large as the field needed to support a second, similar oil drop, how do their electric charges compare with one another?

21. When you hold the north poles of two bar magnets together, the two magnets repel. But each magnet also has a south pole that is attracted toward the north pole of the other magnet. Why don't these attractive and repulsive forces balance one another so that each magnet experiences zero net force?

22. If the forces between electric charges didn't diminish with distance, an electrically charged balloon wouldn't cling to an electrically neutral wall. Why not?

23. The electric field around an electrically charged hairbrush diminishes with distance from that hairbrush. Use Coulomb's law to explain this decrease in the magnitude of the field.

24. When a positively charged cloud passes overhead during a thunderstorm, which way does the electric field point?

25. The cloud in Exercise 24 attracts a large negative charge to the top of a tree in an open meadow. Why is the magnitude of the electric field larger on top of this tree than elsewhere in the meadow?

26. Corona discharges can occur wherever there is a very strong electric field. Why is there a strong electric field around a sharp point on an electrically charged metal object?

27. As you move the north pole of a permanent magnet toward an aluminum baking tray, the tray temporarily becomes magnetic. A pole appears at its surface, facing the approaching north pole. Is the tray's pole north or south?

28. The tray in Exercise 27 also develops a pole at its far surface, away from the approaching north pole. Is that second pole of the tray north or south?

29. Suppose you have a coil of wire that's connected to a source of alternating current. Current flows through the coil first one way and then the other, over and over again. Because the current reverses periodically, north and south poles appear alternately at one end of this coil. If you approach that end with a piece of aluminum, it will be repelled. Why?

30. Some decorative light bulbs have a loop-shaped filament that jitters back and forth near a small permanent magnet. When electric charge moves through the filament, it's attracted to or repelled by the magnet. The filament wire itself isn't magnetic, so why does the permanent magnet push or pull on the filament when the light is turned on?

31. The needle of an automobile speedometer is connected to a spring and a small copper disk. The spring pulls the needle toward zero, but a torque on the disk can pull the needle toward higher speeds. That torque is exerted on the disk by a spinning magnet near the disk's surface. The magnet spins with the car wheels so that the faster the car goes, the faster the magnet spins and the higher the speedometer reads. Why does the moving magnet exert a torque on the copper disk?

32. One of the ways in which a coin-operated vending machine checks to make sure that the coins you feed it are genuine is to roll them past a strong magnet. Why do coins made of good electric conductors such as copper slow down as they pass the magnet?

33. When you move a strong permanent magnet rapidly to the right just above the surface of a stationary aluminum table, what forces will that magnet experience?

34. If you move a permanent magnet across a piece of copper, you will induce electric currents in that copper. If you try the same with a piece of glass, no currents will flow. Why don't currents flow in the glass as the magnet moves by?

Problems

1. You remove two socks from a hot dryer and find that they repel with forces of 0.001 N when they're 1 cm apart. If they have equal charges, how much charge does each sock have?

2. If you separate the socks in Problem 1 until they're 5 cm apart, what force will each sock exert on the other?

3. If you were to separate all of the electrons and protons in 1 g (0.001 kg) of matter, you'd have about 96,000 C of positive charge and the same amount of negative charge. If you placed these charges 1 m apart, how strong would the attractive forces between them be?

4. If you place 1 C of positive charge on the earth and 1 C of negative charge 384,500 km away on the moon, how much force would the positive charge on the earth experience?

5. How close would you have to bring 1 C of positive charge and 1 C of negative charge for them to exert forces of 1 N on one another?

Cases

1. Lightning occurs when large amounts of electric charge flow between the clouds and the ground.
   a. Lightning is most likely to strike when a cloud and the ground beneath it have accumulated large opposite charges. Why is lightning unlikely when they have large like charges?
   b. The atmosphere does a great deal of work in creating the separated charge that produces lightning. Show that it takes work to move positive charge from the negatively charged ground to the positively charged cloud overhead.
   c. When enough opposite charge has accumulated on the ground and in the cloud above it, lightning will strike. One indication that this dangerous charge accumulation has occurred is that your hair begins to stand up. Explain this effect.
   d. A sharp lightning rod reduces any local build-up of electric charge and prevents nearby lightning strikes. How does the lightning rod get rid of local electric charge and allow it to flow gradually to the clouds overhead?

2. A Van de Graaff generator is an electrostatic device that uses a moving nonconductive belt to carry electric charge into a hollow metal sphere. This sphere is insulated from the ground and can accumulate charge until enormous voltages are reached. Small Van de Graaff generators are exciting novelties while large ones are used in research and industry.
   a. A typical Van de Graaff generator uses a rubber belt to carry negatively charged electrons from its base to the sphere. As the belt passes through the sphere, a metal brush touches it. Electrons leave the belt and flow through a wire to the surrounding sphere. Why do they flow onto the outside of the sphere rather than staying near the belt?

b. As negative charge builds up on the sphere, the belt motor begins to strain. What makes it progressively more difficult for the motor to move the belt?
   c. The amount of charge that can build up on the sphere depends on the sphere's surface. Why is it important that there be no sharp points on the sphere?
   d. As you move your hand close to the negatively charged sphere, your hand becomes positively charged and a spark may leap from the sphere to your hand. Why does your hand acquire a positive charge, and from where does it come?
   e. If you insulate yourself from the ground and put your hand on the sphere, negative charge will accumulate on you as well as on the sphere. Explain why your hair stands up.

3. As freshly laundered clothes tumble about in a hot dryer, they rub against one another. Sliding friction transfers electric charges from one piece of clothing to the other so that some items become positively charged and others negatively charged. These charge accumulations produce static cling. Fabric softener is a soaplike chemical that binds to wet fabric fibers and lubricates them. This lubrication is what softens the clothes. But fabric softener also attracts moisture, which in turn reduces static cling.
   a. As you unload the dryer you find several socks clinging to a shirt. How do the charges on the socks and shirt compare to one another?
   b. As you pull the socks off the shirt, you do work on the socks. Into what form is your energy being transformed?
   c. Suppose that your socks are covered with positive charge. As you pull them away from the shirt, what happens to the voltage of the charges on the socks?
   d. The farther each sock is from the shirt, the less that sock and shirt attract one another. Explain.

e. After removing them from the shirt, you find that the socks repel one another. Explain.

f. The next time you do your clothes you decide to add fabric softener. Your clothes emerge from the dryer with a thin layer of moisture on them, and this moisture permits charge to move freely about the clothes. Why does this mobility prevent the build-up of separated charge?

4. The spark lighters used in gas stoves and grills are based on small piezoelectric crystals. When you compress a piezoelectric crystal, it produces a charge separation. One side becomes positively charged while the other becomes negatively charged. In a typical spark lighter, a spring-loaded mass strikes the crystal and produces a huge charge separation. Wires attached to the crystal allow these charges to approach one another across an air gap and create a spark.

a. Separating charge takes energy. Show that the spring-loaded mass does work on the crystal while compressing it.

b. Why does the amount of electric charge on each side of the crystal affect the likelihood that a spark will jump from one wire to the other?

c. To encourage a spark, the wire ends that make up the air gap are usually thin and pointed. Why do sharp points make it easier for charge to begin moving through the air?

d. If the distance separating the two wires becomes too great, the spark lighter will stop working. Why does the distance separating the wires matter at all?

5. A nerve cell is an electrostatic device that operates very differently from a wire. It's a fluid-filled tube that's surrounded by another fluid. Both fluids contain ions (electrically charged atoms) of sodium, potassium, and chlorine. Each sodium or potassium ion has one elementary unit of positive charge while each chlorine ion has one elementary unit of negative charge. The wall of the nerve cell separates the two fluids and normally prevents these ions from passing through it.

a. When the nerve cell is in its resting state, the fluid outside it has slightly more sodium plus potassium ions than chlorine ions. The fluid inside it has slightly more chlorine ions than sodium plus potassium ions. What is the net electric charge of each of these two fluids?

b. How do you know that the resting nerve cell has electrostatic potential energy?

c. To produce its electrostatic potential energy, the nerve cell pumps sodium ions out of the cell. Show that the nerve cell must do work on the sodium ions during this transfer.

d. Which part of the nerve cell has a positive voltage? Which part has a negative voltage?

e. When the nerve cell "fires," it abruptly allows sodium ions to pass through its walls. Which way do they move and what happens to the voltages of the various parts of the cell?

6. Your local market has an electric eye that rings a bell as you walk in the door and block a light beam. The light beam is emitted at one side of the door and normally strikes a small photoconductor on the other side of the door.

a. When the light beam is turned off or blocked, no light strikes the photoconductor. A battery pumps negative charges onto one side of the photoconductor and positive charges onto the other. These opposite charges exert attractive forces on one another so why don't they move together?

b. Because the photoconductor is in the dark, the battery soon stops sending charges to it. What makes it stop?

c. When the light beam reaches the photoconductor, charges can move through it. What has happened inside the photoconductor that allows it to conduct electricity?

d. When the photoconductor is exposed to light, the battery can continue to send charges to it. How has light made it possible for the battery to keep sending charges?

e. You bend down to play with the electric eye. You block the light beam with your back and shine your own flashlight onto the photoconductor. The bell turns off, as though you were out of the way. But when you shine red light from your bicycle taillight onto the photoconductor, it doesn't respond. Why doesn't red light (which has relatively low-energy photons) affect the photoconductor?

7. A magnetic resonance imaging (MRI) machine uses an enormous and extremely strong magnet to study a patient's body. The magnet, which has its north pole at the patient's head and its south pole at the patient's feet, is actually a coil of superconducting wire through which electric charges flow.

a. This fancy electric system seems unnecessary. Why can't the technicians simply put a large number of north magnetic poles near the patient's head and an equal number of south magnetic poles near the patient's feet?

b. The needle of your magnetic compass has its north magnetic pole painted red and its south pole painted white. If you stand a few meters from the MRI machine, at the end where the patient's head is, why does the white end of the compass turn toward the patient's head?

c. The compass is a magnetic dipole, with no net magnetic pole. So why do you feel it pulled toward the patient's head more and more strongly as you get closer to the magnet?

d. Aluminum isn't normally magnetic, but as you carry a large aluminum tray toward the magnet, you find that the magnet repels the aluminum. Explain.

e. You eventually manage to get the aluminum tray up to the magnet. As long as the tray doesn't move, it experiences no magnetic forces. But when you drop it, it falls past the magnet remarkably slowly. What slows down its fall?

Electrodynamics

Electricity and magnetism are related to one another through change and motion. For example, moving electric charges give rise to magnetism, and changing magnetic fields give rise to electricity. In this chapter, we'll examine several objects that make use of the relationships between electricity and magnetism to perform useful tasks. Since the word "dynamics" covers change and motion, these relationships are part of a field known as "electrodynamics." Magnetism was left out of the title to keep it short and to recognize the fact that most magnetism is produced by electricity.

EXPERIMENT: A Nail and Wire Electromagnet

To explore the relationships between electricity and magnetism, try building a simple electromagnet. For this project, you'll need a large steel nail or bolt, a meter or so of insulated wire, a fresh 1.5-volt AA battery, and some small steel objects such as paper clips. The wire's metal conductor should be at least 0.65 mm in diameter (22 gauge or larger) to carry the current you'll send through it without becoming too hot.

Wind the wire tightly around the bolt to form a coil. You should complete more than 50 turns of wire, all in the same direction. The exact number of turns isn't important, and you can make several layers. Be sure that the two ends of the wire are still accessible and remove the insulation from each end so that you can connect them to the battery.

Now test your electromagnet. Connect one uninsulated end of the coil to each terminal of the battery. You can either hold the wires on the terminals with your fin-

> **WARNING**
> The electromagnet that you'll construct in this experiment **will become hot during use.** Be prepared to drop the electromagnet if it becomes uncomfortably hot. **Don't work near flammable materials.**

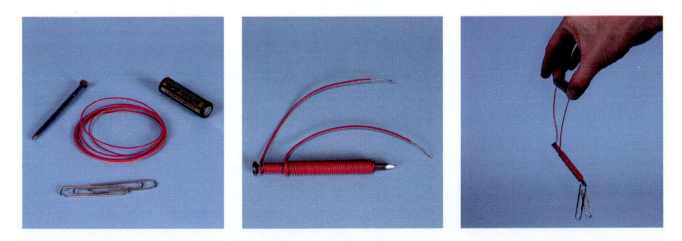

gers or use tape. A 1.5-volt battery can't give you a shock unless your skin is broken, but you should be prepared for the wire to get hot as current flows through it. If it gets too hot to hold, let go and make sure that the wire detaches from the battery so that it doesn't start a fire. Don't use a battery larger than AA or the wire may get dangerously hot.

While current is flowing through the wire, the nail will act as a strong magnet, an electromagnet. Try picking up steel objects with this electromagnet. As you touch each object with the nail, it should stick to the nail's surface. Your electromagnet will temporarily magnetize the steel object and attract it. Try to *predict* what will happen when you touch this magnetized steel object to a second object. *Observe* what happens and see if you can *verify* your prediction. Can you think of a way to *measure* how strong the magnet is? What will happen when you stop the flow of electric current through the coil of the electromagnet? Why does the coil get hot while current is flowing through it?

Chapter Itinerary

This chapter will examine (1) *flashlights,* (2) *electric power distribution,* and (3) *tape recorders.* In *flashlights,* we'll look at how a current of electric charges conveys energy from several batteries to a light bulb. In *electric power distribution,* we'll see how electricity and magnetism are used to transport energy from power plants to cities far away. And in *tape recorders,* we'll explore the ways in which the magnetic structure of certain materials makes it possible to store sound information on strips of specially coated plastic. While there are many other electromagnetic objects that we encounter daily, from doorbell buzzers to light switches to electric stoves, these three topics are representative of most of the basic electromagnetic phenomena.

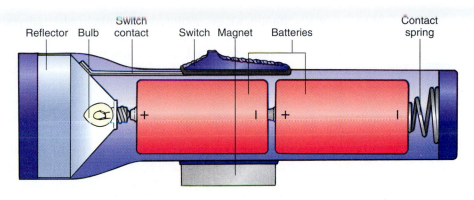

Reflector Bulb Switch contact Switch Magnet Batteries Contact spring

Flashlights

There isn't much to a typical flashlight; you flip a switch and out comes a beam of light. Its few parts are all visible when you open it to replace its batteries. But a flashlight isn't a mechanical device, it's an electric one. It contains an electric circuit, and most of its components are involved in the flow of electricity. Understanding how flashlights work means understanding how electric circuits work and how electricity carries power from batteries to a bulb. As we'll see, flashlights aren't as simple as they appear.

Questions to Think About: *Why are some flashlights brighter than others? Why is it important that all of the batteries point in the same direction? What is the difference between old batteries and new? What makes a flashlight suddenly become dim or bright when you shake it?*

Experiments to Do: *Find a flashlight that uses two or more removable batteries. Turn it on. What did the switch do to make the flashlight produce light? With the flashlight turned on, slowly open the battery compartment. The lamp will probably become dark. You should be able to turn the flashlight on and off by opening and closing the battery compartment. Why does this method work?*

Replace the flashlight's batteries with older or newer ones and compare its brightness. Turn one or more of the batteries around backward and see how that change affects the flashlight. What happens when you put a piece of paper or tape between two of the batteries? What happens when you carefully clean the metal surfaces of each battery with a pencil eraser before putting them in the flashlight?

Electricity and the Flashlight's Electric Circuit

A basic flashlight has only three components—a battery, a bulb, and a switch—connected together by metal strips. The battery transfers energy to the bulb, which then turns it into light. The switch controls the energy transfer so that it only occurs when the switch is closed (the flashlight is on). But what process carries the energy from battery to bulb, and how does the switch start or stop that energy transfer? To answer these questions, we must first understand electricity and electric circuits, so that's where we'll begin.

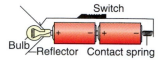

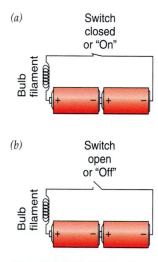

Fig. 9.1.1 A flashlight contains one or more batteries, a light bulb, a switch, and several metal strips to connect them all together. When the switch is turned on (as shown), the components in the flashlight form a continuous loop of conducting materials. Electrons flow around this loop counterclockwise.

(a)   Switch closed or "On"

Bulb filament

(b)   Switch open or "Off"

Bulb filament

Fig. 9.1.2 (*a*) When the flashlight's switch is on, it closes the circuit so that current can flow continuously from the batteries, through the bulb's filament, and back through the batteries. It follows this circuit over and over again. (*b*) When the flashlight's switch is off, it opens the circuit so that current stops flowing.

In the previous chapter, we encountered one form of electricity: static electricity. That term refers to stationary distributions of separated charge, such as the separated charge on a balloon and your hair after you rub them against one another. But there's a second form of electricity that involves moving charge: **dynamic electricity.** It's this form of electricity that powers flashlights, stereos, and microwave ovens. Since most of our discussions in this and subsequent chapters will focus only on the dynamic form of electricity, we'll leave out the word "dynamic" and just call it "electricity."

When you turn a flashlight on, electricity transfers energy from the batteries to the bulb. An electric current flows through both parts, carrying the energy with it. The charged particles that make up this current are electrons, and these electrons follow a circular path that takes them through the batteries and bulb over and over again (Fig. 9.1.1). As long as the flashlight is on, these electrons continue to travel around this circular path, receiving energy from the batteries and delivering it to the bulb.

The circular path taken by electrons in a flashlight is called an **electric circuit.** The flashlight needs this circuit so that its electrons don't accumulate somewhere along the path. Although a few electrons could flow from the batteries to the bulb without having to return, they would begin to repel any additional electrons that tried to join them. By giving the electrons a circular path, the flashlight allows them to carry energy from the batteries to the bulb and then return to the batteries to pick up more energy. This circular path or circuit concept is present in all electric devices and explains the need for at least two wires in the power cord of every home appliance: one wire carries electrons to the appliance to deliver energy, and the other wire carries those electrons back to the power company to receive some more.

But what role does the switch play in all of this? It works by making or breaking the flashlight's circuit (Fig. 9.1.2). The switch forms part of a conducting path that connects the batteries to the bulb. There are two of these paths, with electrons heading to the bulb through one and returning through the other. One of the paths is solid metal and always complete. The second can be interrupted by the switch. When the flashlight is on, the switch completes that conducting path so that electrons can flow continuously around the **closed circuit** (Fig. 9.1.2*a*). A closed circuit appears on the left side of Fig. 9.1.3.

However, when you turn the flashlight off, the switch breaks that conducting path so that electrons can't flow around the **open circuit** (Fig. 9.1.2*b*). Although one path still connects the batteries and bulb, electrons can't travel all the way around the loop. With the two halves of the switch separated by a gap, charge accumulates on the metal strips and current stops flowing through the flashlight. Since no more energy reaches the bulb, it soon becomes dark. An open circuit appears on the right side of Fig. 9.1.3.

There's one other type of circuit worth mentioning. A **short circuit** forms when the two paths between the batteries and the bulb accidentally come in contact with one another (Fig. 9.1.4). When this occurs, the electrons have a shortened path through which they can flow. Because the bulb is supposed to extract energy from the electrons, it's designed to impede their flow and to convert their electrostatic potential energy into thermal energy and light. We call this opposition to the flow of electricity **electric resistance.** Since the shortened path offers less resistance to their movement, most of the electrons flow through it and bypass the bulb. The bulb grows dim or goes out altogether.

Since the bulb is the only part of the flashlight that's meant to get hot, a short circuit leaves the electrons without a safe place to get rid of their energy. The electrons deposit their energy in the batteries and the metal paths, which become hot.

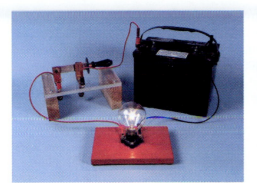 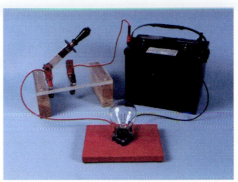

Fig. 9.1.3 When the switch is on (left), current flows around the closed circuit and carries power from the battery to the light bulb. But when the switch is off (right), no current flows through the open circuit.

Since short circuits can start fires, flashlights and other electric equipment are designed to avoid them.

Check Your Understanding #1: Cutting the Power

If you remove just one of the wires from an automobile's battery, the vehicle will not start at all. Why doesn't the other wire supply any energy?

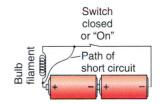

Fig. 9.1.4 When an unwanted conducting path allows current to bypass the flashlight's filament, it forms a short circuit. Because it has no proper place for electrons to deposit their energy, the short circuit becomes hot.

The Electric Current Through the Flashlight

The charged particles that flow through the flashlight's circuit are negatively charged electrons. They pass by in huge numbers, with each electron completing the journey every few minutes. Each electron carries with it a certain amount of energy, so if you want to determine the power reaching the light bulb, you must know how many electrons pass through the bulb each second and how much energy each one delivers. Recall from Section 2.2 that *power* is the amount of energy flowing through a system each second and is measured in watts (abbreviated W). Since the bulb needs a certain amount of energy each second to glow brightly, it makes sense to study the power flowing through the flashlight's circuit.

Rather than counting the electrons individually as they pass, we measure the circuit's **current**—that is, the charge flowing past a particular point in the circuit per unit of time. The SI unit of current is the **ampere** (abbreviated A) and corresponds to 1 C (1 coulomb) of charge passing by the designated point each second. One coulomb is an enormous amount of charge so that when a current of 1 A flows through a wire, there are roughly 6.25×10^{18} or 6,250,000,000,000,000,000 electrons passing through it each second.

Even though each electron only carries a tiny amount of energy, there are so many electrons flowing in a 1-A current that they transfer a substantial amount of power. Since each coulomb of 1-V electrons carries 1 J of energy, a 1-A current of these electrons carries 1 J per second or a power of 1 W. By itself, each 1-V electron carries only 1.602×10^{-19} J or 0.0000000000000000001602 J of energy, but these tiny amounts of energy add up when lots of electrons are involved.

However, current has a direction. When you turn on the flashlight in Fig. 9.1.1, electrons begin to flow around the circuit counterclockwise. They flow from the battery chain's negative terminal, through the switch, through the bulb's filament, and into the battery chain's positive terminal. Given this counterclockwise flow of electrons, it would seem sensible to say that current flows counterclockwise around the circuit.

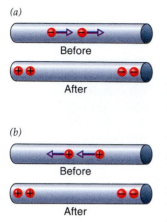

Fig. 9.1.5 A current of negatively charged particles flowing to the right through a piece of wire (*a*) can't easily be distinguished from a current of positively charged particles flowing to the left (*b*). The end result of both processes is an accumulation of positive charge on the left end of the wire and negative charge on its right end.

Unfortunately, electrons are negatively charged. Rather than have current carry a negative charge with it, scientists and engineers decided long ago to have current point in the direction of positive charge flow. Because of this convention, current officially flows clockwise around the circuit in Fig. 9.1.1. In making this definition of the direction of current, the scientists and engineers chose to pretend that positive charges are carrying the current. These positive charges don't really move; they're simply a useful fiction.

This fiction actually works extremely well, as illustrated by a simple example. When negatively charged electrons flow to the right through a neutral piece of wire, its right end becomes negatively charged and its left end becomes positively charged (Fig. 9.1.5*a*). But exactly the same thing would happen if a current of fictitious positively charged particles were to flow to the left through that same piece of wire (Fig. 9.1.5*b*). Without very special equipment, you can't tell whether negative charges are flowing to the right or positive charges are flowing to the left because the end results are essentially indistinguishable.

We, too, will adopt this fiction and pretend that current is the flow of positively charged particles. In this and subsequent chapters, we'll stop thinking about electrons and imagine that electricity is carried by positive charges moving in the direction of the current. There are only a few special cases in which the electrons themselves are important, and we'll consider those situations separately when they arise.

Check Your Understanding #2: Don't Touch That Pipe

When you walk across wool carpeting in rubber-soled shoes, you accumulate a large number of negative charges on your body, particularly if the air is dry. These negative charges have considerable electrostatic potential energy. When you bring your hand near a large piece of metal, the negative charges will leap through the air as a spark in order to reach the metal. Which way is current flowing in this spark?

Batteries, Bulbs, and Metal Strips

To explain how a flashlight works, we'll have to look closely at how power is transferred through its circuit. But first we must understand more about the flashlight's individual components: its batteries, bulb, and metal strips. Let's start with the batteries.

A battery is a device that pumps positive charges from its negative terminal to its positive terminal, against the direction in which current naturally flows. Since the battery must pull each charge away from the negative terminal and push it onto the positive terminal, it takes work to transfer each charge. The battery uses its chemical potential energy to do this work and thus converts it into electrostatic potential energy. The battery's positive terminal acquires a positive net charge, and its negative terminal acquires a negative net charge.

As this separated charge increases, the electrostatic force acting on a transferred charge becomes stronger and the work required to transfer it becomes larger. Eventually, the battery's chemical forces are no longer able to transfer additional charges and the pumping stops. The charges are then in equilibrium.

A typical alkaline battery pumps charges until the voltage of its positive terminal is 1.5 V above the voltage of its negative terminal. This voltage difference measures the amount of electrostatic potential energy that each unit of charge would release if it were permitted to flow from the positive terminal to the negative terminal. Since the voltage difference is 1.5 V (equivalent to 1.5 joules-per-coulomb), each coulomb of charge would release 1.5 J of energy in moving to the negative terminal.

An alkaline battery gets this energy through an electrochemical reaction in which powdered zinc attached to its negative terminal reacts with manganese dioxide paste attached to its positive terminal. This reaction resembles controlled combustion. In effect, the battery "burns" zinc to obtain the energy needed to pump charges from one of its terminals to the other.

As the battery consumes its store of chemical potential energy, its ability to pump charges diminishes. When its chemicals are nearly exhausted, the random effects of thermal energy in the battery begin to decrease its overall voltage. It also loses its ability to pump large currents and wastes much of its energy as thermal energy. An aging battery can pump less current than a fresh one and gives that current less voltage. Ultimately, less power reaches the flashlight's bulb and it grows dim.

When two batteries are connected together in a chain, so that the positive terminal of one battery touches the negative terminal of the other, the two batteries work together to pump charges from the chain's negative terminal to its positive terminal (Fig. 9.1.6). Each battery pumps charges until its positive terminal is 1.5 V above its negative terminal, so the chain's positive terminal is 3.0 V above its negative terminal. Since only *differences* in voltage matter in a flashlight's circuit, it will be convenient for us to define one part of that circuit as being at 0 V. If we define the battery chain's negative terminal as being at 0 V (Fig. 9.1.6), then its positive terminal must be at 3.0 V.

The voltages of the two batteries add because each battery transfers a certain amount of energy to each charge it pumps. Some flashlights use more than two batteries in the chain. The more batteries in the chain, the more voltage the current receives as it flows through the chain and the greater the voltage difference between the chain's positive terminal and its negative terminal. A flashlight that uses six batteries in its chain has a positive terminal that is 9 V above its negative terminal. A 9-V alkaline battery actually contains a chain of six miniature 1.5-V batteries, arranged so that their voltages add up to 9 V (Fig. 9.1.7).

If you reverse one of the batteries in the chain, the reversed battery will extract voltage from the current passing through it (Fig. 9.1.8). The chain will still pump charge from its negative terminal to its positive terminal, but its overall voltage will be reduced. If the chain has three batteries, two will add voltage to the charge while one will extract voltage. The overall voltage of the chain will be only 1.5 V.

Since current is flowing backward through the reversed battery, that battery extracts energy from the passing charge and turns at least some of it into chemical potential energy. The reversed battery recharges. Battery chargers use a similar technique, pushing current backward through a battery, from its positive terminal to its negative terminal, in order to put chemical potential energy back into a rechargeable battery. However, normal alkaline batteries are "nonrechargeable," meaning that they turn most of the recharging current's energy into thermal energy instead of chemical potential energy. Nonrechargeable batteries may overheat and explode during recharging.

A bulb's behavior is roughly opposite that of a battery. Its filament is a coil of tungsten wire that allows current to flow naturally, from its positive end to its negative end. In traveling through the filament, each charge releases its electrostatic potential energy, and that energy becomes thermal energy and light.

If the filament were a perfect conductor of electricity, each charge passing through the fine wire coil would speed up as its electrostatic potential energy be-

Fig. 9.1.6 When two 1.5-V batteries are connected in a chain, their voltages add so that the chain's positive terminal has a voltage that is 3.0 V higher than the chain's negative terminal. If the chain's negative terminal is chosen to be at 0 V, then the chain's positive terminal is at 3.0 V.

Fig. 9.1.7 A 9-V battery actually contains six small 1.5-V cells, connected in a chain. Positive charges that enter the chain at the battery's negative terminal pass through all six cells before arriving at the battery's positive terminal.

Fig. 9.1.8 When one battery in a chain of three is reversed, the reversed battery's voltage is subtracted from the sum of the others. The chain's positive terminal has a voltage only 1.5 V higher than its negative terminal. The reversed battery recharges.

came kinetic energy. But the filament has a large electric resistance and significantly impedes the flow of electric current. The charges collide with the filament's tungsten atoms and heat it white hot. Their ordered electrostatic energy becomes thermal energy in the tungsten atoms, and the filament emits light.

What about the metal strips connecting the flashlight's batteries and bulb? Because these thick pieces of metal have small electric resistances, they carry current easily and consume only a tiny fraction of the power flowing from the battery to the bulb. In some flashlights, the thick metal case itself helps to carry current. In general, the less electric resistance in the wires carrying current to and from the bulb, the more power reaches the bulb. That's why it's so important to use thick metal strips in the connections.

But a poor connection can spoil this efficient transfer of power. If there is dirt or grease on a battery terminal or worn materials in the switch, the current will have to pass through a large electric resistance. When it does, it will waste a substantial fraction of its power. The bulb will receive less power than it should and will emit relatively little light. Improving the connection, either by shaking the flashlight or by cleaning the metal surfaces, will increase the current flow through the circuit, reduce the wasted power, and brighten the flashlight.

Check Your Understanding #3: Car Batteries

A lead–acid automobile battery provides 12 V between its negative terminal and its positive terminal. It actually contains six individual batteries, connected in a chain. How much voltage does each individual battery produce?

Voltage, Current, and Power in Flashlights

To complete this discussion of flashlights, let's see how the electric current conveys power from the batteries to the bulb. We'll consider a flashlight with two batteries, so that the voltage of the battery chain's positive terminal is 3.0 V above that of its negative terminal. Let's suppose that when this flashlight is turned on, a current of 1.0 A flows through its circuit.

The bulb consumes electric power because the current passing through it experiences a drop in voltage. This **voltage drop,** the voltage before the bulb minus the voltage after it, measures the energy each unit of charge loses while passing through the bulb. Multiplying the voltage drop by the current passing through the bulb gives you the power consumed by the bulb. This observation can be written as a word equation:

$$\text{power consumed} = \text{voltage drop} \cdot \text{current}, \qquad (9.1.1)$$

in symbols:

$$P = V \cdot I,$$

and in everyday language:

When a big current loses a lot of voltage, something will get hot.

Since the voltage drop across the flashlight bulb is 3.0 V and the current through it is 1.0 A, it's consuming 3.0 W of power.

The battery chain produces electric power because the current passing through it experiences a rise in voltage. This **voltage rise,** the voltage after the batteries minus the voltage before them, measures the energy each unit of charge gains while passing through the batteries. Multiplying the voltage gain by the current passing through the batteries gives you the power provided by the batteries. This observation can be written as a word equation:

$$\text{power provided} = \text{voltage rise} \cdot \text{current}, \qquad \textbf{(9.1.2)}$$

in symbols:

$$P = V \cdot I,$$

and in everyday language:

> *It takes a great deal of power to raise the voltage of a large electric current.*

Since the voltage rise across the battery chain is 3.0 V and the current through it is 1.0 A, it's providing 3.0 W of power.

For this particular flashlight, the bulb has been designed to become white hot when it receives 3.0 W of power. It's also designed to allow 1.0 A of current to flow through it when the voltage drop across it is 3.0 V. The bulb's electric resistance and the voltage of the battery chain control the amount of current flowing in the flashlight's circuit and the amount of power transferred from the batteries to the bulb. If you use the wrong bulb in a flashlight, one designed for a different number of batteries, it will receive too much or too little power. If it receives too much, it will burn out quickly. If it receives too little, it will glow dimly. Powerful flashlights that use many batteries have bulbs that are designed to consume large amounts of power and emit intense beams of light.

There are two more important components of the circuit: the strips connecting the battery chain to the bulb and the switch. In effect, these connecting parts are also devices that require power to operate. The power they consume keeps the current in them from coming to a stop, and they convert this power into thermal energy. But because the strips and switch are short and thick, they have little electric resistance and consume little power. Most of the 3.0 W provided to the circuit by the battery chain reaches the bulb. Only when the connections are poor do they extract significant amounts of power from the circuit and reduce the current, voltage drop, and power delivered to the bulb.

➤ Check Your Understanding #4: Current Trends in Music

A large battery powers your portable radio. Current enters the radio through one wire and leaves through another. Which wire has a higher voltage?

➤ Check Your Figures #1: When You Turn the Ignition Key

When you start a car, a 200-A current flows through its starter motor. If that current experiences a voltage drop of 12 V, how much power is being consumed?

➤ Check Your Figures #2: When You Turn the Ignition Key Again

When you restart the car, a smaller current of 150 A flows through the car's starter motor. The battery is supplying power to this current, providing it with a voltage rise of 12 V. How much power is the battery providing?

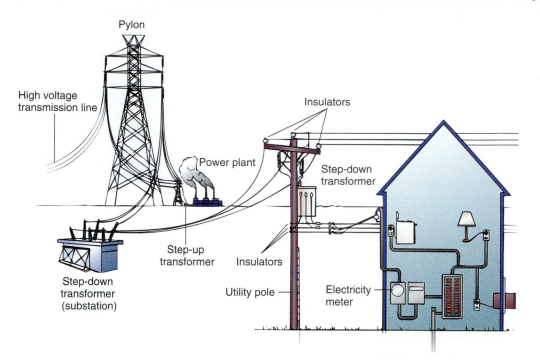

Electric Power Distribution

WARNING

<u>Electricity is danger-</u>
<u>ous, particularly</u>
<u>when it involves</u>
<u>high voltages.</u> The
principal hazard is
that an electric cur-
rent will pass
through your body in
the vicinity of your
heart and disrupt its
normal rhythm.
While very little cur-
rent is needed to
cause trouble, your
skin is such a poor
conductor that it nor-
mally keeps harmful
currents from pass-
ing through you.
However, large volt-
ages can propel dan-
gerous currents

Electricity is a particularly useful and convenient form of ordered energy. Because it's delivered to our homes and offices as a utility, we barely think about it except to pay the bills. The wires that bring it to us never plug-up or need cleaning and work con-tinuously except when there's a power failure, blown fuse, or tripped circuit breaker.

Just how does electricity get to our homes? In this section, we'll look at the prob-lems associated with distributing electricity far from the power plant at which it's gen-erated. To understand these problems, we'll examine the ways in which wires affect electricity and see how electric power is transferred and modified by devices called transformers.

Questions to Think About: *Why do power distribution systems use alternating current? What is the purpose of high-tension wires? Why does the power company place large electric devices on the utility poles near homes or on the ground near neighborhoods? What are the advantages and disadvantages of 120-V versus 220-V electric power?*

Experiments to Do: *Experiments with electric power distribution are rather danger-ous, but you can observe the way in which your local electric power is distributed. If your area is connected to a major electric power network, you'll be able to find an en-tire hierarchy of power conversion facilities. The power should travel from the power plant to your town on high-tension wires, normally located overhead on tall pillars or pylons. These wires should end at a large power conversion facility, where enormous devices transfer power to the lower voltage power lines that fan out across your town. In some places, these wires are overhead, but in others they are underground.*

But this power is still not ready for household use. It goes through at least one more level of conversion before it reaches individual homes. Transformers do this conversion. You may find transformers as boxes or cylinders on utility poles or on the ground outside homes. In cities, transformers are often located inside buildings, out of sight. Even though you may have trouble finding these transformers, they're there. In this section, we'll see why they're necessary.

Direct Current Power Distribution

Batteries may be fine for powering flashlights, but they're not very practical for lighting homes. Early experiments that placed batteries in basements were disappointing because the batteries ran out of energy quickly and needed service and fresh chemicals much too frequently.

A more cost-effective source for electricity was coal- or oil-powered electric generators. Like batteries, generators do work on the electric currents flowing through them and can provide the electric power needed to illuminate homes. But while generators produce electricity more cheaply than batteries, early ones were large machines that required fresh air and attention. These generators had to be built centrally, with people to tend them and chimneys to get rid of the smoke.

This was the approach taken by the American inventor Thomas Alva Edison (1847–1931) in 1882 when he began to electrify New York City. Each of the Edison Electric Light Company's generators acted like a mechanical battery, producing direct current that always left the generator through one wire and returned to it through another. Edison placed his generators in central locations and conducted the current to and from the homes he served through copper wires. However, the farther a building was from the generator, the thicker the copper wires had to be. That's because wires impede the flow of current and making them thicker allows them to carry current more easily.

Wire thickness is important because a current loses kinetic energy as it collides with atoms in the metal. For the current to keep moving forward, it must convert electrostatic potential energy into kinetic energy. The current's voltage gradually diminishes during its trip, so the wire has a voltage drop between one end and the other. A forward electric field is associated with this voltage drop; that field pushes the current forward against the slowing effects of collisions (Fig. 9.2.1).

Because of the competition between collisions and electric field, the charges in the wire's current move forward at a certain average speed. You can increase this average speed, and thus the current flowing through the wire, by increasing the electric field. Raising the voltage drop through the wire increases this electric field, so it's not surprising that the current in the wire is proportional to the voltage drop between its ends.

Another way to increase the current in the wire is to make the wire thicker, so that more charges can move through it at the same time. With more charges moving, they don't have to move as quickly to maintain a large current and the wire exhibits a smaller electric resistance. Resistance incorporates not only thickness, but also length and composition. Overall, the current flowing through a wire is inversely proportional to its electric resistance.

We can combine these two observations together as a word equation:

$$\text{voltage drop} = \text{current} \cdot \text{electric resistance,} \qquad \textbf{(9.2.1)}$$

in symbols:

$$V = I \cdot R,$$

through your skin and put you at risk. While your body usually has to be part of a closed circuit for you to receive a shock, don't count on the absence of an identifiable circuit to protect you from injury—circuits have a tendency to form in surprising ways whenever you touch an electric wire. Be especially careful whenever you're near voltages of more than 50 V, even in batteries, or when you're near any voltages if you're wet or your skin is broken.

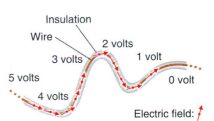

Insulation
Wire
2 volts
3 volts   1 volt
5 volts   0 volt
4 volts
Electric field: ↑

Fig. 9.2.1 This wire is part of a current-carrying electric circuit. The current experiences a voltage drop as it flows through the wire because the current must convert some electrostatic potential energy into kinetic energy in order to continue moving forward. This additional kinetic energy makes up for that lost to thermal energy through collisions with metal atoms.

❏ **German physicist Georg Simon Ohm (1789–1854)** served as a professor of mathematics, first at the Jesuits' college of Cologne and then at the polytechnic school of Nürnberg. His numerous publications were undistinguished, with the exception of one pamphlet on the relationship between current and voltage. This extraordinary document, written in 1827, was initially dismissed by other physicists, even though it was based on good experimental evidence and explained many previous observations by others. In despair, Ohm resigned his position at Cologne, and it was not until the 1840s that his work was accepted. He was finally appointed professor of physics at Munich only two years before his death.

and in everyday language:

Long, skinny jumper cables lose so much voltage carrying current to your car that it probably won't start.

Equation 9.2.1 is called **Ohm's law,** after its discoverer Georg Simon Ohm ❏. The SI unit of electric resistance, the volt-per-ampere, is called the **ohm** (abbreviated Ω, the Greek letter omega) in his honor. Despite its simplicity, Ohm's law is extremely useful in physics and electrical engineering. It applies to an enormous number of systems so that nearly everything can be characterized by an electric resistance. Once an object's electric resistance is known, the voltage drop through it can be calculated from the current flowing through it, or the current flowing through it can be calculated from the voltage drop through it. (For another useful form of Eq. 9.2.1, see ❏.)

> **Ohm's Law**
> The voltage drop through a wire is equal to the current flowing through that wire times the wire's electric resistance.

❏ **Equation 9.2.1** can be rearranged algebraically as

$$current = \frac{voltage\ drop}{electric\ resistance}.$$

This form makes it clear that the current flowing through a wire depends both on the voltage drop through it and on its electric resistance. Increasing the voltage drop increases the current, while decreasing the wire's resistance increases the current.

In the case of a wire conducting current from a generating plant to a home, we care most about how much power the wire wastes as thermal energy. We can determine this wasted power by combining Ohm's law with Eq. 9.1.1:

$$
\begin{aligned}
power\ consumed &= voltage\ drop \cdot current \\
&= (current \cdot electric\ resistance) \cdot current \\
&= current^2 \cdot electric\ resistance \qquad \textbf{(9.2.2)}
\end{aligned}
$$

The wire's wasted power is proportional to the square of the current passing through it! This relationship became all too clear to Edison when he tried to expand his power distribution systems. The more current he tried to deliver over a particular wire, the more power it lost as heat. Doubling the current in the wire quadrupled the power it wasted.

Edison tried to combat this loss by lowering the electric resistances of the wires. He used copper because only silver is a better conductor of current. He used thick wires to increase the number of moving charges. And he kept the wires short so that they didn't have much chance to waste power. This length requirement forced Edison to build his generating plants within the cities he served. Even New York City contained many local power plants. (For an interesting tale about the early days of electric power, see ❏.)

Edison also tried to save power by delivering small currents at high voltages. Since the delivered power is proportional to the voltage drop times the current, Edison could reduce the current flowing through the wires by raising the voltages. Although less current flowed through each home, it underwent a larger voltage drop so the power delivered was unchanged. However, high voltages are dangerous because they tend to create sparks as current jumps through the air. They also produce nasty shocks when current flows through your body. While high voltages could be handled safely outside a home, they could not be brought inside. Edison used the highest voltages that safety would allow.

Alternating Current and Transformers

The real problem with distributing electric power via direct current (DC) is that there's no easy way to transfer power from one circuit to another. Because the generator and the light bulbs must be part of the same circuit, safety requires the entire circuit to use low voltages and large currents. DC power distribution therefore wastes much of its power in the wires connecting everything together.

However, alternating current (AC) makes it easy to transfer power from one circuit to another, so that different parts of the alternating current power distribution system can operate at different voltages with different currents. Since the wires that carry the power long distances operate at high voltages and low currents, they waste little power.

In alternating current, the direction of current flow through a circuit reverses periodically. At one moment, the current leaves the power plant through one wire and returns through a second, while a moment later it leaves through the second wire and returns through the first.

These reversals in direction are accomplished smoothly; the current slows to a stop and then gradually begins to flow backward. During each reversal there is actually a moment in which no current is flowing at all. In the United States, these reversals occur every 120th of a second so that the current completes 60 full cycles of reversal (back and forth) each second (60 Hz). In Europe, the reversals occur every 100th of a second, and the current completes 50 full cycles of reversal each second (50 Hz).

While these current reversals seem needlessly complicated, they have little effect on most household devices. Toasters (Fig. 9.2.2) and incandescent light bulbs still experience voltage drops when alternating currents pass through them and still consume electric power. In fact, power consumption is used to define an effective voltage for alternating current. AC voltage is defined to be equal to the DC voltage that causes the same power consumption. Thus 120-V AC power delivers the same power to a toaster as 120-V DC power.

Edison was steadfastly opposed to alternating current, which he considered to be dangerous and exotic. The champion of alternating current was Nikola Tesla (1856–1943), a Serbian–American inventor, who was backed financially by the American inventor and industrialist George Westinghouse (1846–1914). The advantage that Tesla and Westinghouse saw in alternating current was that its power could be *transformed*, or passed via electromagnetic action, from one circuit to another by a device called a **transformer.**

❏ Love Canal is the United States' most famous toxic waste dump. The dump was created in the 1920s at an abandoned section of canal constructed in 1892 by William T. Love. Love intended his canal to connect the upper and lower Niagara Rivers, so that the descending water could be used to generate DC electric power for the citizens of Niagara Falls, NY. The advent of AC power transmission systems in 1896 made the canal less useful, and it was never finished.

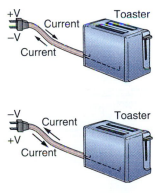

Fig. 9.2.2 The current passing through this toaster reverses directions periodically as the voltages of the electric outlet's two terminals reverse.

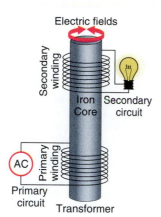

Fig. 9.2.3 A transformer transfers power from its primary circuit to its secondary circuit. As the current in the primary circuit changes during the alternating current's cycle, the magnetic field in the transformer changes. This changing magnetic field produces an electric field that delivers power to the current in the secondary circuit and extracts power from the current in the primary circuit. Overall, power is transferred from the power company in the primary circuit to the light bulb in the secondary circuit.

A transformer uses two of the electromagnetic principles we encountered in the previous chapter: (1) an electric current creates a magnetic field and (2) a changing magnetic field creates an electric field. Individually, these effects allow electricity to produce magnetism and magnetism to produce electricity. Together, they allow electricity to produce electricity.

Since a current creates a magnetic field, a changing current in one circuit creates a changing magnetic field around that circuit. But this changing magnetic field, in turn, creates an electric field. Such an electric field can propel a current in a second circuit, located near the first. Thus, through electric and magnetic processes, a changing current in one circuit can cause a current to flow in a second circuit. If the changing current in the first circuit reverses directions smoothly and periodically, as it does in alternating current, the current in the second circuit will also reverse directions in just the same manner.

That's how a transformer works. Most transformers consist of two wire coils, the *primary coil* and the *secondary coil,* wrapped together around a central iron core (Fig. 9.2.3). When a current flows through the primary coil, it magnetizes the iron core and creates a magnetic field around it. If the current through the primary coil changes, the changing magnetic field around the core induces currents in the secondary coil.

One measure of the core's magnetization is the **magnetic flux** emerging from its ends. This flux takes the form of closed loops or "lines of force" that pass through the core and then return through the air outside it (Fig. 9.2.4). In a transformer, these flux loops run through both the primary and secondary coils before circling around the outsides of both coils. This flux is related to the magnetic field in an interesting way: the field points along the magnetic flux lines, and its magnitude is proportional to how tightly those flux lines are packed together. Where the flux lines are closely spaced, as they are near the poles in Fig. 9.2.4, the magnetic field is relatively strong. Where the flux lines are farther apart, the magnetic field is relatively weak.

When the current in the primary coil changes, the core's magnetization changes and so does the number of magnetic flux loops passing through it. A change in the magnetic flux through the iron core creates an electric field around it (Fig. 9.2.3) that can push on currents inside the two wire coils. This electric field pushes current forward through the secondary coil and transfers power to it. That current experiences a voltage rise as it flows through the secondary coil. The electric field pushes current backward through the primary coil and extracts power from it. That current experiences a voltage drop as it flows through the primary coil. Overall power is being transferred from the primary circuit to the secondary circuit (Fig. 9.2.5). And because the circuits aren't actually connected, no current can ever flow between them. The transformer transfers power, not electric charge.

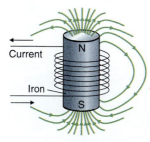

Fig. 9.2.4 The magnetic flux around the iron core of a transformer while current flows through its primary coil. The flux forms complete loops.

> **Check Your Understanding #2: Use Only with AC Power**
> If you send direct current through the primary coil of a transformer, no power will be transferred to the secondary circuit. Explain.

Changing Voltages

The work that the electric field does on charges in the secondary coil depends on the coil's *turns*—the number of times its wire circles the core. The more turns the coil has, the farther the electric field pushes current in that coil and the more work it does on each charge that flows in that current. As a result, the current's voltage rise is proportional to the number of turns in the coil.

When a transformer's primary and secondary coils have the same number of turns, the voltage drop through the primary coil is equal to the voltage rise through the secondary coil. As a result, the energy that one charge loses in going through the primary coil is picked up by another charge as it goes through the secondary coil. Since all of the power removed from the primary circuit is transferred to the secondary circuit, the currents passing through the two coils must also be equal.

But when a transformer's primary and secondary coils have different numbers of turns, the transfer of power is more complicated. If the secondary coil has fewer turns than the primary coil, then the voltage rise in the secondary coil is less than the voltage drop in the primary coil and we have a "step-down" transformer (Fig. 9.2.6). Higher voltage power enters the step-down transformer's primary coil while lower voltage power leaves its secondary coil. Since all of the power removed from the primary circuit must enter the secondary circuit, the secondary circuit must carry a larger current than the primary circuit.

Similarly, if the secondary coil has more turns than the primary coil, then the voltage rise in the secondary coil is more than the voltage drop in the primary coil and we have a "step-up" transformer (Fig. 9.2.7). Lower voltage power enters the step-up transformer's primary coil while higher voltage power leaves its secondary coil. The secondary circuit carries a smaller current than the primary circuit.

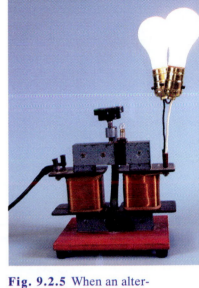

Fig. 9.2.5 When an alternating current flows through the primary coil on the left, it produces a changing magnetization in this transformer's ring-shaped iron core. This changing magnetization produces an electric field that pushes an alternating current through the secondary coil on the right and lights the bulbs.

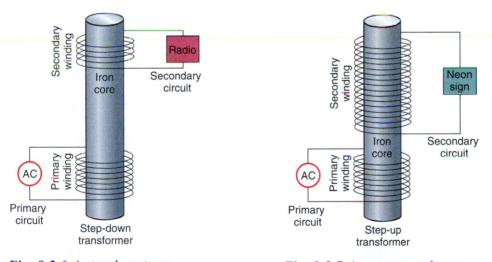

Fig. 9.2.6 A step-down transformer has more turns in its primary coil than in its secondary coil. The voltage rise through the secondary coil is less than the voltage drop through the primary coil, hence the name "step-down." However, the current flowing through the secondary circuit is larger than that flowing through the primary circuit.

Fig. 9.2.7 A step-up transformer has more turns in its secondary coil than in its primary coil. The voltage rise through the secondary coil is more than the voltage drop through the primary coil, hence the name "step-up." However, the current flowing through the secondary circuit is smaller than that flowing through the primary circuit.

common transformers

transformer imperfections

remaining power losses

The voltage and current in the secondary circuit are related to those in the primary circuit by the ratio of secondary coil turns to primary coil turns. The specific relationships are

$$\text{secondary voltage} = \text{primary voltage} \cdot \frac{\text{secondary turns}}{\text{primary turns}}$$

and

$$\text{secondary current} = \text{primary current} \cdot \frac{\text{primary turns}}{\text{secondary turns}}.$$

The first relationship recognizes that the work done on each charge is proportional to the number of times it circles the transformer's core in one of the coils. The second relationship ensures that the power extracted from the primary circuit is the same as that given to the secondary circuit.

Check Your Understanding #3: High-Intensity Lamps
In the United States, household voltage is typically 120 V. Some high-intensity lamp bulbs operate on 12-V power, which they obtain from a step-down transformer. What is the ratio of turns in the coils of this transformer?

Alternating Current Power Distribution

We now have enough tools to understand power transmission. Here's the problem: to transmit electric power efficiently over long distances—from the power plant where it's produced to the city where it's consumed—that power should travel as a small current of very-high-voltage charges. Its high voltage allows this small current to carry lots of power while wasting little of it heating the wires. But to deliver electric power safely to homes, the power should travel as a large current of low-voltage charges. These requirements conflict!

While it's impossible to meet both requirements simultaneously with direct current, transformers make it possible to satisfy them both with alternating current. We can use a step-up transformer to produce the very-high-voltage current suitable for cross-country transmission and a step-down transformer to produce the low-voltage current that's appropriate for delivery to homes (Fig. 9.2.8).

At the power plant, the generator pushes a huge current through the primary circuit of a step-up transformer, producing a voltage drop of about 5000 V through its primary coil. The current flowing through the secondary cir-

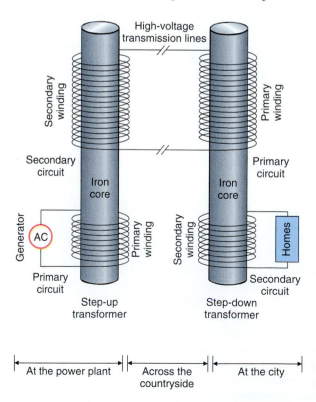

Fig. 9.2.8 Power is transmitted cross-country by stepping it up to very high voltage near the power plant, transmitting it as a small current at very high voltage, and stepping it back down to low voltage near the homes that are to be served. The secondary circuit for the first (left) transformer is also the primary circuit for the second (right) transformer.

Fig. 9.2.9 This giant transformer transfers millions of watts of power from the very-high-voltage cross-country circuits above it to the medium-high-voltage neighborhood circuits to its left. Fans keep the transformer from overheating.

Fig. 9.2.10 The three metal cans on this utility pole are transformers. They transfer power from the medium-high-voltage neighborhood circuits above them to the low-voltage household circuits at the lower right.

cuit is only about 1/100th that in the primary circuit, but the voltage rise through the secondary coil is much higher, typically about 500,000 V.

This transformer's secondary circuit is extremely long, extending all the way to the city where the power is to be used. Since the current in this circuit is modest, the power wasted in heating the wires is within tolerable limits.

Once it arrives in the city, this very-high-voltage current passes through the primary coil of a step-down transformer, where it experiences a huge voltage drop (Fig. 9.2.9). The voltage rise through the secondary coil of this transformer is only about 1/100th as large, but the current flowing through the secondary circuit is about 100 times that in its primary circuit.

Now the voltage is reasonable for use in a city. Before entering homes, this voltage is reduced still further by other transformers. The final step-down transformers can frequently be seen as oil-drum sized metal cans hanging from utility poles (Fig. 9.2.10) or as green metal boxes on the ground (Fig. 9.2.11). Current enters the buildings at between 110 V and 240 V, depending on the local standards. While 240-V electricity wastes less power in the home wiring, it's more dangerous than 110-V power. The United States has adopted a 120-V standard while Europe has a 220-V standard.

Fig. 9.2.11 This transformer transfers power from a medium-voltage underground circuit to a low-voltage underground circuit used by nearby homes. It handles 50 kV·A or 50,000 W of power.

Check Your Understanding #4: High-Tension Wires

If a power utility were able to replace an existing 500,000-V transmission line with one operating at 1,000,000 V, how would that affect the power lost to heat in the wires?

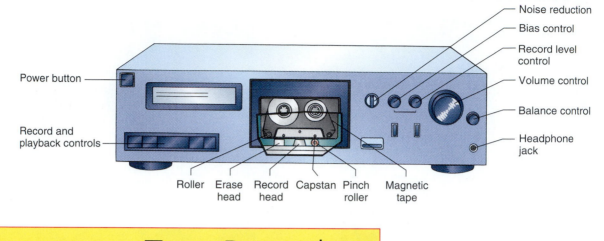

Noise reduction
Bias control
Record level control
Volume control
Balance control
Headphone jack

Power button

Record and playback controls

Roller   Erase   Record   Capstan   Pinch   Magnetic
          head   head   roller   tape

Tape Recorders

Tape recorders save audio or video information on thin plastic tapes. It's already remarkable that audio or video information can be represented electronically, so it's even more surprising that something as simple as a plastic strip can store this information almost indefinitely. Magnetic recording tape and the devices that use it become better every year, yet the basic concepts behind magnetic recording have changed relatively little since the early days of wire recorders and reel-to-reel tapes.

Questions to Think About: *What does it mean to represent sound electrically or as the magnetization of a magnetic tape? How can an electric current representing sound magnetize a tape? Why is it important that the magnetized tape move past the read head during playback? Does it matter how fast the tape moves during recording or playback? Why must you keep magnetic tapes and credit cards away from strong magnetic fields?*

Experiments to Do: *Take a look at a cassette tape recorder. While the actual recording structures are too small to see, the transport mechanisms are quite visible. The tape leaves the left spool of the cassette, travels past several components along the cassette's open end, and is wound back up on the right spool. Among the devices that you should be able to see along the tape's path are two or more shiny metal recording heads and a metal pin near a rubber roller. The two shiny metal heads are the erase head and the record/playback head. On a tape player, there's no erase head and all you'll see is one shiny metal playback head. The metal pin with its opposing rubber pinch roller is the capstan—a precision shaft that spins steadily and pulls the tape along at a constant speed.*

    While it's hard to record sound on tape without a tape recorder, you should have no trouble erasing it. Find an unimportant cassette and make sure that it has something recorded on it. Move a strong permanent magnet across its open end and gently touch the exposed magnetic tape. Now insert the cassette in a player, rewind it slightly, and play it back. You should find a two-second quiet spot, which you "erased" with the magnet. The tape recorder does a better job of erasing because it uses a powerful fluctuating magnetic field. It also uses magnetic fields to record and play back information on the tape. In this section, we'll look at how the tape recorder magnetizes and then reads the tape, and why the tape stays magnetized.

Representing Sound Electrically

Before looking at how a tape recorder records and retrieves sound information from magnetic tape, we must first study how sounds are represented electronically. As we saw in Section 7.2, sound consists of compressions and rarefactions of the air, which move outward from their source at about 300 m/s. In a region of compression, the air's density and pressure are somewhat higher than average, and in a region of rarefaction, its density and pressure are somewhat lower than average.

A tape recorder measures these pressure fluctuations in order to recreate their sound. Its microphone produces an electric current that's proportional to how much the local air pressure differs from the average value. When the air pressure is higher than average, current flows in one direction through the recording circuit, and when the air pressure is lower than average, current flows in the other direction. In this manner, the tape recorder represents anything you can hear as a fluctuating electric current.

When it's time to recreate the sound, the tape recorder uses a speaker to compress and rarefy the air. The speaker produces a local air pressure shift that's proportional to the current in the tape recorder's playback circuit. Now the tape recorder is using a fluctuating electric current to create something you can hear. Because the pressure fluctuations emerging from the speaker are identical to those detected previously by the microphone, the sound created by the tape recorder is indistinguishable from the sound it recorded earlier.

The tape recorder's task is to record the current produced by its microphone and then to recreate that current for its speaker. But while microphones and speakers are themselves quite interesting, this section will focus on recording and playback rather than on measuring and reproducing the sound.

Check Your Understanding #1: Current Music

When the only sound reaching a tape recorder's microphone is a pure 500-Hz tone, the local air pressure is fluctuating up and down 500 times each second. Describe the current in the recording circuit.

❏ In 1898, while a research engineer for the Copenhagen Telephone Company, **Danish engineer and inventor Valdemar Poulsen (1869–1942)** developed a method for recording sound by magnetizing a steel wire. In his original recorder, the wire was stretched taut across his laboratory and the recording and playback devices were trolleys that traveled on that wire. The recording trolley would magnetize the wire to record sound and the playback trolley would measure the wire's magnetization to reproduce the sound. Poulsen would run along with the recording trolley and yell into the recording microphone. He would then listen to a reproduction of his voice as the playback trolley traveled along the wire.

Electrons and Magnetic Atoms

Magnetic tape is a thin strip of plastic, coated with a film of tiny permanent magnets. Each of these permanent magnets has a north magnetic pole at one end and a south magnetic pole at the other. During recording, a strong magnetic field can swap the poles of these particles, altering the tape in a way that can be detected during playback. This controlled modification of the tape, together with its detection at playback, is the basis for magnetic recording. (Early recorders didn't use tape; see ❏.)

There are only a few materials that are noticeably magnetic and fewer still that can become permanent magnets. To be magnetic, a material must contain naturally occurring magnetic dipoles. For it to be a permanent magnet, those dipoles must stay aligned with one another. Most magnetic materials contain tiny dipoles that are randomly oriented so that they cancel one another. Such materials generally appear nonmagnetic when viewed from outside. But there are a few special materials in which the tiny dipoles can stay aligned with one another to give the materials large overall magnetic poles. When their dipoles are aligned together, these materials become permanent magnets.

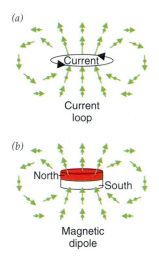

Fig. 9.3.1 The magnetic field around a loop of electric current (a) is similar to that around a dipole magnet having equal north and south poles (b).

Since isolated magnetic poles have never been observed, magnetic materials create their dipoles with moving electric charge. Anytime electric charge moves in a circle or loop, it acts as a magnetic dipole (Fig. 9.3.1). The simplest and most important example of this effect is an electron, which has a spin and behaves like a tiny spinning ball that's covered with negative charge. This circulating electric charge makes even a stationary electron magnetic. If the electron orbits an atom, it has additional magnetism due to that orbital motion.

But quantum physics severely limits the electron's allowed orbits and spins, and therefore its magnetism. In particular, its spin can only produce two different dipole magnetic fields. If the electron is spin-up, effectively rotating counterclockwise when viewed from above, it's a dipole with its south pole on top. If the electron is spin-down, effectively rotating clockwise when viewed from above, it's a dipole with its north pole on top.

But when many electrons join together in an atom, they often form pairs that spin or orbit in opposite directions so that their magnetic fields cancel. The amount of magnetism in an atom depends on how many of its electrons remain unpaired. While most of the electrons do pair up, this pairing brings the electrons particularly close to one another. A few of the least tightly bound electrons often spread out into different orbitals (quantum orbits) to lower their electrostatic potential energies. Adopting the same spin, either up or down, helps to keep these electrons in different orbitals because the Pauli exclusion principle then forbids them from traveling together. Most atoms have a few unpaired electrons and are therefore magnetic.

Check Your Understanding #2: Helium Balloons Aren't Magnetic

A helium atom has two electrons that spin and orbit in opposite directions. How does this explain why helium atoms are nonmagnetic?

Magnetic Solids and Magnetic Tape

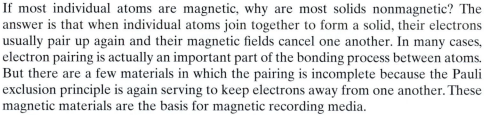

recording media

If most individual atoms are magnetic, why are most solids nonmagnetic? The answer is that when individual atoms join together to form a solid, their electrons usually pair up again and their magnetic fields cancel one another. In many cases, electron pairing is actually an important part of the bonding process between atoms. But there are a few materials in which the pairing is incomplete because the Pauli exclusion principle is again serving to keep electrons away from one another. These magnetic materials are the basis for magnetic recording media.

The three major types of magnetic solids are distinguished by the ordering of their individual magnetic atoms (Fig. 9.3.2). Iron, cobalt, nickel, and chromium

Fig. 9.3.2 (a) The atomic magnets in ferromagnetic solids point in the same direction. (b) In antiferromagnetic solids, they cancel one another completely. (c) In ferrimagnetic solids, there is partial cancellation.

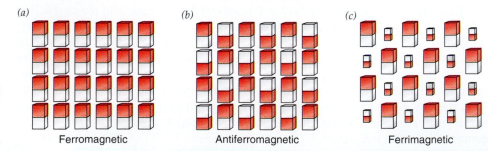

Ferromagnetic   Antiferromagnetic   Ferrimagnetic

dioxide are **ferromagnetic** solids, meaning that each atom's magnetic dipole is aligned with those of its neighbors. Chromium is an **antiferromagnetic** solid, in which the atoms' magnetic dipoles point in alternating directions and completely cancel one another. Ferrite (γ-iron oxide) is a **ferrimagnetic** solid, meaning that its atoms' magnetic dipoles alternate back and forth but, because they have different strengths, don't cancel one another completely. Ferromagnetic and ferrimagnetic materials can be used in magnetic recording media. Many other metals, including aluminum and copper, have no magnetic dipoles in them at all.

But despite the ferromagnetic ordering of its magnetic atoms, a piece of iron usually doesn't appear magnetic. That's because it contains many individual **magnetic domains**—separate regions in which the atoms' magnetic dipoles are aligned with one another—and these domains normally point in different directions. The magnetic fields produced by these isolated magnetic domains tend to cancel one another, so that the iron appears to be nonmagnetic.

To see why domains form, consider what would happen if they didn't. A piece of iron would then have a huge north magnetic pole at one end and an equally huge south magnetic pole at the other (Fig. 9.3.3a). These poles would exert torques on any nearby magnets, including the magnetic atoms *inside* the iron! This twisting action is so strong that it separates the magnetic atoms into magnetic domains (Fig. 9.3.3b) and turns the poles of each domain so that they all point in different directions. The iron effectively *demagnetizes* itself.

But when a piece of iron is placed in a magnetic field, its magnetic character suddenly becomes obvious. Magnetic domains that are aligned with the field grow at the expense of other domains (Fig. 9.3.4a,b) and the iron develops large magnetic poles. The iron becomes *magnetized*. The electrons simply change the directions of their spins, leaving the iron atoms exactly where they were. This process explains why a permanent magnet attracts iron; the magnet magnetizes the iron (Fig. 9.3.4b) and then the two magnets attract one another.

But when you remove the permanent magnet, iron's domains change size again and it quickly demagnetizes, reverting to its nonmagnetic state. Iron is a **soft mag-**

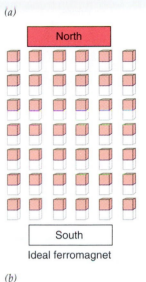

(a)

North

South

Ideal ferromagnet

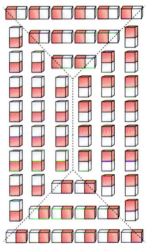

(b)

Magnetic domains

Fig. 9.3.4 (a) Because its magnetic domains balance one another, a piece of iron is normally nonmagnetic. (b) But when the iron is exposed to a magnetic field, domains that are aligned with the field grow at the expense of other domains and the iron becomes a magnet. Since iron is a soft magnetic material, its domains return to normal (a) when the magnetic field is removed. (c) However, the domains in a hard magnetic material can't change sizes easily so that the material remains permanently magnetized.

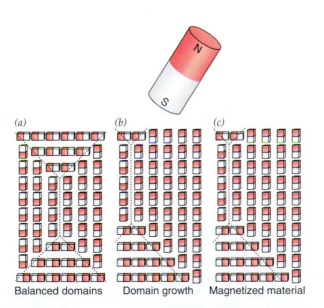

(a)   (b)   (c)

Balanced domains   Domain growth   Magnetized material

Fig. 9.3.3 (a) An ideal ferromagnet's individual magnetic atoms all point in the same direction, so that it behaves as though it had two giant magnetic poles. (b) However, a real ferromagnet's strong magnetic field twists the magnetic atoms out of alignment, so that they separate into magnetic domains with different orientations.

erasing tape

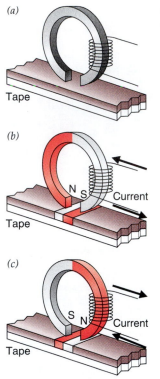

(a)

Tape

(b)

N S Current

Tape

(c)

S N Current

Tape

Fig. 9.3.5 (*a*) A tape head is a ring of soft magnetic material with a gap in it and a coil of wire wound around it. A current in the coil temporarily magnetizes the ring (*b, c*) and permanently magnetizes any hard magnetic material near its gap.

Fig. 9.3.6 This stereo recording head has two soft iron rings, the shiny rectangles above and below the horizontal dividing line. The vertical gaps in these rings are too thin to see.

netic material, meaning that it can't retain a magnetization on its own. Soft magnetic materials aren't useful for magnetic recording media, but they're essential in transformers and generators and play an important role in a tape recorder's recording and playback heads.

Magnetic tapes use **hard magnetic materials,** materials that can remain magnetized on their own. There are several ways to obtain magnetic hardness. Some hard magnetic materials have microscopic structures and defects that impede the resizing of domains. These stiff magnetic domains make a material both hard to magnetize and hard to demagnetize. For example, the domains in common steel have some difficulty resizing, so that steel doesn't return completely to its nonmagnetic state after exposure to a magnetic field. That's why steel nails, screwdrivers, and compass needles remain weakly magnetized after exposure to a strong magnet.

Actually, the permanent magnets used on magnetic tapes are so small that each one has only a single domain and therefore can't hide its magnetic dipole by itself. However, a group of these particles can appear nonmagnetic by orienting their dipoles in opposite directions so that they cancel. Magnetizing a group of these particles means aligning their dipoles together, while demagnetizing the group means orienting their dipoles randomly.

These tiny particles are usually cigar shaped, and their magnetic poles tend to be located at their ends. Swapping the poles is difficult and requires a strong magnetic field, which is why these particles are good for magnetic recording. Since recording and playback are sensitive only to magnetization along the tape's length, the magnetic particles are aligned in that direction when they're coated onto the plastic tape.

Recording Sound onto Magnetic Tape

We are now ready to look at how a tape recorder records sound information on a magnetic tape. This is done by a miniature electromagnet in the recording head, a tiny ring of soft magnetic material with a coil of wire wrapped around it (Figs. 9.3.5, 9.3.6, and 9.3.7). When a current flows through the coil, the coil's magnetic field temporarily magnetizes the ring.

The ring is actually incomplete; it has a tiny gap just at the point where it touches the magnetic tape (Fig. 9.3.5*a*). When current flows one way through the coil, the gap has a north pole on its left end and a south pole on its right end (Fig. 9.3.5*b*). When the current reverses, the poles also reverse (Fig 9.3.5*c*). During recording, sound is represented as a fluctuating current in the coil and therefore a fluctuating magnetic field in the ring's gap.

Before it encounters the magnetic recording head, the coating on the magnetic tape is demagnetized. Although the individual particles in the coating are always magnetic, they're arranged so that roughly half of their north poles point forward and half point backward (Fig. 9.3.8*a*). The recording head's job is to reorient those tiny magnets so that their poles all point in the same direction (Fig. 9.3.8*b*). The particles don't actually flip over; their poles simply interchange.

Fig. 9.3.7 This thin film magnetic recording head is so small that it can only be viewed through a microscope. The head's iron ring is the shiny white object near the top, encircled by the coil of wire that connects to the circuit at the bottom.

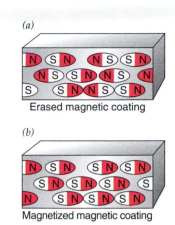

(a)

Erased magnetic coating

(b)

Magnetized magnetic coating

Fig. 9.3.8 In an erased magnetic coating (*a*), roughly half the particles have north poles on the left and half on the right. In a magnetized portion of coating (*b*), all of the particles have their north poles aligned on the same side.

As the magnetic tape moves steadily past the recording head, the gap in the recording head magnetizes regions of the tape's magnetic coating in one direction or the other (Fig. 9.3.9). The depth to which the coating is magnetized depends on the strength of the magnetic field in the gap, which is proportional to the current in the coil. Since that current represents sound and air pressure, the magnetization depth is greatest during loud parts and least during quiet parts. The direction of magnetization, north poles forward or backward, depends on the direction of current flow through the coil, which depends on whether the recorder is trying to represent a compression or a rarefaction of the air.

There are two more important details about magnetic recording. First, the recording head expects to start with a blank tape, so the entire coating must be demagnetized by an erase head located just upstream from the recording head. The erase head is essentially a recording head with an extra wide gap and a large, high-frequency alternating current in its coil. As the magnetic coating passes through the rapidly reversing magnetic field in the gap, the magnetic alignments of its particles flip back and forth many times. About half end up oriented in one direction and half in the other.

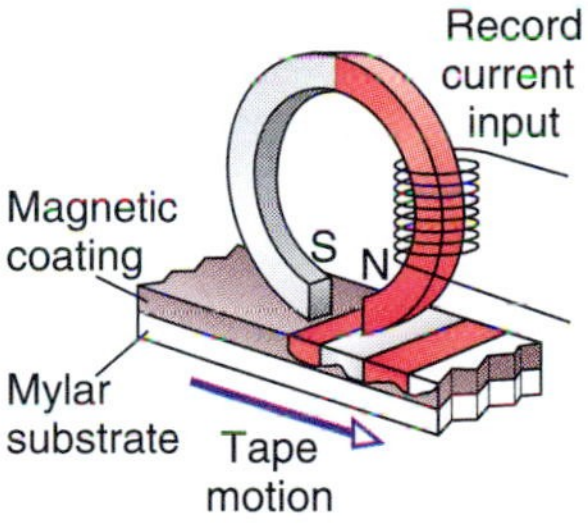

Second, the recording head has trouble flipping the magnetic alignments of particles during quiet portions of the sound. To solve this problem, the tape recorder sends an additional high-frequency alternating current through the coil in the recording head. This *bias current* flips the particles' magnetic alignments back and forth during the recording process. While the bias current doesn't give the tape any particular overall magnetization, it allows the other current in the coil, the current representing sound, to magnetize parts of the coating even during quiet passages. Tape recorders adjust the bias current according to the magnetic hardness of the particles in the coating, either type I (normal bias), type II (high bias), or type IV (metal bias).

Record current input

Magnetic coating

Mylar substrate   Tape motion

Fig. 9.3.9 The recording head magnetizes regions of the tape's magnetic coating as the tape moves past the head. The depth of this magnetization is proportional to the amount of current in the coil.

➤ Check Your Understanding #4: No More Rock and Roll
If you bring a strong magnet near a recorded magnetic tape, that tape may be erased. Why?

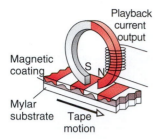

Fig. 9.3.10 As the magnetized tape passes across the gap in the playback head's ring, the ring is magnetized and the magnetic flux passing through the coil changes. Current representing the recorded sound is induced in the coil.

noise reduction

stereo recording

tape transport

Playback

The tape recorder recreates sound by passing the magnetic tape across a playback head. The playback head is similar to the recording head, and the two are often the same device. Fancy tape recorders have a separate playback head located downstream from the recording head, so that you can listen to the tape as you are recording it. Separate recording and playback heads also outperform a single head that must carry out both duties.

As the magnetized tape passes across the gap of the playback head, it temporarily magnetizes the ring (Fig. 9.3.10). The deeper the magnetization of the coating, the more the ring is magnetized and the more magnetic flux passes through the coil of wire wrapped around it. As the tape moves past the head, this flux changes and induces currents in the coil. This current is amplified and sent to a speaker, where it reproduces the compressions and rarefactions of the air that occurred in the original sound.

Check Your Understanding #5: Feel Those Good Vibrations
Is the induced current in the coil of a playback head larger during a quiet passage or a loud passage?

Epilogue for Chapter 9

In this chapter, we studied movement and change, and how they introduce relationships between electricity and magnetism. In *flashlights,* we examined circuits to see how a current of electric charges can transfer power from batteries to a light bulb. We found that both voltage and current contribute to this power transfer. In *electric power distribution,* we saw how alternating electric current makes it possible to transfer power from one circuit to another by way of a transformer. We learned that transforming electric power to extremely high voltage and low current minimizes the power wasted between power plants and cities.

And in *tape recorders,* we found that certain materials are intrinsically magnetic and saw that this magnetism can be used to store the information needed to reconstruct sound. We explored the ways of making permanent magnets and using those permanent magnets to construct recording tapes.

Explanation: A Nail and Wire Electromagnet

When you connect the wire from one terminal of the battery to the other, a current begins to flow from the positive terminal to the negative terminal through the wire. (In reality, negatively charged electrons move from the battery's negative terminal, through the wire, to its positive terminal, but we've adopted a fiction that positive charges are heading the other way.) This current produces a magnetic field around the wire. Because the wire is coiled around the nail, this magnetic field passes through the nail and causes its magnetic domains to resize until most of the nail's magnetic atoms are aligned with the field. Without any current in the wire, the magnetic domains in the steel point in many different directions, so the nail appears nonmagnetic. However, with the current orienting the domains, they together produce a large magnetization. The nail becomes magnetic and exerts large magnetic forces on other nearby objects.

Chapter Summary

How Flashlights Work: A flashlight produces light when a current flows through its electric circuit. This circuit consists of a chain of batteries, a light bulb, a switch, and several metal strips, all connected in a continuous loop. The current obtains energy as it passes through the battery, and it releases this energy as it passes through the filament of the light bulb, heating that filament white hot.

The switch provides a way to control the flow of current in the circuit. When the switch is off, it breaks the circuit and prevents current from flowing completely around the circuit. Without a steady current to carry power from the batteries to the bulb, the flashlight is dark. When the switch is on, it completes the circuit so that power can reach the bulb and the flashlight becomes bright.

How Electric Power Distribution Works: To minimize power losses in the transmission lines between power plants and cities, power distribution systems use alternating current and transformers. Near the power plant, relatively low-voltage, high-current electric power is converted into very-high-voltage, low-current power for transmission on cross-country power lines. Because the power consumed by these high-tension wires depends on the square of the currents they carry, the power losses are greatly reduced by this technique. When the power arrives at the city, it's converted into medium-voltage, high-current power for distribution to neighborhoods. Finally, in neighborhoods, step-down transformers convert this power to low-voltage, very-high-current power for distribution to individual homes and offices.

How Tape Recorders Work: During recording, the air pressure fluctuations (sound) detected by a microphone become fluctuations in the current through a coil of wire. This coil is wrapped around a tiny ring of soft magnetic material, which is temporarily magnetized by the coil's current. The ring is broken by a narrow gap, and the ring's magnetization creates poles on opposite sides of that gap. Demagnetized magnetic tape is drawn across the gap and is magnetized by its magnetic fields. The tape's magnetization now represents the original sound.

During playback, the magnetized tape is drawn across the gap in the ring of a playback head (often the same as the recording head). As each magnetized region travels across the gap, the coating's magnetic field temporarily magnetizes the ring. This changing magnetization induces current in the coil wrapped around the ring, and this current is then amplified and sent to a speaker, where it reproduces the original sound.

Important Laws and Equations

1. Power Consumption: The electric power consumed by a device is the voltage drop across it times the current flowing through it, or

$$\text{power consumed} = \text{voltage drop} \cdot \text{current}. \quad (9.1.1)$$

2. Power Production: The electric power provided by a device is the voltage rise across it times the current flowing through it, or

$$\text{power provided} = \text{voltage rise} \cdot \text{current}. \quad (9.1.2)$$

3. Ohm's Law: The voltage drop through a wire is equal to the current flowing through that wire times the wire's electric resistance, or

$$\text{voltage drop} = \text{current} \cdot \text{electric resistance} \quad (9.2.1)$$

Check Your Understanding—Answers

Section 9.1 FLASHLIGHTS

1. Removing a single wire from the battery breaks the circuit and prevents a steady flow of electric current through the car's electric system.

Why: Removing the negative wire from the battery stops electrons from leaving the battery. Removing the positive wire prevents electrons from returning to the battery. Either way, no steady current will flow, so the car will not operate.

2. The current flows from the metal toward your hand.

Why: The negative charges on your body flow toward the metal piece. Because current is defined as the flow of positive charges, it points in the direction opposite the flow of negative charges. Thus the current flows toward your hand. In this case, electricity doesn't involve a complete circuit. However, the charges only move for a brief instant. For the charges to move continuously, they would have to be recycled and there would have to be a circuit.

3. Each individual battery produces 2 V.

Why: The voltages of the individual batteries add so that it takes six of the 2-V batteries to produce a 12-V chain. If one of the batteries becomes weak, through damage, loss of fluid, or consumption of its chemical potential energy, the voltage of the chain will drop below 12 V. Pushing current backward through a lead–acid battery recharges it.

4. The wire through which current enters the radio.

Why: The radio is consuming power, so the current passing through it is experiencing a voltage drop. The current has a higher voltage when it enters the radio than when it leaves.

Section 9.2 ELECTRIC POWER DISTRIBUTION

1. It would roughly double.

Why: Doubling the length of a wire is like placing two identical wires one after the next. If each wire uses 1 unit of power, then two wires should use roughly 2 units of power. Electric resistance is proportional to the length of a wire and inversely proportional to the cross-sectional area of that wire. Shortening and thickening a wire reduces its resistance.

2. When direct current flows through the transformer's primary coil, it creates a constant magnetic field around the iron core. Since that field doesn't change, it doesn't create any electric fields and doesn't induce current in the transformer's secondary coil.

Why: The current through the primary coil must change so that the magnetic field in the coils will change and current will be induced in the secondary coil. Transferring power from one circuit to another is so useful that there are many DC powered devices that switch their power on and off to mimic alternating current, so that they can use transformers.

3. About 10 primary coil turns for every secondary coil turn.

Why: To make the voltage rise across the secondary coil equal to 1/10th the voltage drop across the primary coil, the step-down transformer must have only 1 secondary coil turn for each 10 primary coil turns. The secondary circuit will have 10 times the current of the primary circuit.

4. It would reduce the amount of heat produced to only 25% of the previous value.

Why: A 1,000,000-V transmission line would be able to carry the same power as the 500,000-V transmission line with only half the current. Since the power wasted by the transmission lines themselves is proportional to the square of the current, halving the current reduces the power waste to 25%.

Section 9.3 TAPE RECORDERS

1. It's an alternating current, reversing its direction 1000 times each second.

Why: The 500-Hz tone causes the air pressure to change up and down through 500 complete cycles each second. Each time the pressure goes above average, the current through the circuit flows in one direction. Each time the pressure goes below average, the current flows in the other direction. That makes 1000 reversals in current each second.

2. While the individual electrons are magnetic, their magnetic fields exactly cancel one another.

Why: Since the electrons orbit and spin in opposite directions, all magnetism associated with those motions are also in opposite directions. They therefore sum to zero.

3. The particle's magnetic poles will interchange.

Why: Although the particle itself can't move, its magnetic poles can. When the repulsion between the two north poles becomes strong enough, the poles of the magnetic particle will interchange and it will then present its south pole to the north pole of the other magnet.

4. The strong magnetic field magnetically realigns the particles in the tape's coating.

Why: The magnetic particles in the tape can respond to a strong magnetic field by interchanging their magnetic poles. When this happens, the recorded information is lost and is replaced by meaningless magnetization. That's why magnets must be kept away from magnetic recording media.

5. The current reaches its largest values during a loud passage.

Why: When loud passages are recorded, they produce deep regions of magnetization in the tape's coating. This extensive magnetization causes large changes in the magnetic flux through the ring of the playback head and induces large currents in the coil around that ring.

Check Your Figures—Answers

Section 9.1 FLASHLIGHTS

1. 2400 W.

Why: The voltage drop across the starter motor is 12 V (each coulomb of charge loses 12 J of energy in passing through it) and the current through it is 200 A (200 C of charge pass through it each second). We can use Eq. 9.1.1 to determine the power the motor is consuming:

$$\text{power consumed} = \text{voltage drop} \cdot \text{current}$$
$$= 12 \text{ V} \cdot 200 \text{ A} = 2400 \text{ W}.$$

2. 1800 W.

Why: The voltage rise across the battery is 12 V (each coulomb of charge gains 12 J of energy in passing through it) and the current through it is 150 A (150 C of charge pass through it each second). We can use Eq. 9.1.2 to determine the power the battery is providing:

$$\text{power provided} = \text{voltage rise} \cdot \text{current}$$
$$= 12 \text{ V} \cdot 150 \text{ A} = 1800 \text{ W}.$$

Section 9.2 ELECTRIC POWER DISTRIBUTION

1. The current through you would double and the power you're consuming would increase by a factor of 4.

Why: Since the voltage drop across you would double, Eq. 9.2.1 indicates that the current through you should also double. That's because the average speed of the charges inside you would double. From Eq. 9.2.2 we see that doubling the current through you increases the power you're consuming by a factor of 4. It's not surprising that 220-V electric power delivers worse shocks than 110 V.

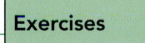

Exercises

1. The electric switch for the light in your room has two wires attached to it. One day the switch breaks so that the light won't turn off. What should the switch be doing with the electric circuit it's in, and what is it now doing instead?

2. The cord running from the electric outlet to your table lamp has two wires inside it. Current flows to the lamp through one wire and returns from the lamp through the other. If a heavy metal file cabinet cuts into the cord and electrically connects those two wires to one another, will the lamp become brighter or dimmer? Explain your answer.

3. The power source for an electric fence pumps charge from the earth to the insulated fence wire. The earth can conduct electricity. When an animal walks into the wire, it receives a shock. Identify the circuit through which the current flows.

4. A bird can perch on a high-voltage power line without getting a shock. Why doesn't current flow through the bird?

5. The positive terminal of a battery has a net positive charge. You are normally electrically neutral. If you touch the positive terminal of the battery, does any current flow between the terminal and your finger? If so, for how long?

6. The two prongs of a power cord are meant to carry current to and from a lamp. If you were to plug only one of the prongs into an outlet, the lamp wouldn't light at all. Why wouldn't it at least glow at half its normal brightness?

7. The rear defroster of your car is a pattern of thin metal strips across the window. When you turn the defroster on, current flows through those metal strips. Why are there wires attached to both ends of the metal strips?

8. When current is flowing through a car's rear defroster (Exercise 7), the voltage at each end of the metal strips is different. Which end of each strip has the higher voltage, the one through which current enters the strip or the one through which current leaves, and what causes the voltage drop?

9. You're given a sealed box with two terminals on it. You use some wires and batteries to send an electric current into the box's left terminal and find that this current emerges from the right terminal. If the voltage of the right terminal is 6 V higher than that of the left terminal, is the box consuming power or providing it? Is it more likely to contain batteries or light bulbs? How can you tell?

10. Suppose that you have two 1.5-V batteries, some wires, and a light bulb. One of the batteries, however, is unlabeled and there is nothing on it to indicate which is its positive terminal and which is its negative terminal. The other battery is properly labeled. How can you determine which terminal of the unlabeled battery is positive and which is negative?

11. One time-honored but slightly unpleasant way in which a hobbyist determines how much energy is left in a 9-V battery is to touch both of its terminals to his tongue briefly. He experiences a pinching feeling that's mild when the battery is almost dead but one that's startlingly sharp when the battery is fresh. (Don't try this technique yourself.) What is the circuit involved in this taste test? How is energy being transferred?

12. Electric power is often sold on the basis of kilowatt-hours (kWh). A house that uses an average of 1000 W for an hour has consumed 1 kWh. Explain why a kilowatt-hour is a unit of energy.

13. If you overload an extension cord so that it's carrying too much electric current, why does it become hot?

14. A fuse limits the current that can flow through an electric circuit to the value printed on its label. Why can putting a 50-A fuse in a circuit rated to carry only 30 A lead to a fire?

15. Your friends are installing a loft in their room and are using thin speaker wires to provide power to an extra outlet. If they only draw a small amount of current from the outlet, the voltage drop in each of the wires will remain small. Why?

16. When your friends from Exercise 15 plug a large home entertainment system into the outlet, it doesn't work properly because the voltage rise provided by the extra outlet is only 60 V. The power company provides a voltage rise of 120 V, so where is the missing voltage?

17. Many older buildings use a single electric circuit to provide power to a room's lights and its electric outlets. One wire delivers current to the lights and outlets and another wire returns that current to the power company. When you plug a hair dryer into an outlet in such a room and turn it on, the room lights dim. Why do the lights dim?

18. The electric company makes sure that the voltage drop across an incandescent light bulb is always 120 V. But as the bulb's filament warms up, its electric resistance increases. What happens to the current passing through the filament?

19. During a particular power surge, the voltage drop across an incandescent light bulb increases by 10%. If the bulb's electric resistance doesn't change, what effect will this surge have on the current passing through the bulb's filament?

20. Why will the power surge in Exercise 19 cause the bulb to consume approximately 20% more power than normal?

21. Spot welding is used to fuse two sheets of metal together at one small spot. Two copper electrodes pinch the sheets together at a point and then run a huge electric current through that point. The two sheets melt and flow together to form a spot weld. Why does this technique only work with relatively poor conductors of electricity such as stainless steel and not with excellent conductors such as copper?

22. The primary coil of a transformer makes 240 turns around the iron core, and the secondary coil of that transformer makes 80 turns. If the primary voltage is 120 V, why will the secondary voltage be only 40 V?

23. If a 3-A current is passing through the primary coil of the transformer in Exercise 22, what current is passing through the secondary coil of that transformer?

24. A steel paper clip is normally not magnetic. However, when you touch one end of the paper clip to the north end of a permanent magnet, the paper clip becomes magnetic. Which end of the paper clip becomes a south pole?

25. Aluminum-coated plastic films and tapes have a mirrorlike surface and are frequently used for decorations. Why can't aluminum-coated tape be used to record sound in a tape recorder?

26. Why does a magnet stick to the surface of a refrigerator?

27. If you hold a permanent magnet the wrong way in an extremely strong magnetic field, its magnetization will be permanently reversed. What happens to the magnetic domains inside the permanent magnet during this process?

28. A strong magnet can pick up a whole chain of steel paper clips. Explain how the first paper clip can pick up the second paper clip if the two aren't normally magnetic.

29. Junk yards use large direct current electromagnets to pick up scrap iron and steel. Why don't they use alternating current in those electromagnets?

30. When you have the magnetic strip on an ID or credit card read, it's always done by moving the card past a playback head. Why must the card be moving for the head to read it?

31. Suppose two magnetic tapes have pure tones recorded on them. One has a low-frequency tone while the second has a high-frequency tone. Explain how the magnetizations of the two tapes differ from one another.

32. Why must the recording particles in a magnetic tape be made from a hard magnetic material rather than a soft one?

33. Why must the ring of a magnetic recording head be made from a soft magnetic material rather than a hard one?

34. What does exposing the magnetic strip of an ID or credit card to a strong magnetic field do to that strip that makes it unreadable?

35. If you sprinkle iron powder onto the magnetic strip of an ID or credit card, the powder will stick and form a pattern of stripes. Why does the powder stick to the magnetic strip?

36. Why does iron powder form a pattern of stripes on the magnetic strip of an ID or credit card (see Exercise 35)?

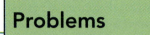

Problems

1. An automobile has a 12-W reading lamp in the ceiling. This lamp operates with a voltage drop of 12 V across it. How much current flows through the lamp?

2. The rear defroster of your car operates on a current of 5 A. If the voltage drop across it is 10 V, how much electric power is it consuming as it melts the frost?

3. Your portable FM radio uses two 1.5-V batteries in a chain. If the batteries send a current of 0.05 A through the radio, how much power are they providing to the radio?

4. You have two flashlights that operate on 1.5-V "D" batteries. The first flashlight uses two batteries in a chain while the second uses five batteries in a chain. Each flashlight has a current of 1.5 A flowing through its circuit. What power is being transferred to the bulb in each flashlight?

5. You have two flashlights that have 2-A currents flowing through them. One flashlight has a single 1.5-V battery in its circuit, while the second flashlight has three 1.5-V batteries connected in a chain that provides 4.5 V. How much power is the battery in the first flashlight providing? How much power is each battery in the second flashlight providing?

6. How much power is the bulb of the first flashlight in Problem 5 consuming? How much power is the bulb of the second flashlight consuming?

7. A 1.5-V alkaline "D" battery can provide about 40,000 J of electric energy. If a current of 2 A flows through two "D" batteries while they're in the circuit of a flashlight, how long will the batteries be able to provide power to the flashlight?

8. A radio-controlled car uses four "AA" batteries to provide 6 V to its motor. When the car is heading forward at full speed, a current of 2 A flows through the motor. How much power is the motor consuming at that time?

9. Your car battery is dead, and your friends are helping you start your car with cheap jumper cables. One cable carries current from their car to your car, and a second cable returns that current to their car. As you try to start your car, a current of 80 A flows through the cables to your car and back, and a voltage drop of 4 V appears across each cable. What is the electric resistance of each jumper cable?

10. If you replace the cheap cables in Problem 9 with cables having half their electric resistance, what voltage drop will appear across each new cable if the current doesn't change?

11. Each of the two wires in a particular 16-gauge extension cord has an electric resistance of 0.04 Ω. You're using this extension cord to operate a toaster oven, so a current of 15 A is flowing through it. What is the voltage drop across each wire in this extension cord?

12. How much power is wasted in each wire of the extension cord in Problem 11?

13. The two wires of a high-voltage transmission line are carrying 600 A to and from a city. The voltage between those two wires is 400,000 V. How much power is the transmission line delivering to the city?

14. In bringing electricity to individual homes, the power in Problem 13 is transferred to low-voltage circuits so that the current passing through homes experiences a voltage drop of only 120 V. How much total current is passing through the homes of the city?

Cases

1. An uninterruptible power supply (UPS) allows a computer system to operate during a brief power outage. Most personal computer UPS systems use an electronic device to convert 12-V DC power from a battery into 120-V AC power for the computer. The UPS circuitry first converts the 12-V DC into 12-V AC and then uses a transformer to increase the voltage to 120 V, as required by the computer. In Europe, the final voltage is 220 V.

a. Why can't the transformer directly convert 12-V DC into 120-V DC?

b. An electronic switching system converts the 12-V DC from the batteries into 12-V AC for the transformer. What is different about the current flowing through the two wires attached to the battery and the current flowing through the two wires attached to the transformer?

c. How does power move from the 12-V AC side of the transformer to the 120-V AC side of the transformer?

*d. Energy is conserved. If there is an average current of about 20 A in the 12-V primary circuit of the transformer, what is the average current in the 20-V secondary circuit of that transformer?

e. The length of time that the UPS can deliver power to the computer is limited by the battery's chemical potential energy. Suppose you wish to extend that time by adding a second 12-V battery to the system. How should you connect the two batteries together so that the current arriving at the pair of batteries will still experience a 12-V rise in voltage and yet be able to extract power from both batteries?

2. Each electric outlet in a home is connected to the power company through a fuse or circuit breaker. When properly installed, current flows from the power company, through the fuse or circuit breaker, through the outlet and the appliances connected to it, and then back to the power company—a complete circuit. The fuse or circuit breaker limits the current that the circuit can carry.

a. If you plug too many appliances into a socket that has nothing to limit the current it can supply, the outlet and home wiring will become hot. Why?

b. A fuse contains a short wire with a relatively high electric resistance. Why will this wire melt first if too much current is drawn through the circuit?

c. Why will current stop flowing through the appliances when the fuse melts and opens the circuit?

d. Why is a short circuit that bypasses the appliances very likely to melt the fuse and open the circuit?

e. A circuit breaker contains a heater wire and a thermostat. The current flowing through the circuit protected by the circuit breaker passes through both the heater wire and the thermostat. If the breaker is supposed to prevent too much current from passing through the circuit, should the thermostat open or close the circuit when it determines that the heater wire has become very hot?

3. A recycling plant is trying to separate metals from other trash automatically. It grinds the trash up into small pieces and then sends these pieces past the poles of a strong electromagnet.

a. Iron and steel scraps in the trash are attracted to the poles of the electromagnet but aluminum scraps are not. Why do these metals behave so differently?

b. If they're moving fast enough, aluminum scraps experience a magnetic drag force that separates them from nonmetallic trash. What is the origin of this force?

c. To look for metal hidden inside bundles of paper, the recycling plant exposes the bundles to an alternating magnetic field (one that reverses directions many times a second). If it finds that something inside a bundle creates another alternating magnetic field in response, it knows that the bundle contains metal. How can metal inside the trash create this second alternating magnetic field?

4. A home burglar alarm uses various sensors to detect an intruder. These sensors are connected to the alarm's control unit by wires. The control unit determines when a break-in has occurred, sounds an alarm, and notifies the authorities.

a. Each sensor is attached to the control unit via two wires rather than just one. It's very hard to signal a break-in electrically with only one wire. Why do two wires make it so much easier for a sensor to indicate that a break-in has occurred?

b. One of the simplest sensors is just a thin strip of metal foil that runs along the edge of a window. If a burglar breaks the window, the foil strip will be severed. How can the control unit determine if the strip has broken, using its two wires?

c. A more sophisticated sensor can tell when a door is opened. It uses a magnet attached to the door and a small device called a reed switch attached to the doorframe. This switch contains two iron strips that are arranged head to tail in a line, but normally don't quite touch—they're bent slightly apart and must bend together to make contact. When the magnet's north pole is near the end of one of the iron strips, that strip becomes magnetic. Why?

d. The first iron strip, now magnetic, attracts the other strip and they pull together and touch. Why does the first strip attract the second strip?

e. When the door is closed, the magnet is near the reed switch. But as the door opens, the magnet moves away from the reed switch and the two iron strips spring apart. How can the control unit tell when the door opens?

5. A ground fault interrupter is a device that senses when some of the current flowing out one side of an electric outlet isn't returning through the other side of that outlet. The only way such an imbalance can occur is if some current is returning to the electric company through the ground. Because that accidental current path might include your body, the interrupter shuts off current as soon as it senses trouble.

a. Inside the interrupter, the two wires from the electric company pass together around the iron core of a single transformer, forming two identical primary coils. If the alternating current passing through each coil is equal in magnitude but opposite in direction, how does the magnetization of the transformer's core change with time?

b. If the alternating current passing through each coil isn't equal (perhaps because some current is escaping from a hair dryer plugged into the outlet), how does the magnetization of the transformer's core change with time?

c. The transformer's core has a secondary coil wrapped around it. If all of the current flowing to the hair dryer through one wire returns from it through the other wire, current passing through this secondary coil will experience no change in voltage. But if some of the current flowing to the hair dryer doesn't return, current in the secondary coil will experience a change in voltage. Explain.

d. The current from the secondary coil is used to trigger a switch that disconnects the outlet from the electric company. This switch has manual "test" and "reset" buttons. To test the ground fault interrupter, you press the "test" button. What can this button do to simulate a real current accident?

6. A typical audio speaker, such as that in a radio, consists of a permanent magnet, a coil of wire, and a movable speaker cone. The coil of wire is connected electrically to the radio's amplifier so that the radio controls how much current flows through the coil. The coil is attached to the speaker cone so that if the coil moves, the speaker cone moves and creates sound waves in the air.

a. When the amplifier sends an electric current through the coil, the coil is either attracted or repelled by the permanent magnet. Its copper wire isn't normally magnetic, so why does the coil experience forces from a magnet?

b. If the radio reverses the direction of current through the coil 500 times each second (250 complete cycles each second, or 250 Hz), what will happen to the speaker cone?

c. Pushing the speaker cone forward quickly compresses the air in front of the cone and requires energy from the radio. Show that the speaker cone does work on the air while compressing it.

d. When you talk in front of a speaker, the sound of your voice moves the speaker cone and the coil. The coil moves back and forth near the permanent magnet. What happens to the mobile electric charges in the wire coil?

e. If you connect two speakers together, so that the two wires from one speaker's coil connect to the two wires from the other speaker's coil, you will form a complete circuit. What will happen to the first speaker when you talk into the second speaker?

7. The needle of a magnetic compass is a permanent magnet. When the compass is level, the needle rotates easily about a vertical axis so that its north magnetic pole points toward the earth's north geographic pole.

a. What type of magnetic pole is located near the earth's north geographic pole?

b. The compass needle is sensitive to any nearby steel or iron. Whenever a piece of iron is held near the compass, one end of its needle turns toward that iron. Why is either end of the needle attracted to pieces of iron?

c. The needle is also disturbed by nearby electric currents. Why?

d. Some compasses have mechanisms that can prevent their needles from turning. If you lock a compass's needle in place and pass its north pole near the north pole of a very strong permanent magnet, you will spoil the compass. When you release the needle, it will point toward the earth's south geographical pole. What will have happened to the needle?

CHAPTER 10

Electronics

Electric currents and magnetized metals can do more than simply power light bulbs or record music. They can also represent many different things, from sounds, to video images, to computerized information. In this chapter, we'll see how electricity and magnetism give rise to the field of electronics.

Electronic devices are tools that manipulate electric currents in sophisticated ways. They first appeared in the early 20th century with the development of vacuum tubes and have been maturing ever since. The invention of transistors has accelerated the electronic revolution so that it now permeates every facet of modern society.

EXPERIMENT: Listening to Yourself Talk

In this chapter, we'll discuss some of the components used in electronic devices and the ways in which these devices perform their tasks. One such task is the amplification of audio signals—electronic representations of sound. You can observe this amplification by playing with an audio system.

This activity is most instructive if the audio system has a microphone and a microphone input, so that you can hear your voice played through the audio system. Many tape recorders have microphones built into them or have microphone inputs that you can use. So do many computers. Even if you don't have a real microphone, you can use your headphones as microphones because the speakers inside them can also work as microphones. A speaker becomes a microphone when you do work on

it with your voice. If you don't have any way to introduce your voice into the audio system, you'll have to make do with prerecorded sound or radio.

Turn on the audio system and connect the microphone (or headphone) to the microphone input. Talk into the microphone. If you have all the switches set properly—you must be in "record mode" if you're using a tape recorder—then your voice should be reproduced through the audio system's speakers or headphones.

Predict what will happen if you adjust each knob on the audio system and then *observe* the result. Were you able to *verify* your predictions? Did the base and treble knobs affect its sound? How are these knobs and buttons controlling the behavior of this electronic device? What characteristics of the audio system would you *measure* to see how well it performs, if you had the right tools?

While this electronic reproduction of sound may seem trivial in these days of CDs, DVDs, and digital stereo, it involves processes that would have astounded Thomas Edison or any other person in the 1800s.

Chapter Itinerary

To understand the remarkable properties of electronic devices, we'll examine two of them: (1) *audio amplifiers* and (2) *computers.* In *audio amplifiers,* we'll look at how small electric currents can be amplified into much larger currents and examine the components that make this amplification possible. In *computers,* we'll see how electricity and magnetism can represent both numbers and information and investigate the basic tools a computer uses to process them.

While this brief survey can't go beyond the most basic issues in electronics, it should provide a good basis on which to build. Once you understand how a few important electronic components work, it's not so hard to see how to combine them in ways that no one else has ever imagined. Like building blocks that control the movements of charged particles, electronic components can be put together to do almost anything.

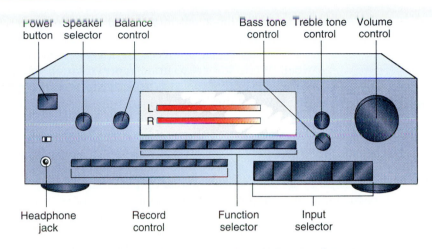

Power button · Speaker selector · Balance control · Bass tone control · Treble tone control · Volume control · Headphone jack · Record control · Function selector · Input selector

SECTION 10.1 # Audio Amplifiers

Audio amplifiers are electronic devices that reproduce and enlarge small electric currents. They're commonly found in radios, stereos, and musical instruments, where they scale up small currents representing sound and provide the large powers needed by speakers.

Early audio amplifiers were built with vacuum tubes, relatively bulky devices that wasted power and aged quickly. Transistors have made audio amplifiers much more practical. They perform the task of amplification so efficiently that most of the electric power used in an amplifier is delivered to the speakers. And because transistors are tiny and reliable, transistor amplifiers operate almost indefinitely without maintenance and are routinely built into games, toys, appliances, automobiles, and even greeting cards.

Questions to Think About: Why do audio amplifiers need electric power to operate? What does an audio amplifier's power rating mean? Why does an audio amplifier become warm as it operates? How does the volume control work? What do the treble and bass knobs on an audio amplifier do?

Experiments to Do: You almost certainly own something with an audio amplifier in it. A component stereo system has a large one. Find an amplifier and try to look inside it, perhaps through its cooling holes. You should see lots of small objects with wires protruding from their ends or sides. Are any of these familiar? Adjust the treble and bass controls and listen to how they change the sound. Do the same with the volume control. Can you explain what these controls are doing to the currents delivered to the speakers?

What an Audio Amplifier Does

There are many devices that produce audio signals, including microphones, musical instruments, tape recorders, and computers. A **signal** is a common term for any elec-

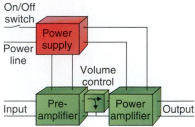

Fig. 10.1.1 A block diagram of a typical audio amplifier.

power supplies

Fig. 10.1.2 The electronic components on this "printed circuit" card are connected to one another by narrow metal strips. Because photographic techniques are used to form such strips on the surfaces of fiberglass cards, printed circuit cards are inexpensive to produce and extremely reliable.

tric representation of information, and an **audio signal** is an electric current that represents sound.

The audio signal produced by a typical microphone or instrument is weak. While it's an accurate representation of the sound, it doesn't have the power that speakers need to produce that sound at a reasonable volume. Something must first enlarge the audio signal; it needs to be *amplified*.

Devices that enlarge various characteristics of signals are called **amplifiers.** An audio amplifier is an amplifier that's designed to boost signals in the frequency range that we hear or feel (20 Hz to 20,000 Hz). It has two separate circuits—an input circuit and an output circuit—and it uses the small current passing through its input circuit to control a much larger current passing through its output circuit. In this manner, the amplifier provides more power to its output circuit than it consumes from its input circuit.

There are many different kinds of audio amplifiers, but most have the same principal components: a preamplifier, a power amplifier, and a power supply (Fig. 10.1.1). The preamplifier is designed to boost the small input signal from a tape deck, a CD player, or a microphone, to a level that the power amplifier is designed to use. The preamplifier is part of two circuits. Its input signal arrives through one circuit, and its output signal leaves through the other circuit. The preamplifier makes sure that the current reaching the power amplifier is within the proper range, protecting the power amplifier and improving its performance.

Just how much the preamplifier must boost the input signal depends on the input device. Some input devices, such as tape decks or CD players, send relatively large currents to the preamplifier, and their signals need little preamplification. Other devices, such as phonographs, produce such tiny currents that the preamplifier must boost their signals significantly. Because of the different preamplification requirements, a typical amplifier has several kinds of inputs, some for large signals and others for small signals.

The power amplifier resembles the preamplifier, except that the power amplifier starts with a relatively large input signal and creates an extremely powerful output signal, capable of delivering watts of power to a speaker system. A power amplifier doesn't have the subtlety of a preamplifier and is usually built of much larger electronic components.

Between the preamplifier and the amplifier is the volume control. The volume control is a device that splits the preamplifier's output current into two parts. One part is simply returned to the preamplifier directly, while the other part passes through the power amplifier's input. Since the power amplifier only amplifies current that flows through its input, its output signal strength is proportional to the fraction of preamplifier current it receives. When you turn up the volume, you increase that fraction.

Both the preamplifier and the power amplifier provide more power to their output circuits than they consume from their input circuits. They must obtain this power from somewhere, which is why they're connected to a power supply. This power supply does work on the current passing through it and is either a battery or a power-line operated device that imitates a battery.

While a typical audio amplifier contains hundreds of electronic components, often mounted together on *printed circuit cards* (Fig. 10.1.2), many of those components are relatively unimportant. They are there to ensure that the amplifier boosts all audio signals equally, regardless of strength or frequency. If we're willing to settle for less ideal performance, we can construct an audio amplifier from just a handful of parts. We'll do just that, after we first study those parts.

Check Your Understanding #1: Speak Up Please

Why can't you just connect the two wires of a microphone directly to the two wires of a speaker and have the speaker reproduce your voice loudly as you talk into the microphone?

Fig. 10.1.3 Resistors impede the flow of electric current, converting some of that current's power into thermal energy. The larger resistors can handle more thermal power without overheating.

Resistors

Our first electronic component is a **resistor,** which is simply two wires connected by a poor conductor of electricity (Figs. 10.1.3 and 10.1.4). Since current flows through a poor conductor only if there's an electric field pushing it forward, a current-carrying resistor must have a voltage drop through it. In accordance with Ohm's law (see Section 9.2), that voltage drop is proportional to the current in the resistor.

But the voltage drop through a resistor is also related to the resistor's electric resistance—to how poor a conductor it is. The larger its resistance, the less current flows through the resistor at a particular voltage drop. Some are relatively good conductors ("low" resistance) while others are relatively poor conductors ("high" resistance).

As we saw in Section 9.2, a resistor's electric resistance is defined as the voltage drop through it divided by the current flowing through it and is measured in ohms (abbreviated Ω). A resistor with a few ohms of resistance is nearly a good conductor while a resistor with a few million ohms of resistance is nearly an insulator. A k in front of the Ω means thousands (kΩ or kiloohms), and an M in front of the Ω means millions (MΩ or megaohms). Thus a 100-kΩ resistor has a resistance of 100,000 Ω and a 10-MΩ resistor has a resistance of 10,000,000 Ω. A resistor's resistance is marked on its cylindrical body, often in the form of brightly colored stripes. Ten different colors represent the digits 0 through 9, as well as various powers of ten, so that a resistor with stripes brown (1), black (0), and red (×100) is a 1000 Ω resistor.

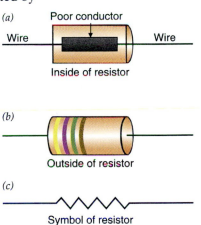

Fig. 10.1.4 (*a*) A resistor is two wires with a poor conductor of electricity between them. (*b*) It's usually encased in a cylindrical shell, with colored stripes to indicate its resistance. (*c*) In a schematic diagram of an electronic device, the resistor is represented by a zigzag line.

resistor types and labeling

Check Your Understanding #2: A Waste of Energy

Your radio produces sound when its amplifier allows current from its power supply to flow through its speaker. During a quiet moment, the amplifier sends only a small current through the speaker, and this current retains most of its electric power. How can the radio use up the current's remaining power before returning it to the power supply for reuse?

Fig. 10.1.5 Capacitors store separated electric charge. Each of these capacitors contains two conducting surfaces separated by a thin insulating layer.

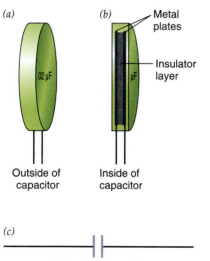

(a)   *(b)*   Metal plates

.02 μF   μF   Insulator layer

Outside of capacitor   Inside of capacitor

(c)

Symbol for capacitor

Fig. 10.1.6 (*a*) A capacitor is usually a disk or cylinder with two protruding wires. Its capacitance is printed on its surface. Inside (*b*), the wires are connected to two conducting plates that are separated by a thin insulating layer.

capacitor types

Capacitors

Our second electronic component is a **capacitor,** a device that stores separated electric charge (Fig. 10.1.5). It consists of two conducting plates separated by a thin insulating layer (Fig. 10.1.6). When one plate is positively charged and the other is negatively charged, the opposite charges attract one another. This attraction allows the plates to store large quantities of separated charge, while leaving the capacitor as a whole electrically neutral.

You can charge a capacitor's plates by transferring charge from its negative plate to its positive one. The work you do during this transfer is stored in the capacitor as electrostatic potential energy and is released when you let the separated charge get back together. A charged capacitor acts like a battery when connected to a circuit because it pushes charge through the circuit from its positive plate and collects that charge from the circuit with its negative plate.

Since (positive) charge has more electrostatic energy on the positive plate than on the negative plate, the voltage of the positive plate is higher than the voltage of the negative plate. The voltage difference between the plates is proportional to the separated charge on them. The more separated charge the capacitor is holding, the larger the voltage difference between its plates.

This voltage difference also depends on the structure of the capacitor. Enlarging the plates allows the like charges on each plate to spread out, so that they repel one another less strongly. Thinning the insulating layer between the plates allows the opposite charges on the two plates to move closer together, so that they attract one another more strongly. Both of these changes lower the separated charge's electrostatic potential energy and, consequently, the voltage difference between the plates.

Because these changes allow the capacitor to store separated charge more easily, they increase its **capacitance**—that is, the separated charge it holds divided by the voltage difference between its plates. The SI unit of capacitance is the coulomb-per-volt, also called the **farad** (abbreviated F). A capacitor with a farad of capacitance stores an incredible amount of separated charge, even at a low voltage difference, while a capacitor with a billionth of a farad of capacitance is much more typical. The Greek letter μ in front of the F means millionths (μF or microfarads), the letter n in front of the F means billionths (nF or nanofarads), and the letter p in front of the F means trillionths (pF or picofarads). A capacitor's capacitance is marked on its wrapper, often in an abbreviated form.

Check Your Understanding #3: Charging Up

Your stereo's amplifier obtains power from the power line. Since the power line delivers no power during the moments when its alternating current reverses direction, the stereo's power supply must store energy. How can it do this?

Diodes

Our third electronic component is a **diode,** a one-way device for current. A diode allows current to flow through it in one direction but not in the other. Although diodes

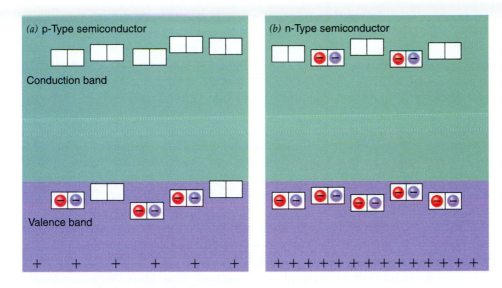

(a) p-Type semiconductor

Conduction band

Valence band

(b) n-Type semiconductor

Fig. 10.1.7 (a) p-Type semiconductor has missing electrons (and positive atomic nuclei) and can conduct electricity through its partly filled valence band. (b) n-Type semiconductor has extra electrons (and positive atomic nuclei) and can conduct electricity through its partly filled conduction band.

have taken many forms over the years, most modern diodes are built from semiconductors such as silicon. We discussed semiconductors in Section 8.2 and found that their properties are in between those of metals and insulators.

A semiconductor diode is made by joining together two different semiconductors. These two semiconductors aren't pure; they don't have perfectly filled valence levels and perfectly empty conduction levels. Instead, they're **doped** with atomic impurities that either create a few empty valence levels (**p-type semiconductor,** Fig. 10.1.7a) or place a few electrons in the conduction levels (**n-type semiconductor,** Fig. 10.1.7b). These empty valence levels or conduction level electrons allow p-type and n-type semiconductors to conduct electricity. The doping atoms bring with them just the right amount of positive charge in their nuclei to keep both p-type and n-type semiconductors electrically neutral.

But when a piece of p-type semiconductor touches a piece of n-type semiconductor, something remarkable happens: a **p-n junction** forms at the place where the two meet (Fig. 10.1.8). To reduce their potential energies, higher-energy conduction

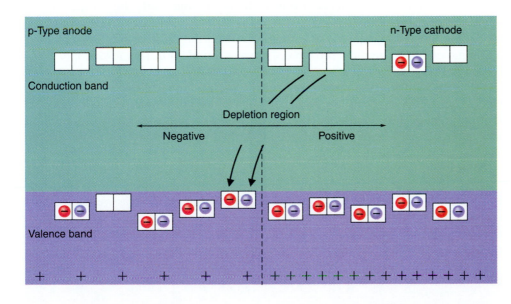

p-Type anode

Conduction band

Depletion region

Negative   Positive

Valence band

n-Type cathode

Fig. 10.1.8 When p- and n-type semiconductors touch, conduction level electrons flow from the n-type semiconductor to the p-type semiconductor, creating a thin, electrically polarized depletion region at the junction.

level electrons from the n-type semiconductor flow across the p-n junction and fill in empty lower-energy valence levels in the p-type semiconductor. This electron flow creates separated charge. The n-type semiconductor acquires a positive net charge because it now has fewer electrons than positive charges. The p-type semiconductor acquires a negative net charge because it now has more electrons than positive charges. Electrostatic forces from this separated charge oppose the further flow of electrons across the junction and gradually bring that flow to a halt. Everything is then in equilibrium.

Near the p-n junction, there is now a **depletion region**—an area in which electron flow has emptied all of the conduction levels and filled all the valence levels. With no conduction level electrons or empty valence levels left, the depletion region can't conduct electricity and charge can't move across the p-n junction. The depletion region is an insulator, and the two pieces of semiconductor have become a diode.

Recalling our theater analogy from Section 8.2, the p-n junction is analogous to a theater with two halves. In the left or "p-type" half, the balcony is empty, and even the ground floor has some empty seats. In the right or "n-type" half, the ground floor is filled and there are even a few people in the balcony. Since these two halves touch, people in the right balcony notice the empty seats in the left ground floor, and a few of them near the center of the theater cross from the right balcony to the left ground floor to take advantage of the better seats. Near the center of the theater, the ground floor is now filled and the balcony is empty, forming a depletion region in which no one can move left or right. The theater can't conduct people!

Returning to the actual diode, let's now look at what happens when we attach wires to each semiconductor half and try using a battery to push electrons across the p-n junction. If we push electrons leftward, adding them to the n-type side and removing them from the p-type side, the depletion region becomes thinner (Fig. 10.1.9a). We're adding electrons to the n-type conduction levels and pushing them toward the p-n junction. We're also removing electrons from the p-type valence levels and pulling them away from the p-n junction.

The extra electrons on the n-type side and the missing electrons on the p-type side create a voltage difference between the diode's two halves. When that difference reaches about 0.6 V (for a silicon diode), the depletion region becomes so thin that conduction level electrons in the n-type material begin to flow leftward across the junction into empty valence levels in the p-type material. The p-n junction is conducting electric current.

In the theater analogy, we're adding people on the right to the n-type balcony and removing them on the left from the p-type ground floor. The new people in the n-type balcony can move about the empty seats and migrate toward the center of the theater. Similarly, the empty seats in the p-type ground floor allow people to shift about so that empty seats become available near the center of the theater. At that point, people in the n-type balcony can cross over to the p-type ground floor and there is a net leftward flow of people through the theater. The theater is conducting people from right to left.

But what happens when we try to send electrons backward through the diode, pushing them into the p-type side and removing them from the n-type side (Fig. 10.1.9b). In that case, the depletion region becomes thicker as we fill in empty valence levels in the p-type side and remove conduction level electrons from the n-type side. The widening depletion region prevents charge from moving and no current flows across the p-n junction. It remains an insulator.

In the theater analogy, we're removing people on the right from the n-type balcony and adding them on the left to the p-type ground floor. Soon the n-type

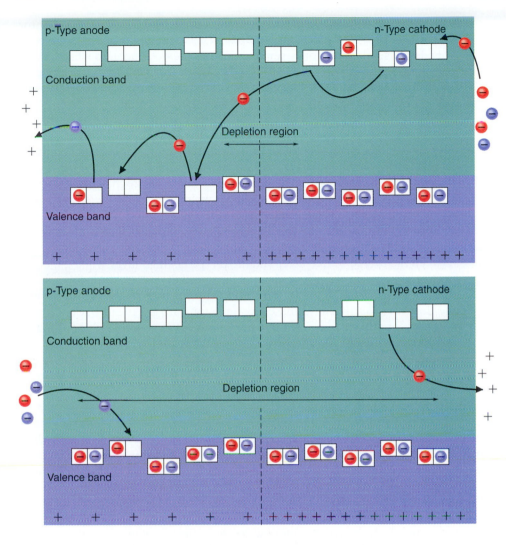

Fig. 10.1.9 (*a*) When you add electrons to the n-type side of a p-n junction and remove them from the p-type side, the depletion region thins and current can flow across the junction. (*b*) When you add electrons to the p-type side and remove them from the n-type side, the depletion region thickens and no current can flow across the junction.

balcony is virtually empty and the p-type ground floor is essentially full. The entire theater is now a depletion region and behaves like the insulator theater of Section 8.2. No one can move and the theater can't conduct people.

Since it allows current to flow in one direction but not the other, the p-n junction is a diode. For historical reasons, the diode's p-type side is called the *anode* and its n-type side is called the *cathode*. Current, which is the flow of positive charge, can only pass through a diode from its anode to its cathode.

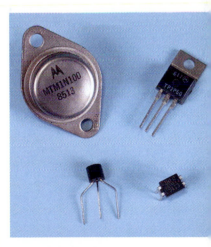

Check Your Understanding #4: It's a One-Way Street

What will happen if you include a p-n junction (a diode) in the AC circuit that connects the power company to your table lamp?

Transistors

Our fourth and final electronic component is a **transistor,** a device that allows a tiny electric charge to control the flow of a large electric current (Fig. 10.1.10). Invented in 1948 by three American physicists, William Shockley (1910–1989), John

Fig. 10.1.10 These MOSFETs make it possible for small electric charges to control large electric currents. The larger transistors can handle more electric power and thermal power without overheating.

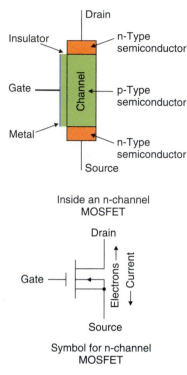

Inside an n-channel
MOSFET

Symbol for n-channel
MOSFET

Fig. 10.1.11 In an n-channel MOSFET, current tries to flow through a narrow channel of weakly doped p-type semiconductor. Normally, the channel has no conduction level electrons or empty valence levels to carry current. But when positive charge is placed on the gate, it attracts conduction level electrons into the channel. The channel becomes an n-type semiconductor and conducts current.

Bardeen (1908–1991), and Walter Brattain (1902–1987), transistors are now the basis for nearly all electronic equipment. Like diodes, transistors are built from doped semiconductors—semiconductors such as silicon to which chemical impurities have been added. But unlike diodes, which simply prevent current from flowing backward through a single circuit, transistors allow the current in one circuit to control the current in a second circuit.

While there are many types of transistors, the simplest and most important type is the *field effect transistor.* Actually, even here there are several types, so we'll focus on one that is widely used in audio, video, and computer equipment: the *n-channel metal-oxide-semiconductor field effect transistor* or *n-channel MOSFET.* Despite its complicated name, the n-channel MOSFET is a relatively simple device, consisting principally of three semiconductor layers and a nearby metal surface (Fig. 10.1.11). The three layers are called the *source,* the *channel,* and the *drain,* while the metal surface is the *gate.*

The three semiconductor layers form two p-n junctions. One junction is between the n-type source layer and the p-type channel layer while the other junction is between the p-type channel layer and the n-type drain layer. Each junction behaves like a diode, permitting current to flow through it in only one direction. But because these two junctions are arranged back to back, they prevent any current from flowing between the source and the drain. If you were to include this transistor in a circuit and tried to send current through it from the drain to the source, no current would flow. But there is a way to turn on the transistor so that current *can* flow through it. To see how this works, we need to study electrons in the three layers.

The channel is a p-type semiconductor, so it naturally has a few empty valence levels. The source and drain are an n-type semiconductor, so they naturally have some conduction level electrons. Contact between the three layers allows conduction level electrons to move out of the source and drain and into the channel's empty valence levels (Fig. 10.1.12a). The channel is lightly doped so that it has relatively few empty valence levels and these are entirely filled by electrons from the source and drain. The channel is therefore one big nonconducting depletion region, which is why no current can flow through it.

But if more electrons were somehow to enter the channel layer, they would have to go into its conduction levels. The channel layer would then have extra conduction level electrons, just like the n-type source and drain layers. In fact, the channel would then behave like an n-type semiconductor and the p-n junctions would vanish. Current would be able to flow through the transistor, from its drain to its source.

Drawing extra electrons into the channel from outside the transistor is the task of the metal gate. Separated from the channel by an incredibly thin insulating layer, the gate controls the channel's ability to carry current. When a tiny positive charge is placed on the gate through a wire, it attracts conduction level electrons into the channel and the channel begins to conduct current (Fig. 10.1.12b). The more positive charge there is on the gate, the more conduction level electrons enter the channel and the more current can flow. In effect, the transistor behaves like an adjustable resistor with a resistance that decreases as the positive charge on its gate increases.

We can now understand the n-channel MOSFET's name. "n-Channel" refers to the channel's n-type behavior when its gate is positively charged and the channel is carrying current. Although the channel is chemically p-type, it becomes electrically n-type when extra electrons are drawn into it and it acquires a negative net charge. "Metal-oxide-semiconductor" indicates that the metal gate is separated from the

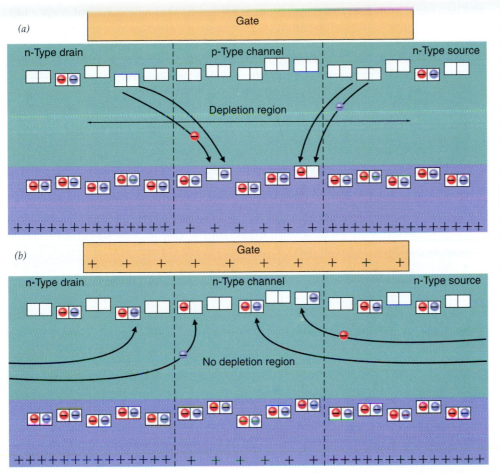

Fig. 10.1.12 (*a*) When an n-channel MOSFET is formed from three layers of n-, p-, and n-type semiconductor, conduction level electrons fill the empty valence levels in the central p-type channel and produce a vast, insulating depletion region. (*b*) But when positive charge is placed on the nearby gate, it attracts electrons from outside the MOSFET into the conduction levels and turns the entire structure into a conducting n-type semiconductor.

semiconductor channel by a thin insulating layer of oxide. This insulator is easy to puncture, which is why computers and other electronic devices are so easily damaged by static electricity. "Field effect transistor" indicates that the electric field from charge on the gate is what draws electrons into the channel and controls the current flow through the transistor.

► Check Your Understanding #5: Power and Control

Widening the channel of an n-channel MOSFET allows it to handle more current between its source and drain. However, the enlarged transistor needs more positive charge on its gate to control that current. Explain.

A Simple Audio Amplifier

Figure 10.1.13 shows a simple audio amplifier, built from the components we've just studied. The figure is a **schematic diagram**—that is, a conceptual drawing that uses symbols to represent each electronic component. The symbols used for the various components appeared in previous figures, and the lines between them represent wires connecting them together.

The amplifier has only five components: an n-channel MOSFET, two resistors,

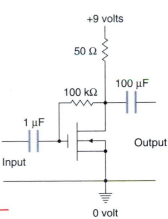

Fig. 10.1.13 A simple audio amplifier can be built with one n-channel MOSFET, two resistors, and two capacitors. A 9-V battery powers the device.

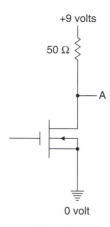

+9 volts

50 Ω

A

0 volt

Fig. 10.1.14 The voltage at A depends on the resistance of the MOSFET. The lined triangle at the bottom signifies connection to ground (often the earth itself).

and two capacitors. It draws power from a 9-V battery (or an equivalent power supply) and amplifies a tiny alternating current in its input wires into a large alternating current in its output wires.

To understand how the amplifier works, let's first remove everything but the MOSFET and the 50-Ω resistor (Fig. 10.1.14). Any current that passes through one of these components must also pass through the other. When the MOSFET doesn't conduct current, no current flows through the 50-Ω resistor and it experiences no voltage drop. So the voltage at A is 9 V. But if the transistor does conduct current, a voltage drop will appear through the 50-Ω resistor and the voltage at A will decrease.

The transistor will conduct current only if positive charge is put on its gate. That can be done by connecting the gate to A with a 100-kΩ resistor (Fig. 10.1.15). Since A is at +9 V, it's positively charged and pushes charge toward anything at lower voltage. Current flows slowly through the resistor from A to the gate. But as positive charge accumulates on the gate, the transistor begins to conduct current and the voltage at A drops. The charge at A gradually loses voltage until it can no longer flow through the resistor to the gate.

At that point, the amplifier has reached a stable situation with the voltage at A near 5 V and a modest amount of charge on the transistor's gate. The 100-kΩ resistor has provided feedback to the transistor. If the transistor conducts too little current, charge flows onto its gate and makes it conduct more. If the transistor conducts too much current, charge flows off its gate and makes it conduct less.

The amplifier is now exquisitely sensitive to small changes in the charge on the transistor's gate. If you add just a tiny bit more positive charge, down goes the voltage at A. If you remove just a tiny bit of positive charge, up goes the voltage at A. The amplifier's input signal adds or subtracts positive charge from the gate and the amplifier's output signal emerges from A.

The amplifier has two input wires. Current from a microphone or other source flows into the amplifier through one wire and returns through the other. But the input signal is not connected directly to the gate. Instead, it's connected to the gate through a capacitor (Fig. 10.1.16). The capacitor allows the voltage on the transistor's gate to be different from the voltage on the top input wire. This flexibility is important in most audio amplifiers.

While the capacitor doesn't allow charge to flow from one plate to the other, it does respond to currents. As positive charge flows onto one side of the capacitor through the upper input wire, this charge attracts negative charge onto the other side of the capacitor, drawing it away from the gate. The capacitor remains electrically neutral but the gate becomes more positively charged. Overall, charge flowing into the capacitor from the input wire has caused an equal amount of charge to flow from the capacitor to the gate.

When an alternating current flows back and forth between the two input wires, charge moves on and off the transistor's gate. As a result, the voltage at A varies up and down. Even a tiny alternating current on the input wires creates a large changing voltage at A.

All that remains is to use the changing voltage at A for some purpose. The amplifier has two output wires. Current from a speaker or other device flows into the amplifier through one wire and returns through the other. The amplifier provides power to this current. But the output signal isn't connected directly to A. Instead, it's connected to A through a capacitor (Fig. 10.1.17). Current flowing into one side of this capacitor causes a similar current to flow out the other side of the capacitor. The capacitor allows the voltage at A to be different from the voltage on the top output wire. Once again, this flexibility is useful in amplifiers.

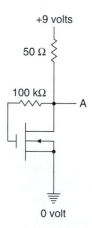

+9 volts

50 Ω

100 kΩ

A

0 volt

Fig. 10.1.15 The 100-kΩ resistor transfers positive charge to the gate until the voltage at A drops to about 5 V.

Tiny alternating currents through the two input wires can produce large alternating currents through the two output wires. This amplifier works remarkably well, given its simplicity. If you connect a microphone to the input wires and a speaker to the output wires, the speaker will do a surprisingly good job of reproducing the sound in the microphone.

However, our simple amplifier isn't perfect. It distorts the sound somewhat and it doesn't handle all frequencies or amplitudes of sound equally. It also wastes a large amount of electric power heating the 50-Ω resistor. Fancier amplifiers carefully correct for these problems. Many of them use feedback to make sure that their output signals are essentially perfect replicas of their input signals, only larger. They sense their own shortcomings and correct for them.

But perfect replication of the input signal isn't always desirable. Sometimes you want to boost the volume for part of the sound. The treble and bass controls on an amplifier allow you to selectively change the volumes for the high- and low-frequency portions of the sound, respectively. These controls often use resistors and capacitors to select and modify particular ranges of audio frequencies. Resistors slow the movement of charge, and capacitors store charge. Together, these devices can respond differently to alternating currents of different frequencies. Such resistor–capacitor "filters" are common in audio equipment.

An amplifier's power rating indicates the peak power it can deliver to the speakers during a loud passage. The average power should never reach this value. But even during what seems like only a moderately loud passage, there will be moments during which the current must become extremely large. Different tones can work together to create a sudden high or low pressure at the microphone and the speaker must follow this pressure. Even when a 100-W stereo is only producing 20 W of sound on average, there may be brief moments when it delivers 100 W of power to the speakers. If the amplifier is asked to deliver more power than it's capable of producing, the audio signal it sends to the speakers will be distorted and the sound will be unpleasant.

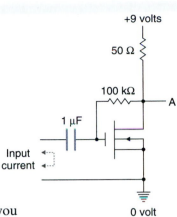

Fig. 10.1.16 Because current flowing back and forth through the two input wires affects the charge on the transistor's gate, it also affects the voltage at A.

Check Your Understanding #6: Sound Control

When you connect a microphone to the input of an MOSFET-based amplifier, the microphone sends current back and forth through the input wires. As a result, charge moves onto or off what critical control element in the amplifier?

 vacuum tubes, bipolar transistors

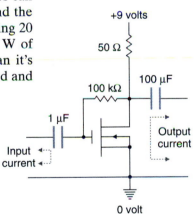

Fig. 10.1.17 The amplifier causes currents to flow back and forth through the two output wires. The alternating current in the output wires is a good replica of the alternating current in the input wires, only larger.

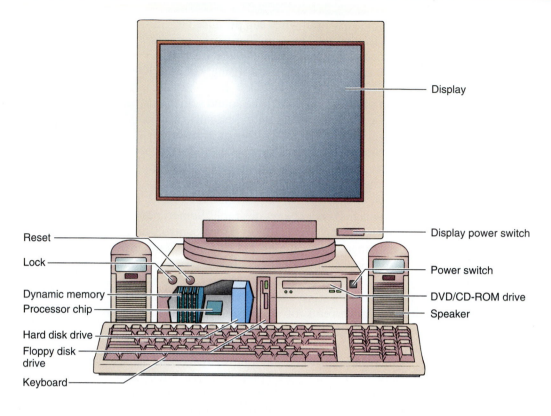

Display

Display power switch

Power switch

DVD/CD-ROM drive

Speaker

Reset

Lock

Dynamic memory

Processor chip

Hard disk drive

Floppy disk drive

Keyboard

SECTION 10.2   Computers

Computers are devices that process numbers and other information according to a collection of programmed rules. They aren't really intelligent: they just do exactly what they're told extremely rapidly. It's not clear whether computers make our lives easier or more frantic, but they certainly let us do things we couldn't otherwise do. Computers are not only found on desktops and at web sites but are also embedded inside everything from microwave ovens to automobiles. These embedded computers analyze information gathered by sensors and determine how the machines should respond. They've become so inexpensive that small, dedicated computers now do tasks once done by specialized electronic devices better and more cheaply.

Questions to Think About: *Why are most computers electronic rather than mechanical? How can electricity represent numbers? How can numbers be used to represent letters, words, and symbols? If people use computers to write computer programs, how are the first programs for a new computer written?*

Experiments to Do: *Computers are so common that you'll have no trouble finding one to study. Turn it on or start it up. The computer will begin to act on a series of rules that are stored in its memory until it eventually pauses for you to type or point. It will then act on what you type according to still more rules. Try describing its process of starting up, waiting for you to type something, and then acting on what was typed in* *terms of a series of rules like "draw rectangular box" or "read typed character from keyboard." How many of these simple rules do you think are contained in your favorite word processing package or computer game?*

Unlike a person who might respond differently each time you say hello, the computer responds exactly the same way each time you start it up. If it doesn't, then one of the rules it's following specifies the way in which it should vary its response. With no creativity of its own, the computer obeys its orders to the letter even when appearing unpredictable. (For an interesting note about such programming, see ❏.)

Some of the rules the computer follows affect what it displays on its screen. For example, it may display a character you typed. Does it have to do this or could it display a different character instead? Its rules may also involve storing the characters you type in its memory. In what situations should such storage occur?

Work with the computer for a while and try to think of it as a machine that's simply acting on your input—typing and pointing—according to specific rules and producing appropriate output—displayed material, printing, and sound. How does this view of a computer as a machine that responds to inputs with appropriate outputs apply to the computers that control automobiles or microwave ovens?

Representing Numbers

A computer represents numbers with physical quantities such as charge, current, voltage, and magnetization. There are two common techniques for this representation. The first technique relates the value of a physical quantity to the number itself. For example, a charge of 124 C could represent *124,* or a voltage of −2.313 V could represent *−2.313.* This representation technique is called **analog,** and computers that use it are *analog computers.* But while analog computers are frequently used in electronic devices such as in the noise reduction part of your tape deck, they are rarely used for calculations.

Most of the devices we think of as computers use a second representation technique called **digital,** in which numbers are first decomposed into digits and these digits are then individually represented by specific values of a physical quantity. For example, *124* could be decomposed into the decimal digits 1, 2, and 4, and then these decimal digits could be represented as three separate charges, currents, or voltages. Because each charge, current, or voltage only has to represent the integers from 0 to 9, its value doesn't have to be very accurate. If a voltage of 4 V is supposed to represent the digit 4, then 3.9 V or 4.1 V will still be understood to mean 4.

We usually break numbers into ones, tens, hundreds, thousands, and so on—the powers of 10—because we work in **decimal.** But we could also break numbers into ones, twos, fours, eights, sixteens, and so on. Instead of using the powers of 10, as in decimal, we would be using the powers of 2. This system for representing numbers with the powers of 2 is called **binary.**

In decimal, *124* is written as 124, meaning that *124* contains 1 hundred (10^2), 2 tens (10^1), and 4 ones (10^0). When these pieces are added together, $100 + 20 + 4$, you obtain *124.* In binary, *124* is written as 1111100, meaning that *124* contains 1 sixty-four (2^6), 1 thirty-two (2^5), 1 sixteen (2^4), 1 eight (2^3), 1 four (2^2), 0 twos (2^1), and 0 ones (2^0). When these pieces are added together, $64 + 32 + 16 + 8 + 4$, they again total *124.* This apparently complicated way to represent even a fairly small number is actually quite useful. The number has been broken into pieces that have

❏ In an odd twist of fate, the first computer program was written long before the first computer was built. That program was penned by **British mathematician Augusta Ada King, countess of Lovelace (1815–1852),** as part of her work with fellow **mathematician Charles Babbage (1791–1871)** on his "Analytical Engine." While Babbage's Engine wasn't actually built until 1991, Ada Lovelace is rightly considered to be the first computer programmer. The tradition of women in programming continued when **American mathematician and rear admiral Grace Murray Hopper (1906–1992)** became one of the first programmers on a real computer, Harvard's Mark I in 1944. Hopper was the first to apply the word *bug* to an unexpected computer failure, when she referred to some moths that had infested the Mark I.

only two possible values. There either *is* a thirty-two in the number being represented or there *isn't*. The only two symbols you need when representing a number in binary are 0 and 1.

Because *124* is 1111100 in binary, you could represent *124* by the charge, current, or voltage of seven separate objects. The first five objects would represent 1's while the last two would represent 0's. For example, if the seven objects were capacitors, the first five might hold separated charge while the last two might hold no charge. A device that needed to know what number the capacitors represented would measure their charges. Finding separated charge in the first five capacitors (11111) and no separated charge in the last two (00), it would determine that the capacitors represent 1111100 binary or *124*.

Binary is useful because fast electronic devices that turn currents on and off are relatively easy to build. It's much harder to build fast devices that deliver the specific currents, charges, or voltages needed for decimal or even analog representation. Analog representation is also susceptible to electronic imperfections and noise because an analog device that tries to represent *124* with a voltage of 124 V might accidentally produce 123 V or 125 V. Imagine a bank computer that couldn't tell $124 from $123 or $125!

Although representing *124* in binary takes at least seven separate quantities, there is no confusion in the number being represented. Computers that decompose numbers into binary in this manner are called *digital computers*. Because they can reliably represent numbers of any size, digital computers are much more precise and accurate than analog computers.

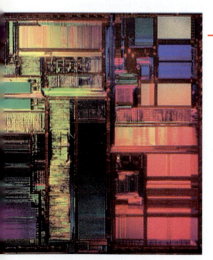

Fig. 10.2.1 This microscope photograph shows a Pentium integrated circuit microprocessor. Aluminum strips connect millions of MOSFET transistors and other components that have been formed by photographic techniques on the surface of a thin wafer of silicon.

> Check Your Understanding #1: New Math
> What number does binary 10000001 represent?

Storing Numbers and Information

Inside a digital computer, numbers are represented in binary form. The number of binary digits used to represent a number determines how large that number can be. Each binary digit is called a **bit,** and using more bits allows you to represent larger numbers. By associating these numbers with other things—words, sounds, colors, names, and so on—the computer can use bits to represent information. The more detailed that information, the more bits are needed to represent it.

Eight bits can be used to represent any number from *0* (which is 00000000) to *255* (which is 11111111). Since there are fewer than 256 of many common objects, these objects can be identified by groups of eight bits. For example, the keyboard symbols have all been assigned numbers between *0* and *255,* with *65* denoting the letter "A". Since the eight bits 01000001 represent *65,* they also specify an A. Groups of eight bits are so common and useful that they are called **bytes.**

There are several ways for a computer to store a bit. In the computer's main memory (often called random access memory or RAM), each bit is a tiny capacitor that uses the presence or absence of separated electric charge to denote a 1 or a 0. The computer stores a bit by producing or removing separated charge and recalls that bit by checking for that charge.

Each capacitor is built right at the end of its own tiny n-channel MOSFET. This

MOSFET controls the flow of charge to or from the capacitor. To store or recall a bit, the computer places positive charge on the gate of the MOSFET so that the MOSFET becomes electrically conducting. The memory system can then transfer charge to or from the bit's capacitor.

Storing the bit is relatively easy; the computer simply sends the appropriate charge through the MOSFET to the capacitor. But recalling the bit is harder because the charge on the capacitor is extremely small. Sensitive amplifiers in the memory system detect any charge flowing through the MOSFET from the capacitor and report what they find to the computer. Since this reading process removes charge from the capacitor, the memory system must immediately store the bit again.

Unfortunately, these tiny capacitors can't hold separated charge forever because it leaks out through their surroundings. Computer memory that uses charged capacitors to store bits is called *dynamic memory* and must be *refreshed* (read and restored) hundreds of times each second to ensure that a 1 doesn't accidentally switch to a 0 or vice versa.

The capacitors and MOSFETs used in dynamic memory and other computer components are formed on the surface of a thin wafer of silicon, creating an **integrated circuit** or "computer chip" (Figs. 10.2.1, 10.2.2, and 10.2.3). Careful processing, using photographic and chemical techniques, produces the tiny structures needed to make capacitors and MOSFETs that are less than 1 micron on a side. Since individual memory chips often store more than 64 million bits, they have more than 64 million capacitors and MOSFETs on their surfaces.

While dynamic memory is inexpensive, the time it takes to store and retrieve bits is relatively long and a computer with only dynamic memory will run rather slowly. That's why computers also use a faster but more costly type of memory, *static memory,* that employs a different technique for storing and retrieving bits. We'll discuss static memory later on, after we've studied the logic components from which it's built.

But using only static memory isn't cost effective either. Instead, a typical computer uses a large amount of cheap dynamic memory and a smaller amount of expensive static memory. The computer puts numbers it needs frequently in the faster static memory. When it needs numbers that are stored in the slower dynamic memory, it transfers them temporarily into the static memory, so that it can refer to them again quickly if necessary. This arrangement, in which numbers and information move from slow memory to fast memory on demand, is called *caching* and the fast memory is called a *cache.*

Numbers that are needed even less frequently than those in dynamic memory, or that must be retained even when the computer is off, are usually stored magnetically. Hard disks, floppy disks, and computer tapes all use the magnetic recording concepts we discussed in Section 9.3. The recording surface consists of a thin layer of tiny magnetic particles. In a hard disk, this layer rests on an extremely smooth aluminum disk (Fig. 10.2.4). In a floppy disk, it appears on both sides of a stiff plastic disk. And in a computer tape, this layer is coated on a thin strip of plastic.

 magnetic data storage

> Check Your Understanding #2: Like Sending a Letter to Yourself
> Every few thousandths of a second, a computer stops briefly to read and then rewrite every bit in its dynamic memory. What is going on?

Fig. 10.2.2 Each of these integrated circuits contains vast numbers of transistors, resistors, and capacitors. The two tiny chips in the foreground are unmounted.

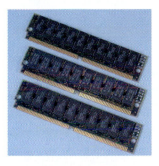

Fig. 10.2.3 These dynamic memory modules use tiny capacitors to store millions of bits.

Fig. 10.2.4 Each of the 3 platters in this hard disk stores many billions of bits. Microscopic read and write heads are mounted at the tips of the gold colored arms so that pivoting those arms can position the heads anywhere on the spinning platters.

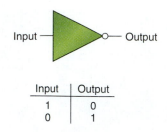

Input	Output
1	0
0	1

Fig. 10.2.5 An inverter, shown here symbolically, produces one output bit that is the inverse of its one input bit.

Processing Numbers and Information

We've seen how numbers and information can be represented and stored as bits, so now let's look at how a computer processes those bits. This processing is done by electronic devices that take groups of bits as their inputs and produce new groups of bits as their outputs. Since their output bits are related to their input bits by the rules of logic, these electronic devices are called *logic elements*.

The simplest logic element is the *inverter,* which has only one input bit and one output bit. Its output is the inverse of its input (Fig. 10.2.5). If an inverter's input bit is a 1, then its output bit is a 0 and vice versa. Inverters are used to reverse an action—turning a light on rather than off or starting a tape rather than stopping it. Inverters are also used as parts of more complicated logic elements.

But inverters aren't just abstract logic elements; they're real electronic devices. They act on electric inputs and create electric outputs. In household computers, inverters and other logic elements represent input and output bits with electric charge. Positive charge represents a 1 and negative charge represents a 0. Thus when positive charge arrives at the input of an inverter, the inverter releases negative charge from its output.

Inverters and other logic elements are usually constructed from both n-channel and p-channel MOSFETs. We've already seen that n-channel MOSFETs conduct current only when their gates are positively charged. p-Channel MOSFETs are just the reverse, conducting current only when their gates are negatively charged. The drain and source of a p-channel MOSFET are made from p-type semiconductor while the channel is made from n-type semiconductor. Since n-channel and p-channel MOSFETs are exact complements to one another, logic elements built from them are called Complementary MOSFET or *CMOS* elements.

A CMOS inverter consists of one n-channel MOSFET and one p-channel MOSFET (Fig. 10.2.6). The n-channel MOSFET is connected to the negative terminal of the computer's power supply and controls the flow of negative charge to the inverter's output. The p-channel MOSFET is connected to the power supply's positive terminal and controls the flow of positive charge to the output. When negative charge arrives at the inverter's input and moves onto the gates of the MOSFETs, only the p-channel MOSFET conducts current and the output becomes positively charged. When positive charge arrives at the input, the n-channel MOSFET conducts current and the output becomes negatively charged.

But a computer needs logic elements that are more complicated than inverters. One such element is the Not-AND or NAND gate. This logical element has two input bits and one output bit, and its output bit is 1 unless both input bits are 1's (Fig. 10.2.7). It's called a Not-AND gate because it's the inverse of an AND gate. An AND gate produces a 0 output unless both input bits are 1's. Simple memoryless logic elements are often called *gates*.

A CMOS NAND gate uses two n-channel MOSFETs and two p-channel MOSFETs (Fig. 10.2.8). The two n-channel MOSFETs are arranged in **series**—one after the next—so that current passing through one must also pass through the other. If either transistor has negative charge on its gate, no current can flow through the series. Components arranged in series must carry the same current, but they may experience different voltage drops.

The two p-channel MOSFETs are arranged in **parallel**—one beside the other—so that current can flow through either one of them to the output. If either transistor has negative charge on its gate, current can flow from one side of the pair to the

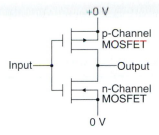

Fig. 10.2.6 When negative charge arrives at the input of a CMOS inverter, its p-channel MOSFET (top) permits positive charge to flow to the output. When positive charge arrives at the input, the n-channel MOSFET (bottom) sends negative charge to the output.

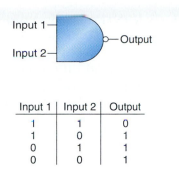

Input 1	Input 2	Output
1	1	0
1	0	1
0	1	1
0	0	1

Fig. 10.2.7 The output bit of a Not-AND or NAND gate, shown here symbolically, is a 1 unless both input bits are 1's.

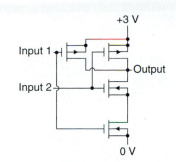

Fig. 10.2.8 A CMOS NAND gate has two input bits. When negative charge arrives through either input, the chain of n-channel MOSFETs (bottom) stops conducting current and one of the two p-channel MOSFETs (top) permits positive charge to reach the output. Only if both inputs are positively charged will negative charge reach the output.

other. Components arranged in parallel share the current reaching them through one wire and pass it on to the second wire. But while one may carry more of the current than the other, both experience the same voltage drop.

If negative charge arrives at either input of the CMOS NAND gate, the series of n-channel MOSFETs will be nonconducting and one of the p-channel MOSFETs will deliver positive charge to the output. But if positive charge arrives at both inputs, both p-channel MOSFETs will be nonconducting and the series of n-channel MOSFETs will deliver negative charge to the output. Thus the CMOS NAND gate has the correct logic behavior.

These two logic elements, inverters and NAND gates, can be combined to produce any conceivable logic element. For example, they can be used to build an adder, a device that sums the numbers represented by two groups of input bits and produces a group of output bits representing that sum. These adders can themselves be used to build multipliers, and multipliers can be built into still more complicated devices. In this fashion, the simplest logic elements can be used to construct an entire computer.

In fact, two NAND gates are all that's required to make one bit of static memory (Fig. 10.2.9). That bit is a *flip-flop,* an element that's capable of adopting two different stable arrangements or *states*. The flip-flop has two inputs that normally receive positive charge and two outputs that normally emit opposite charges—one positive and the other negative. You can set the state of the flip-flop by temporarily delivering negative charge to one of its inputs. The associated output then becomes positively charged and the other output becomes negatively charged. The flip-flop remains like that until you temporarily deliver negative charge to its other input.

Each bit of static memory uses one flip-flop. Since a flip-flop can store its bit extremely quickly and retrieve it without then having to rewrite it, static memory is much faster than dynamic memory. But since each flip-flop requires eight transistors, static memory is much more expensive than dynamic memory.

Actually, a computer isn't built exclusively from NAND gates and inverters. To improve its speed and reduce its size, it uses a few other basic logic elements as well.

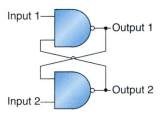

Fig. 10.2.9 A flip-flop built from two NAND gates. Delivering negative charge to one of the two inputs sets the flip-flop's state. Static memory is based on flip-flops.

Like the CMOS NAND gate and inverter, these elements are constructed directly from n-channel and p-channel MOSFETs.

All of these logic elements are wired together in an intricate pattern to create the final computer (Fig. 10.2.10). When you type a character at the keyboard, you send positive or negative charge to various logic elements and they act on that charge, storing a bit here, adding a bit there, and retrieving another bit from somewhere else. Moments later, the character appears on the display and becomes part of a document in your word processing package.

Check Your Understanding #3: Getting It Together

How could you use inverters and NAND gates to create an AND gate, a logic element that has two inputs and one output, with the output being 1 only if both of the inputs are also 1's?

Fig. 10.2.10 The processor chip of this personal computer appears at the top center, covered by a fan and black heat sink—a metal grid that transfers heat from the processor chip to the air. Cards of dyamic memory fit into slots to the right of the processor. Cables on the far right connect the processor to its magnetic disk drives.

The Limits to a Computer's Speed

Computers perform calculations extremely quickly. The simple NAND gate we examined above acts in about one-billionth of a second. That's the time it takes for the MOSFETs to start or stop conducting current and for the charge on the output wire to reach its proper value. Since scientists and engineers are continually improving the performances of MOSFETs, CMOS logic elements get faster every year.

But computer speed also depends on the time it takes for signals to travel between logic elements. While the signals from one element travel to other elements at almost the speed of light, that transfer speed limits how fast the computer can work. Each element must wait for its signals to arrive, and packing the elements closely together shortens the wait.

However, charged particles themselves travel much slower than light, so how can the signals move at nearly light speed? We can understand this by imagining a garden hose at the end of a closed water faucet. If you suddenly open the faucet, fresh water will begin to enter the hose. If the hose is already filled with water, water will begin to pour out of the hose long before fresh water from the faucet moves very far into the hose. In fact, water will begin to pour out of the hose almost as soon as you open the faucet. That's because the increase in water pressure that occurs when you open the faucet travels through the hose at the speed of sound.

Similarly, when charge enters a wire, its electric field pushes like charge ahead of it. Something resembling a pressure rise rushes through the wire at almost the speed of light, and like charge begins to flow out of the far end of the wire long before the original charge arrives there. Since light takes about 3 billionths of a second to travel a meter or about 1 billionth of a second to travel a foot, putting the entire computer on a single, coin-sized integrated circuit is a good idea. The logic elements are then so close together that they can exchange signals extremely quickly.

But there's another problem that slows a computer's calculations. Whenever a logic element changes its output from a 1 to a 0 or a 0 to a 1, it must change the charge on its output wire. If that wire is long and wide, it will have a fair amount of the old charge on it. Removing this old charge and replacing it with the new charge takes time because only so much current will flow through the tiny MOSFETs in the logic element. Thinning and shortening the wires, and otherwise reducing their capacitances, speeds a computer's operation.

speed and power consumption

> **Check Your Understanding #4: Thinner Wires Are Faster Wires**
> Halving the width of the wires between logic elements roughly doubles the speed of the computer. Explain.

Epilogue for Chapter 10

In this chapter we looked at two electronic devices to see how they use electricity and magnetism to perform useful tasks. In *audio amplifiers*, we saw how resistors, capacitors, and transistors are combined to enlarge small electric currents representing sound. We learned that resistors impede the flow of current, that capacitors store separated electric charge, that diodes allow current to flow only in one direction, and that transistors allow the current in one circuit to control the flow of current in another circuit.

In *computers*, we encountered some of these same components again, but this time in a different context. We examined the ways in which numbers and information can be represented using binary bits and electric charge, and then learned how arrangements of transistors can perform complicated logic operations on that representation. We found out what limits a computer's speed and learned why further miniaturization will make computers even faster.

Explanation: Listening to Yourself Talk

The sound of your voice causes pressure fluctuations at the surface of the microphone, and the microphone converts them into current fluctuations in an electric circuit. However, the fluctuating current flowing through the two wires of the microphone doesn't have the power needed to operate a speaker. That's where the amplifier comes in. When the small current from the microphone passes through the amplifier's input circuit, it produces a similar but larger current in its output circuit. This enlarged current flows through the two wires to the speaker and causes it to reproduce the sound of your voice. Adjustments to the audio system's controls, particularly the treble and bass knobs, alter its amplification for specific frequency ranges and can change the volume and tone of the reproduced sound.

Chapter Summary

How Audio Amplifiers Work: An audio amplifier permits a small current flowing through its input circuit to control a much larger current flowing through its output circuit. In a modern amplifier, this control is achieved with transistors, typically n-channel MOSFETs. The input current adds or subtracts charge from the gates of these transistors and changes their electric resistances. These transistors are connected to a power source, such as a battery or power supply, and determine how much power from the power source is transferred to the output wires. As the input signal varies, the transistors replicate its fluctuations in their output signal. However, the amplifier provides the output signal with more power than it consumes from the input signal. The audio amplifier thus provides power amplification.

How Computers Work: A computer processes numbers and other information by operating on them according to preprogrammed logic rules. It assigns numbers to all information and then represents those numbers as groups of binary bits. These bits can only take values of 0 or 1. The computer processes bits using logic elements that follow specific rules relating their inputs to their outputs. These logic elements respond to their inputs without any creativity or intelligence, acting only as they were designed to act.

The electronic components that carry out these logic operations are formed on the surface of an integrated circuit. The 1's and 0's are represented as positive and negative charge, and this charge moves about on the computer chip under the control of n-channel and p-channel MOSFETs. The arrangements of these MOSFETs determine the logic operations that they perform and thus the relationships between their inputs and outputs.

When not in use by the logic elements, bits are stored in several different ways. Static memory, itself a type of logic element, holds bits that are needed most frequently. Dynamic memory, which uses capacitors to store charge, holds bits that are needed less frequently. And bits that are rarely needed are stored as magnetization of a recording medium or on optical surfaces.

Check Your Understanding—Answers

Section 10.1 AUDIO AMPLIFIERS

1. The tiny amount of power provided by the microphone is not sufficient to make the speaker emit loud, powerful sound.

Why: You can't get something for nothing. If you want loud sound, you must provide the power to create that sound. An amplifier allows a tiny signal from a microphone to control the flow of power from a power supply to a speaker.

2. The amplifier can send the current through a resistor.

Why: By passing the current through a resistor, the amplifier converts the current's electric power into thermal power. The current leaves the resistor with only enough electric power to carry it back to the power supply for reuse. Amplifiers often use resistors to get rid of unusable electric power. While such actions may seem wasteful, it's extremely hard to use all of the power supply's electric energy to make sound. Controlling current is most easily done by impeding its flow, and that often means resistors and wasted electric power.

3. It can store the energy as separated charge on a large capacitor.

Why: The stereo's power supply saves energy in large electrolytic capacitors. During the times when the power line is providing plenty of electric power, the power supply uses that power to increase the separated charge in its capaci-

tors. When the power line isn't providing power, the power supply draws energy out of the capacitors. Unfortunately, electrolytic capacitors age quickly and often leak their liquid electrolyte. Once they leak, they can't store energy properly and your stereo will hum.

4. Current will only flow through the circuit half the time and your lamp will be dim.

Why: If you include a diode in an AC circuit, it will prevent current from flowing in one direction. Current will flow through the circuit only during the half of each cycle when the diode's cathode is negatively charged and its anode positively charged. Since the lamp will receive only about half of its normal power, its filament won't reach normal operating temperature. The lamp will glow dimly but the bulb will last an extraordinarily long time. Diodes are often used in this manner to create dim light levels in lamps or appliances.

5. The larger transistor also has a larger gate. With more surface over which to spread its charge, the gate needs more positive charge in order to draw conduction level electrons into the channel.

Why: MOSFETs range in size from remarkably small (less than 1 square-micron) to relatively large (several square-millimeters). The smallest ones are used in computer chips, where millions of MOSFETs are created on a single wafer of silicon only a centimeter square. A tiny charge on the gate of one of these MOSFETs will allow it to conduct current. The largest MOSFETs are used in power control devices such as amplifiers and power supplies. These transistors have large gates and much more charge is needed to allow one of them to conduct current.

6. The gate of a MOSFET.

Why: In all likelihood, the input current adds or removes charges from the gate of a MOSFET in the amplifier's first stage of amplification.

Section 10.2 COMPUTERS

1. 129

Why: Binary 1000001 contains only 1 hundred-and-twenty-eight (2^7) and 1 one (2^0). All the other powers of 2 are not present. Since 128 + 1 is *129,* that is the number represented by this binary value.

2. The computer is making sure that the charge stored on each capacitor in the dynamic memory adequately represents that bit's contents.

Why: Since charge leaks quickly from the capacitors in dynamic memory, the contents of each bit must be refreshed many times a second. This refreshing process, reading each bit and storing it back into memory, slows the computer down slightly.

3. You could connect an inverter to a NAND gate so that the output signal of the NAND gate is the input signal of the inverter. When both of the NAND gate's input signals are 1's, it will produce an output of 0. This 0 will arrive at the inverter, which will invert it and yield an output of 1.

Why: Connecting logic elements together one after the next is the standard method for producing more complicated logic elements. In this case, two elements produce a third.

4. Halving the widths of the wires also halves the amount of charge they store. Since a logic element trying to change the charge on a wire now has to move only half as much charge, it can act twice as fast as before.

Why: The wires in the computer act as capacitors, storing separated charge. The wire itself stores one type of charge (positive or negative) while the other type of charge can be located elsewhere on the computer chip. This stored charge in its wires slows down the computer because the logic elements must provide that charge. Thinning a wire reduces its capacitance and makes it less of a burden on the logic elements.

Exercises

1. Suppose you connected a microphone directly to a large unamplified speaker. Why wouldn't the speaker reproduce your voice loudly when you talked into the microphone?

2. Why can't an audio amplifier operate without batteries or a power supply?

3. You like to listen to old phonograph records but your new stereo amplifier has no input for a phonograph. You connect the phonograph to the stereo's CD player input but find that the volume is extremely low. Why?

4. To correct the volume problem in Exercise 3, you buy a small preamplifier and connect it between the phonograph and the CD player input of the stereo. The volume problem is gone. What is the preamplifier doing to fix the problem?

5. You have a small electronic game that runs on 6 V from four "AA" batteries. While you're on a trip, you want to operate it from the car's 12-V electric system. The game requires a current of 1 A, so you connect the game to the car's positive terminal through a 6-V resistor. Current from the car's positive terminal enters the resistor at 12 V but it's down to 6 V by the time it leaves the resistor and enters the game. The game then returns this current to the car's negative terminal and everything works beautifully. How does the resistor reduce the voltage, and what happens to the power associated with the missing voltage?

6. Why is it important that the resistor in Exercise 5 be rated to handle at least 6 W of electric power?

7. When you turn off the game (see Exercises 5 and 6), the voltage on both sides of the 6-V resistor becomes 12 V. Why did the voltage drop across the resistor change?

8. Suppose a battery is transferring positive charges from one plate of a capacitor to the other. Why does the work that the battery does in transferring a charge increase slightly with every transfer?

9. Two capacitors are identical except that one has a thinner insulating layer than the other. If the two capacitors are storing the same amount of separated electric charge, which one will have the larger voltage difference between its plates?

10. Even though capacitors can store large amounts of separated charge, they normally remain neutral overall. When positive charge flows onto one plate of a capacitor in an electronic device, negative charge will flow onto the other plate to leave the capacitor neutral. If the plates were initially uncharged, what effect will this movement of charge have on the voltages of the two plates?

11. The large electrolytic capacitors used to store energy in an amplifier's power supply sometimes leak. Since its liquid electrolyte acts as one of the plates of a capacitor, a loss of electrolyte effectively reduces the surface area of that capacitor's plates. If the voltage difference between the plates isn't allowed to change, does the amount of separated electric charge stored in the capacitor increase or decrease? Why?

12. In an n-channel MOSFET, the source and drain are connected by a thin strip of *p-type* semiconductor. Why is this device labeled as having an n-channel rather than a *p-channel*?

13. The gate of a MOSFET is separated from the channel by a fantastically thin insulating layer. This layer is easily punctured by static electricity, yet the manufacturers continue to use thin layers. Why would thickening the insulating layer spoil the MOSFET's ability to respond to charge on its gate?

14. MOSFETs are often used as fast switches. They can open or close circuits in a few billionths of a second and never wear out. Suppose you have a battery, a light bulb, an n-channel MOSFET, and some wires. How could you arrange those parts so that adding or removing positive charge from the MOSFET's gate would turn the light bulb on or off, respectively? (Draw a schematic diagram of your circuit.)

15. When your calculator is on, current flows from the battery's positive terminal, through the calculator, and back to the battery. However, on its way back to the battery, the current passes through an n-channel MOSFET that acts as the calculator's real on/off switch. When you push the "on" button, you put positive charge on the MOSFET's gate. When you push the "off" button, you remove that charge. Explain how these movements of charge control the current flowing through the processor.

16. Your calculator (see Exercise 15) turns itself off when you don't use it for a few minutes. How does it use the n-channel MOSFET in its main circuit to do this?

17. A MOSFET doesn't change instantly from a perfect insulator to a perfect conductor as you vary the charge on its gate. With intermediate amounts of charge on its gate, the MOSFET acts as a resistor with a moderate electric re-

sistance. This flexibility allows the MOSFET to control the amount of current flowing in a circuit. Explain why a MOSFET becomes warm as it controls that current.

18. Some light fixtures have a low-light setting that uses a diode. Instead of flowing directly to and from the light bulb, alternating current from the power company must first pass through the diode. Why does the diode halve the average power reaching the light bulb?

19. We combine the three decimal digits 6, 3, and 1 to form 631 in order to represent the number *631*. What does the 6 in 631 mean? What are there 6 of?

20. We combine the three binary bits 1, 0, and 1 to form 101 in order to represent the number 5. What does the leftmost 1 in 101 mean? What is there 1 of?

21. What numbers do the two binary bytes 11011011 and 01010101 represent?

22. How is the number *165* represented in binary?

23. Do the decimal representations 453 and 0453 refer to the same number? Why?

24. Do the binary representations 11101111 and 011101111 refer to the same number? Why?

25. Why are there no 2's in the binary representation of a number? (In other words, why isn't 1101121 a valid binary representation?)

26. Dynamic memory stores bits as the presence or absence of separated charge on tiny capacitors. It takes energy to produce separated charge, and a computer that minimizes this energy will use less electric power. Why does making the insulating layers of the memory capacitors very thin reduce the energy it takes to store each bit in them?

27. How could you use an n-channel MOSFET, a battery, a light bulb, and some wire to determine whether or not there is positive charge on one plate of a small capacitor?

28. The tiny MOSFETs that are used to move charge onto and off the capacitors in dynamic memory are so small that they're never very good conductors. Why do their modest electric resistances lengthen the time it takes to store or retrieve charge from the memory capacitors?

29. Why does the effect described in Exercise 28 limit the speed with which a computer can store or retrieve bits from its dynamic memory?

30. Computer memory is extremely sensitive to static electricity. If a spark were to introduce even a tiny amount of extra charge onto the capacitors in a dynamic memory chip, the bits stored in that chip would be spoiled. Why?

31. Each time a computer logic element places positive charge on a wire and then removes that positive charge, there is a net movement of charge from the computer power supply's positive terminal to its negative terminal. The more often the computer does this, the more energy it consumes. Why?

32. If you connect the output of one inverter to the input of a second inverter, how will the output of the second inverter be related to the input of the first inverter?

33. If two logic elements in a computer are connected by a wire that's twice as long as necessary, the second element will take twice as long to begin responding to the output of the first element. Why?

Cases

1. The bright red, green, and yellow lights that you find on many electronic devices are light emitting diodes or LEDs. Like any other diode, an LED carries current in only one direction. But unlike a normal diode, an LED emits a photon of light whenever an electron shifts from a conduction level in its n-type cathode to a valence level in its p-type anode.

   a. Explain why an LED's brightness is proportional to the electric current flowing through it.

   b. Light carries energy. Use conservation of energy to show that the current passing through an LED must experience a voltage drop.

   c. A photon of green light has more energy than a photon of red light. Why must an LED that produces green photons have a larger voltage drop than an LED that produces red photons?

   d. The power supply in a typical electronic device delivers current at too high a voltage for an LED. If that current were sent directly through the LED and then returned to the power supply, the LED would receive too much power, overheat, and burn out. To prevent such a disaster, the electronic device first sends the current through a resistor and then through the LED. How does the resistor protect the LED?

   e. An LED's color is determined by the energy needed to shift an electron across the band gap in the semiconductor from which the LED is made. Because blue photons have more energy than green photons, development of blue LEDs took a long time. The materials needed to produce blue LEDs are almost true insulators rather than semiconductors. What distinguishes a semiconductor from an insulator?

2. A Leiden jar is an early form of capacitor, consisting of a glass jar with metal coatings on its inside and outside surfaces. Because the two coatings don't touch, the jar can store separated charge on them and they act as the two plates of a capacitor.

   a. A metal post extends upward from the inner coating and rises above the top of the jar. Why does capping this post with a small metal sphere, rather than a sharp metal

point, improve the jar's ability to store separated charge?

   b. The jar is made of thick glass so that the two metal coatings are far apart. Why does transferring even a modest amount of charge from one coating to the other result in a huge voltage difference between the two coatings?

   c. If the glass of the Leiden jar were made thinner, you would have to transfer far more charge from one plate to the other to produce the voltage difference in part b. Why?

   d. If a Leiden jar is storing separated charge and you connect the two coatings with a wire, a bright spark will appear. From where did the energy for this spark come?

   e. Some Leiden jars use removable metal cans in place of the metal coatings. If you disassemble such a jar while it's storing separated electric charge, the voltage difference between the two cans will rise. What provides the additional energy associated with this increased voltage difference?

3. The power supply in your stereo amplifier uses alternating current from the electric company to provide direct current to the amplifier. This supply converts electric power from one form to another through the use of transformers, diodes, capacitors, and transistors.

   a. The power supply provides a relatively small voltage rise to the large current that it sends through the amplifier. The voltages provided by the electric company are too high for the amplifier and the currents are too small. How does a transformer make it possible for the power supply to provide a relatively small voltage rise to a relatively large current that flows between it and the amplifier?

   b. The power supply's transformer provides AC electric power through two wires, but the amplifier needs DC electric power through two wires. To fix this mismatch, the stereo connects these two pairs of wires with four diodes so that even though the currents through the two wires of the transformer reverse, the currents through the two wires of the amplifier don't reverse. How are those four diodes connected between the transformer and amplifier? (Draw a picture and indicate which way current can flow through each diode.)

c. While the diodes (see part b) ensure that current always flows in one direction through the amplifier, the transformer can't provide current when the current from the power company is reversing. To maintain a steady current through the amplifier, the power supply uses a large capacitor. When the transformer's current is large, some of that current is used to transfer charge between the capacitor's plates so that it stores separated charge. When the transformer's current is small, this separated charge is allowed to flow through the amplifier as current. Draw a graph of the voltage difference between the two plates of the capacitor versus time. Mark the times when the amount of separated charge is increasing and when it's decreasing.

d. The transformer, diodes, and capacitor do a pretty good job of sending direct current through the amplifier, but there still tend to be periodic fluctuations in that current. To keep the current stable, the amplifier uses a regulating device. Just before the current enters the amplifier, it passes through an n-channel MOSFET. An electronic sensor measures the current through the amplifier and determines if that current is too high or too low. It then adjusts the charge on the gate of the MOSFET to increase or decrease the MOSFET's electric resistance and lower or raise the current through the amplifier. If the sensor detects that the current is too high, should it increase or decrease the positive charge on the MOSFET's gate? Explain.

4. You enjoy listening to your little portable radio while jogging, but it hasn't been working properly since it got wet in the rain last week. The problem is that it keeps turning itself off. Because the radio makes no "click" sound when you press the "on" or "off" buttons and because it turns itself off automatically after an hour, you know that the switch that controls the radio's power is electronic. It's probably an n-channel MOSFET that's connected in series with the radio's electronics so that current from the battery must pass through both the electronics and the MOSFET before returning to the battery.

a. Why won't any power reach the electronics when the MOSFET isn't conducting current?

b. What must the "on" button do to make the n-channel MOSFET conduct current so that the radio will operate?

c. What must the "off" button or the automatic shutdown do to stop the n-channel MOSFET from conducting current, so that the radio will turn off?

d. Water is a poor conductor of electricity, but with patience you can send charge through it. If water is slowly turning off the radio, what is it probably doing?

e. You discover a small drop of water inside the "off" button, allowing positive charge to flow slowly from the gate of the n-channel MOSFET to the negative terminal

of the battery. You remove the water and the radio works perfectly. Why did the drop cause trouble and why did removing the drop fix the radio?

5. Your new electric guitar and amplifier sound great, although all the neighbors seem to have gone on vacation since you bought it last week. After 6 straight hours of jamming you decide to take a break and figure out how it works.

a. You begin by examining the strings and pickups—the devices that sense the strings' motions and represent them as electric currents. The pickup near each string is a coil of wire wrapped around a small permanent magnet. The permanent magnet magnetizes the steel string so that it induces an alternating current in the pickup coil as it vibrates back and forth. What provides the power for this alternating current?

*b. The electric power provided by the pickup is much too small to drive a speaker, so it must be amplified. The first step is to send it through a preamplifier. Current from the pickup flows through the preamplifier's input circuit and an amplified version of that current flows through the preamplifier's output circuit. The voltage rise at the preamplifier's output is 10 times as large as the voltage drop at its input, and the current passing through its output circuit is 10 times as large as the current passing through its input circuit. The power the preamplifier is providing is how many times as large as the power it's receiving from the pickup?

c. The output of the preamplifier is connected to the main power amplifier. You open a side panel of the amplifier and notice several large MOSFETs bolted to a heat sink. Since the amplifier was running recently (you wisely turned it off and unplugged it before opening it up), the heat sink is quite warm. These MOSFETs have been controlling the flow of current from the amplifier's power supply to your speaker, so their electric resistances have been fluctuating up and down. What has the amplifier been doing to the MOSFETs to make their electric resistances change?

d. Why have the MOSFETs been producing thermal energy?

e. You reinstall the amplifier's side panel and take a look at the speaker cabinet. This cabinet contains several speakers, each of which has a coil of wire that becomes magnetic when current flows through it. A nearby permanent magnet pushes on the magnetic coil, and this force moves a paper cone to create sound. The currents needed for high volume are large, and they flow to and from the speaker through the speaker wires. You notice that the speaker wires are warm—they have been wasting some of the power from your amplifier! You knew it was a mistake to buy cheap thin speaker wire, but it's too late now. You have extra wire, which you can use to reduce the wasted power. Should you add a second wire in parallel to each of the present wires or should you add a second wire in series with each of the present wires? Explain your answer.

CHAPTER 11

Electromagnetic Waves

We have seen that changing electric fields produce magnetic fields and that changing magnetic fields produce electric fields. These relationships between electric and magnetic fields allow them to create one another even in empty space. In fact, they can form electromagnetic waves, in which the two fields recreate one another endlessly and head off across space at an enormous speed. These electromagnetic waves are all around us and are the basis for much of our communications technology, for radiative heat transfer, and for our ability to see.

EXPERIMENT: Boiling Water in an Ice Cup

You can experiment with electromagnetic waves using a microwave oven. As we'll see in Section 11.3, microwaves are a type of electromagnetic wave that falls between radio waves and light. For reasons that we'll discuss in that section, microwaves can transfer energy to water molecules in a liquid but not to water molecules in an ice crystal. That difference allows for some interesting cooking tricks.

Take a block or cube of ice from the freezer and melt a shallow bowl-shaped depression in its top surface. Then put the block on a microwave-safe ceramic plate and return it to the freezer to cool. Once the block and plate are cold and frozen, take them out of the freezer and put them quickly into the microwave. Fill the block's bowl-shaped depression with water and immediately start the microwave.

If you've been quick enough, the water in the depression will still be liquid when the oven starts producing microwaves. The liquid water will absorb power from the microwaves that fill the cooking chamber but the ice will not. The water will become extremely hot and will begin to melt into the block. Why was it important to chill the plate before putting the block into the microwave oven?

If you're lucky and the block is large enough, the water in the depression will reach a full boil before it melts its way through the block. Try to *predict* how the shape of the liquid region will change as the microwaves heat it up. *Observe* what happens and try to *verify* your prediction. *Measure* how quickly the ice melts and

think about how you could use that measurement to calculate how much power the microwave oven is delivering to the water.

Chapter Itinerary

In this chapter, we'll discuss how electromagnetic waves are formed and detected in three common situations: (1) *radio,* (2) *television,* and (3) *microwave ovens.* In *radio,* we'll examine the ways in which charge moving on an antenna can emit or respond to electromagnetic waves and how those waves can be controlled in order to send an audio signal from a radio transmitter to a radio receiver. In *television,* we'll look at how beams of electrons inside a picture tube form a picture on the screen and consider how electromagnetic waves carry the information needed to produce that picture to the television. In *microwave ovens,* we'll see how microwaves affect water molecules and metals. While we can't see the electromagnetic waves that these three devices use, they clearly play important roles in our world.

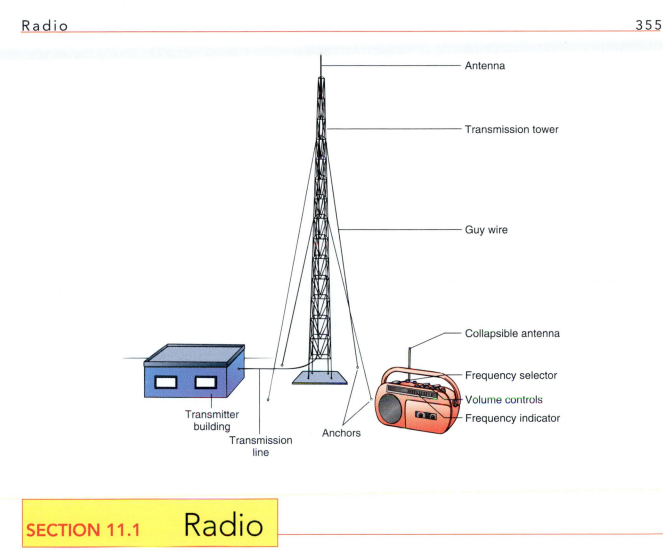

Antenna

Transmission tower

Guy wire

Collapsible antenna

Frequency selector

Volume controls

Frequency indicator

Transmitter
building

Transmission
line

Anchors

Radio

We've seen that electric currents can represent sound and carry speech and music anywhere wires will reach. But how can we send sound to someone who is moving? We
need a way to represent sound that doesn't involve wires. We need radio.

This section describes how radio works. We'll look at how radio waves are transmitted and how they're received. We'll also examine the common ways in which sound
is represented by radio waves so that it can travel through space to a receiver far away.

Questions to Think About: How might the movement of electric charge in one metal
antenna affect electric charge in a second antenna nearby? What about when the second antenna is far away from the first antenna? What does it mean when a radio station claims to transmit 50,000 watts? How does your radio select one channel from
among all the possibilities?

Experiments to Do: Listen to a small AM radio and notice that the volume of the
sound depends on the radio's orientation or location. Radio waves are pushing electric
charges back and forth along the radio's hidden internal antenna. You can sometimes
find an orientation in which the radio is silent because in that orientation the radio
waves are unable to move charges along the antenna. If you put the radio inside a metal
box, it will also become silent. Can you explain why?

You can try similar experiments with a cordless telephone—actually a radio transmitter and receiver. See how far you can go with the handset before you lose contact with the base unit. Notice that the antenna's size and orientation affect its range. What happens to the reception if you stand behind a large metal object?

Antennas and Tank Circuits

A radio transmitter communicates with a receiver via radio waves. These waves are produced by electric charge as it moves up and down the transmitter's antenna. We've already seen that electric charge produces electric fields and that moving charge produces magnetic fields. However, something new happens when charge *accelerates*. Accelerating charge produces a mixture of changing electric and magnetic fields that can reproduce one another endlessly and travel long distances through empty space. These interwoven electric and magnetic fields are known as **radio waves.** When radio waves pass a radio receiver's antenna, electric charge in that antenna begins to move up and down, too. If this motion is vigorous enough, the receiver will be able to detect the transmission.

But before we look at the structure of a radio wave and at how it travels through space, let's start with a much simpler situation. We'll look at how two nearby metal antennas affect one another. Figure 11.1.1 shows a radio transmitter and a radio receiver, side by side. Because of their proximity, electric charge on the transmitter's antenna is sure to affect charge on the receiver's antenna.

To communicate with the nearby receiver, the transmitter sends charge up and down its antenna. This charge's electric field surrounds the transmitting antenna and extends all the way to the receiving antenna, where it pushes charge down and up. Unfortunately, the resulting charge motion in the receiving antenna is weak and the receiver may have difficulty distinguishing it from random thermal motion or from motion caused by other electric fields in the environment. Therefore the transmitter adopts a clever strategy—it moves charge up and down its antenna rhythmically at a particular frequency. Since the resulting motion on the receiving antenna is rhythmic at that same frequency, it's much easier for the receiver to distinguish from unrelated motion.

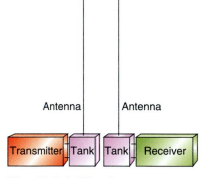

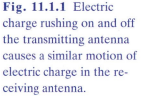

Fig. 11.1.1 Electric charge rushing on and off the transmitting antenna causes a similar motion of electric charge in the receiving antenna.

Using this rhythmic motion has another advantage: it allows the transmitter and receiver to each use a **tank circuit**—a resonant electronic device that behaves like a playground swing for electric charge. A tank circuit consists of only a capacitor and an **inductor**—a coil of wire that becomes magnetic when a current passes through it (Fig. 11.1.2). As we'll see shortly, charge swings or "sloshes" back and forth through the tank circuit at a particular frequency. And just as you can get a child swinging strongly at the playground by giving her a gentle push every swing, so the transmitter can make charge slosh strongly through its tank circuit by giving that charge a gentle push every cycle. By helping the transmitter move larger amounts of charge up and down the antenna, the tank circuit dramatically strengthens the transmission.

A second tank circuit attached to the receiving antenna helps the receiver detect this transmission. Gentle, rhythmic pushes by fields from the transmitting antenna cause more and more charge to move through the receiving antenna and its attached tank circuit. While the motion of charge on this antenna alone may be difficult to detect, the much larger charge sloshing in the tank circuit is unmistakable.

We can understand how a tank circuit works by looking at how charge moves between its capacitor and its inductor. The inductor is basically just an electromagnet whose magnetic field is proportional to the current flowing through it. Magnetic fields contain energy, so the inductor must extract energy from this current as its magnetic field grows stronger and return energy to the current as its magnetic field becomes weaker.

This exchange of energy between the inductor and the current gives rise to an interesting effect: an inductor opposes any change in the current flowing through it. As the current tries to increase, the inductor's rising magnetic field produces an electric field. That electric field pushes backward on the current and opposes the increase. As the current tries to decrease, the inductor's collapsing magnetic field again produces an electric field. However, this time the electric field pushes forward on the current and opposes the decrease. Thus current in an inductor changes relatively slowly.

Let's imagine that a tank circuit starts out with separated charge on the plates of its capacitor (Fig. 11.1.2a). Since the inductor conducts electricity, current begins to flow from the positively charged plate, through the inductor, to the negatively charged plate. But the current through the inductor must rise slowly and, as it does, it creates a magnetic field in the inductor (Fig. 11.1.2b).

Soon the capacitor's separated charge is gone and all of the tank circuit's energy is stored in the inductor's magnetic field (Fig. 11.1.2c). But the current keeps flowing, driven forward by the inductor's opposition to current changes. The inductor uses the energy in its magnetic field to keep the current flowing, and separated charge reappears in the capacitor (Fig. 11.1.2d). Eventually, the inductor's magnetic field decreases to zero and everything is back to its original state—almost. While all of the tank circuit's energy has returned to the capacitor, the separated charge in that capacitor is now upside-down (Fig. 11.1.2e).

This whole process now repeats in reverse. The current flows backward through the inductor, magnetizing it upside-down, and the tank circuit soon returns to its original state. This cycle repeats over and over again, with charge sloshing from one side of the capacitor to the other and back again.

A tank circuit is an electronic harmonic oscillator, equivalent to the mechanical harmonic oscillators we examined in Chapter 7. Like all harmonic oscillators, its period (the time per cycle) doesn't depend on the amplitude of its oscillation. Thus no matter how much charge is sloshing in the tank circuit, the time it takes that charge to flow over and back is always the same.

The tank circuit's period depends only on its capacitor and its inductor. The larger the capacitor's capacitance, the more separated charge it can hold with a given amount of energy and the longer it takes that charge to move through the circuit as current. The larger the inductor's **inductance**—its opposition to current changes—the longer that current takes to start and stop. A tank circuit with a large capacitor and a large inductor may have a period of a thousandth of a second or more, while one with a small capacitor and a small inductor may have a period of a billionth of a second or less.

Inductance is defined as the voltage drop across the inductor divided by the rate at which current through the inductor changes with time. This division gives inductance the units of voltage divided by current per time. The SI unit of inductance is

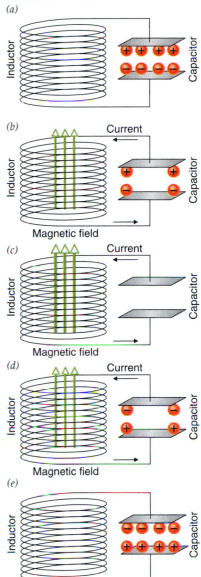

Fig. 11.1.2 A tank circuit consists of a capacitor and an inductor. Energy sloshes rhythmically back and forth between the two components.

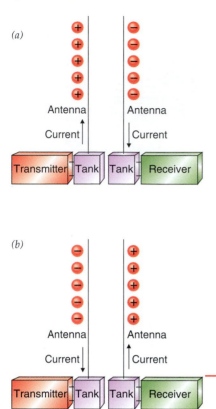

Fig. 11.1.3 (*a*) As current flows *up* the transmitting antenna, it causes current to flow *down* the receiving antenna. (*b*) As current flows *down* the transmitting antenna, it causes current to flow *up* the receiving antenna.

the volt-second-per-ampere, also called the **henry** (abbreviated H). While large electromagnets have inductances of hundreds of henries, a 1-μH (0.000001 H) inductor is more common in radio.

Its resonant behavior makes the tank circuit useful in radio. That's because small, rhythmic pushes on the current in a tank circuit can lead to enormous charge oscillations in that circuit. In radio, these rhythmic pushes begin when the transmitter sends an alternating current through a coil of wire. Fields from this coil push current back and forth through the nearby transmitting tank circuit, causing large amounts of charge to slosh back and forth in it and travel up and down the transmitting antenna. That charge's electric field then pushes rhythmically on charge in the receiving antenna, causing large amounts of charge to travel down and up it and slosh back and forth in the receiving tank circuit (Fig 11.1.3). The receiver can easily detect this sloshing charge.

Energy flows from the transmitter to the receiver via resonant energy transfer—from the transmitter, to the transmitting tank circuit and antenna, to the receiving antenna and tank circuit, and finally to the receiver. This sequence of transfers can work efficiently only if all the parts have resonances at the same frequency. Tuning a radio receiver to a particular station is largely a matter of adjusting its capacitor and inductor so that its tank circuit has the right resonant frequency.

Check Your Understanding #1: No Tanks

Why doesn't the radio transmitter simply push electric charge directly on and off the antenna, without using a tank circuit?

Radio Waves

When the two antennas are close together, charge in the transmitting antenna simply exerts electrostatic force on charge in the receiving antenna. But when the antennas are far apart, the interactions between them are more complicated. Charge in the transmitting antenna must then emit a *radio wave* in order to push on charge in the receiving antenna. Like a water wave, a radio wave is a disturbance that carries energy from one place to another. But unlike a water wave, which must travel in a fluid, a radio wave can travel through otherwise empty space, from one side of the universe to the other.

A radio wave is a type of **electromagnetic wave**—that is, a wave consisting only of a changing electric field and a changing magnetic field. These fields recreate one another over and over again as the wave travels through empty space at the **speed of light**—exactly 299,792,458 m/s (approximately 186,282 miles-per-second).

The radio wave is created when electric charge in the antenna accelerates. While stationary charge or a steady current produces constant electric or magnetic fields, accelerating charge produces fields that change with time. As charge flows up and down the antenna, its electric field points alternately up and down and its magnetic field points alternately left and right. These changing fields then recreate one another again and again, and sail off through space as an electromagnetic wave.

The wave emitted by a vertical transmitting antenna has a **vertical polarization**—that is, its electric field points alternately up and down (Fig. 11.1.4*a*). The distance between adjacent "ups" is its wavelength. For radio waves, that wavelength is usu-

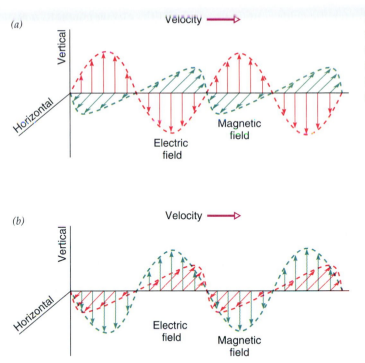

(a)

Velocity

Vertical

Horizontal

Magnetic field

Electric field

(b)

Velocity

Vertical

Horizontal

Electric field

Magnetic field

Fig. 11.1.4 (*a*) In a vertically polarized electromagnetic wave, the electric field points alternately up and down while the magnetic field points alternately right and left. (*b*) In a horizontally polarized wave, the electric field points alternately left and right while the magnetic field points alternately up and down.

ally 1 m or more. The wave's magnetic field is perpendicular to its electric field and thus points alternately left and right.

Had the transmitting antenna been tipped on its side, the wave's electric field would have pointed alternately left and right and the wave would have had a **horizontal polarization** (Fig. 11.1.4*b*). The wave's magnetic field would then point alternately up and down. Whatever the polarization, the electric and magnetic fields move forward together as a traveling wave, so the pattern of fields moves smoothly through space at the speed of light.

If you stood in one place and watched this wave pass, you'd notice its electric field fluctuating up and down at the same frequency as the charge that created it. When the wave passes a distant receiving antenna, it pushes charge up and down that antenna at this frequency. If the receiving tank circuit is resonant at this frequency, the amount of charge sloshing in it should become large enough for the receiver to detect.

A radio station can optimize its transmission by using a transmitting antenna of the proper length. When that length is exactly a quarter of the wavelength of the radio wave it's transmitting, charge sloshes vigorously up and down the antenna in a natural resonance. Surprisingly enough, the antenna is another electronic harmonic oscillator, with a period that depends only on its length. (In fact, the antenna is the top half of a tank circuit, with its tip acting as one plate of a capacitor and its length acting as the top half of an inductor. Objects at the base of the antenna complete the tank circuit.) When the transmitting tank circuit and antenna have resonances at the same frequency, there's a resonant energy transfer from one to the other. As you might expect, these resonant effects help to produce a powerful radio wave.

The transmitting antenna sends the strongest portion of its radio wave out perpendicular to its length. That's not unexpected because the motion of charge on the antenna is most obvious when viewed from a line perpendicular to its length. Thus a vertical antenna sends most of its wave out horizontally, where people are likely to receive it. No wave emerges from the ends of an antenna.

Both electric and magnetic fields contain energy, so as the electromagnetic wave travels through space, it carries energy away from the transmitter. When a radio station advertises that it "transmits 50,000 W of music," it's claiming that its antenna emits 50,000 J of energy per second or 50,000 W of power in its electromagnetic wave. The receiving antenna must absorb enough of this power to detect the wave. But the farther the wave gets from the transmitting antenna, the more spread out and weak it becomes. Trees and mountains also absorb or reflect some of the wave and hinder reception.

For the best reception, a listener should be located where the radio wave is strong and where there's an unobstructed path from the transmitting antenna to the receiving antenna. The receiving antenna should be a quarter wavelength long and it should be oriented along the radio wave's polarization—vertical for a vertically polarized radio wave or horizontal for a horizontally polarized radio wave. Aligning the receiving antenna with the wave's polarization makes certain that the wave's electric field pushes charge *along* the antenna, not *across* it.

To ensure good reception regardless of receiving antenna orientation, many radio stations transmit a complicated *circularly* polarized wave that combines both vertical and horizontal polarizations. To form this wave, they need several quarter-wavelength antennas. For wavelengths under a few meters, these antennas can all be attached inexpensively to a single mast. That's why commercial FM and TV broadcasts, which use short-wavelength radio waves, are usually transmitted with circular polarization. However, commercial AM broadcasts, which use long-wavelength radio waves, are transmitted only with vertical polarization.

Because commercial FM radio waves usually include both polarizations, FM receiving antennas can be vertical or horizontal. Portable FM receivers often use vertical telescoping antennas while home receivers frequently use horizontal wire antennas. All of these antennas are roughly a quarter wavelength long.

directional transmission

A quarter-wavelength antenna for commercial AM radio would have to be about 100 meters long, so straight AM antennas (such as those on cars) are much shorter than optimal. That's why many other AM antennas respond to the radio wave's horizontal magnetic field rather than its vertical electric field. This fluctuating magnetic field induces currents in a horizontal coil of wire. AM receivers often contain hidden or external coil antennas.

Check Your Understanding #2: There's No Place Like Home

Why do cordless telephones only work when they're close to their base units?

Representing Sound: AM and FM Radio

A radio transmitter does more than simply emit a radio wave. It uses that radio wave to *represent sound*. Because sound waves are fluctuations in air pressure and radio waves are fluctuations in electric and magnetic fields, a radio wave can't literally "carry" a sound wave. However, a radio wave can carry information to a radio receiver, instructing the receiver how to reproduce a particular sound. The radio wave is then *representing* the sound.

All the receiver needs to know in order to reproduce the sound is what the air pressure at its speakers should be. To convey this information, the radio station alters its radio wave to represent compressions and rarefactions of the air. There are two common techniques by which a radio wave can represent these pressure changes. One is called amplitude modulation and involves changing the overall strength of the radio wave. The other is called frequency modulation and involves small changes in the frequency of the radio wave.

In the **amplitude modulation** (AM) technique, air pressure is represented by the strength of the transmitted wave (Fig. 11.1.5). To represent a compression of the air, the transmitter is turned up so that more charge moves up and down the transmitting antenna. To represent a rarefaction, the transmitter is turned down so that less charge moves up and down the antenna.

The frequency at which charge moves up and down the antenna remains steady, so only the amplitude of the radio wave changes. The receiver measures the strength of the radio wave and uses this measurement to recreate the sound. When it detects a strong radio wave, it pushes its speaker toward the listener and compresses the air. When it detects a weak radio wave, it pulls its speaker away from the listener and rarefies the air.

In the **frequency modulation** (FM) technique, air pressure is represented by the frequency of the transmitted wave (Fig. 11.1.6). To represent a compression of the air, the transmitter's frequency is increased slightly so that charge moves up and down the transmitting antenna a little *more* often than normal. To represent a rarefaction, the transmitter's frequency is decreased slightly so that the charge moves up and down a little *less* often than normal. These changes in frequency are extremely small—so small that charge continues to slosh strongly in all the resonant components and reception is unaffected. The receiver measures the radio wave's frequency and uses this measurement to recreate the sound. When it detects an increased frequency, it compresses the air and when it detects a decreased frequency, it rarefies the air.

Although the AM and FM techniques for representing sound can be used with a radio wave at any frequency, the most common commercial bands are the AM band between 550 kHz and 1600 kHz (550,000 Hz and 1,600,000 Hz) and the FM band between 88 MHz and 108 MHz (88,000,000 Hz and 108,000,000 Hz). Elsewhere in the spectrum of radio frequencies are many other commercial, military, and public transmissions, including TV, short wave, amateur radio, telephone, police, and aircraft bands. These other transmissions use AM, FM, and a few other techniques to represent sound and information with radio waves.

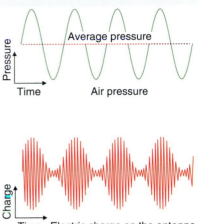

Fig. 11.1.5 When sound is transmitted using amplitude modulation, air pressure is represented by the strength of the radio wave. A compression is represented by strengthening the radio wave and a rarefaction is represented by weakening it.

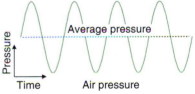

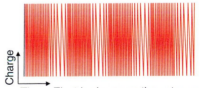

Fig. 11.1.6 When sound is transmitted by frequency modulation, air pressure is represented by changing the frequency of the radio transmitter slightly. A compression is represented by increasing that frequency and a rarefaction by decreasing it.

► Check Your Understanding #3: Another Volume Control
When you are listening to the AM radio in a car and drive through a tunnel, the volume becomes very low. Explain.

reception quality, modulation effects, information rate

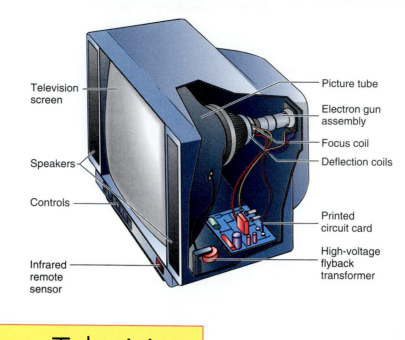

Television screen

Speakers

Controls

Infrared remote sensor

Picture tube

Electron gun assembly

Focus coil

Deflection coils

Printed circuit card

High-voltage flyback transformer

SECTION 11.2   Television

WARNING
Televisions contain dangerously high voltages, even after they have been unplugged. Since a television uses beams of high-energy electrons to create the images that you see, it needs high voltages. To avoid any risk of shock, you should never open a television while it's operating. But even turning the unit off doesn't necessarily make it safe. Because the television uses capacitors to store separated electric charge in its high-voltage power supply, high voltages can persist inside it for

In the previous section, we learned how radio sends sound through space via an electromagnetic wave. In this section, we'll see how television performs the much more complicated feat of sending pictures via an electromagnetic wave. To do this, it breaks each picture up into tiny dots and rapidly transmits information about those dots. As we explore television, we'll look at how a picture is represented by a video signal, how it's transmitted from one place to another, and how a television set turns that signal back into a picture. We'll also look briefly at the ongoing transition from analog to digital television.

Questions to Think About: *Why does it take a few seconds for a television set to warm up? Why are there tiny colored spots on the front surface of a picture tube? What happens when a picture tube "burns out"? If you take a photograph of the TV screen, why will you probably find that only part of the screen appears on the photograph?*

Experiments to Do: *One interesting but possibly costly experiment to do with a television set is to expose it to the field of a strong magnet. The magnet will deflect the stream of electrons flowing through the picture tube and create beautiful colors on the screen. However, just inside the surface of a color picture tube is a metal mask that's easily magnetized. If you hold a magnet to a color TV, this mask may have to be professionally demagnetized. But you can safely try this experiment with a black and white TV.*

A less risky experiment is to cut a narrow slot in a card and to watch a TV through this horizontal slot. Move the card up and down rapidly so that you only see the screen for a brief moment as the slot passes in front of your eye. You should see horizontal dark and light bands across the image. Can you explain why only part of the picture is illuminated? What happens when you hold the slot vertically and move it left and right?

Finally, use a magnifying glass to look at the screen of a color television set. You'll see a pattern of red, green, and blue dots or stripes. How can these simple colored dots create all of the colors that you see when you watch the television?

Creating a Television Picture

A television builds its picture out of tiny colored dots, arranged in a rectangular array on the screen. The number of dots in this array depends on the television standard, which varies with country. For the following discussion, we'll consider only the *NTSC* (National Television Systems Committee) analog color television standard used in the United States, which specifies an array 525 dots high by about 440 dots wide. Other analog television standards differ somewhat in the details but not in the concepts.

However, the world is in transition to new digital transmission standards, and these standards will replace the analog ones during the first decade of the new millennium. We'll examine those new standards later in this section. What we won't have room to examine is the gradual transition from picture tubes to flat panel displays. We'll only look at picture tubes.

While an NTSC analog television lights up its dots one by one, it finishes with all of them so quickly that it appears as though they're all illuminated at once. Your eye responds slowly to changes in light, and you see the whole pattern of dots on the screen as a single picture, a television picture. To create this picture, the television starts at the upper left-hand corner of the screen and scans through the dots horizontally from left to right. Every 1/15,750th of a second, it starts a new horizontal line. Since there are 525 horizontal lines, the television completes the entire picture every 1/30th of a second.

Actually, if the television worked its way from the top of the screen to the bottom in 1/30th of a second, our eyes would sense a slight flicker. To reduce this flicker, the television builds the image in two passes from top to bottom: it illuminates the odd numbered lines during the first pass and the even numbered lines during the second pass. That way, the television scans down the screen once every 1/60th of a second, and there is essentially no flicker at all.

minutes or more after you turn it off and unplug it. Don't open a television until you're sure that it has no more stored electric energy in its capacitors.

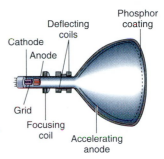

Fig. 11.2.1 The main components of a black and white picture tube. Electrons travel from left to right and illuminate the screen.

> ► **Check Your Understanding #1: Horizontal Hold**
>
> If you take a photograph of a television screen, using an exposure time of 1/250th of a second, the finished photograph will contain only a band of the television image, about a quarter of the screen high. Explain.

Black and White Picture Tubes

These dots of light are produced on the inside surface of a glass picture tube when electrons collide with a **phosphor**—a chemical that emits light when energy is transferred to it (Figs. 11.2.1 and 11.2.2). These electrons are emitted by a hot surface in the neck of the picture tube and accelerate toward positive charge on the phosphor-coated screen. When the electrons strike the phosphor, they transfer energy to it and it glows brightly.

But the television exerts forces on the electrons as they fly through the empty space inside the picture tube. These forces focus the electrons into a narrow beam and then steer that beam to various points on the phosphor screen. This focusing

Fig. 11.2.2 The left side of this small picture tube forms an electron beam that is then steered by the magnetic coils on the right side so that it hits the appropriate locations on the phosphor screen (not shown).

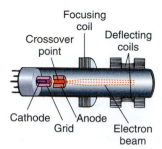

Fig. 11.2.3 The neck of a picture tube, showing how the electron beam comes to a focus once inside the anode and again at the screen.

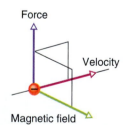

Fig. 11.2.4 Two of the three electron sources in a color picture tube. Each source has its own cathode, grid, and anode and is responsible for illuminating one of the colored phosphors.

Fig. 11.2.5 An electron moving through a magnetic field experiences a force that's at right angles to both its velocity and the magnetic field. A positively charged particle would experience a force in the opposite direction.

and deflecting are done by components in the neck of the picture tube, a region that's shown in detail in Figs. 11.2.3 and 11.2.4. For the moment, let's consider only a black and white picture tube.

The electrons emerge from a hot *cathode,* a device that uses thermal energy to emit electrons directly into space. This cathode is heated by a nearby filament and takes a few seconds to warm up when you turn on the television. If the filament breaks or burns out, the picture tube is ruined.

Surrounding the cathode is a hollow *grid* that is negatively charged. Since this grid repels electrons, most of the electrons leaving the cathode return to its surface. However, the grid has a small hole through which some of the electrons can escape. Once they escape through the grid, these electrons are attracted by a positively charged *anode* in front of them and accelerate forward.

The shape of the electric field between the cylindrical cathode, grid, and hollow anode has an interesting effect on the electrons: it focuses them to an extremely narrow spot inside the anode. Regardless of which way electrons were heading when they left the cathode's surface, they all accelerate toward the same point inside the anode, the *crossover point.*

But the electrons don't stop at the crossover point. Instead, they continue on through the anode and sail off toward the screen. After they have passed through the crossover point, the electrons are heading away from one another. They must be brought back together again so that they all strike the screen at exactly the same spot. This second focusing action is done by a magnetic field.

While it may seem surprising that a magnetic field can influence the path of an electron, we've already seen that moving charge is magnetic. A magnet pushes on charge flowing through the wires of an electromagnet, so why shouldn't that magnet push on charges flying through an empty picture tube?

As it passes through a magnetic field, a moving electron experiences a force that's proportional to its charge and velocity, and to the strength of the magnetic field. Called the Lorentz force, after Dutch physicist Hendrik Antoon Lorentz (1853–1928), this force also depends on the angle between its velocity and the magnetic field. It's strongest when the velocity and field are at right angles to one another and diminishes to zero when they both point in the same or in opposite directions. Oddly enough, the Lorentz force acts at right angles to both the electron's velocity and the magnetic field (Fig. 11.2.5).

A wire coil that circles the neck of the picture tube creates the magnetic field that focuses electrons to a spot on the screen. This field points directly toward the screen. Since the electrons' velocities also point toward the screen, the Lorentz forces on them are small. But if we look down the neck of the tube at the screen (Fig. 11.2.6), we see that these electrons have small outward components of velocity that take them away from the center of the tube and the crossover point they just left behind. These outward components of velocity are at right angles to the magnetic field, so the electrons experience Lorentz forces.

Each electron accelerates toward its right as it flies through the magnetic field so that it travels in a spiral. Viewed down the neck of the tube, this motion appears circular and is called **cyclotron motion,** after the particle accelerator based on it. Remarkably, the time each electron takes to complete one full circle of cyclotron motion doesn't depend on how large that circle is. All of the electrons complete exactly one full circle on their flights from the crossover point to the screen and hit the screen at the same spot!

The television also uses magnetic fields to steer the electron beam to different parts of the screen. A pair of coils above and below the picture tube's neck produces a vertical magnetic field that deflects the electron beam horizontally (Fig. 11.2.7). By adjusting the amount and direction of current in these horizontal deflecting coils, the television can control the horizontal position of the beam spot on the screen.

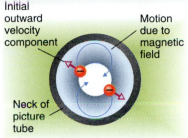

A second pair of coils mounted to the left and right of the tube's neck produces a horizontal magnetic field that deflects the electron beam vertically (Fig. 11.2.8). The amount and direction of current in these vertical deflecting coils determine the vertical position of the beam spot on the screen.

When the electron beam strikes the phosphor coating on the inside of the screen, it transfers energy to that phosphor, which then emits white light (recall this is a black and white television). Creating a bright image takes lots of energy, so the electron beam is accelerated on its way to the screen. A high-voltage power supply ($+15,000$ V to $+25,000$ V) pumps positive charge onto the inside of the screen and the surrounding accelerating anode, and this charge attracts the electrons. By the time they hit, the electrons have enough kinetic energy to make the phosphor glow bright white.

But white isn't the only choice in a television picture. To produce a gray or black spot, the television reduces the current of electrons in the beam. It controls this current by adjusting the charge on the picture tube's grid. The more negative charges on the grid, the harder it is for electrons to pass from the cathode to the anode and the fewer of them that strike the phosphor. The television carefully adjusts this grid charge as it sweeps the electron beam back and forth across the screen and thus creates a complete television picture, one dot at a time.

Fig. 11.2.6 As diverging electrons move down the neck of the picture tube toward the screen, the focusing magnetic field makes them accelerate to their right. They travel in a spiral and return together just as they hit the screen.

► Check Your Understanding #2: Currents in Time

The electron beam is swept from left to right by changing the current through the horizontal deflection coils. If each sweep takes 1/15,750th of a second, how does the current through these deflection coils vary with time?

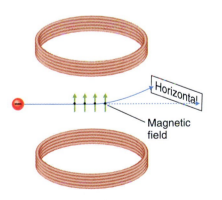

Horizontal deflecting coils

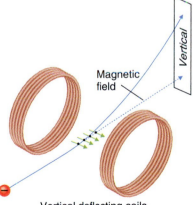

Vertical deflecting coils

Fig. 11.2.7 The horizontal deflecting coils are located above and below the neck of the picture tube. The vertical magnetic field they produce deflects the electron beam either left or right.

Fig. 11.2.8 The vertical deflecting coils are located to the left and right of the picture tube's neck. The horizontal magnetic field they produce deflects the electron beam either up or down.

(a)

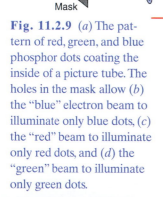

Phosphor pattern

(b)

Blue beam

Mask

(c)

Red beam

Mask

(d)

Green beam

Mask

Fig. 11.2.9 *(a)* The pattern of red, green, and blue phosphor dots coating the inside of a picture tube. The holes in the mask allow *(b)* the "blue" electron beam to illuminate only blue dots, *(c)* the "red" beam to illuminate only red dots, and *(d)* the "green" beam to illuminate only green dots.

Fig. 11.2.10 A color television creates its colored image with millions of tiny colored dots. This is a magnified photograph of those dots.

Color Picture Tubes

Color picture tubes work much like black and white tubes, except that they have three electron beams and phosphors that emit red, green, and blue light. As we'll discuss in the next chapter, mixtures of red, green, and blue light can make you perceive any color. A color television appears full color by carefully mixing these three colored lights.

The inside surface of a color television screen is coated with thousands of tiny phosphor dots (Figs. 11.2.9*a* and 11.2.10). Some of these dots emit red light, some green light, and some blue light. The television directs electrons at these dots through holes in a metal mask. Three separate electron beams, coming from three slightly different angles, pass through the holes and strike the phosphors. Since each beam can only strike one color of phosphor dots (Fig. 11.2.9*b–d*), each beam controls the brightness of one of the three colors.

While some picture tubes use phosphor stripes rather than dots, and a grille rather than a mask, the basic idea is the same. Only one electron beam can hit a particular phosphor stripe. These masks and grilles are carefully aligned inside picture tubes, and because they must stay aligned even when heated by the electron beams, they're made from special thermally compensated metals. These metals are easily magnetized, which is why you shouldn't hold a strong magnet near a color television. The resulting magnetization would then deflect the electron beams and produce distorted images and colors. Fortunately, many modern televisions and computer monitors have built-in degaussing systems that demagnetize their picture tubes when you turn them on.

Check Your Understanding #3: The Blue and the Red

The three electron beams in a color picture tube always converge to a single spot on the screen, even when they're being deflected. How does the momentum transferred to electrons in the "blue" beam by this deflection compare to that transferred to electrons in the "red" beam?

Analog Television Transmission

Representing a picture with a current or an electromagnetic wave is much more complicated than representing sound. Known as a video signal, this representation must indicate the brightness and color of every dot in the picture many times a second. In the NTSC analog standard, the video signal is a fluctuating voltage or radio wave amplitude that denotes the brightness and color of the dots, one at a time. In the newer digital standards, the video signal carries compressed digital information that allows a special purpose computer to recreate the images.

The basic structure of an NTSC black and white video signal is relatively simple. Dot by dot, line by line, it indicates how bright the screen should be as the television scans its way through the picture (Fig. 11.2.11). To indicate when each new line begins, the video signal briefly shifts outside its normal range and takes a value beyond the one that indicates full blackness. The television responds to this *horizontal sync signal* by steering its electron beam back to the left edge of the screen and starting a new line. To indicate when a new picture begins, the video signal extends beyond blackness for a longer time (not shown in Fig. 11.2.11) and this *vertical sync signal* causes the television to steer its electron beam back to the top of the screen.

An NTSC color video signal contains the same brightness and synchronization signal, so that it's compatible with black and white televisions. But it also contains a separate color signal. These two signals, called *luminance* and *chrominance,* respectively, are combined into the same video signal through some extremely clever electrical engineering. But rather than examining that accomplishment directly, let's take a look at why combining the signals was so difficult. It's a matter of bandwidth.

A video signal is sent by varying the voltage on a wire or the amplitude of a radio wave—essentially the AM technique we discussed in the previous section. The more pieces of video information that are sent each second, the faster these voltage or amplitude variations must occur. As they travel through wires or space, these rippling variations themselves have frequencies and wavelengths and thus complicate the simple picture of a voltage on a wire or a pure radio wave in space.

A video signal on a wire has a *range* of frequencies present in it, from zero frequency up to some limit that depends on how much information is being sent each second. Similarly, a video signal traveling via a radio wave has a range of frequencies present in it, stretching from well below the official frequency of the wave, the *carrier frequency,* to well above it. Once again, the more information being sent each second, the wider the range of frequencies in the signal, a range that's known as the signal's **bandwidth.**

Competing television stations must be assigned widely different carrier frequencies to avoid having any of their frequencies overlap. An NTSC video signal requires 6.0 MHz of bandwidth: 1.25 MHz below the carrier frequency and 4.75 MHz above it. As a result, broadcast television channels are assigned carrier frequencies 6 MHz apart. Now we can understand why the NTSC color standard is so remarkable: 6 MHz isn't very much bandwidth, yet it's enough to display sharp, full-color images that move smoothly across the screen.

Consider radio stations for comparison. By international agreement, an AM radio station may use 10 kHz of bandwidth, 5 kHz above and below its carrier frequency. To stay within that bandwidth, the audio signal can't contain frequencies above 5 kHz. While this restricted frequency range is bad for music, it allows competing stations to function with carrier frequencies only 10 kHz apart, so that 106 different stations can operate between 550 kHz and 1600 kHz.

An FM radio station may use 200 kHz of bandwidth, 100 kHz on each side of its carrier frequency. This luxurious allocation permits FM radio to represent a very broad range of audio frequencies, in stereo, which is why an FM radio station can do a much better job of sending music to your radio than an AM station can.

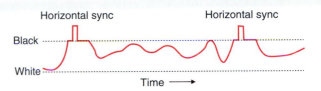

Fig. 11.2.11 A black and white video signal represents the brightness of the picture line by line. A new horizontal line is indicated when the video signal briefly extends beyond black.

Check Your Understanding #4: Beating the Bandwidth

Video tape recorders store video and audio signals on magnetic recording tape. The bandwidth available to the recording process is limited. Older VCR standards such as VHS, Beta, and 8 mm allow less than 6-MHz bandwidth for recording while some of the modern standards such as Super-VHS, Super-Beta, and Hi-8 mm approach or exceed 6 MHz. How does this recording bandwidth affect picture quality?

Cable Television Transmission

Cable television is similar to broadcast television except that it sends electromagnetic waves through a cable rather than empty space. Television cable consists of an insulated metal wire inside a tube of metal foil or woven metal mesh. This wire-inside-a-tube arrangement is called **coaxial cable** because its two metal components share the same centerline or axis. Electromagnetic waves can propagate easily through a coaxial cable, following its twists and turns from the company that produces the waves to the television receiver that uses them. The fact that wires are assisting these waves in their travels makes them more complicated than waves in empty space. However, they still involve electric and magnetic fields and still propagate forward at nearly the speed of light.

Because the electromagnetic waves inside a coaxial cable don't interact with those outside it, the whole electromagnetic spectrum inside the cable can be used for television channels. Since the cable can handle frequencies up to about 1000 MHz, it can carry about 170 channels. Adding more channels is just a matter of running more additional cables.

However, coaxial cables must now compete with optical fiber cables that guide light from one place to another. We'll examine optical fibers in Section 13.2. Like radio waves, light is an electromagnetic wave and can be amplitude modulated to represent information. But light's frequency is extremely high; the frequencies of visible light range from 4.5×10^{14} Hz to 7.5×10^{14} Hz. If we were to allocate television channels 6 MHz apart throughout the visible spectrum, there would be about 50 million channels!

► **Check Your Understanding #5: Networking**

Computers often communicate with one another through networks of wire or optical fiber cables. How does this network resemble cable television?

Digital Television

Despite its sophisticated engineering, the NTSC analog standard wastes most of its bandwidth. That's because it transmits large amounts of repetitive and nonessential information. Television images usually don't change much from picture to picture and even within a single picture, there are often regions containing simple color patterns that don't need to be specified dot-by-dot. Since analog television transmits every dot of every picture, it can't avoid this waste. But the new digital transmission standards can. By compressing out unnecessary information, these new standards allow the images conveyed by a digital television signal to have far more detail than analog television, while keeping the amount of transmitted information at a manageable level.

The new *ATSC* (Advanced Television Systems Committee) digital television standards include a variety of resolutions, picture shapes, and update rates. Its most spectacular picture resolution is 1080 lines tall by 1920 dots wide. Often called HDTV (High Definition Television), this standard has roughly eight times the resolution of the NTSC analog standard, so the pictures look amazingly clear. Moreover, because the transmission is digital, it's essentially free of visual noise—the pictures are flawless.

To send these HDTV images as often as 60 times a second is a demanding process. Much of it is computer science; the picture and sound compression/decompression techniques are similar to those used by web sites to display images or present audio programs. Where electromagnetic waves enter the story is in the actual transmission of those compressed images.

Instead of representing brightness and color as the amplitude of an electromagnetic wave, digital transmission converts the pictures into a stream of numbers and then represents those numbers via the amplitude of an electromagnetic wave. The numbers are divided not into binary (base-2) digits like most computers, but into octal (base-8) or hexadecimal (base-16) digits. This means that instead of having the radio wave amplitude take only 2 different values, 0 and 1, the amplitude takes either 8 or 16 different values. By packing more information into each digit this way, the transmission becomes more efficient.

The choice of octal versus hexadecimal depends on the quality of the electromagnetic path between transmitter and receiver. Since the television has to distinguish reliably between the 8 or 16 possible values for each digit, a noisy transmission path isn't suitable for hexadecimal. Outdoor broadcasts are more susceptible to electromagnetic noise than cable broadcasts, so outdoor broadcasts use the octal representation while cable broadcasts use the hexadecimal one. In fact, to improve the reliability of the outdoor broadcasts, one-third of the information in each octal digit is used to detect and correct transmission errors.

Actually, all of the digital video signals contain some error detection and correction information that allows a television to repair most transmission errors perfectly. The acceptable limit for irreparable errors is just one mistake per second, so the pictures are nearly perfect: no snow, no ghosts, and no fuzzy images.

The ATSC standards make remarkably efficient use of transmission bandwidth. To avoid undermining analog television during the transition period, new digital transmissions must fit into the same 6-MHz channels used by analog television. That means that a digital video signal must keep all of its frequencies within 6 MHz and must therefore be careful how fast it modulates the amplitude of its radio wave. Using a number of sophisticated techniques, digital transmission manages to pack information into that 6 MHz of bandwidth to within about 20% of the theoretical limit.

Check Your Understanding #6: More Is Better

If a cable system were so well built that its televisions could easily distinguish between 32 different wave amplitudes, how much faster could a station send information to its viewers than it can in the hexadecimal transmission scheme?

picture tubes, sidebands and channels

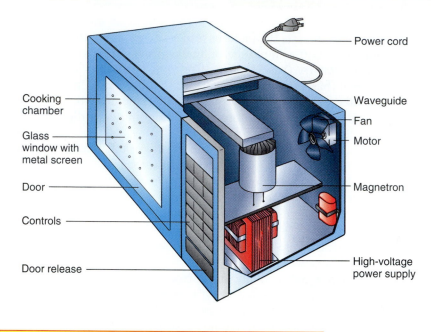

Power cord

Cooking chamber

Glass window with metal screen

Door

Controls

Door release

Waveguide

Fan

Motor

Magnetron

High-voltage power supply

Microwave Ovens

In addition to carrying sounds and pictures from one place to another, electromagnetic waves can carry power. One interesting example of such power transfer is a microwave oven. It uses relatively high-frequency electromagnetic waves to transfer power directly to the water molecules in food, so that the food cooks from the inside out. This section discusses both how those waves are created and why they heat food.

Questions to Think About: *Why do microwave ovens tend to cook food unevenly if you don't move the food during cooking? How can part of a frozen meal become boiling hot while another part remains frozen? Why must you be careful with metal objects placed inside the oven? Why do some objects remain cool in the microwave oven while other objects become extremely hot? How does microwave popcorn work?*

Experiments to Do: *A microwave oven transfers power primarily to the water in food. You can see this effect by placing completely water-free food ingredients such as salt, baking power, sugar, or salad oil on a microwave-safe ceramic dish in a microwave oven. Cook the ingredients **briefly.** You will find that the ingredients and dish remain cool. Add just a little water to the collection. What happens when you cook them this time?*

Now try cooking a very cold ice cube. The cube should come directly from the freezer on an ice-cold plate, so that its surface is solid and dry. What happens? If ice contains water and water is what absorbs power in a microwave oven, why doesn't the ice absorb power and melt?

What Are Microwaves?

When we studied incandescent light bulbs in Section 6.2, we discussed the *wavelengths* of electromagnetic waves. While looking at radio and television, we mostly considered the *frequencies* of electromagnetic waves. It's time to bring these two concepts together. A typical electromagnetic wave has both a wavelength *and* a frequency. Like Fig. 6.2.2, Fig. 11.3.1 shows the approximate wavelengths of many types of electromagnetic waves. But it also shows their frequencies.

Although an electromagnetic wave is characterized by both its wavelength and its frequency, the product of these two values is always the speed of light. To see why this relationship is so, consider a passing train. You can determine the train's speed by multiplying the length of a train car by how often those cars pass you. If each car is 10 m long and 6 cars pass you per second, then the train is moving 60 m/s. Now consider a passing electromagnetic wave. You can determine the wave's speed by multiplying the distance between equivalent peaks in the electric field (the wavelength) by how often those peaks pass you (the frequency). If the distance between peaks is 299.792458 m and 1,000,000 peaks pass you per second, then the wave is moving 299,792,458 m/s, which is the speed of light in empty space. This relationship can be written as a word equation:

$$\text{speed of light} = \text{wavelength} \cdot \text{frequency},\qquad\textbf{(11.3.1)}$$

in symbols:

$$c = \lambda \cdot \nu,$$

and in everyday language:

The higher the frequency of an electromagnetic wave, the shorter its wavelength.

Since wavelength and frequency aren't independent, you can calculate one from the other.

Radio and television broadcasts use the low-frequency, long-wavelength portion of the electromagnetic spectrum. Commercial AM radio is at frequencies of 550 kHz to 1600 kHz (wavelengths of 545 m to 187 m), commercial FM radio is at frequencies of 88 MHz to 108 MHz (wavelengths of 3.4 m to 2.8 m), and commercial television is at frequencies of 54 MHz to 806 MHz (wavelengths of 5.6 m to 0.4 m). Most of these waves have wavelengths longer than 1 m and are called radio waves. But some are shorter than 1 m and are called **microwaves.** Microwaves extend from wavelengths of 1 m down to 1 mm.

microwave communications

Check Your Understanding #1: Going Nowhere Fast

When an electromagnetic wave tries to travel through a transparent material, the material slows it down somewhat. The wave's frequency doesn't change, so what happens to its wavelength?

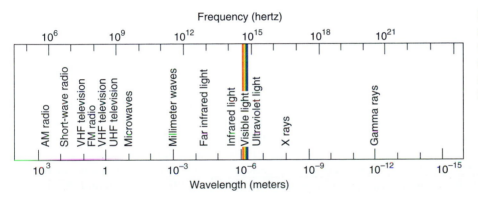

Fig. 11.3.1 The electromagnetic spectrum. Microwaves have wavelengths between about 1 m and 1 mm, corresponding to frequencies from 300 MHz up to 300 GHz.

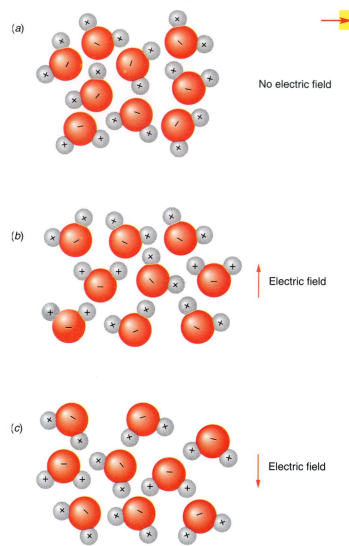

(a)

No electric field

(b)

Electric field

(c)

Electric field

Fig. 11.3.2 The water molecules in liquid water are randomly oriented when there's no electric field. But a field tends to orient them with their positive ends in the direction of the field.

microwaves and food safety

→ **Check Your Figures #1: Shopping for Food**
The red light used by many grocery store checkout stations to scan product codes is produced by a helium–neon laser. This light is an electromagnetic wave with a wavelength of approximately 633 nm. What is its frequency?

Heating Food with Microwaves

To explain how a microwave oven heats food, let's begin by looking at water molecules. Water molecules are electrically polarized—that is, they have positively charged ends and negatively charged ends. This polarization comes about because of quantum mechanics and the tendency of oxygen atoms to pull electrons away from hydrogen atoms. The water molecule is bent, with its two hydrogen atoms sticking up from its oxygen atom like Mickey Mouse's ears. When the oxygen atom pulls the electrons partly away from the hydrogen atoms, its side of the molecule becomes negatively charged while the hydrogen atoms' side becomes positively charged. Water is thus a *polar molecule*.

In ice, these polar water molecules are arranged in an orderly fashion with fixed positions and orientations. But in liquid water, the molecules are randomly oriented (Fig. 11.3.2). Their arrangements are constrained only by their tendency to bind together, positive end to negative end, to form a dense network of coupled molecules. This binding between the positively charge hydrogen atom on one water molecule and the negatively charged oxygen atom on another molecule is known as a hydrogen bond.

If you place liquid water in a strong electric field, its water molecules will tend to rotate into alignment with the field. That's because a misaligned molecule has extra electrostatic potential energy and accelerates in the direction that reduces its potential energy as quickly as possible. In this case, the water molecule will experience a torque and will undergo an angular acceleration that makes it rotate into alignment. As it rotates, the molecule will bump into other molecules and convert some of its electrostatic potential energy into thermal energy.

A similar effect occurs at a crowded party when everyone is suddenly told to face the front of the room. People brush against one another as they turn and sliding friction converts some of their energy into thermal energy. If the people were told to turn back and forth repeatedly, they would become quite warm. The same holds true for water. If the electric field reverses its direction many times, the water molecules will turn back and forth and become hotter and hotter.

A microwave's fluctuating electric field is well suited to heating water. A microwave oven uses 2.45-GHz (2.45-gigahertz or 2,450,000,000-Hz) microwaves to twist food's water molecules back and forth, billions of times per second. As the water molecules turn, they bump into one another and heat up. The water absorbs the microwaves and converts their energy into thermal energy. This particular microwave frequency was chosen because it was not in use for communications and because it cooks food uniformly. If the frequency were higher, the microwaves would be absorbed too strongly by food and wouldn't penetrate deep into large items. If the frequency were lower, the microwaves would pass through foods too easily and wouldn't cook efficiently.

This twisting effect explains why only foods or objects containing water or other polar molecules cook well in a microwave oven. Ceramic plates, glass cups, and plastic containers are water-free and usually remain cool. Even ice has trouble absorbing microwave power because its crystal structure constrains the water molecules so they can't turn easily.

But while ice melts slowly in a microwave oven, the liquid water it produces heats quickly. This peculiar heating behavior explains why it's so easy to burn yourself on frozen food heated in a microwave oven. Portions of the food that defrost first absorb most of the microwave power and overheat while the rest of the food remains frozen solid. You never know whether your next bite will break your teeth or sear the roof of your mouth. To address this problem, many microwave ovens have defrost cycles in which microwave heating is interrupted periodically to let heat flow naturally through the food to melt the ice. Once the frozen parts have melted, all of the food can absorb microwaves.

Check Your Understanding #2: Microwave Popcorn

A popcorn kernel contains moist starch trapped inside a hard, dry hull. You can scorch this hull by cooking the corn in hot oil but not when you cook it in a microwave oven. How can the microwave oven pop the corn without risk of overheating the hull?

Metal in a Microwave Oven

Contrary to popular lore, metal objects and microwave ovens aren't always incompatible. In fact, the walls of the oven's cooking chamber are metal, yet they cause no trouble when exposed to microwaves during cooking. Like most metal surfaces, the walls reflect microwaves. They do this by acting as both receiving and transmitting antennas. Electric fields in the microwaves cause mobile charges in the metal surfaces to accelerate and absorb the original microwaves. But as these charges accelerate, they emit new microwaves. The emitted microwaves have the same frequencies as the original ones, but they travel in new directions. The original microwaves have been reflected by the surface.

The cooking chamber walls reflect the oven's microwaves and keep them bouncing around inside. Even the metal grid covering the window reflects microwaves. That's because charge has enough time during a microwave cycle to flow around each hole in the grid and compensate for the hole's presence. As long as the wavelength of an electromagnetic wave is much larger than the holes in a metal grid, the wave reflects perfectly from the grid. At 12.2 cm, the wavelength of the oven's microwaves is much larger than the holes in its window grid so those waves reflect. In fact, if there's nothing inside the oven to absorb the microwaves, they'll bounce around inside it until they return to their source, a vacuum tube called a magnetron (Fig. 11.3.3), and eventually cause it to overheat.

While metal surfaces help confine the microwaves inside the oven, cooking your food and not you, extra metal inside the microwave can cause trouble. If you wrap food in aluminum foil, the foil will reflect the microwaves and the food won't cook. However, food placed in a shallow metal dish cooks reasonably well because microwaves enter the open top, pass through the food, reflect, and pass through the food again.

Sometimes metal's mobile charges do more than just reflect microwaves. If enough charge is pushed onto the sharp point of a metal twist-tie or scrap of aluminum foil, some of it will jump right into the air as a spark. This spark can start a fire, particularly when the twist-tie is attached to something flammable, like a plastic or paper bag. As a rule of thumb, never put a sharp metal object in the microwave oven.

Some metal objects heat up in a microwave oven. When microwaves push charge back and forth in a metal, the metal experiences an alternating current. If the metal has a substantial electric resistance, this alternating current will experience a voltage drop and heat up the metal. While thick oven walls and cookware have low resistances and remain cool, thin metal strips quickly overheat. Metallic decorations on porcelain dinnerware are particularly susceptible to damage in a microwave oven, so that warming up coffee in Grandma's gold-rimmed teacup is a recipe for almost instant disaster. When putting metal into a microwave oven, make sure that it is thick enough to conduct electricity well.

Resistive heating in conducting objects can actually be useful at times. Since microwave ovens cook food inside and out at the same time, the food's surface never gets particularly hot and the food doesn't brown or become crisp. To get around this problem, some foods come with special wrappers that conduct just enough current to become very hot in a microwave oven. These wrappers provide the high surface temperatures needed to brown the food.

Another peculiar feature of microwave ovens is that they don't always cook evenly. That's because the amplitude of the microwave electric field isn't uniform throughout the oven. As the microwaves bounce around the cooking chamber, they pass through the same spot from several different directions at once. When they do, their electric fields combine with one another. At one location, the individual electric fields may point in the same direction and reinforce one another, so that food there heats up quickly. But at another location, those fields may point in opposite directions and cancel one another. Food at this second spot doesn't cook well at all.

If nothing is moving in the microwave oven, the pattern of microwaves inside it doesn't move either. There are then regions in which the electric field has very large amplitudes and regions in which the amplitudes are very small. The larger the amplitude of the electric field, the faster it cooks food.

To heat food uniformly in such a microwave oven, you must move the food around as it's cooking. Many ovens have turntables inside that move the food automatically. Another solution to this problem is to stir the microwaves around the oven with a rotating metal paddle. The pattern of microwaves inside the chamber changes as the paddle turns and the food cooks more evenly. Still other microwave ovens use two separate microwave frequencies to cook the food. Because these two frequencies cook independently, it's unlikely that a portion of the food will be missed by both waves at once.

microwave oven safety

Fig. 11.3.3 This oven's magnetron microwave source is located in the middle of the picture, just to the left of its cooling fan. Microwaves travel to the cooking chamber through the metal rectangle on top of the oven. The high-voltage transformer at the bottom right provides power to the magnetron.

You place a thick metal divider into your microwave oven, so that it divides the cooking chamber exactly in half. The oven sends its microwaves into the right half of the chamber. If you put food in the left half of the chamber, will it cook?

magnetrons

Epilogue for Chapter 11

This chapter examined three common devices that are based on electromagnetic waves. In *radio*, we saw that electromagnetic waves can be created by accelerating electric charge and that these waves can be detected by looking for their effects on other electric charges. We also examined the techniques that are used to send audio signals through space by having them control either the amplitude or the frequency of electromagnetic waves.

In *television*, we looked at how electric and magnetic fields create, focus, and steer beams of electrons toward the screen of a picture tube. We studied the dot-by-dot way in which a picture is represented and examined the analog and digital techniques used to send those pictures via electromagnetic waves.

In *microwave ovens*, we explored the ways in which electromagnetic waves can interact directly with polar water molecules and can transfer energy to those molecules. We also saw how interactions between microwaves and a metal object can lead to reflection, sparking, or heating.

Explanation: Boiling Water in an Ice Cup

The water molecules in ice are held rigidly in place by ice's crystalline structure. Unable to move or turn, these water molecules can't follow the microwave radiation's rapidly changing electric field. As a result, ice is unable to absorb much energy from the microwave radiation.

In contrast, molecules in liquid water can move about relatively freely and turn back and forth with the changing electric field of the microwaves. These water molecules absorb energy from the microwave radiation and become hotter. The liquid water's temperature rises.

While heat tends to flow out of the hot water and into the ice, water is not a good conductor of heat. Convection stops working once the water temperature rises above 4 °C and the hotter water begins to float on the colder water beneath it. In favorable circumstances, the water can begin to boil before it has melted the ice cup that contains it.

Chapter Summary

How Radio Works: A radio transmitter creates a radio wave when electric charge accelerates up and down its antenna. To get as much charge moving as possible, the transmitter attaches a tank circuit to the antenna and slowly adds energy to that tank circuit until an enormous amount of charge is flowing up and down the antenna. If the antenna is a quarter-wavelength long, it's also resonant at the transmission frequency and boosts the amount of charge sloshing up and down.

The radio receiver detects this radio wave when it causes charge to accelerate up and down the receiving antenna. If both the receiving antenna and the receiver's tank circuit are resonant at the transmission frequency, large amounts of charge will slosh back and forth in the receiver's tank circuit and the receiver will detect the transmission.

This radio wave can represent sound using either the AM or FM technique. In the AM technique, the strength of the wave is increased or decreased to represent compressions or rarefactions of the air, respectively. In the FM technique, the precise frequency of the transmission is increased or decreased to represent those compressions or rarefactions.

How Television Works: A black and white television creates an image on its picture tube by scanning a beam of electrons rapidly across the inside of the screen. It starts at the top of the screen and scans left to right while gradually moving the beam down toward the bottom of the screen. The current of electrons in the beam determines the brightness of any particular spot. That beam is focused to a tiny point on the screen by electric and magnetic fields and is scanned magnetically from left to right and top to bottom. A color television tube scans three separate electron beams across the inside surface of the screen, where they illuminate blue, red, or green phosphor patches through holes in a metal mask.

The electromagnetic wave that carries the television signal through space fits within a 6-MHz range of frequencies. Getting an analog video signal to fit in this narrow bandwidth is already an accomplishment, but digital compression and transmission techniques are allowing even higher resolution pictures to be transmitted within that same bandwidth.

How Microwave Ovens Work: A microwave oven uses microwaves to cook food. These microwaves bounce around the cooking chamber, where they transfer energy to water molecules in the food. Because a water molecule is polar, having a positive end and a negative end, it tends to align with an electric field. The microwave's fluctuating electric field causes the tightly packed water molecules to twist back and forth rapidly, and the ensuing collisions heat the water and cook the food.

Important Laws and Equations

1. Relationship Between Wavelength and Frequency: The product of the frequency of an electromagnetic wave times its wavelength is the speed of light, or

$$\text{speed of light} = \text{wavelength} \cdot \text{frequency}. \quad (11.3.1)$$

Check Your Understanding—Answers

Section 11.1 RADIO

1. The amount of charge that the transmitter can move directly on and off the antenna is too small to create a strong radio signal.

Why: The tank circuit is useful because it allows the transmitter to move much more charge. Just as a tuning fork is inefficient at emitting sound waves by itself, so a radio antenna is inefficient at emitting radio waves by itself. You can make the tuning fork much louder by coupling it to an object that resonates at its frequency. Similarly, you can make the radio antenna emit a much stronger radio wave by coupling it to a tank circuit that resonates at its frequency.

2. When the cordless telephone is too far from its base unit, their electromagnetic waves become so spread out that they have trouble communicating.

Why: The powers emitted by the base unit and the handset are small, so that their waves are relatively difficult to detect. As long as the handset and base unit are nearby, they are able to detect each other's waves. But when the distance between them becomes too great, the waves become too spread out to detect and the handset and base unit lose contact with one another.

3. The tunnel blocks most of the radio wave. Since only a small fluctuating wave reaches your radio, the radio produces only small fluctuations in air pressure with its speaker.

Why: An AM radio has trouble distinguishing between a distant transmission representing loud music and a nearby transmission representing soft music. In both cases, the receiver detects only small variations in the current moving up and down its antenna. That's why you must turn up the volume of an AM radio as you move farther from the transmitting antenna or as you enter a tunnel.

Section 11.2 TELEVISION

1. During the 1/250th of a second exposure, the television will only complete about a quarter of its scan from the top of the screen to the bottom of the screen.

Why: While the television's scanning process easily fools your eye into seeing a complete image, a camera records exactly what it sees. If the exposure time is too short, the television will only be able to build part of the screen image during the exposure, and the photograph will contain only a horizontal band of the picture.

2. This current changes smoothly and steadily from flowing in one direction around the coils to flowing in the other direction around the coils. Every 1/15,750th of a second, it restarts this process.

Why: The current flowing through the coils resembles a series of ramps when graphed as a function of time. It increases steadily up to a point and then suddenly returns to its starting value. As the current increases steadily, the electron beam is deflected less to the left and more to the right. Then the current is suddenly returned to its original value and the beam snaps back to the left. Because magnetic fields from this changing current can cause objects to move, some televisions emit an annoying tone at 15,750 Hz.

3. They are exactly the same.

Why: The deflection process doesn't prevent the beams from coming together at a single point on the screen. Instead, it gives electrons in each beam exactly the same impulse, so that they all experience the same change in momentum and travel together to a specific point on the screen.

4. The higher the recording bandwidth, the finer the horizontal detail in the picture.

Why: A VCR that can record very high frequencies in the video signal will capture the finest details in each horizontal line of the picture. Recent high-bandwidth VCRs are able to record more detail than is currently present in broadcast television.

5. Both computer networks and cable television systems send information as electromagnetic waves in wires or glass fibers.

Why: The information that moves through a computer network travels as electromagnetic waves in cables. Sometimes those cables are coaxial cables; sometimes they are optical fibers; and sometimes they consist of twisted pairs of wires (which can also carry and guide electromagnetic waves). A computer transmits information through the cable by creating and modulating an electromagnetic wave, and another computer receives the information by detecting that modulated wave.

6. 25% faster.

Why: In base-16 or hexadecimal, each digit carries the equivalent of 4 binary bits of data. In base-32, each digit carries the equivalent of 5 binary bits. That's a 25% increase in data flowing to the televisions. Unfortunately, cable networks aren't perfect enough for base-32 transmission.

Section 11.3 MICROWAVE OVENS

1. The wavelength gets shorter.

Why: Even when an electromagnetic wave slows down while going through material, the relationship between its wavelength, frequency, and speed remains valid. Since the frequency can't change (the wave peaks keep passing by at the same rate), the wavelength has to shrink when the speed decreases.

2. The microwave oven transfers heat to water molecules in the starch so that the hull never becomes hotter than the material inside it.

Why: A corn kernel cooked in oil is heated by contact with the hot oil and pot. You can easily overheat the outer hull and burn it. But microwaves transfer heat to the water molecules inside the kernel. The hull can't overheat because the hottest thing it touches is the starchy insides of the kernel. When the pressure of steam inside the kernel becomes high enough, the hull breaks and the kernel "pops."

3. No.

Why: The metal divider would reflect microwaves and keep them from entering the left half of the oven.

Check Your Figures—Answers

Section 11.3 MICROWAVE OVENS

1. About 4.74×10^{14} Hz.

Why: Since the product of frequency and wavelength for an electromagnetic wave is equal to the speed of light, its fre-quency is equal to the speed of light divided by its wave-length:

$$\frac{299{,}792{,}458 \text{ m/s}}{0.000000633 \text{ m}} = 4.74 \times 10^{14} \text{ Hz.}$$

Exercises

1. The ignition system of an automobile produces sparks to ignite the fuel in the engine. During each spark process, charges suddenly accelerate through a spark plug wire and across a spark plug's narrow gap. Sometimes this process in-troduces noise into your radio reception. Why?

2. To diminish the radio noise in a car (see Exercise 1), the ignition system uses wires that are poor conductors of elec-tricity. These wires prevent charges from accelerating rapidly. Why does this change improve your radio reception?

3. The electronic components inside a computer transfer charge to and from wires, often in synchrony with the com-puter's internal clock. Without packaging to block electro-magnetic waves, the computer will act as a radio transmit-ter. Why?

4. When a television signal travels through a coaxial ca-ble, charge moves back and forth on both the central wire and the surrounding tube. Show that both electric and mag-netic fields are present in the coaxial cable.

5. A tank circuit consists of an inductor and a capacitor. Give a simple explanation for why the magnetic field in the inductor is strongest at the moment the separated charge in the capacitor reaches zero.

6. Suppose you include an inductor in an electric circuit that includes a battery, a switch, and a light bulb. Current leaving the battery's positive terminal must flow through the switch, the inductor, and the light bulb before returning to the battery's negative terminal. The current in this circuit increases slowly when you close the switch, and it takes the light bulb a few seconds to become bright. Why?

7. When you open the switch of the circuit in Exercise 6, a spark appears between its two terminals. As a result, the cir-cuit itself doesn't open completely for about half a second, during which time the bulb gradually becomes dimmer. The bulb's behavior indicates that the current in the circuit dimin-ishes slowly, rather than stopping abruptly when you open the switch. Why does the current diminish slowly?

8. The metal wires from which most tank circuits are made have electric resistances. Why do these resistances prevent charge from sloshing back and forth forever in a tank circuit, and what happens to the tank circuit's energy as time passes?

9. FM radio stations must operate at very different fre-quencies to avoid interfering with each other's broadcasts. Why?

10. Cordless telephones often have a switch that allows you to change the radio frequencies they use to communi-cate with their base units. Why is this feature more impor-tant in cities than it is in rural areas?

11. While a particular AM radio station claims to transmit 50,000 W of music power, that's actually its *average* power. There are times when it transmits more power than this and times when it transmits less. Explain.

12. When your receiver is too far from an AM radio sta-tion, you can only hear the loud parts of the transmission. When it's too far from an FM station, you lose the whole sound all at once. Explain the reasons for this difference.

13. When an AM radio station announces that it's trans-mitting at 950 kHz, that statement isn't quite accurate. Explain why it may also be transmitting at 948 kHz and 954 kHz.

14. The Empire State Building has several television an-tennas on top, added in part to increase its overall height. These antennas aren't very tall. Why do short antennas, lo-cated high in the air, do such a good job of transmitting television?

15. If you tried to communicate with a friend using radios that operated at 2.45 GHz, you'd probably hear lots of noise instead of your friend's voice. Explain.

16. A Penning ion trap suspends an ion (a charged atom or molecule) in empty space using electric and magnetic fields. In this trap, like charges on metal surfaces above and below the ion keep it from moving up or down, and a vertical mag-netic field confines the ion horizontally by making it move in a circle. Why does the ion move in a circle?

17. A television's high-voltage power supply transfers electrons from the phosphor screen to the cathode for reuse. This supply does considerable work on the electrons it transfers. How is that work related to the light emitted by the screen?

18. The force that a magnetic field exerts on a moving charged particle does no work on that particle. Use the definition of work to explain why there's no work done.

19. A cyclotron is a particle accelerator invented in 1929 by American physicist Ernest O. Lawrence. It uses electric fields to do work on charged particles as they follow circular paths in a strong magnetic field. Lawrence's great insight was that all the particles take the same amount of time to complete one circle, regardless of their speed or energy. That fact allows the cyclotron to do work on all the particles at once as they circle together. Why is Lawrence's discovery also crucial to the sharpness of a television picture?

20. If the inside of the picture tube's screen didn't have a thin layer of aluminum on it, the electrons would accumulate there. Why would that be a problem?

21. X rays are produced by a vacuum tube in which electrons from a hot, negatively charged cathode accelerate toward a positively charged metal anode. Collisions between the electrons and the metal atoms produce X rays. Why is it important to shield the X-ray tube from magnetic fields?

22. The sun emits a stream of energetic electrons and protons called the *solar wind*. These particles frequently get caught up in the earth's magnetic field, traveling in spiral paths that take them toward the north or south magnetic poles. When they head northward and collide with atoms in the earth's upper atmosphere, those atoms emit light we know as the *aurora borealis,* or northern lights. These particles also interfere with radio reception. Why do they emit radio waves?

23. If you hold a strong permanent magnet near the face of a color television picture tube, the image will be distorted in both shape and color. What causes this distortion? (*Warning: The distortion may not go away when you remove the magnet, so don't try this on a good television set!*)

24. When you put your hand near a picture tube that has just been turned on, negative charges flow from your hand to the glass screen. What attracts these charges to the screen?

25. Porous, unglazed ceramics can absorb water and moisture. Why are they unsuitable for use in a microwave oven?

26. Why are most microwave TV dinners packaged in plastic rather than aluminum trays?

27. Why is it so important that a microwave oven turn off when you open the door?

28. Why can operating a microwave oven for a long time with only salad oil inside it damage the magnetron?

29. When you're listening to FM radio near buildings, reflections of the radio wave can make the reception particularly bad in certain locations. Compare this effect to the problem of uneven cooking in a microwave oven.

30. Dish-shaped reflectors are used to steer microwaves in order to establish communications links between nearby buildings. Those reflectors are often made from metal mesh. Why don't they have to be made from solid metal sheets?

Problems

1. The frequency of the radio wave emitted by a cordless telephone is 900 MHz. What is the wavelength of that wave?

2. Citizens band (CB) radio uses radio waves with frequencies near 27 MHz. What are the wavelengths of these waves and how long should a quarter-wavelength CB antenna be?

3. The electromagnetic waves in blue light have frequencies near 6.5×10^{14} Hz. What are their wavelengths?

4. Amateur radio operators often refer to their radio waves by wavelength. What are the approximate frequencies of the 160-m, 15-m, and 2-m wavelength amateur radio bands?

5. The radio waves used by cellular telephones have wavelengths of approximately 0.36 m. What are their frequencies?

Cases

1. Microwave ovens allow for some interesting cooking and funny disasters.

a. Baked Alaska is a dessert in which a hot, baked meringue contains cold, frozen ice cream. The reverse is

Frozen Florida, a dessert in which cold, frozen meringue contains boiling hot liqueur. Frozen Florida is prepared by taking a frozen meringue ball (cooked egg whites) containing liquid liqueur (a water–alcohol mixture that remains liquid at low temperature) out of the freezer and putting it in a microwave oven briefly. Why does the liqueur get hot while the meringue remains frozen?

b. If you try to cook an egg in a microwave oven, the egg may explode. Compare this result to the process of cooking popcorn in a microwave oven.

c. Some prepared foods come with browning sheets that contain very thin metallic layers. These sheets become hot in a microwave oven and help to brown the surface of the food. Why does a thin metallic film become so hot?

d. If you wrap a piece of food in sturdy metal screening with holes about 2 millimeters on a side, the food won't cook in a microwave oven. Why can't the microwaves get through the screen's holes the way light can?

2. As a DJ at a very small local radio station, you often find yourself involved in technical issues for the station's two channels, one of which is AM at 1020 kHz and the other of which is FM at 89.5 MHz. The station is in the process of building a new transmitting system.

a. To save money, the director wants to use a single antenna for both channels. You warn him that the AM channel needs a taller antenna than the FM channel. Why is that true?

b. The director had planned to put the antennas next to the station, which is in a valley at the base of a small mountain. You suggest putting them at the top of the mountain, despite the extra cost of wires. Why is altitude important, particularly for the FM antenna?

c. The AM channel must be careful not to "overmodulate" the radio wave during very loud passages because it distorts the sound people hear in their radios. You explain this effect as due to moments when the transmitter actually turns itself completely off. Why would the transmitter stop transmitting any wave at all?

d. The FM channel must also avoid overmodulation during loud passages because it will get in trouble with the FCC. Other FM stations in your area will also be angry at your station for spoiling the reception of their transmissions. How can your FM station affect those other FM stations when they operate at different carrier wave frequencies?

3. Your garage door opener is triggered by a radio transmitter. You press the button in your car to open the door.

a. When you press the button, the transmitter moves charge onto and off a small antenna. Why doesn't it just put charge on the antenna and leave the charge there?

b. If you're close enough to the garage door, it will open. Why must the transmitter and receiver be close together?

c. When you press the button of your transmitter, the radio in your car still sounds normal. Why doesn't the transmitter interfere with radio reception in your car?

d. In the early days of garage door openers you could drive around town opening other people's garages. To make things more secure, the transmitters now send codes to the receivers using AM modulation. Each transmitter turns itself on and off quickly in a pattern that's recognized only by the proper receiver. Graph the charge on a transmitter's antenna as a function of time during such an AM transmission.

4. When you talk on a cellular telephone, two radio waves connect you with the telephone network. Your telephone transmits one of these waves, and a stationary base unit sends the other to you. Base units are located a few kilometers apart and each one provides service to the telephones in its vicinity—its "cell." A connection between a telephone and a base unit occupies one of 832 channels currently allocated to cellular communications. Each channel corresponds to radio waves of two frequencies: a lower frequency wave (in the range 825.03–849.96 MHz) that's emitted by the telephone and a higher frequency wave (in the range 870.03–894.96 MHz) that's emitted by the base unit. The waves of different channels are spaced 0.030 MHz (30 kHz) apart and those of your telephone and its base unit are 45 MHz apart.

*a. When your telephone and its base unit are connected on Channel 1, your telephone's wave has a frequency of 825.03 MHz, and its base unit's wave has a frequency of 870.03 MHz. What are the wavelengths of those two waves?

*b. What is the proper length for the telephone's antenna if it's to have a natural resonance at the frequency of the radio wave it's transmitting?

c. Why will there be a problem if 900 people in a conference center all try to use their cellular telephones at once?

d. The base unit usually has several antennas, so that it can interact with a cell phone located anywhere around it through any wave polarization. However, a minimal base unit could get by with a single antenna oriented vertically. A single antenna oriented horizontally would leave certain people on the ground unable to interact with this base unit at all. Where would these people be located, and why wouldn't they be able to interact with this horizontal antenna?

e. If the base unit were using a single vertical antenna, your telephone's antenna would also have to be vertical to work well. Why?

f. If you're in a valley or a metal building, your telephone may lose contact with the base unit. Why?

g. Your telephone modulates its carrier wave in order to send your voice to the base unit. Why must it be careful not to produce waves that have frequencies that are more than 15 kHz away from that of its carrier wave?

h. If you talk while traveling, a computer system makes sure that your telephone is always communicating with the nearest base unit. Why is it important that every telephone and base unit transmit the weakest radio waves that still permit reliable communication between them?

5. The Global Positioning System or GPS is a system of earth-orbiting satellites that provide position information to anyone with a GPS radio receiver. Each satellite transmits two microwaves, one at 1.57542 GHz and the other at 1.2276 GHz. These waves are modulated by carefully timed pulses so that by measuring exactly when it receives those pulses from four or more satellites, a GPS receiver can determine its position on earth to within 100 m. Recent changes in U.S. Government policy will allow GPS receivers to locate their positions even more accurately over the next few years.

a. Because each satellite carries a time-standard cesium atomic clock, its pulses are emitted at precisely known times. The GPS receiver knows exactly when it receives the pulses, so it can tell exactly how far it is from the satellite. How?

b. Because a GPS receiver only receives microwaves from satellites that are above the horizon, there must be at least 17 satellites in orbit at once. Why can't the receiver detect satellites that are below the horizon?

*****c.** What are the wavelengths of the two microwaves?

*****d.** One of the reasons for using microwaves in the GPS is that microwaves can be received by small antennas with well-defined locations. How long should an antenna be to receive the lower frequency GPS microwave?

e. Geologists are using the GPS to study motions of the earth's crust that are as small as a few millimeters. To do this, they actually study the electric and magnetic fields of the microwaves. Why would two GPS receivers located a few centimeters apart detect somewhat different electric and magnetic fields if they studied the microwave from one particular satellite at precisely the same moment?

6. A cordless microphone uses a 925-MHz carrier wave to carry sound information from a performer to the audio system of a theater.

a. The tank circuits in the microphone and in the theater's receivers have carefully been tuned to precisely the same resonant frequencies. Why is this adjustment important?

b. The frequencies of the waves emitted by the microphone include those slightly above and below 925 MHz. Explain.

c. Because the performer moves and tips the microphone during the performance, the theater uses two receivers at different locations. This ensures that the radio wave from the microphone is always strong at one of the receivers. How is it possible for the wave to be weak at one of the receivers?

d. Each receiver has two antennas oriented at right angles to one another. Why does this ensure that the receiver can detect the radio wave regardless of how the microphone's antenna is tipped?

CHAPTER 12

Light

While radio waves and microwaves are useful for communications and energy transfer, there's another portion of the electromagnetic spectrum that we find far more important: light. Light consists of very-high-frequency, very-short-wavelength electromagnetic waves. Light's frequencies are so high that normal antennas can't handle it. Instead, it's absorbed and emitted by the individual charged particles in atoms, molecules, or materials. Because of its special relationship with the charged particles in matter, light is important to physics, chemistry, and materials science. Moreover, it's one of the principal ways by which we interact with the world around us.

EXPERIMENT: Splitting the Colors of Sunlight

We see light because it stimulates cells in our eyes. This stimulation is an example of light's ability to influence chemistry. And because our eyes are able to distin-

guish between different wavelengths of light, we perceive colors. While sunlight normally appears uncolored, that's because it contains a rich mixture of wavelengths that our eyes interpret as whiteness. But there are situations in which sunlight becomes separated into its constituent colors.

You can observe this separation of colors by looking at sunlight passing through a cut crystal glass or bowl. Hold the cut crystal object in direct sunlight and observe the light that it redirects toward your eyes or projects onto a white sheet of paper nearby. While some of the light you see will still be white, you should see colors as well.

Turn the object slowly in your hand and observe how the colors change. You will see gradual progressions from one color to the next. *Predict* the order of colors you'll see and then *observe* the actual sequence. Were you able to *verify* your predictions? Is the sequence of colors always the same? How does this sequence relate to the colors of the rainbow? What is the relationship between this sequence and the wavelengths of light? Can you *measure* the relative spacings of the classic rainbow colors: red, orange, yellow, green, blue, indigo, and violet?

Chapter Itinerary

In this chapter, we'll examine three sources of light: (1) *sunlight,* (2) *fluorescent lamps,* and (3) *lasers.* In *sunlight,* we'll see how sunlight travels to our eyes and how its passage through the atmosphere, raindrops, and soap bubbles can separate it into its constituent colors. In *fluorescent lamps,* we'll explore the ways in which atoms and molecules emit and absorb light, and how different atoms and molecules can be used to produce light of different colors. In *lasers,* we'll look at how atoms and molecules can duplicate or amplify the light passing through them and thus produce intense beams of highly ordered light. In the process of studying these three light sources, we'll also learn about three different types of light: thermal light, atomic resonance light, and coherent light.

Sunlight

For thousands of years, people have marked the passage of time by the rising and setting of the sun over the horizon. The sun first appears as a red disk in the east every morning, rises white in the blue sky, and then sets once again as a red disk in the west. The sunlight that we see takes about 8 minutes to travel the 150,000,000 km from the sun to our eyes and provides most of the energy and heat that make life on earth possible. While the light in sunlight is really just another electromagnetic wave, and could be considered part of the previous chapter, it's so important to everyday life that it deserves special attention. And so we'll begin by looking at how sunlight interacts with our world.

Questions to Think About: Why is the sun red at sunrise and sunset? Why is the sky blue during the day? Why can you occasionally see sunbeams when sunlight enters an otherwise dark room through a small window? Why do we see colors when sunlight passes through cut crystal or a soap bubble?

Experiments to Do: Sunlight is actually composed of many different electromagnetic waves. These waves differ in frequency and wavelength like the radio waves from your two favorite stations. But you don't need a machine to help you distinguish between various wavelengths of light; you can use your eyes. Take a look at a soap bubble on a bright, sunny day. You see the bubble because it reflects light. In fact, the clear bubble appears colored, even though the sunlight hitting it is white. That's because the bubble separates sunlight according to wavelength and sends only certain wavelengths toward your eyes.

Sunlight and Electromagnetic Waves

Electromagnetic waves can have any wavelength, from thousands of kilometers to a fraction of the width of an atomic nucleus. The radio waves and microwaves that we examined in the previous chapter have wavelengths longer than 1 mm. In this chapter, we'll turn our attention to shorter wavelength radiation. In particular, we'll study electromagnetic waves with wavelengths between 400 nm and 750 nm (recall that 1 nm or 1 nanometer is 10^{-9} m). These are the electromagnetic waves that we perceive as **visible light** and the principal components of sunlight.

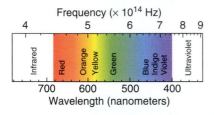

Because the electromagnetic waves in sunlight have such short wavelengths, their frequencies lie between 10^{14} Hz and 10^{15} Hz (Fig. 12.1.1). As one of these waves of sunlight passes by, its electric field fluctuates back and forth almost 1,000,000,000,000,000 times each second. Since producing microwaves, which have much longer wavelengths and much lower frequencies, already requires specialized components and tiny antennas, what can possibly emit or absorb light waves? The answer is the individual charged particles in atoms, molecules, and materials. These tiny particles can move extremely rapidly, often vibrating about at frequencies of 10^{14} Hz, 10^{15} Hz, or even more. As these charged particles accelerate back and forth, they emit light waves. Similarly, passing light waves cause individual charged particles in atoms, molecules, and materials to accelerate back and forth, thereby absorbing the light waves as well.

Fig. 12.1.1 The visible portion of the spectrum of sunlight. Each wavelength of visible light has a particular frequency and is associated with a particular color. At the ends of the visible spectrum are invisible infrared and ultraviolet lights.

Sunlight originates at the outer surface of the sun, in a region called the *photosphere*. There, atoms and other tiny charged systems (mostly atomic ions and electrons) jostle about at 5800 °C. Since these charged particles accelerate as they bounce around, they emit electromagnetic waves.

Because the sun's surface emits light through the random, thermal motions of its charged particles, the distribution of wavelengths it emits is determined only by its temperature. It emits a black body spectrum, like the incandescent light bulbs we discussed in Section 6.2. Because the photosphere's temperature is 5800 °C, the jostling motions are extremely rapid and most of the sunlight falls in the visible portion of the electromagnetic spectrum (Fig. 12.1.2).

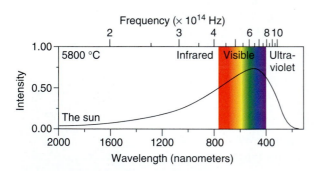

However, not all sunlight is visible. On the long-wavelength, low-frequency side of visible light is **infrared light.** We can't see infrared light with our eyes but we feel it when we stand in front of a hot object. In sunlight, infrared light is produced by charges that are accelerating back and forth more slowly than average.

Fig. 12.1.2 Sunlight comes from the sun's photosphere, where the temperature is 5800 °C. This light has a black body distribution of wavelengths, with much of its intensity concentrated in the visible portion of the overall electromagnetic spectrum.

On the short-wavelength, high-frequency side of visible light is **ultraviolet light.** We can't see ultraviolet light either, but we're aware of its presence because it induces chemical damage in molecules. It causes sunburns and encourages skin to tan. In sunlight, ultraviolet light is produced by charges that are accelerating back and forth more rapidly than average.

Check Your Understanding #1: The Rosy Glow of Candlelight

Why does a burning candle emit reddish or yellowish light?

Sunlight's Passage to the Earth

Sunlight travels from the sun to the earth at the speed of light. But what sets the speed of light? Actually, it's one of the fundamental constants of nature, with a defined value of 299,792,458 m/s in empty space. While one could argue that the speed of light is set by the relationships between the electric and magnetic fields, that observation simply passes the buck. If you were then to ask what sets the relationships between the electric and magnetic fields, the answer would be the speed of light.

Rather than justifying why sunlight travels as fast as it does in empty space, let's look at what happens to it when it enters a region that's not empty. After all, sunlight eventually reaches the earth's atmosphere and, when it does, several interesting things happen.

First, the sunlight slows down as its electric and magnetic fields begin to interact with the electric charges and magnetic poles in the atmosphere. Light polarizes the molecules it encounters, a process that delays its passage and reduces its speed. Since most transparent materials respond much more strongly to light's electric field than to its magnetic field, we'll concentrate on only electric effects.

The factor by which light slows down in a material is known as the material's **index of refraction.** Light travels particularly slowly through materials that are easy to polarize, and some of them have indices of refraction of 2 or even 3. However, because air near sea level is only slightly polarizable, its index of refraction is just 1.0003. While this reduction in light's speed is too small to notice, we do notice the polarized air particles that cause it. These polarized air particles are what make the sky blue (Fig. 12.1.3).

The particles in air consist of individual atoms and molecules, small collections of atoms and molecules, water droplets, and dust. As a wave of sunlight passes through one of these particles, the particle becomes polarized. Its electric charges accelerate back and forth as the sunlight's electric field pushes them around and they reemit a new electromagnetic wave of their own.

This new wave draws its energy from the original wave. In effect, the particle acts as a tiny antenna, temporarily receiving part of the electromagnetic wave and

Fig. 12.1.3 During the day, the sky above Monument Valley appears blue because the earth's atmosphere scatters mostly blue sunlight toward us. At night, the scattered sunlight is gone and the atmosphere is a clear window through which to observe the stars. The time-lapse photograph on the right shows these stars as nearly vertical streaks in the night sky.

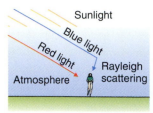

Fig. 12.1.4 As sunlight passes through the atmosphere, some of its blue light undergoes Rayleigh scattering from particles in the air. We see this redirected blue light as the diffuse blue sky. The remaining light reaches our eyes directly from the sun and tends to be reddish, particularly at sunrise and sunset.

immediately retransmitting it in a new direction. This process, where a tiny particle redirects the path of a passing light wave, is called **Rayleigh scattering,** after the English physicist Lord Rayleigh (John William Strutt, 1842–1919) who first understood it in some detail.

While most sunlight travels directly to our eyes, some of it undergoes Rayleigh scattering and reaches us by more complicated paths. We see the direct light as coming from the brilliant disk of the sun, but the scattered light gives the entire sky a fairly uniform blue glow (Fig. 12.1.4). But why is this glow blue?

The sky's blue color comes about because the tiny air particles that Rayleigh scatter sunlight are too small to make good antennas for that light. We observed in Section 11.1 that an antenna works best when it is a quarter as long as the wavelength of its electromagnetic wave. The air particles make particularly bad antennas for longer-wavelength red light, so that very little red sunlight undergoes Rayleigh scattering on its way through the atmosphere. But the air particles are not such bad antennas for shorter-wavelength blue light. Some of the blue sunlight does Rayleigh scatter and reaches our eyes from all directions. We see this Rayleigh scattered light as the blue glow of the sky.

Rayleigh scattering not only makes the sky blue; it also makes the sunrises and sunsets red. As the sun rises or sets, its light must travel long distances through the earth's atmosphere in order to reach your eyes. Its path is so long that most of the blue light Rayleigh scatters away miles to your east or west and all you see is the remaining red light. Sometimes the whole local sky appears reddish because there simply isn't any blue light left to scatter toward you. Sunrises and sunsets are particularly colorful when extra dust or ash is present in the atmosphere to enhance the Rayleigh scattering. Air pollution, forest fires, and volcanic eruptions tend to create unusually red sunrises and sunsets.

In contrast, clouds and fog appear white because they're composed of relatively large water droplets. These droplets are larger than the wavelengths of visible light and scatter all of sunlight's wavelengths equally well. Although this scattering is often so effective that you can't see the sun's disk through a cloud, it doesn't give the cloud any color. The cloud simply looks white.

▶ Check Your Understanding #2: Seeing the Blues
The air in a dark, smoky room often looks bluish when illuminated by white light. What creates this bluish appearance?

Rainbows

refraction angles

Sometimes water droplets do separate the colors of sunlight. When sunlight shines on clear, round raindrops as they fall during a storm, these raindrops can create a rainbow. To understand how clear spheres of water can bend sunlight's path and separate it according to wavelength, we must understand three important optical effects: refraction, reflection, and dispersion.

Let's begin by looking at what happens when a wave of sunlight passes directly through a raindrop. Because water is more polarizable than air, the wave slows down inside the raindrop and its cycles bunch together (Fig. 12.1.5). While this bunching effect reduces the light's wavelength inside the drop, the light's frequency remains unchanged. The cycles don't disappear as they go through the raindrop, they just move more slowly.

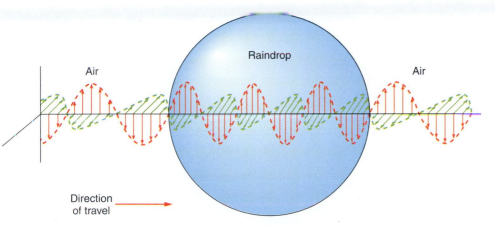

Fig. 12.1.5 As an electromagnetic wave enters a material, its speed decreases and the waves bunch up together. Its wavelength decreases.

If a narrow wave of sunlight is aimed directly through the center of the raindrop, it will follow a straight path and emerge essentially unaffected from the other side (Fig. 12.1.6*a*). But if that wave strikes the raindrop near the top, it will bend as it enters the water (Fig. 12.1.6*b*). Because the lower edge of the wave will reach the water first and slow down, the upper edge will overtake it and the wave will bend downward. The wave will head more directly into the water.

As the wave in Fig. 12.1.6*b* leaves the raindrop, its upper edge emerges first and speeds up while the lower edge lags behind. The wave bends downward even further and heads less directly into the air and away from the water.

This bending of sunlight at the boundaries between materials is called **refraction.** It occurs whenever sunlight changes speeds as it passes through a boundary at an angle. If sunlight slows down at a boundary, it bends to head more directly into the new material. If sunlight speeds up at a boundary, it bends to head less directly into the new material. The amount of the bend increases as the speed change increases.

However, part of the sunlight striking a boundary doesn't pass through the boundary at all. Instead, it reflects because of an **impedance mismatch**—an abrupt change in opposition to the light wave's passage. In general, **impedance** is the measure of a system's opposition to the passage of a current or a wave. For an electromagnetic system, impedance measures how much voltage or electric field is needed to produce a particular current or magnetic field. In other words, electric impedance measures how hard it is for electric activity to produce magnetic activity. Impedance effects are common in nature (see ❏) and also apply to mechanical waves and currents. When sound and water waves encounter impedance mismatches, they also partly reflect.

The impedance of empty space is high because an electric field there has nothing to aid it in producing a magnetic field. But inside most materials, the electric field has help. The electric field polarizes the material, which then helps to create the magnetic field. Because of this assistance, the impedance of most materials is much less than that of empty space. Since air is almost empty space, the boundary between air and water is an impedance mismatch for light.

Passing through an impedance mismatch upsets the balance between a light wave's electric and magnetic fields. To compensate for this imbalance, part of the incoming wave experiences **reflection** off the boundary. Thus some sunlight reflects each time it enters or leaves a water droplet. The fraction of light that reflects depends on the severity of the impedance mismatch but is typically 4% between air

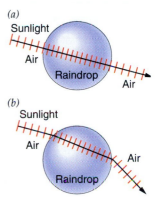

Fig. 12.1.6 A side view of two narrow waves of sunlight entering and leaving raindrops. The lines drawn across each light wave represent upward electric field maxima and are bunched together as light slows down in water.

❏ When electromagnetic waves travel through wires, they reflect from impedance mismatches. These mismatches occur whenever the relationship between electric and magnetic fields changes and must carefully be avoided in television wiring. If you don't provide an impedance matching device when connecting an antenna wire (300 Ω) to a video cable input (75 Ω), you'll have reflections in the wires and ghost images on the screen.

❑ Sand appears white because it redirects sunlight in all directions. One explanation for this effect is that the sand grains act as tiny antennas that respond to and reemit light's electromagnetic waves. A second explantion is that the sand grains present the sunlight with thousands of air–sand boundaries from which to reflect. However, both explanations are descriptions of exactly the same physics—the charged particles in the sand grains are electrically polarized by the waves passing through them. These waves are randomly redirected without being absorbed, and they give the sand its white appearance.

Fig. 12.1.7 A rainbow forms when water droplets reflect sunlight back toward your eyes. Because the different wavelengths of light follow slightly different paths, we see the different colors coming from slightly different directions and observe bands of color.

and most transparent materials, including water (for reflection from sand, see ❑). In contrast, metals polarize so easily that their impedances are essentially zero and they reflect light almost perfectly.

There is one more important point about sunlight's passage through water: red light travels about 1% faster through water than violet light does. That's because higher-frequency violet light polarizes the water molecules a little more easily than lower-frequency red light, and that increased polarization slows down the violet light. This frequency dependence of light's speed in a material is called **dispersion.** Dispersion affects refraction. The more light slows as it enters a raindrop, the more it bends at the boundary. Since violet light slows more than red light, violet light also bends more and the different colors of sunlight follow somewhat different paths through the raindrop.

A rainbow is created when raindrops separate sunlight according to color (Fig. 12.1.7). To see the rainbow, you stand with the sun at your back and look up at the sky. When sunlight hits the raindrops, they redirect some of that light back toward you. Since each raindrop redirects light only in a narrow range of angles, you can't see light from every raindrop. Only the raindrops in a narrow arc of the sky redirect light toward you. This arc appears brightly colored because raindrops at the inner edge of the arc send violet light toward you while raindrops at the outer edge of the arc send red light toward you. In between, you see all the colors of the rainbow.

Figure 12.1.8 shows how a raindrop redirects different colors of light in different directions. While there are many possible paths light can take through the raindrop, this path is the one that produces rainbows. Sunlight enters near the top of the raindrop and bends inward. Violet light bends more than red light, so the sunlight begins to separate according to color. Some sunlight is also reflected from the raindrop, but doesn't contribute to rainbows.

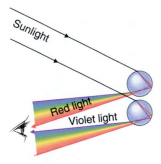

Fig. 12.1.8 As sunlight passes through spherical raindrops, its colors separate. Violet light bends more at each air/water boundary than does red light, and the two emerge from the raindrops heading in different directions. You see red light coming toward you from the upper raindrops and violet light from the lower raindrops.

When the light inside the raindrop strikes the back surface, most of it leaves the drop and is lost. However, a small fraction of the light reflects from that surface and continues to travel through the raindrop. When this light reaches the raindrop's front surface, most of it leaves the drop. Violet light bends more strongly than red light as they reenter the air, so the different colors of light leave the drop heading in different directions. Since violet light is redirected more upward than red light, you see violet light coming toward you from the lower raindrops. Red light is redirected more downward so you see it coming toward you from the upper raindrops. Thus the upper arc of the rainbow is red while the lower arc is violet.

➤ Check Your Understanding #3: The Look of Diamonds

A diamond pendant sparkles with color when you look at it in sunlight. From where do the colors come?

Soap Bubbles

Soap bubbles also separate sunlight into its various colors (Fig. 12.1.9), but they use a wave phenomenon called interference. We encountered **interference** with microwave ovens, when we noted that microwaves passing through a particular location from several different directions sometimes help and sometimes hinder the cooking at that location. When the electric fields there add together, they help one another and cooking proceeds quickly. This mutual assistance is called **constructive interference** (Fig. 12.1.10a). But when their electric fields cancel one another, cooking is slow and the effect is called **destructive interference** (Fig. 12.1.10b).

Both forms of interference occur when sunlight reflects from the outer skin of a soap bubble. As each wave of sunlight hits that thin film of soapy water, the film's front surface reflects about 4% of the wave and the film's back surface reflects another 4% (Fig. 12.1.11). Since both reflections travel in the same direction, the reflected light that you see reaches your eyes via two different paths, one from each reflection. If these two waves arrive **in phase**—that is, with their electric fields synchronized and assisting one another—you see the particularly bright reflection of constructive interference. If the two waves arrive **out of phase**—with their electric fields canceling one another—you see the particularly dim reflection of destructive interference.

Whether you see constructive or destructive interference depends on the wavelength of the sunlight. The back surface reflection has to travel twice through the soap film, so it's delayed relative to the front surface reflection. If the delay is just long enough for the wave to complete an integral number of cycles, then the two reflected waves are in phase with one another as they head toward your eye and you see a bright reflection. If the delay allows the back surface reflection to complete an

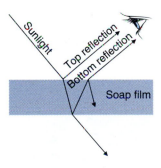

Fig. 12.1.11 Sunlight reflects from both the front and back surfaces of a soap film. The back surface reflection is delayed relative to the front surface reflection because it must pass twice through the soap film itself. If the two reflected waves arrive in phase, you see a bright reflection. If they are out of phase, you see a dim reflection.

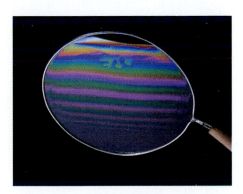

Fig. 12.1.9 Light reflected by the front and back surfaces of this soap film interferes with itself and gives the film its colorful appearance. Since the colors are determined by film thickness and the film's thickness increases in the downward direction, the film displays horizontal bands of color.

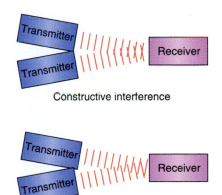

Constructive interference

Destructive interference

Fig. 12.1.10 (a) When waves from two separate paths arrive in phase at the receiver, constructive interference produces a particularly large effect on the receiver. (b) When waves arrive out of phase, destructive interference produces a particularly weak effect on the receiver.

extra half cycle, then the two reflected waves are out of phase with one another and you see a dim reflection.

Sunlight contains many different wavelengths of light, and these wavelengths behave differently during the reflection process. You see a colored reflection, consisting mainly of those wavelengths of light that experience constructive interference. Because the delay experienced by the back surface reflection depends on the thickness of the soap film, you can actually determine the film's thickness by studying its colors.

Check Your Understanding #4: A Slick-Looking Oil Slick

A thin layer of oil or gasoline floating on water appears brightly colored in sunlight. From where do these colors come?

Sunlight and Polarizing Sunglasses

polarizing materials

(a)

(b)

Fig. 12.1.12 (*a*) Ordinary sunglasses simply darken the scene. (*b*) Polarizing sunglasses block horizontally polarized light, the main component of glare.

All sunglasses absorb some of the sunlight passing through them, but the best ones absorb horizontally polarized light much more strongly than vertically polarized light. These polarizing sunglasses dramatically reduce glare by eliminating most of the light reflected from horizontal surfaces.

When light strikes a transparent surface at right angles, about 4% of that light is reflected, regardless of its polarization. But when light strikes a horizontal surface at a shallow angle, horizontally polarized light reflects much more strongly than vertically polarized light. That's because horizontally polarized light's horizontal electric field pushes electric charges back and forth along the surface. Charges shift relatively easily in that direction and the surface becomes even easier to polarize as the angle becomes shallower. As a result, the surface reflects more horizontally polarized light at shallow angles than it does at steeper angles.

But vertically polarized light's vertical electric field pushes electric charges up and down vertically. At shallow angles, this field acts to lift the charges in and out of a horizontal surface. Since the charges can't leave the surface, the surface becomes harder to polarize as the angle becomes shallower. A horizontal surface doesn't reflect much vertically polarized light, and there is even a special angle, *Brewster's angle,* at which no vertically polarized light reflects at all.

Sunlight is an even mixture of vertically and horizontally polarized waves. Since horizontally polarized waves reflect most strongly from horizontal surfaces, sunglasses that absorb horizontally polarized light will prevent you from seeing most of this reflected light (Fig. 12.1.12). That's why polarizing sunglasses are so effective at reducing glare.

Check Your Understanding #5: Looking into the Reflecting Pool

When you look into a pool of water, you see mostly a reflection of the sky. But when you wear polarizing sunglasses, you see into the water clearly. Explain.

why the sky looks darker

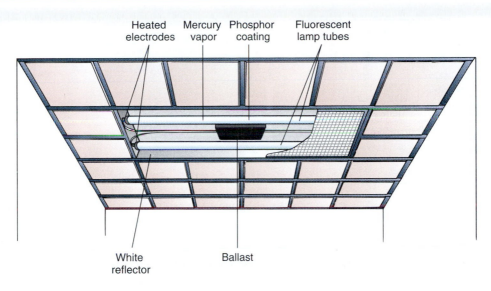

Heated electrodes | Mercury vapor | Phosphor coating | Fluorescent lamp tubes

White reflector | Ballast

SECTION 12.2 Fluorescent Lamps

Energy efficiency is crucial to modern lighting. While incandescent lamps provide pleasant, warm illumination, most of the power they consume is wasted as invisible infrared light. Fluorescent and other gas discharge lamps produce far more visible light with the same amount of electric power and now dominate office, industrial, and street lighting. In this section we'll explore several types of discharge lamps—fluorescent, mercury vapor, sodium vapor, and metal–halide lamps—that all share a common theme: current flow through a gas.

Questions to Think About: *Why do different fluorescent lamps often have slightly different colors? Why is the light from a neon sign red? Are neon atoms red? Why do most fluorescent lamps take a few seconds to turn on? Why are streetlights so dim when they first turn on? Why are some highway lights orange?*

Experiments to Do: *Examine the discharge lamps around you, particularly the white fluorescent tube lamps. Where do their lights originate? In a fluorescent lamp, it comes from the white phosphor coating on the tube's inner surface. What about in a neon lamp or a mercury vapor streetlight? Compare the colors of various lamps, including several different fluorescent lamps. Are their lights identical?*

Both fluorescent and neon lamps start almost immediately. But watch how long it takes a streetlight to start. While fluorescent and neon lamps remain cool during operation, the mercury, sodium, and metal-halide lamps used in street lighting get hot. Watch a streetlight warm up. Does its color change? If the streetlight loses power even for a moment, it must wait about 5 minutes before it can start again. Why must it wait?

How We See Light and Color

Before examining fluorescent lamps, let's look at how our eyes recognize color. While it might seem that they actually measure the wavelengths of light, that's not

Fig. 12.2.1 The red-sensitive cells in our retina detect light near 600 nm, the green-sensitive cells near 550 nm, and the blue-sensitive cells near 450 nm. The red-sensitive cells also respond near 440 nm, so that we see violet.

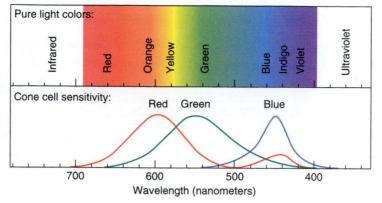

the case. Instead, our retinas contain three groups of light-sensing *cone cells* that respond to three different ranges of wavelengths. One group of cone cells responds to light near 600 nm and lets us see red, another responds to light near 550 nm and lets us see green, and a third responds to light near 450 nm and lets us see blue (Fig. 12.2.1). These cone cells are most abundant at the center of our vision. While our retinas also contain *rod cells,* which are more light sensitive than cone cells, rod cells can't distinguish color. They are most abundant in our peripheral vision and provide us with night vision.

Having only three types of color-sensing cells doesn't limit us to seeing just three colors. We perceive other colors whenever two or more types of cone cells are stimulated at once. Each type of cell reports the amount of light it detects, and our brains interpret the overall response as a particular color.

While light of a certain wavelength will stimulate all three types of cone cells simultaneously, the cells don't respond equally. If the wavelength is 680 nm, the red cone cells will respond much more strongly than the green or blue cells. Because of this strong red response, we see the light as red.

Other wavelengths of light stimulate the three types of cells somewhat more evenly. Light with a wavelength of 580 nm is in between red and green light. Both the red-sensitive and the green-sensitive cone cells respond about equally to this light and we see it as being yellow.

But we also see yellow when looking at an equal mixture of 640-nm light (red) and 525-nm light (green). The 640-nm light stimulates the red-sensitive cone cells and the 525-nm light stimulates the green-sensitive ones. Even though there is no 580-nm light entering our eyes, we see the same yellow color as before.

In fact, mixtures of red, green, and blue light can make us see virtually any color. For that reason, these three are called the **primary colors of light.** A color television uses red, green, and blue phosphors to produce full-color pictures. While the television only produces mixtures of three different lights, it controls those mixtures carefully and we see all the possible colors.

Check Your Understanding #1: Mixed Up Light

If you look at a mixture of 70% red light and 30% green light, what color will you see?

The Shortcomings of Incandescent Lamps

When all three of our color-sensing cells respond about equally, we see white light. That's because our vision evolved under a single incandescent light source: the sun. Sunlight stimulates the red-, green-, and blue-sensitive cells in our eyes about evenly, so any other source of "white light" must do the same.

While an incandescent light bulb makes a good attempt at producing white light, it suffers from two serious drawbacks. First, because its filament can't reach the temperature of the sun's surface (5800 °C), it emits a black body spectrum that is much redder than sunlight. Second, because most of its electromagnetic radiation is invisible infrared light, it doesn't use energy efficiently to produce useful visible light.

The best alternatives to incandescent lamps are gas discharge lamps. Some discharge lamps do excellent jobs of producing white light, and all of them are far more energy efficient than incandescent light bulbs. The most common and familiar of these lamps are fluorescent lamps. Instead of heating a filament until it glows white hot, a fluorescent lamp uses electrons to excite atoms, which then give off light. A fluorescent lamp remains fairly cool as it operates, so it wastes little power heating the room or giving off infrared radiation.

Check Your Understanding #2: Making Whiter Light

Some photographic lamps simulate sunlight by placing a blue filter in front of an incandescent bulb. This filter absorbs some of the red light so that the lamp appears whiter. Does this filter increase or decrease the lamp's energy efficiency?

Fluorescent Lamps: Atoms and Light

At the heart of a fluorescent lamp is a narrow glass tube that's sealed at both ends. This tube contains argon, neon, and/or krypton gases at a pressure and density only 0.3% that of the atmosphere outside the tube. The tube also contains a few drops of liquid mercury metal, some of which evaporates to form mercury vapor. About one in every thousand gas atoms inside the tube is a mercury atom, and it's these mercury atoms that create the light.

The tube lights up when an electric current passes through it. It has metal electrodes at each end so that current can enter the gas through one electrode and leave through the other. But as long as the gas contains no mobile electric charges, it can't conduct electricity. Something must inject electric charges into the gas so that electricity can move through it.

Fluorescent lamps use two techniques to introduce charges into the gas. Most heat their electrodes, so that thermal energy ejects electrons from their surfaces. Others use high voltages to squeeze like charges onto the electrodes until those charges push one another into the gas. But regardless of technique, the result is a gas that conducts electricity.

The lamp then starts a **discharge**—that is, a current flowing through the gas. This current consists mostly of electrons, accelerating from the tube's negatively charged electrode to its positively charged electrode (Fig. 12.2.2). While these electrons collide frequently with various gas atoms, they have so little mass that they usually just bounce off the gas atoms without losing much energy. Like a Ping-Pong ball re-

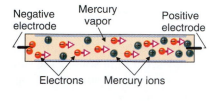

Fig. 12.2.2 A fluorescent tube sends electrons through a low-pressure gas. These electrons collide with mercury atoms in the gas, transferring energy to those mercury atoms and causing them to emit ultraviolet light. Positively charged mercury ions, created by particularly energetic collisions, keep the electrons from repelling one another to the walls of the tube.

bounding from an elephant, the electron does most of the bouncing and then continues on its way.

However, every so often an electron will collide with a mercury atom and something different will happen; the mercury atom will rearrange internally and absorb part of the electron's kinetic energy. The electron will rebound with less energy than it had before and the mercury atom will go on to emit light.

To understand how the mercury atom manages to turn the electron's kinetic energy into light, we need to look at the structure of atoms. As we learned in the section on xerographic copiers, quantum physics permits the mercury atom's 80 electrons to orbit its nucleus only in certain allowed orbitals. Because of the Pauli exclusion principle, only one electron of each spin (up and down) can be in a given orbital, so the atom's electrons spread out into at least 40 different orbitals.

To minimize its potential energy, the mercury atom's 80 electrons normally fill the 40 orbitals that are closest to its nucleus. This arrangement is called the **ground state** of the atom. Like all isolated atoms near room temperature, the mercury atoms in a fluorescent lamp are almost always in their ground states.

But when a mercury atom is struck hard by an electron, its electrons may rearrange. One or more of its electrons may move into a formerly unoccupied orbital. This new arrangement of electrons is called an **excited state** of the atom, indicating that at least one of the electrons is not in its ground state orbital.

The mercury atom doesn't stay in this excited state long. It has too much potential energy and begins a series of rearrangements that quickly returns it to its ground state. In each rearrangement, an electron makes a transition from one orbital to another and the atom radiates light. Electronic transitions in which light is emitted or absorbed are called **radiative transitions.** We've seen that hot objects emit light, a process called **incandescence,** but here the mercury atom emits light without heat, a process called **luminescence.** While radiative transitions are too complicated for us to study in much detail, we can make some useful observations about atoms and light.

First, let's see why an electron that remains in its orbital *doesn't* emit light. After all, an orbiting object is accelerating and accelerating charge should emit light. The problem lies in our view of the orbiting electron. It isn't simply a negatively charged ball orbiting a positively charged nucleus (Fig. 12.2.3*a*). It's actually a negatively charged wave extending uniformly all the way around the nucleus (Fig. 12.2.3*b*). This wave doesn't change with time but remains steady. Since the wave's electric charge doesn't shift back and forth, it doesn't emit an electromagnetic wave.

The electron's wave nature is a consequence of quantum physics. Like all objects in our universe, electrons travel as waves when they move from place to place, and it's only when you go looking for them that you find them at particular locations (see ❏). When an electron orbits an atomic nucleus in a single orbital, it moves as a wave that fills the whole orbital evenly and doesn't change with time. Because of its unchanging form, an electron in an orbital doesn't emit any electromagnetic radiation at all.

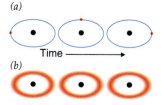

Fig. 12.2.3 (*a*) In the classical picture of an electron in an atomic orbital, the electron is a tiny object that moves around the nucleus as time passes. (*b*) In the quantum picture, the electron is a diffuse wave that doesn't change in any measurable way as time passes.

But when that electron begins a transition to an empty, lower-energy orbital, its wave begins to change with time. The wave moves rhythmically back and forth during the transition so that by the time the transition is complete, the electron will have emitted a *particle* of light. Like an electron, light travels as a *wave* but behaves as a *particle* when you try to locate it. While it's being emitted or absorbed by an atom, light exhibits its particulate nature.

Each time the mercury atom undergoes a radiative transition, the atom emits a single particle of light, a photon, which then heads off through space as a wave. This photon carries with it the energy released by the mercury atom as its electron shifted to a lower-energy orbital. The photon's energy is equal to the frequency of its wave times a fundamental constant of nature known as the **Planck constant,** first used by German physicist Max Planck (1858–1947) in 1900 to explain the light spectrum of a hot object. This relationship between energy and frequency can be written as a word equation:

$$\text{energy} = \text{Planck constant} \cdot \text{frequency}, \qquad (12.2.1)$$

in symbols:

$$E = h \cdot \nu,$$

and in everyday language:

Ultraviolet light and X rays can injure your skin and tissues because each particle of high-frequency light carries a great deal of energy.

The Planck constant is so small, only 6.626×10^{-34} J · s, that light's particulate nature is difficult to observe. Because a photon of ultraviolet light with a frequency of 10^{15} Hz has an energy of only 6.626×10^{-19} J, even a weak ultraviolet beam contains an incredible number of photons. With so many photons hitting your skin, you can't tell that they're arriving as particles. However, an ultraviolet photon is energetic enough to damage a molecule in your skin, contributing to a sunburn and inducing your skin to tan as a defensive response. X rays have even higher frequencies, and their energetic photons can cause more severe damage.

A mercury atom can only emit a photon corresponding to the energy difference between two of its **states**—that is, its possible arrangements of electrons. This effect severely limits the spectrum of light that a mercury atom can emit. Since all mercury atoms have identical states, they all emit the same characteristic spectrum of light. Each type of atom has its own unique light spectrum.

Returning to the fluorescent lamp, we see that electrons moving through the tube can transfer energy to mercury atoms during collisions. Each time a collision leaves an atom in an excited state, that atom returns to the ground state through a series of radiative transitions. With each transition, a photon carries away some of the atom's excess energy until it returns to the ground state. The photons that the mercury atoms emit are characteristic of mercury and are determined by the energy differences between its orbitals.

❏ In 1927, **American physicists Clinton Joseph Davisson (1881–1958) and Lester H. Germer (1896–1972)** showed that electrons travel as waves by observing interference effects when electrons reflected from different atomic layers in a crystal of nickel metal. When the various electron waves arrived at a detector in phase, the detector found many electrons. When the waves arrived out of phase, the detector found few electrons. Their work was aided by a fortuitous accident in which air entered the experiment's glass vacuum tube. While carefully eliminating oxygen from their nickel sample, they managed to perfect its crystalline structure, making it possible to observe the interferences.

► Check Your Understanding #3: Colorful Atoms

Many fireworks involve brilliantly colored lights. How do the atoms in burning chemicals produce particular colors of light?

► Check Your Figures #1: The Particles in a Radio Wave

How much energy is carried by a photon from a 1000-kHz AM radio station?

The Phosphor Coating in a Fluorescent Tube

But the fluorescent tube has a problem. While its mercury atoms emit light, most of that light is ultraviolet. The final radiative transition that returns each mercury atom to its ground state releases a large amount of energy and produces a photon with a wavelength of 254 nm. This light can't go through the glass walls of the tube, and you couldn't see it if it did. So the fluorescent lamp converts it into visible light, using phosphor powder on the inside of the glass tube.

As we saw in Section 11.2, phosphors are solids that luminesce when something transfers energy to them. Their behavior is similar to that of an atom: an energy transfer shifts the phosphor from its ground state to an excited state and it then undergoes a series of transitions that return it to its ground state. Some of those transitions are radiative ones and emit light.

In a television, the phosphor is excited by a beam of electrons. In a fluorescent lamp, the phosphor is excited by ultraviolet light. This excitation or energy transfer is actually a radiative transition, but one in which the photon is absorbed by the phosphor as one of its electrons makes a transition from a lower-energy level to a higher-energy level. (Recall from Section 8.2 that in solids the paths taken by electrons are called levels rather than orbitals.) The light's electric field pushes the electron's wave back and forth rhythmically until it shifts to the new level. The photon disappears and the phosphor receives its energy.

Once the phosphor is in an excited state, its electrons begin to make transitions back to their ground state levels. Most of these transitions radiate visible light, the light that you see when you look at the lamp. However, some of these transitions radiate invisible infrared light or cause useless vibrations in the phosphor itself. Despite this wasted energy, phosphors are relatively efficient at turning ultraviolet light into visible light, a process called **fluorescence.**

The phosphors in a fluorescent lamp are carefully selected and blended to fluoresce over a broad range of visible wavelengths. While this light doesn't have the same spectrum as sunlight, it appears white because it stimulates the red-, green-, and blue-sensitive cells in our eyes about equally. The blending of phosphors is necessary because, like atoms, each phosphor fluoresces with a characteristic spectrum of light that's determined by the energy differences between its levels. Several different phosphors are needed to create the right balance between red, green, and blue lights. While some advertising and novelty lamps use brightly colored, unblended phosphors, fluorescent lighting tubes come in six standard color blends: cool white, deluxe cool white, warm white, deluxe warm white, white, and daylight.

When they were first introduced, fluorescent tubes used the daylight phosphor. This phosphor emits too much blue light and makes everything look cold and medicinal. Phosphors have improved over the years so that the most common phosphors today, cool white and warm white, look much more pleasant. Light from the "cool" phosphors resembles daylight, while that from the "warm" phosphors resembles incandescent lighting. Household fluorescent lamps often use deluxe cool white and deluxe warm white, which are even more pleasant versions of white but which are slightly less energy efficient.

Check Your Understanding #4: Sorry, We Forgot the Coating

If there were no phosphor coating on the inside of a fluorescent tube, what would you see when the lamp operated?

A Few Practical Issues

A fluorescent tube needs a substantial electric field inside it to keep its electrons moving forward. Since this electric field is proportional to the voltage drop through the tube, longer tubes need higher voltages. Power line voltages (110 V to 240 V) are appropriate for tubes up to 3 m in length, but the longer and often colored fluorescent tubes used in artwork or advertising require much higher voltages.

To keep electrons from pushing one another into its walls, a fluorescent tube must also contain positively charged mercury ions. The discharge naturally produces these ions during particularly energetic collisions. The result, a gaslike mixture of positively charged ions and negatively charged electrons, is called a **plasma.** All operating discharge lamps contain plasmas. When you look at the rich red glow of a neon sign, you're observing a plasma. Neon atoms in the discharge are emitting their characteristic spectrum, which is mostly red light.

cylindrical tubes

Since hot electrodes are needed to inject electrons into the plasma, a typical fluorescent lamp runs current through filaments at each end of its tubes during start-up (Fig. 12.2.4). Once the discharge is operating, the electrodes are kept hot by electrons that hit them at the ends of their journeys through the tube. Unfortunately, these filament/electrodes are fragile. They're damaged by a process called **sputtering,** in which positive mercury ions from the plasma collide with the filament/electrodes and chip away their tungsten atoms. Because sputtering is particularly severe during start-up, a typical filament/electrode breaks after a few thousand starts. That's why you shouldn't turn a fluorescent lamp on and off more than once every few minutes.

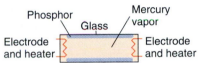

Fig. 12.2.4 In a hot-electrode fluorescent tube, the electrodes are actually filaments and are heated by running currents through them.

> **Check Your Understanding #5: Getting Red-y**
> Why do the ends of some fluorescent lamps glow red during starting?

Mercury, Metal–Halide, and Sodium Lamps

While a low-pressure mercury discharge emits mostly ultraviolet light, a high-pressure mercury discharge emits more visible light than ultraviolet light. This change occurs because ultraviolet light becomes trapped in densely packed mercury atoms and only visible light is able to escape from the discharge.

Known as **radiation trapping,** this effect occurs because mercury atoms absorb 254-nm photons just as well as they emit them. The same radiative transition that causes a mercury atom to emit a 254-nm photon can also run backward to absorb that photon. In a dense gas of mercury, whenever one mercury atom tries to emit a 254-nm photon, another mercury atom snaps it up. So while the discharge keeps pouring energy into the mercury atoms, they can't get rid of it as 254-nm photons. Instead, they emit most of their energy through radiative transitions between other excited states. Since this light is much less likely to be caught by other mercury atoms, it emerges from the lamp as bluish visible light.

When you first turn on a high-pressure mercury lamp, most of the mercury is liquid and the pressure is low. The lamp starts like a small fluorescent tube without phosphor, so you see very little light. But the tube is designed to heat up during operation so that the liquid mercury evaporates to form a dense gas. As the gas pressure rises, the tube's color charges until it emits brilliant, blue-white light.

To make a lamp that's a little less bluish, some high-pressure mercury lamps contain additional metal atoms. These atoms are introduced into the lamps as metal–iodide compounds, making them metal–halide lamps. Sodium, thallium, indium, and scandium iodides all help to strengthen the red end of the spectrum and give metal–halide lamps a warmer color.

Pure sodium lamps resemble mercury lamps, except that they use sodium atoms. Sodium is a solid at room temperature, so both low- and high-pressure sodium lamps must heat up before they begin to operate properly. A low-pressure sodium lamp is extremely energy efficient because its 590-nm light comes directly from a sodium atom's strongest radiative transition. Many highways are illuminated by the yellow-orange glow of low-pressure sodium lamps.

But this monochromatic illumination is unpleasant and permits no color vision at all. While it may be acceptable on a highway, you wouldn't want it near your home. That's why people buy high-pressure sodium lamps for home use (Fig. 12.2.5).

Remarkably enough, the 590-nm emission itself smears out at high pressure to cover a wide range of wavelengths, from yellow-green to orange-red. This spreading occurs because of the many collisions suffered by the densely packed sodium atoms as they try to emit 590-nm light. These collisions distort the atomic orbitals so that the photons emerge with somewhat shifted energies. Overall, a high-pressure sodium lamp emits remarkably little light exactly at 590 nm because the atoms trap that light. There is actually a hole in the lamp's spectrum right at 590 nm.

High-pressure discharge lamps suffer from a problem not found in low-pressure lamps: they're difficult to start when hot. It's much harder to initiate a discharge in a high-pressure gas than in a low-pressure gas, so they all start at low pressure and then evolve to high pressure. If the discharge in a high-pressure mercury, sodium vapor, or metal–halide lamp is interrupted, the lamp must cool down before it can be restarted.

fluorescent fixtures

Check Your Understanding #6: Slow Glowing

At dusk, a mercury streetlight turns on. It glows dimly at first and gradually increases in brightness. What is happening during this warm-up period?

Fig. 12.2.5 The active component of a high-pressure sodium vapor lamp is a small translucent tube. As the lamp warms up, sodium metal in the tube evaporates to form a brilliant yellow discharge. The dense vapor of sodium atoms in the tube traps the 590 nm light so that the lamp emits a richer spectrum of wavelengths and a less monochromatic glow than a low-pressure sodium lamp.

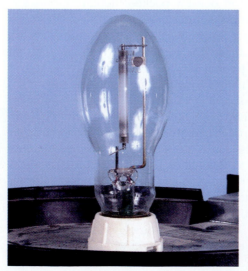

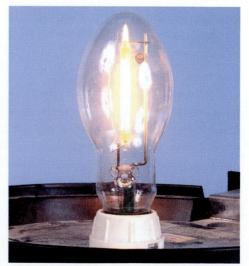

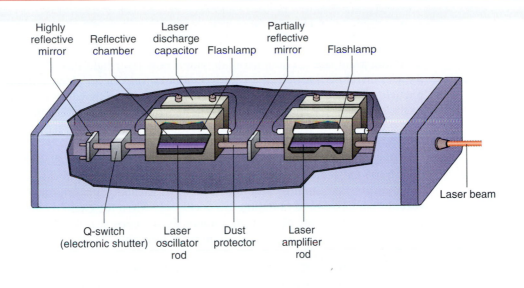

Highly reflective mirror · Reflective chamber · Laser discharge capacitor · Flashlamp · Partially reflective mirror · Flashlamp

Laser beam

Q-switch (electronic shutter) · Laser oscillator rod · Dust protector · Laser amplifier rod

Lasers

SECTION 12.3

Few devices have inspired our imaginations more than lasers. Since their invention in the late 1950s, lasers have found countless uses, from cutting metal and clearing human arteries to surveying land and playing compact discs. But lasers are more than just novel applications of old ideas. Instead, they bring together quantum and optical physics to produce a new type of light. This light is radically different from that produced by incandescent and fluorescent lamps, and its properties make it particularly useful for many applications. In this section we'll examine the nature of this new light and the ways in which lasers produce it.

Questions to Think About: Why is laser light usually brightly colored? Why does laser light often appear as a narrow beam? In movies, "lasers" are often shown to emit bright streaks of light that can be dodged if one jumps quickly enough; is that view realistic?

Experiments to Do: While lasers are household objects, the ones in CD or DVD players and laser printers are relatively inaccessible. If you don't own a laser pointer, look at a store barcode scanner. The scanning system contains a gas or solid-state laser that emits a very narrow beam of bright red light. This system also contains a rotating mirror or holographic disk that directs the beam as a pattern of thin stripes onto anything that passes through the scanner. A light sensor inside the window watches for this moving beam of light to travel across a label. If you look down at the laser light emerging from the scanner's window or observe the spot of a laser pointer on the wall, you'll see both the purity of its color and its strange speckled character; the light appears to consist of tiny light and dark speckles. These speckles are caused by interference effects, which are extremely pronounced in the ordered laser light. This light also appears unusually bright and, as when looking at the sun, you should keep your gaze brief to avoid eye injury.

Lasers and Laser Light

To understand lasers, you must understand how laser light differs from the normal light emitted by hot objects or by individual atoms in an electric discharge. Each particle of normal light, each photon, is emitted willy-nilly without any relationship to the other light particles being emitted nearby. Because of this light's independent and unpredictable character, it's called *spontaneous light* and its creation is called **spontaneous emission of radiation.**

But theoretical work by Albert Einstein and others in the 1920s and 1930s predicted the existence of a second type of light, *stimulated light,* that can be created when an excited atom or atomlike system duplicates a passing photon. While this **stimulated emission of radiation** can occur only when the excited atom is capable of emitting the duplicate photon spontaneously, the copy that it produces is so perfect that the two photons are absolutely indistinguishable. Together, these two photons form a single electromagnetic wave.

To get a slightly better picture for how such stimulation occurs, think about an isolated atom in an excited state. That atom will eventually return to its ground state, but it must emit one or more photons in order for this to happen. That atom waits around in the excited state until it begins a spontaneous radiative transition. During the transition, one of the atom's electrons accelerates back and forth and the atom emits a photon.

But if a similar photon passes through the atom as it's waiting in the excited state, that photon's electric field can stimulate the radiative transition process through sympathetic vibration. The field pushes and pulls on the atom's electrons and makes them accelerate back and forth. While this effect is small, it may be enough to trigger the emission of light. If the atom does emit light, the photon it will produce will be a perfect copy of the stimulating photon.

When this stimulated emission process was first discovered, people immediately recognized that it made light amplification possible. If enough excited systems could be assembled together, a single passing photon could be duplicated exactly over and over again. Instead of a single particle of light, you would soon have thousands, or millions, or even trillions of identical light particles.

However, implementing this idea had to wait until the late 1950s, when the technical details for how to actually achieve light amplification were worked out. In 1960, the first laser oscillators were constructed. These were devices that emitted intense beams of light, in which each particle of light was identical to every other particle of light. A single particle of light had been duplicated by the stimulation process into countless copies.

When individual excited atoms or atomlike systems emit light through spontaneous emission, the particles of light head off separately as many independent electromagnetic waves (Fig. 12.3.1a). Light consisting of many independent electromagnetic waves is called **incoherent light.**

But when that same collection of excited atoms or atomlike systems emit light by stimulated emission, all of the light particles are *absolutely identical* and form a single electromagnetic wave (Fig. 12.3.1b). Light consisting of many identical photons and a single electromagnetic wave is called **coherent light.** Because of its single wave nature, coherent light exhibits remarkable interference effects. These effects are easily seen in the coherent light emitted by lasers.

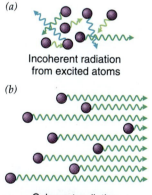

(a)

Incoherent radiation
from excited atoms

(b)

Coherent radiation
from excited atoms

Fig. 12.3.1 (*a*) Photons of incoherent light are created independently and have somewhat different wavelengths and directions of travel. (*b*) Photons of coherent light are produced by stimulated emission and are identical to one another in every way.

Check Your Understanding #1: A Light Dusting of Photons

If you were to measure the electric field in the light from a flashlight, you would find that it fluctuates about randomly. Why is the light's electric field so disorderly?

Light Amplifiers and Oscillation

Producing coherent light requires amplification. You must start with only one particle of light and duplicate it many times. The basic tool for this duplication of light is a **laser amplifier** (Fig. 12.3.2). When weak light enters an appropriate collection of excited atoms or atomlike systems—the **laser medium**—that light is amplified and becomes brighter. The new light has exactly the same characteristics as the original light, but it contains more photons.

However, when we think of lasers, we rarely imagine a device that duplicates photons from somewhere else. We usually picture one that creates light entirely on its own. To do that, the laser must produce the initial particle of light that it then duplicates to produce others. A **laser oscillator** is a device that uses the laser medium itself to provide the seed photon, which it then duplicates many times (Fig. 12.3.3). If a laser medium is enclosed in a pair of carefully designed mirrors, it's possible for the stimulation process to become self-initiating and self-sustaining. However, the mirrors must be curved properly and must have the correct reflectivities. One mirror must normally be extremely reflective, while the other must transmit a small fraction of the light that strikes its surface.

When the laser medium is placed between these two mirrors, there is a chance that a photon, emitted spontaneously by one of the excited systems, will bounce off a mirror and return toward the laser medium. As that returning photon passes through the laser medium, it's amplified. Because the photon was emitted by one of the excited systems, it has the right wavelength to be amplified by other excited systems. (For a discussion of a photon's properties, see ❏.)

By the time the original photon leaves the laser medium, it has already been duplicated many times. This group of identical photons then bounces off the second mirror and returns for another pass through the laser medium. It continues to bounce back and forth between the mirrors until the number of identical photons in the collection is astronomical.

Eventually there are so many identical photons that the laser medium is no longer able to amplify them. The laser medium has only so much stored energy and only so many excited systems in it. If the laser medium continues to receive additional energy, it may continue to amplify the light somewhat. But if it doesn't receive more energy, light amplification will eventually cease.

To let the light out of this laser oscillator, one of its mirrors is normally *semitransparent*—that is, some of the photons that strike the surface of the mirror travel through it rather than reflecting. This transmission creates a beam of outgoing light, a *laser beam*. The laser beam continues to emerge from the mirror as long as the amplification process can support it. Because this laser beam consists of duplicates of one original photon, it's coherent light. For technical reasons, many lasers duplicate more than one original photon simultaneously, so that their laser beams are a little less coherent than they might be. However, with suitable fine-tuning, one original photon can usually be made to dominate the laser beam.

When you focus a flashlight's beam with a lens, its independent photons won't end up exactly together at the focus of the lens. That's because these photons leave the flashlight heading in somewhat different directions. But since virtually all of the photons in a laser beam are identical, they can all focus together to an extremely small spot. That's why a laser printer employs a laser; a laser beam can illuminate a very small spot on the photoconductor drum that is used in the xerographic process to produce a printed image.

Fig. 12.3.2 A laser amplifier uses excited atoms or atomlike systems to increase the number of light particles leaving the laser medium. The incoming light is duplicated by stimulated emission.

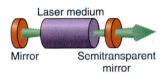

Fig. 12.3.3 A laser oscillator is a laser amplifier enclosed in mirrors. Oscillation occurs when the laser medium spontaneously emits one photon in just the right direction. This photon bounces back and forth between the two mirrors and is duplicated many times. Some of the light is extracted from this laser by making one of the mirrors semitransparent.

❏ While it might seem that a photon should have an exact wavelength and frequency, and travel in only one direction, that's not the case. Photons travel as electromagnetic waves and spread in more than a single direction. And because each photon has a beginning and an end, its wave contains more than a single wavelength or a single frequency. Thus while lasers can produce some of the most perfect electromagnetic waves imaginable, those waves still spread outward slightly and still have a range of wavelengths and frequencies.

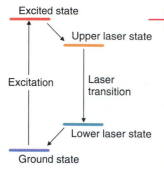

Fig. 12.3.4 An ideal laser system passes through four different states during the laser's operation.

types of lasers

→ **Check Your Understanding #2: More of a Good Thing**

If you take the laser beam from a particular laser oscillator and send it through a similar laser amplifier, what will happen to the laser beam?

How a Laser Medium Works

Obtaining the excited systems needed to amplify light is a critical issue for lasers. Ideally, a laser involves four different states of an atom or atomlike system: the ground state, an excited state, the upper laser state, and the lower laser state. The reason for having four separate states should become clear in a moment.

Let's consider an atom that acts as an ideal laser amplifier (Fig. 12.3.4). The atom starts in its ground state. A collision or the absorption of a photon shifts it to the excited state, giving it the energy it needs to amplify light. The atom then shifts to the upper laser state, either by emitting a photon or as the result of a collision. This preliminary shift is important because it prevents the excited atom from returning directly back to the ground state and avoiding the amplification process. Once it has shifted to the upper laser state, the atom is stuck there and will wait around long enough to amplify light.

When a suitable photon passes through the atom, that photon stimulates the emission of a duplicate photon and the atom undergoes a radiative transition to the lower laser state. So far, so good. However, if the atom remains in the lower laser state, it might absorb a photon of the laser light and return to the upper laser state. To prevent this sort of radiation trapping, the atom must quickly shift to the ground state, either by emitting a photon or as the result of another collision. The atom is then ready to begin the cycle all over again.

This four-state cycle, or something close to it, is found in most lasers. In each case, something provides the energy needed to shift atoms or atomlike systems in the laser medium from their ground states to their excited states. This transfer of energy into the laser medium to prepare it for amplifying light is called *pumping*. How a particular laser medium is pumped depends on the laser.

The most common pumping mechanisms are electronic and optical. In electronic pumping, currents of charged particles use their kinetic or electrostatic energies to excite the medium's atoms or atomlike systems from their ground states to their excited states. In optical pumping, intense light is shone on the laser medium, causing a similar excitation.

The most common examples of electronic pumping are diode lasers. Found in laser pointers and CD or DVD players, these lasers are closely related to the diodes we discussed in Section 10.1. When current flows through a diode from its anode to its cathode, electrons shift from conduction levels in the n-type semiconductor to empty valence levels in the p-type semiconductor. As they shift, the electrons release energy and normally produce heat.

But in specially designed diodes, the high-energy electrons join briefly with positive charges in the p-type semiconductor to form excited atomlike systems. These systems radiate away their energy as visible light. Red, yellow, green, and blue light-emitting diodes or LEDs are common in electronic devices. However, these excited atomlike systems can also amplify light so that carefully structured light-emitting diodes can act as laser oscillators and emit intense coherent light (Fig. 12.3.5).

As for optical pumping, the most important examples are ion-doped solid-state lasers. These lasers are based on atomic ions embedded in transparent solids. Common ions are titanium (Ti), neodymium (Nd), and erbium (Er), and they are often

Fig. 12.3.5 This tiny semiconductor chip is a diode laser that emits an intense beam of coherent light when a current flows through it.

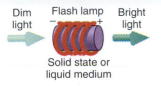

Fig. 12.3.6 In an optically pumped laser, intense light from a flash lamp, an arc lamp, or even another laser transfers energy to the laser medium. Atoms or atom-like systems inside the medium store this energy and use it to amplify light.

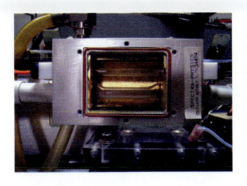

Fig. 12.3.7 This flash-lamp-pumped laser amplifier contains a purple neodymium:YAG rod. The rod is in the bottom half of the opened, gold-lined amplifier box and is protected by a glass tube. Light from a long flashlamp in the top half of the box excites the neodymium ions so that they can amplify infrared light passing horizontally through the rod.

embedded in sapphire, yttrium aluminum garnet (YAG), or glass. Ti:sapphire, Nd:YAG, and Er:glass lasers are important to modern research, technology, and optical communication systems. When these laser media are exposed to extremely bright light, their ions become excited, and they can act as laser oscillators or amplifiers (Figs. 12.3.6 and 12.3.7).

Check Your Understanding #3: Not So Fast, 007

A secret agent in a movie uses a tiny, hand-held laser to burn a hole through a thick metal plate. From the point of view of power, why is such a laser essentially impossible to make?

special lasers

Epilogue for Chapter 12

In this chapter we explored the creation and movement of light. In *sunlight,* we looked at how light is scattered during its passage through the atmosphere and at how it bends and reflects when it moves from one material to another. We also examined interference effects that occur when a light wave follows more than one path to a particular destination and learned how polarizing sunglasses can diminish glare.

In *fluorescent lamps,* we examined electric discharges in gases. We saw that atoms excited by collisions with charged particles can subsequently emit light through radiative transitions. We studied the primary colors of light to see how the phosphors on the inside surface of a fluorescent tube are able to produce a reasonable facsimile of white sunlight. And in *lasers,* we looked at the difference between common incoherent light and the unusual coherent light emitted by a laser. We saw that lasers use stimulated emission to duplicate photons, so that a small number of initial photons can be amplified into an enormous number.

Explanation: Splitting the Colors of Sunlight

Light bends as it passes through the cut facets of the crystal glass or bowl. Because of dispersion, the angle of each bend depends slightly on the wavelength of the light involved, so that the different colors of sunlight follow slightly different paths through the crystal. When the light emerges from the crystal, its various wavelengths head in somewhat different directions; you see colors. Color sequences progress from long wavelength to short wavelength, or vice versa. You see red, orange, yellow, green, blue, indigo, and violet, or the reverse—the colors of the rainbow.

Chapter Summary

How Sunlight Works: Sunlight originates at the 5800 °C outer surface of the sun, when electrically charged particles there accelerate back and forth rapidly and emit electromagnetic waves. This sunlight travels at the speed of light through empty space until its reaches the earth's atmosphere. There it slows slightly and some of it Rayleigh scatters. Short-wavelength light is Rayleigh scattered more strongly than long-wavelength light, so the sky appears blue.

As the sunlight passes through various objects, it slows down and its colors separate. When sunlight passes through falling raindrops, its different wavelengths follow different paths and create rainbows. As sunlight reflects from thin films such as soap bubbles, its waves are divided and may follow several different paths to the same destination. These light waves then interfere with one another so that some waves appear strong and bright while others are weak and dim. Interference depends on the wavelengths of light so thin films appear brightly colored.

How Fluorescent Lamps Work: A fluorescent lamp emits visible light when the phosphor coating on the inside of its tube is exposed to ultraviolet light produced inside the tube by a mercury vapor discharge. Electrons are injected into the tube by heated electrodes at each end, and these electrons carry current through the tube. When the electrons collide with mercury atoms, these atoms are excited and begin to make transitions that return them to their ground states. While making these transitions, the mercury atoms emit light. The strongest emission is ultraviolet light at 254 nm. This light excites the phosphor coating and causes it to emit visible light.

How Lasers Work: Lasers amplify light through the process of stimulated emission. In this process, energy is transferred to atoms or atomlike systems contained in a laser medium. While these excited systems might emit light spontaneously, they can be stimulated into emitting duplicates of a passing photon. When a photon with just the right wavelength passes through an excited system, that system is likely to give up its stored energy by emitting an exact copy of the initial photon. In a laser oscillator, two mirrors cause light to bounce back and forth through the laser medium. An initial photon is duplicated endlessly to produce coherent light. One of the mirrors is semitransparent so part of the light emerges from the laser as a laser beam.

Important Laws and Equations

1. Relationship Between Energy and Frequency: The energy in a photon of light is equal to the product of the Planck constant times the frequency of the light wave, or

$$\text{energy} = \text{Planck constant} \cdot \text{frequency}. \quad (12.2.1)$$

Check Your Understanding—Answers

Section 12.1 SUNLIGHT

1. Charged particles in the hot flame accelerate back and forth and emit electromagnetic waves that include the low-frequency end of the visible spectrum.

Why: The accelerating charged particles in hot objects emit light. The hotter the object, the faster the charged particles move and accelerate and the higher the frequencies of light they emit. A candle isn't hot enough to emit the whitish light of the sun, so it emits mostly reddish or yellowish light.

2. Rayleigh scattering from the microscopic smoke particles scatters more blue light than red light and gives the air a bluish glow.

Why: When smoky air in a dark room is illuminated by a bright spotlight or another white light source, the tiny particles in the air Rayleigh scatter some of the light. When you look at the air against a dark background you can see this glow. Since Rayleigh scattering affects blue light most strongly, the air appears bluish.

3. A diamond exhibits dispersion so that different frequencies of sunlight follow somewhat different paths through the diamond's polished facets. The different colors of sunlight emerge from the diamond traveling in slightly different directions, so that you can see them individually.

Why: One of the delightful aspects of a diamond is its strong dispersion. It bends violet light much more than red light so that sunlight is separated into its different colors as it passes through the stone. Cleverly cut facets help to isolate the individual colors.

4. From interferences between light reflected by the top and bottom surfaces of the floating layer of oil.

Why: A thin film of almost anything on water will appear colored because of interference. Part of each light wave striking the film reflects from the top surface, and part reflects from the bottom surface. These two reflected waves interfere with one another in a wavelength-dependent manner and make the layer appear brightly colored.

5. Most of the light that reflects from the water is horizontally polarized. The sunglasses block horizontally polarized light so that you see mostly light from within the pool of water.

Why: Sunlight reflecting from a horizontally polarized surface is mostly horizontally polarized waves. By blocking horizontally polarized light, polarizing sunglasses virtually eliminate this reflected light, and you see mostly light from below the surface. If you turn the sunglasses sideways, they will block the wrong polarization of light and you will see mostly the reflected light. Try it.

Section 12.2 FLUORESCENT LAMPS

1. Orange.

Why: Your red-, green-, and blue-sensitive cone cells have the same response to this light mixture as they would have to pure 600-nm light. Such light appears orange, so you see orange when viewing the mixture.

2. It decreases the lamp's energy efficiency.

Why: Although the bluish filter changes the spectrum of wavelengths so that the amount of blue light leaving the lamp increases relative to the red light, it does this by absorbing some of the red light. The filter gets hot, and less electric power used by the bulb leaves as visible light.

3. When fire adds energy to an atom and shifts it to an excited state, it emits photons in order to return to its ground state. Each photon's energy, frequency, and color are determined by energy differences between the atom's orbitals.

Why: Each color in fireworks is produced by a particular type of atom. Some atoms, such as strontium and lithium, emit red light as they return to their ground states. Barium atoms emit green light, copper atoms emit blue light, and sodium atoms emit yellow-orange light.

4. You would see only a dim (blue-white) glow from the lamp because most of the light produced by the mercury discharge is invisible ultraviolet light.

Why: Without its phosphor coating, a fluorescent tube produces very little visible light. The phosphor coating is needed to convert the discharge's ultraviolet light into visible light. But even if you could see ultraviolet light, you would see very little of it leaving the tube. The tube is made of glass, which absorbs virtually all ultraviolet light with wavelengths shorter than 350 nm. Even in a normal fluorescent tube, any ultraviolet not converted to visible light by the phosphors is absorbed by the glass tube.

5. To start many lamps, their electrodes must be heated red-hot. You can often see light emitted by these hot-filament electrodes.

Why: In most fluorescent fixtures, the filaments are heated to high temperatures just before the discharge starts. Thermal energy then ejects electrons from the filament so that they can carry current through the gas.

6. The small high-pressure discharge tube is heating up, vaporizing more mercury.

Why: When the streetlight first turns on, the pressure of mercury atoms inside it is low and it emits very little visible light. As the discharge heats the tube, more mercury atoms evaporate until eventually all of the mercury inside the tube is gaseous.

Section 12.3 LASERS

1. The flashlight produces incoherent light, with many independent light waves contributing their individual electric fields to the overall electric field. If you were to measure that overall field, you would find it very disorderly.

Why: Because the photons of incoherent light are independent, their individual electric fields don't fluctuate together.

At any place and time, the individual electric fields will add in a complicated, random fashion. As time passes, this overall field will fluctuate randomly. But coherent light, in which each photon is the same as all the others, has a much more orderly electric field because all of the photons contribute equally.

2. It will become even brighter.

Why: A laser oscillator normally emits as intense a beam of light as it can, based on the amount of stored energy in the laser medium. This beam of light can be amplified further by sending it through a separate laser amplifier. Most high-powered lasers use a laser oscillator and one or more laser amplifiers to create particularly bright beams of light.

3. The light in such a laser beam would carry an enormous amount of power. Something must transfer that power to the laser medium, an impossible task in a tiny, hand-held unit.

Why: The power in a laser beam comes from the laser medium. The laser medium must have gotten it from somewhere else. Since batteries aren't up to delivering thousands of watts of electric power, it's unlikely that powerful hand-held lasers will ever be developed. Even if a suitable power source were available, these lasers would tend to overheat. The conversion of power to light is not perfectly efficient, and most of the energy ends up as thermal energy in the laser's components. Lasers require cooling to remove this wasted energy.

Check Your Figures—Answers

Section 12.2 FLUORESCENT LAMPS

1. 6.626×10^{-28} J.

Why: Since the Planck constant is $6.626 \cdot 10^{-34}$ J · s and the frequency of the radio wave is 10^6 Hz or 10^6 cycles/second,

the energy per photon is given by Eq. 12.2.1 as

energy $= (6.626 \times 10^{-34} \text{ J} \cdot \text{s}) \cdot (10^6 \text{ cycles/s}) = 6.626 \times 10^{-28}$ J.

This energy is so small that it's virtually impossible to observe the particulate character of radio waves.

Exercises

1. How long should an antenna be to receive or transmit red light well?

2. How long should an antenna be to receive or transmit violet light well?

3. How rapidly should a charged particle on the sun's surface move back and forth to emit a wave of red light?

4. How rapidly should a charged particle on the sun's surface move back and forth to emit a wave of violet light?

5. Figure 12.1.1 is accurate only for light waves traveling in empty space. To make this figure suitable for light passing through water, how would it have to change?

6. Rayleigh scattering also occurs for infrared and ultraviolet light. Which of these two types of light experiences the strongest Rayleigh scattering?

7. When astronauts aboard the space shuttle look down at the earth, its atmosphere appears blue. Why?

8. When astronauts walked on the surface of the moon, they could see the stars even though the sun was overhead. Why can't we see the stars while the sun is overhead?

9. If you shine a flashlight horizontally at a glass full of water, the glass will redirect the light beam. How?

10. A laser light show uses extremely intense beams of light. When one of these beams remains steady, you can see the path it takes through the air. What makes it possible for you to see this beam even though it isn't directed toward you?

11. To make the beams at a laser light show (see Exercise 10) even more visible, they're often directed into mist or smoke. Why do such particles make the beams particularly visible?

12. Why can you see your reflection in a calm pool of water?

13. Use the concepts of refraction, reflection, and dispersion to explain why a diamond emits a spray of colored lights when sunlight passes through its cut facets.

14. Diamond has an index of refraction of 2.42. If you put a diamond in water, you see reflections from its surfaces. But if you put it in a liquid with an index of refraction of 2.42, the diamond is invisible. Why is it invisible, and how is this effect useful to a jeweler or gemologist?

15. Why is a pile of granulated sugar white while a single large piece of rock candy (solid sugar) is clear?

16. Basic paper consists of many transparent fibers of cellulose, the main chemical in wood and cotton. Why does paper appear white, and why does it become relatively clear when you get it wet?

17. Before the era of paints, people would whiten their walls with whitewash, a mixture of lime particles and water. Lime is a mineral and a large crystal of lime is clear. So why does a wall coated with lime particles appear white?

18. When snow is packed into ice on the surface of a road, it appears white. But when the snow melts into water and refreezes, it can be nearly invisible "black ice." What makes packed snow ice so much more visible than black ice?

19. On a rainy day you can often see oil films on the surfaces of puddles. Why do these films appear brightly colored?

20. When two sheets of glass lie on top of one another, you can often see colored rings of reflected light. How do the nearby glass surfaces cause these colored rings?

21. The colored rings described in Exercise 20 change size when you squeeze the glass sheets closer together. Why do they change size?

22. If you're wearing polarizing sunglasses and want to see who else is wearing polarizing sunglasses, you only have to turn your head sideways and look to see which people now have sunglasses that appear completely opaque. Why does this test work?

23. Why is it easier to see into water when you look directly down into it than when you look into it at a shallow angle?

24. When you stand at the edge of a lake and try to look into the water, you mostly see a reflection of the sky. Why is it much easier to see into the lake when you're wearing polarizing sunglasses?

25. Light near 480 nm has a color called cyan. What mixture of the primary colors of light makes you perceive cyan?

26. Fancy makeup mirrors allow users to choose either fluorescent or incandescent illumination to match the lighting in which they'll be seen. Why does the type of illumination affect their appearances?

27. While many disposable products no longer contain mercury, a potential pollutant, fluorescent tubes still do. Why can't the manufacturers eliminate mercury from their tubes?

28. When white fabric ages it begins to absorb blue light. Why does this give the fabric a yellow appearance?

29. To hide yellowing (see Exercise 28), fabric is often coated with fluorescent "brighteners" that absorb ultraviolet light and emit blue light. In sunlight, this coated fabric appears white, despite absorbing some blue sunlight. Explain.

30. Increasing the power to an incandescent bulb makes its filament hotter and its light whiter. Why doesn't increasing the power to a neon sign change its color?

31. If the low-pressure neon vapor in a neon sign were replaced by low-pressure mercury vapor, the sign would emit almost no visible light. Why not?

32. A fluorescent tube is most efficient when the 254-nm photons emitted by its mercury atoms reach the phosphor coating without being absorbed and reemitted too many times by those atoms. Each absorption and reemission risks wasting energy as heat. Use this fact to explain why a fluorescent tube operating above 40 °C is relatively inefficient.

33. Use the information in Exercise 32 to explain why fluorescent tubes are made relatively narrow.

34. A compact disc player uses a beam of laser light to read the disc, focusing that light to a spot less than 1 micron (10^{-6} m) in diameter. Why can't the player use a cheap incandescent light bulb for this task, rather than a more expensive laser?

35. Explain why the electromagnetic wave emitted by a radio station is coherent—a low-frequency equivalent of coherent light.

36. If a radio transmitter and antenna can emit coherent electromagnetic radiation (see Exercise 35), why aren't there similar transmitters and antennas for coherent light?

37. Does a radio transmitter form its coherent electromagnetic radiation (see Exercises 35 and 36) by duplicating one photon over and over again, the way a laser does?

38. One of the most accurate atomic clocks is the hydrogen maser. This device uses excited hydrogen molecules to duplicate 1.420-GHz microwave photons. In the maser, the molecules have only two states: the upper maser state and the lower maser state (which is actually the ground state). To keep the maser operating, an electromagnetic system constantly adds excited state hydrogen molecules to the maser and a pump constantly removes ground state hydrogen molecules from the maser. Why does the maser require a steady supply of new excited state molecules?

39. Why must ground state molecules be pumped out of a hydrogen maser (see Exercise 38) as quickly as possible to keep it operating properly?

40. While some laser media quickly lose energy via the spontaneous emission of light, others can store energy for a long time. Why is a long storage time essential in lasers that produce extremely intense pulses of light?

41. Astronomers occasionally detect stimulated emission in the electromagnetic waves coming from excited gases far out in space. There are obviously no mirrors out in space, so this radiation isn't coming from laser oscillators. What must be happening to produce this stimulated radiation?

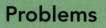

Problems

1. The yellow light from a sodium vapor lamp has a frequency of 5.08×10^{14} Hz. How much energy does each photon of that light carry?

2. If a low-pressure sodium vapor lamp emits 50 W of yellow light (see Problem 1), how many photons does it emit each second?

3. A particular X ray has a frequency of 1.2×10^{19} Hz. How much energy does its photon carry?

4. A particular light photon carries an energy of 3.8×10^{-19} J. What are the frequency, wavelength, and color of this light?

5. If an AM radio station is emitting 50,000 W in its 880-kHz radio wave, how many photons is it emitting each second?

Cases

1. The color of the sky depends on the weather and the air.
  a. On a very humid day, when the air contains many water droplets that are about 1 micron (0.000001 m) in diameter, the air appears hazy and white. Why?
  b. On a very clear, dry day when there is no dust or water in the air, the sky is very dark blue. Why?
  c. Why is the sky black on the airless moon?
  d. When you look at the surface of a calm lake, you see a reflection of the sky. The water is not a metal, yet you see a reflection. What change occurs as light passes from air to water that causes some of the light to be reflected?
  e. Mirages appear on sunny days in the desert when very hot air from the earth's surface bends light from the horizon upward so that you see it as though it were coming from the ground in the distance. Hot air has fewer air molecules in it than colder air at the same pressure. Why does light bend slightly as it moves from colder air to hotter air?

2. The electronic flash in a typical camera is based on a xenon flashlamp, a tube filled with xenon gas at high pressure with an electrode at each end. This flashlamp is electrically connected to a small but powerful capacitor to form a circuit so that if the flashlamp were to conduct current, it would allow current to flow from one plate of the capacitor to the other.
  a. Before you take a picture, the camera places separated electric charge on the two plates of the capacitor until a voltage drop of about 300 V appears across the xenon flashlamp. The flashlamp, however, conducts no current. Why not?

  b. When you take a picture, the shutter opens and the camera causes a small high-voltage transformer to inject a few electrons into the gas in the flashlamp. The lamp suddenly allows current to flow from one plate of the capacitor to the other and the lamp "flashes." Why does this introduction of electrons into the flashlamp cause it to "flash"?
  c. The flashlamp will only last for a certain number of flashes because each flash damages the electrodes. Why does the flash damage the electrodes?
  d. The flashlamp uses high-pressure xenon rather than low-pressure xenon. Why does high-pressure xenon give a more uniform spectrum of light than low-pressure xenon?
  e. The flashlamp uses xenon gas rather than sodium gas, in part because xenon emits light over a very broad range of wavelengths and does a good job of simulating sunlight. But why wouldn't a high-pressure sodium vapor flashlamp be practical, even if you didn't care that it was orange in color?

3. A paint is a plastic that contains dye molecules and small clear particles. While the dye molecules absorb certain wavelengths of light and give the paint its color, the particles are often completely clear. Nonetheless, these randomly shaped and oriented particles are very important to the paint's appearance.
  a. The speed of light is different in the particles than it is in the plastic surrounding them. When light strikes the surface of one of these particles, why does some of it reflect?

b. A dyeless paint appears white because light bounces around among its particles and emerges heading in all directions. Sketch a few of the paths light could take in the paint and show how it could reach the surface beneath the paint.

c. Lead carbonate is a clear, toxic chemical that was put in paints before 1930 to make them white. The speed of light in a particle of lead carbonate isn't that much slower than in plastic. How does this fact help explain why "lead paints" weren't very good at covering dark surfaces, requiring lots of layers of paint before the surfaces really looked white?

d. Modern paints contain particles of clear, nontoxic titanium dioxide. Light travels very slowly in this material. Explain why titanium dioxide based paints look so white, even after only a single layer of paint.

e. Red paint usually contains a mixture of titanium dioxide particles and red dye molecules. The red dye molecules absorb any light that isn't red but they don't affect red light at all. How would the appearance of this paint change if it didn't contain any titanium dioxide particles?

4. Interior house paints contain dye molecules that absorb certain wavelengths of light and give the paints their colors.

a. Dye molecules only absorb light of certain wavelengths for the same reason as atoms. Explain that reason briefly.

b. The dyes in most house paints eventually convert the light energy they absorb into thermal energy. However, fluorescent or neon paints emit light of a new color. Why is the light emitted by fluorescent paints always longer in wavelength than the light they absorb?

c. A paint's appearance depends strongly on the light that illuminates it. Why are a white wall and a red wall indistinguishable when they're both illuminated by 650-nm light?

d. Two paints that contain different dyes can look indistinguishable when illuminated by sunlight, even though they absorb slightly different portions of the visible light range. In what way can their absorptions differ without your being able to see any difference between the paints?

e. Two paints that look indistinguishable in sunlight may look very different when illuminated by a fluorescent lamp. Why does the difference between the paints only appear for certain illuminations?

f. If you're trying to match new paint to old paint and want the match to work regardless of illumination, you'll have to find a new paint that uses exactly the same dye molecules as the old paint. Why?

5. While fluorescent lamps can be dimmed, it's not as easy as dimming an incandescent light bulb. Fluorescent lamps tend to turn off when they don't run at full power, so a dimmed fluorescent fixture must make sure that the discharge continues to operate, even at low power.

a. Why can a fluorescent fixture be dimmed only if it continuously heats the filaments of its fluorescent tubes?

b. While an incandescent light bulb can be dimmed by reducing the voltage drop across its filament, reducing the voltage drop across a fluorescent tube will spoil its discharge. With the voltage drop across the tube reduced, the average energy of its electrons would be substantially lower than normal and its mercury atoms would emit almost no ultraviolet light at all. Why is there a minimum energy that an electron must have in order to cause a mercury atom to emit ultraviolet light?

c. Reducing the voltage drop across the fluorescent tube would virtually stop the production of positive mercury ions in the vapor. That would make the tube's filaments last longer, but the discharge wouldn't operate. With no positively charged particles inside the tube, what would happen to the electrons as they flowed from one electrode to the other?

d. Rather than reducing the voltage drop across its fluorescent tubes, a dimmed fluorescent fixture turns the discharge on and off about 120 times a second. Fortunately, this rapid pulsed operation doesn't shorten the tube life. Why don't we see this flickering light output?

e. Current passing through the electronic switching system of the dimmed fluorescent fixture experiences a small voltage drop. Why does the switching system get warm?

6. In 1965, American physicists Arno Penzias and Robert W. Wilson discovered that space is filled with thermal radiation, left over from the Big Bang that formed the universe. While this thermal radiation was once extremely hot, the universe's expansion has cooled it to only 3 K. It now consists mostly of microwaves that are extremely difficult to detect.

a. Satellite dish antennas are designed to detect coherent microwave radiation—a low-frequency equivalent of coherent light. Why are microwaves emitted by a normal microwave transmitter and antenna (e.g., a magnetron) coherent?

b. The microwaves reaching us from space are thermal radiation and effectively come from individual charged particles that moved back and forth to produce them. Why is this microwave radiation incoherent?

c. A normal satellite dish antenna will have trouble detecting microwaves from the Big Bang because they are incoherent. Why won't the charges on the antenna of a satellite dish respond strongly to these incoherent microwaves?

d. To detect the incoherent microwaves, Penzias and Wilson used a maser amplifier—a device that amplifies microwave photons via stimulated emission. How did introducing this amplifier make it possible to detect the microwave photons with a microwave antenna and receiver?

CHAPTER 13

Optics

Many of the devices around us perform useful tasks by manipulating light. Cameras record images of the objects in front of them while magnifying glasses allow us to see details we'd miss with our eyes alone. Still other gadgets, such as CD or DVD players, use light in ways that have nothing to do with vision. But all of these objects manipulate light using similar techniques—the techniques of optics.

While optical tools such as lenses and prisms have been around for hundreds of years, advances of modern technology have accelerated the development of optics. Just as the invention of transistors has sped the growth of the electronics industry, so the invention of lasers has sped the growth of the optics industry. The two fields are not far apart in many ways, and there is hope that one day computers will be as much optical devices as they are electronic.

EXPERIMENT: Focusing Sunlight

There are many household devices that manipulate light, and one of the most familiar is a magnifying glass. A magnifying glass bends light rays toward one an-

other as they pass through it. In this chapter, we'll see how a simple converging lens of this sort can magnify an object or cast its image onto a sheet of film. For the moment, we'll simply use it to bring light rays together.

On a clear day with the sun roughly overhead, take a magnifying glass and a dark piece of paper outdoors to a place where there is nothing flammable around. You will use light from the sun to heat the paper until it smokes, so be prepared to extinguish a fire if necessary. Hold the magnifying glass just above the paper and slowly lift the lens upward toward the sun. Make sure that the lens is turned so that its surface faces the sun. You should see a bright circle appear on the dark paper, and that circle should become smaller and smaller as you lift the lens. The sun's light rays are bending inward as they pass through the lens and are converging together as they head toward the paper. Be careful not to let this bright circle of light touch your skin because it will burn you. If the spot isn't round, try tilting the lens differently.

When the lens is just the right height above the paper, the circle of light will reach a minimum size. It will appear as a brilliant white disk that is actually an image of the sun itself. If the magnifying glass is wide enough and the sun is bright enough, the paper will begin to smoke. The image of the sun is then so bright that it is able to heat the paper to its ignition temperature. If the lens is quite large, the paper may begin to burn with a flame.

You can project images of other brightly illuminated objects on a sheet of white paper. Try to *predict* the sizes and orientations of those images. *Observe* what happens. Did you *verify* your predictions? Try to *measure* the distance to an object by seeing how far the lens must be from the paper to get a sharp image.

Chapter Itinerary

This process of bringing light together to form a small spot or an image of a distant object is a common theme in optics. In this chapter, we'll examine several systems that are based on this sort of manipulation of light: (1) *cameras* and (2) *optical recording and communication*. In *cameras*, we'll see how lenses bend light to form images and how those images are used to create photographs. In *optical recording and communication*, we'll explore the roles of lasers in optics while investigating several novel optical effects.

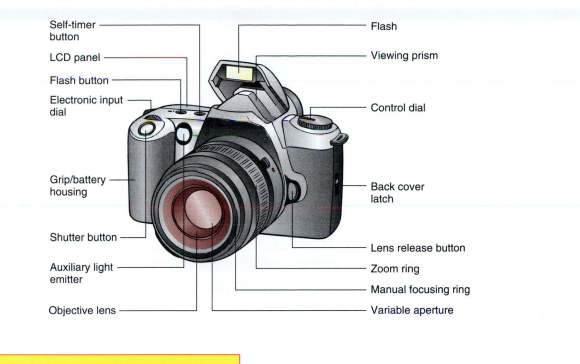

Self-timer button

LCD panel

Flash button

Electronic input dial

Grip/battery housing

Shutter button

Auxiliary light emitter

Objective lens

Flash

Viewing prism

Control dial

Back cover latch

Lens release button

Zoom ring

Manual focusing ring

Variable aperture

SECTION 13.1 Cameras

In the two centuries since their invention, cameras have become extremely easy to use. What started as a hobby for a few dedicated enthusiasts has evolved into an everyday activity. But despite all of the technological improvements, photography still employs many of the same principles it did in the 1800s. Cameras still use lenses to project images onto light-sensitive surfaces, and photographers still have to worry about getting the exposure right, focusing properly, and avoiding the blur of rapid motion. In this section, we'll explore some of the principles that make these cameras work.

Questions to Think About: Why are expensive camera lenses so complicated, with so many separate pieces of glass? Why does a longer lens seem to bring the objects nearer to you? What does a camera's aperture do? How does the camera's shutter speed affect the photograph?

Experiments to Do: The basic activity of a camera—projecting an image of the scene in front of you onto the film—requires nothing more than a simple magnifying glass. In fact, you can use almost any simple lens that's bowed outward in the middle, including the eyeglasses of a farsighted person with no astigmatism. Stand in a darkened room, opposite a window. Hold a sheet of white paper so that it's parallel to the window and move the lens back and forth in front of the paper. Make sure that the lens itself is parallel to the window.

You should find a distance at which the lens casts a clear image of the window onto the sheet of paper. You'll find that the scene appears upside down and backwards and that the image becomes fuzzy if you move the lens toward or away from the paper. You'll also find that, when the window's image is sharp, the outside scene's image is fuzzy and vice versa. Which way must you move the lens to bring more distant objects into focus?

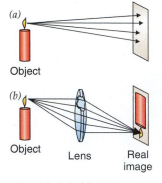

Cut a hole in a piece of cardboard and use it to cover all but the center portion of the lens. The image will now be substantially darker, but it will also be sharper over a wider range of distances from the paper. If the hole is narrow enough, virtually everything focuses at once. Evidently, the diameter of the lens and its ability to focus on several things at once are closely related. Why?

Lenses and Real Images

When you take a picture of the scene in front of you, the lens of your camera bends light from that scene into a real image on a light-sensitive surface. A **real image** is a pattern of light, projected in space or on a surface, that exactly reproduces the pattern of light in the original scene. Since the real image that's projected looks just like the scene you're photographing, recording the light in that image is equivalent to recording the appearance of the scene itself.

While that light-sensitive surface was once always photographic film, digital cameras are gradually replacing film with electronic image sensors. Though the following discussion refers only to film, it could equally well refer to electronic image sensors instead. The two light-sensing surfaces are essentially interchangeable.

Real images don't occur without help. When light from a candle falls directly on a sheet of photographic film, it produces only diffuse illumination (Figs. 13.1.1*a* and 13.1.2*a*). You can't tell by looking at the film what the candle looks like because the light that leaves the candle travels in all directions and is as likely to hit the top of the film as it is to hit the bottom.

That's why a camera needs a **lens,** a transparent object that uses refraction to form images. The light passing through a lens bends twice, once as it enters the glass or plastic and again as it leaves. In a camera lens, this bending process brings much of the light from one point on the candle back together at one point on the film. As you can see in Figs. 13.1.1*b* and 13.1.2*b*, the real image that forms is upside down and backward. This inversion of the real image relative to the object always happens when a single lens creates a real image.

The curved shape of the camera lens allows it to form a real image. Light passing through the upper half of the lens is bent downward while light passing through the lower half is bent upward. Because the camera lens bends light rays toward one another, it's a **converging lens.** You can see how it forms an image in Fig. 13.1.1*b* by following some of the rays of light leaving one point on the candle.

The upper ray from the candle flame travels horizontally toward the top of the

Fig. 13.1.1 (*a*) Without a lens, light from a candle uniformly illuminates a sheet of photographic film. (*b*) When a lens is introduced between the candle and the film, it brings light from each point on the candle back together on the film's surface, forming an upside down and backward real image of the candle. The distance between the lens and the film must be chosen correctly, or the image will be blurry.

Fig. 13.1.2 (*a*) When candlelight falls directly on a sheet of paper, it produces no image. (*b*) But a lens inserted between the candle and paper forms an inverted real image of the flame on the paper.

(*a*)

(*b*)

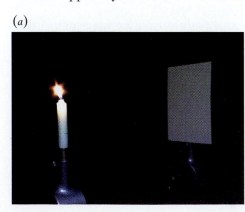

lens. As it enters the lens and slows down, this ray of light bends downward. It bends downward again as it leaves the lens and travels downward toward the bottom of the film.

The lower ray from the candle flame travels downward toward the bottom of the lens and bends upward as it enters the lens. It bends upward again as it leaves the lens and travels horizontally toward the bottom of the film.

These two rays of light reach the film at the same point. They are joined there by many other rays from the same part of the candle flame, so that a bright spot forms on the film. Overall, each part of the candle illuminates a particular spot on the film, so the lens creates a complete image of the candle on the film.

However, the lens can bring the light back together to form a sharp image on the film only if the lens and film are separated by just the right distance (Fig. 13.1.3). If the film is too close to the lens, then the light doesn't have room to come together. If the film is too far from the lens, then the light begins to come apart again before reaching the film. In either case, the image on the film is blurry. The candle's real image is only *in focus* at one distance from the lens.

If the candle moves toward or away from the camera lens, the distance between the lens and the film must also change (Fig. 13.1.4). When the candle is far away, all of its light rays that pass through the lens arrive traveling almost parallel to one another and the inward bend caused by the lens makes those rays converge together quickly. The rays come into focus relatively near the lens and that's where the film must be (Fig. 13.1.4*a*). The candle's image on the film is much smaller than the candle itself because the light rays have only a short distance over which to move up or down after leaving the lens.

When the candle is nearby, its light rays that pass through the lens are diverging rapidly and the inward bend caused by the lens is just barely enough to make those rays converge at all. As a result, the rays come into focus relatively far from the lens (Fig. 13.1.4*b*). The candle's image on the film is quite large because the light rays have considerable distance over which to move up or down after leaving the lens.

Because distant and nearby objects form real images at different distances from the camera lens, they can't both be in focus on the same sheet of film. When you take a picture of a person standing in front of a mountain, only one of them can be in sharp focus. However, if you're willing to compromise a little bit on sharpness, a lens can sometimes form acceptable images of both objects.

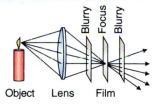

Fig. 13.1.3 The real image is only in focus when the film is just the right distance from the lens. If the film is too near or far from the lens, the image is blurry.

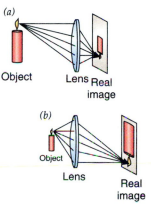

Fig. 13.1.4 (*a*) The light from a distant candle is traveling in almost the same direction, and the lens focuses it easily. The real image forms close to the lens. (*b*) The light from a nearby candle diverges, and the lens has more difficulty bending it back together. The real image forms far from the lens. If the candle is too close to the lens, no real image forms at all.

> Check Your Understanding #1: Seeing the Lights

If you hold a magnifying glass at the proper distance above a sheet of white paper, you will see an image of the overhead room lights on the paper. Explain.

Focusing and Lens Diameter

A disposable camera is little more than a box with a lens. The lens projects a real image of the scene in front of it onto the film. Light in the real image exposes the film and, once developed, the film displays the image permanently. While it also has a shutter that starts and stops the exposure, a flash to provide extra light, and a mechanism for advancing the film, there's little else to this simple camera.

(a)

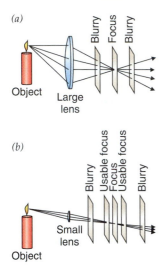

(b)

Fig. 13.1.5 (*a*) A large lens collects lots of light but its focus is critical. The image is blurry except at the actual focus. (*b*) A small lens collects less light but its image is relatively sharp anywhere near the focus.

Fig. 13.1.6 The aperture of this lens can be reduced by closing its internal diaphragm, dimming its image but increasing its depth of focus.

However, there are limitations to the disposable camera design. One of the most severe limitations is that you can't focus it—the camera has a fixed distance between the lens and film. Nonetheless, it manages to form relatively sharp real images on the film, even when there are objects at various distances from the camera. These simple cameras work because they use narrow (small-diameter) lenses. A narrow lens gathers less light than a wide lens but doesn't require *focusing*—adjusting the distance between the lens and the film.

Because a wide (large-diameter) lens brings rays together from many different directions, you must focus it (Fig. 13.1.5*a*); if the film is even slightly too near or too far from the lens, the film's image will be blurry. But a narrow lens forms a reasonably clear image even without focusing. Any rays from one part of the scene that succeed in passing through the narrow lens must already be fairly close together. Their initial closeness means that these converging rays illuminate only a small part of the film even when the film isn't exactly the right distance from the lens (Fig. 13.1.5*b*). Since the film can't record every minute detail anyway, the image that forms on it doesn't have to be in absolutely perfect focus. As a result, a camera with a narrow lens and no focus adjustment manages to take pretty good pictures.

Unfortunately, these simple cameras collect very little light and need extremely light-sensitive film. This high-speed film doesn't produce as sharp photographs as low-speed film. Furthermore, the pictures produced by simple cameras lack fine details; while everything is almost in focus, most things are a bit fuzzy if you look carefully or make enlargements.

More sophisticated cameras use wider lenses that gather more light and expose the film much more rapidly. They also automatically adjust the distance between the lens and the film. They identify the object you are photographing and position the camera lens so that it projects a sharp, real image on the film. As the camera focuses, you can usually see components in the lens moving backward or forward to arrive at the correct distance from the film.

Even a camera with a wide lens can take advantage of the narrow lens trick for focusing. Its lens contains an internal diaphragm that reduces its **aperture** or effective diameter. The *diaphragm* is a ring of metal strips with a central opening. These strips can swing in or out, changing the diameter of the diaphragm's opening and with it the aperture of the lens (Fig. 13.1.6).

When its lens aperture is narrow, the sophisticated camera imitates a simple camera. Nearly everything is essentially in focus simultaneously. In such a situation, the camera has a large *depth of focus*. Actually, the sophisticated camera can bring the most important object into perfect focus, so it produces pictures that are superior to those from simple cameras. However, narrowing the aperture of the lens also reduces the amount of light reaching the film. The scene in front of the camera must either be very bright or the exposure of the film must be very long. You don't get something for nothing.

While widening the aperture of a large lens makes full use of its light-gathering capacity, focusing then becomes crucial. Even a small error in the lens-to-film distance produces a blurry picture, so the depth of focus is very small. This trade-off, between light gathering and depth of focus, is a continual struggle for photographers. However, photographers sometimes take advantage of the small depth of focus in wide lenses to blur the background or foreground of a photograph deliberately. At other times photographers choose long exposures at narrow apertures in order to bring an entire scene into sharp focus. And to capture fast motion while retaining this large depth of focus, photographers use a flash to brighten the scene and shorten the exposure.

Check Your Understanding #2: Portrait Photos

While taking a photograph of your friend, with the diaphragm of your large camera lens wide open, you notice that the background is blurry. Explain.

Focal Lengths and F-Numbers

Lenses are characterized by two quantities: focal length and f-number. The **focal length** of a lens is the distance between the lens and the real image it forms *of a very distant object*. For example, if a real image of the moon forms 100 mm behind a particular lens, then that lens has a focal length of 100 mm. The focal lengths of camera lenses range from less than 10 mm in many digital cameras to about 2 m in film cameras used for nature photography.

When light from a scene passes through a short focal length lens, it comes to a focus near that lens and produces a relatively small image on the film. Because a long focal length lens permits the light passing through it to spread out more before coming to a focus, it produces a larger real image on the film.

The "normal" lens for a particular camera has a focal length that allows all of the objects in your central field of vision to fit onto the piece of film. When you hold the finished photograph about 30 cm from your eyes, the objects in the picture appear about the same size they did when the photograph was taken. The focal length of a camera's normal lens is about 1.5 times the horizontal width of its light-sensitive surface (Table 13.1.1).

A wide-angle lens has a shorter focal length than the normal lens (Fig. 13.1.7). The image it projects onto the film is smaller, and most of the objects in your entire field of vision appear in the photograph. A telephoto lens has a longer focal length than the normal lens. The image it projects onto the film is larger, with only objects at the center of the scene appearing in the photograph.

In addition to indicating where the image of a distant object forms, the focal length of the camera lens relates the object distance to the image distance. The **object distance** is the distance between the lens and the object you're photographing. The **image distance** is the distance between the lens and the real image it forms (Fig. 13.1.8). The relationship is called the **lens equation** and can be written in a word equation:

Fig. 13.1.7 The 24-mm lens (top) forms a small image, close to the lens. The 300-mm lens (bottom) forms a large image, far from the lens.

$$\frac{1}{\text{focal length}} = \frac{1}{\text{object distance}} + \frac{1}{\text{image distance}}, \qquad \textbf{(13.1.1)}$$

in symbols:

$$\frac{1}{f} = \frac{1}{o} + \frac{1}{i},$$

and in everyday language:

The farther away an object is, the closer to the lens its image forms.

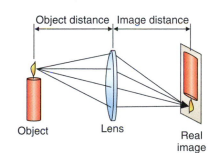

Fig. 13.1.8 The relationship between the object distance, the image distance, and the focal length of the lens is given by the lens equation.

Table 13.1.1 Several Cameras, the Widths of the Film or Image Sensor They Use, and Their Normal Lenses

TYPE OF CAMERA	SENSOR WIDTH	NORMAL LENS
Digital camera	8 mm	12 mm
35-mm Camera	36 mm	50 mm
$2\frac{1}{4}$-inch Medium format camera	$2\frac{1}{4}$ inches	80 mm
5-inch Portrait camera	5 inches	180 mm

> **The Lens Equation**
> One divided by the focal length of a lens is equal to the sum of one divided by the object distance and one divided by the image distance.

According to the lens equation, the image distance for a distant object is equal to the focal length of the lens. That agrees with our earlier discussion of focal length. But when the object is nearby, the image distance becomes larger than the focal length. That's why a camera lens moves away from the film as you focus closer. And when the object distance becomes less than the focal length, the image distance becomes negative and no real image forms at all. That's why you can't focus on an object that's too close to the lens.

A lens's **f-number** characterizes the brightness of the real image it forms on the film, with smaller f-numbers indicating brighter images. The f-number is calculated by dividing the lens's focal length by its diameter. Since long focal length lenses naturally produce larger and dimmer images on the film, the f-number takes into account both the light-gathering capacity of the lens and its focal length. Increasing a lens's diameter increases its light-gathering capacity and decreases its f-number. Increasing a lens's focal length decreases the brightness of its real image and increases the f-number. Doing both at once, increasing the lens diameter and focal length equally, leaves the brightness and f-number unchanged.

Most sophisticated cameras use large-diameter lenses so that their f-numbers are generally less than 4. Since it's difficult to fabricate a lens that's larger in diameter than its focal length, the smallest practical f-number is about 1. And because long focal length lenses need large apertures to keep their f-numbers small, some telephoto lenses are huge.

The diaphragm inside a lens allows you to decrease the lens's aperture and thus increase its f-number. A factor of 2 increase in f-number corresponds to a factor of 2 decrease in the lens's effective diameter and a factor of 4 decrease in the lens's light-gathering area. Thus when you double the f-number of the lens, you must compensate by quadrupling the film's exposure time. Although closing the aperture increases the lens's depth of focus, it requires a longer exposure.

Check Your Understanding #3: Bright and Sharp Photographs

On bright, sunny days, your automatic camera takes photographs with large depths of focus, while on dark, overcast days its pictures have much smaller depths of focus. What causes this difference?

Check Your Figures #1: The Image of an Apple

If the distance from an apple to a converging lens is twice the focal length of that lens, where will the real image of the apple form?

Improving the Quality of a Camera Lens

A high-quality camera lens isn't a single piece of glass or plastic. Instead, it's composed of many separate elements that function together as a single lens. This complexity improves the quality of the real image. To begin with, dispersion in a single element lens causes different colors of light to bend differently and focus at different distances behind that lens. Known as *chromatic aberration,* this problem can be fixed by using lens elements made of different types of glass, with different amounts of dispersion. These elements compensate for one another so that the overall lens, known as an *achromat,* has very little dispersion and almost no color focusing problems.

other lens flaws

After correcting for color and other technical image problems, a sophisticated camera lens may contain more than ten individual elements. For the purposes of the lens equation, this complicated lens has an effective center from which to calculate object and image distances. But having so many separate elements creates reflection problems; each time light passes from air to glass or vice versa, some of it reflects. To avoid fogging the photographs with this bouncing stray light, the individual elements are *antireflection coated* with thin layers of transparent materials. The best coatings use interference effects to cancel out the reflected light waves and give the lens only a weak violet reflection.

A zoom lens is a complicated lens that can change the size of the real image it projects onto the film (Fig. 13.1.9). By carefully rearranging the individual lens elements within the overall lens, the zoom lens can adjust its effective focal length as well as the distance between the effective lens and the film. As the zoom lens changes from short focal length to long focal length, the image it projects on the film becomes larger. This effect allows you to compose the picture so that the scene fills the photograph completely. A lens that can change its focal length while retaining the same f-number and still keep the real image in focus on the film is a truly remarkable achievement.

Fig. 13.1.9 A zoom lens contains many optical elements. This complicated structure allows the photographer to change the lens's focal length and f-number and corrects for color and focus imperfections that are present in simpler lens systems.

> Check Your Understanding #4: Not Such a Good Picture

If you use a large magnifying glass to form a real image of an overhead fluorescent light fixture, you will find that the corners of the light fixture are blurry and have rainbow colors in them. Why is the image quality so poor?

The Viewfinder and Virtual Images

When you peer through the viewfinder of a sophisticated film camera (Fig. 13.1.10), you're looking at the same real image that will be projected onto the film during the exposure. The light you see travels through the camera's main lens, reflects off a mirror, and projects onto a translucent screen inside the top of the camera. You're simply inspecting this screen and the real image through a magnifying lens in the eyepiece. During the exposure, the mirror flips out of the way, the shutter opens, and the real image projects briefly onto the film.

Since the screen and real image are only an inch or two from your eye, you can't focus on them without the help of the eyepiece lens. The eyepiece lens is converging, but in this case it doesn't form a real image. Instead, it forms a **virtual image**—an image located at a negative image distance—that is, on the wrong side of the lens!

The screen displaying the scene that you're photographing is so close to the eyepiece lens that the object distance is less than that lens's focal length. According to the lens equation, the image distance should be negative and it is; the image is located on the screen side of the eyepiece lens (Fig. 13.1.11, p. 422). You can't put your fingers in the light and project this image on your skin because the image is virtual rather than real.

Fig. 13.1.10 The mirror in the center of this reflex camera directs light from the lens (removed for this photograph) onto the focusing screen above it. During the exposure, the mirror swings upward to permit light from the lens to strike the film at the back of the camera.

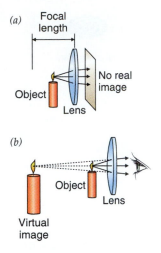

Fig. 13.1.11 (*a*) Light from an object very near a converging lens diverges after passing through the lens and no real image forms. (*b*) Your eye sees a virtual image that is large and far away.

Fig. 13.1.12 This magnifying glass creates an enlarged virtual image located far behind the printed text. You can't touch the image or put your fingers in its light, but you can see it clearly with your eyes.

However, you can see this image through the eyepiece. It's located farther away than the screen itself, so your eye can comfortably focus on it. And the image is magnified—the eyepiece lens is acting as a magnifying glass (Fig. 13.1.12). This lens provides magnification because when you look at the screen through it, the screen image covers a wider portion of your field of vision. This magnification increases as the eyepiece lens's focal length decreases. That's because a shorter focal length eyepiece lens must be quite close to the screen, where it can bend light rays coming from a smaller region so that they fill your field of vision. The eyepiece lens in a typical camera has been chosen so that the screen fills a comfortable portion of your visual field, allowing you to examine the virtual image in great detail and adjust the lens and camera settings until you have just the right picture in your view. Then all you have to do is take the photograph.

Check Your Understanding #5: Real and Virtual Images

As you move a magnifying glass slowly toward the photograph in front of you, you see an inverted image that grows larger and nearer to your eye. This image eventually becomes blurry, and then a new upright and enlarged image appears on the photograph's side of the lens. What's happening?

film, shutter, focusing, projectors, film advance

Cover

Previous track

Next

Stop

Play/pause

Compact disc

LCD display

Cover latch

Line output

Volume

Headphone jack

Tone control

Optical Recording and Communication

Using light to convey information is as old as signal fires and as natural as sight itself. But advances in light sources, optical materials, and electronics have radically increased the possibilities for optical information systems. Optics and information go so well together that they're partly responsible for the current information revolution. The introduction of compact disc players in the early 1980s transformed the music industry virtually overnight, and optical fibers are knitting our world together at an astonishing pace. In this section, we'll look at how optical devices use light to manipulate information.

Questions to Think About: Why are CDs and DVDs so colorful? Where is their information stored? How can CD or DVD players ignore fingerprints, dust, and other surface contamination during playback? Why is a laser involved in recording and playing back a CD or DVD? What limits the duration of a CD's audio or a DVD's video? Why are CDs and DVDs so free of noise? How can a glass fiber direct light in a curving path? How can light travel through kilometers of glass fiber without becoming dim?

Experiments to Do: Hold a prerecorded CD or DVD by its edges and look at its unlabeled surface. Beneath the clear plastic face is a smooth, shiny layer that reflects a rainbow of colors. Tiny pits in this layer cause this coloration. These pits form a spiral track around the center of the disc, and adjacent arcs of this track are so closely spaced that light waves reflecting from them interfere, as from a soap film. The reflective layer is so thin that you can see through it if you hold it in front of a bright light. Can you see the reflections of dust particles on the unlabeled surface? How deep below the plastic surface is the shiny layer?

Digital Recording

CDs sound better than audiotapes because they represent sound in digital form—as a series of numbers. And because CD players can read those numbers perfectly,

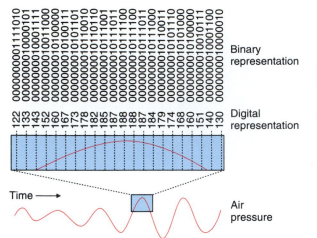

Binary representation

Digital representation

Time →

Air pressure

Fig. 13.2.1 Sound can be represented as a series of numbers. Each number corresponds to the air pressure at a particular moment in time.

digital sound issues

even from dirty or damaged discs, the sound is virtually free of noise. In contrast, audiotapes represent sound in analog form, as the depth and direction of magnetization in a special surface. Any imperfections in that magnetization appear directly in the reproduced sound.

Like CDs, DVDs also represent sound in digital form. However, many DVDs hold video information as well. Like digital television that we discussed in Section 11.2, the video information on a DVD is highly compressed by squeezing out repetitive and nonessential data. Nonetheless, the pictures it represents are crystal clear and free of visual noise.

Both CDs and DVDs represent sound's compressions and rarefactions of the air as numbers (Fig. 13.2.1). These numbers are essentially air pressure measurements taken over and over again during the recording process, and the player uses them to reproduce the recorded sound. While even a brief sound involves thousands of numbers, this digital representation allows for precise control of that sound at all times.

In CDs, the air pressure measurements are made 44,100 times each second for two independent audio channels—to permit stereo recording. These measurements are recorded on the disc in binary form, using 16 bits for each measurement. Since the air pressure can go down as well as up, these bits represent the positive and negative integers from -32,768 to 32,767, which in turn represent how much the air pressure is above or below the average pressure. Pressure measurements with 16 bits of precision are sufficient to reproduce both loud and soft music with almost perfect fidelity.

Audio DVDs can choose from several measurement rates, bits per measurement, and numbers of channels. A typical DVD might have five audio channels: left-front, center-front, right-front, left-rear, and right-rear. The three front channels might have 96,000 pressure measurements per second at 24 bits per measurement, and the two rear channels might have 48,000 measurements per second at 20 bits per measurement. While all of these samples, bits, and channels involve far more information than is stored on a CD, a DVD compresses that information before storing it. In contrast, a CD's information is uncompressed.

In either case, air pressure measurements aren't simply recorded one after the next on a disc's surface. Instead, these numbers are extensively reorganized before they're stored. This reorganization allows the player to reproduce the sound perfectly even if the disc can't be read completely. As we'll see shortly, reading these discs is a technological *tour de force* and is subject to various imperfections. To be sure that the sound (and video) can be reproduced completely and without interruption, the numbers are recorded in an encoded manner. They appear redundantly so that even if one copy of a number is illegible, there is still enough legible information along the same arc of the spiral track to completely recreate that missing number. This duplication of information reduces the playing time of both CDs and DVDs but is essential for reliability.

Its encoding scheme leaves a CD or DVD almost completely immune to all but the most severe playback problems. In principle, you can damage or obscure a 2-mm wide swath of the disc, from its center to its edge, and the player will still be able to reproduce the sound (and video) perfectly. But damage along an arc of the spiral track is far more threatening to the data. If the player can't read a long stretch of a single arc, It won't be able to recover the information. That's why you should always clean a CD or DVD from its center outward to its edge.

Check Your Understanding #1: Cordless Clarity Away from Home

Some cell phones transmit your voice in digital form. In general terms, how does this digital transmission work?

The Structure of CDs and DVDs

Standard CDs and DVDs are 120 mm in diameter and 1.2 mm thick. One side of a CD is clear and smooth but the other side contains a sandwich of layers: a thin film of aluminum, a protective lacquer, and a printed label (Fig. 13.2.2a). In contrast, a DVD is laminated from two 0.6-mm thick clear plastic discs, with one, two, or four reflective layers of aluminum, gold, or silicon stacked up in between them (Fig. 13.2.2c). The more layers in the DVD, the more information it holds.

The reflective layers are the recording surfaces. These layers are so thin that they actually transmit a small amount of light. In the gold or silicon DVD layers, this semitransparency is essential because it allows the optical system that reads information to send light through the semitransparent layer to the aluminum layer

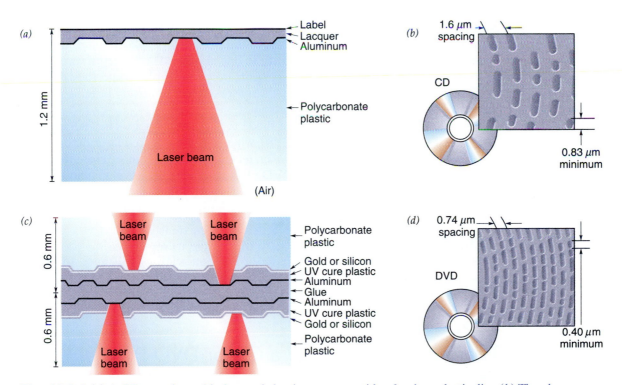

Fig. 13.2.2 (*a*) A CD contains a thin layer of aluminum on one side of a clear plastic disc. (*b*) The aluminum layer has tiny pits that are detected by a 780-nm laser beam. (*c*) A DVD contains up to four aluminum, gold, or silicon layers sandwiched between two clear plastic discs. The gold or silicon layers are semireflective. (*d*) Pits in those layers are detected by a 650-nm laser beam. The beam can focus either on the semireflective layer or, by passing through that layer, on the aluminum layer beyond it.

beyond it. But even the aluminum layers transmit some light. While aluminum's electrons accelerate in response to the light's electric field and normally reflect that light completely, there aren't enough electrons in these 50-nm to 100-nm thick layers to do the job and some light gets through.

The reflective layers aren't perfectly smooth. Instead, each has a narrow spiral track formed in its surface (Fig. 13.2.2b,d). This track is a series of microscopic pits, as little as 0.83 μm long on a CD or 0.40 μm long on a DVD. The spacing between adjacent arcs in the spiral track are only 1.6 μm apart on a CD and just 0.74 μm apart on a DVD. The lengths of the pits and the flat "lands" that separate them represent numbers. The player examines these pits and lands as the disc turns and converts their lengths into numbers, sound, and video.

The pit lengths and the spacings between arcs weren't chosen arbitrarily. Since electromagnetic waves are unable to detect structures much smaller than their wavelengths (see Section 11.3), the laser beam's wavelength limits the smallest features on a disc. In a CD player, that beam's wavelength is 780 nm in air and 503 nm in polycarbonate plastic—short enough to detect the pits of a CD easily. In a DVD, the laser beam's wavelength is between 635 nm and 650 nm in air and between 410 nm and 420 nm in plastic—just short enough to detect the pits in a DVD. The wavelength reduction inside the disc occurs because polycarbonate plastic has an index of refraction of 1.55, meaning that the light's speed in that plastic is reduced from its vacuum speed by a factor of 1.55. Its wavelength is reduced by the same factor.

The player detects a pit by bouncing light from the disc and determining how much of it reflects. As the focused laser beam passes over a pit, the reflection becomes dim, in part because the curved pit scatters light in all directions and in part because of interference effects. Light that's reflected back from a pit travels farther than light that's reflected from the flat region around it, so electric and magnetic fields in the two waves are shifted relative to one another. The pit depth was chosen so that the two reflected waves are approximately out of phase and they destructively interfere. Overall, the player's light sensors detect relatively little light when the laser beam is located over a pit.

A CD or DVD player uses a laser diode to produce its light. The 780-nm standard for CD players was adopted in 1980, when 780-nm infrared laser diodes were reliable but still fairly expensive. However, technology has advanced since then, and the 635-nm to 650-nm standard for DVD players reflects the development of inexpensive red laser diodes. New standards are likely to follow technology, so optical recording systems based on blue laser diodes are probably just around the corner.

Check Your Understanding #2: The Blue Light Special
When disc players begin to use blue lasers, optical discs will have to be redesigned. If the new laser light has a wavelength of 400 nm in air or vacuum, how should they change the depth of the pits in the reflective layer?

The Optical System of a CD or DVD Player

A CD or DVD player's optical system measures the lengths of the tiny pits as they move by on a spinning disc. That reading process requires incredible precision. Not only must the player focus its spot of laser light exactly on the reflective layer, but it must also follow the spiral track as it moves by. The disc itself is neither perfectly flat nor perfectly round, so the player must continuously adjust its reading unit during playback. The optical system must keep its laser beam focused on the reflective layer (autofocusing) and must follow the track as it passes (autotracking). These two automatic processes are beautiful examples of the use of feedback.

The basic structure of a typical CD or DVD player is shown in Fig. 13.2.3. Light from a laser diode passes through several optical elements on its way to the disc's reflective layer. It comes to a tight focus on that layer, where it illuminates only a single track. Some light reflects from the layer and returns through the optical elements. Finally, the reflected light turns 90° at a special mirror called a polarization beam splitter and focuses on an array of light detectors. The player measures the electric currents flowing through the detectors and uses those measurements both to obtain data from the disc and to control the focusing and tracking systems.

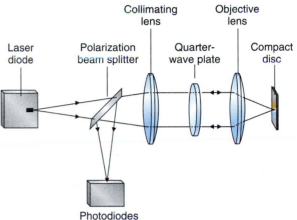

Let's examine this optical system one element at a time. After leaving the laser diode, light passes through a *polarization beam splitter*. This device analyzes the light's polarization. As we saw in Section 12.1, different polarizations of light reflect differently when they strike a transparent surface at an angle. In this case, polarized light from the laser passes through the 45° surface, but light of the other polarization reflects. The beam splitter is specially coated to separate the two polarizations almost perfectly.

Light from the laser diode diverges rapidly as it passes through the beam splitter. It's not that the laser diode is broken or poorly designed; it's that a light wave emerging from a small opening naturally spreads outward, like ripples on a pond. This spreading is known as **diffraction** and occurs whenever a light wave is truncated by passage through an opening. The smaller the opening, the worse is the spreading. Because the emitting surface of the laser diode is essentially a very small opening, the laser beam spreads rapidly as it heads away from the diode. The player uses a converging lens located after the beam splitter to stop this spreading. At that point, the light beam is already wide enough that diffraction causes little additional spreading. The beam leaves the lens *collimated,* meaning that it maintains a nearly constant diameter after passing through the lens.

The laser light then passes through a *quarter-wave plate.* This remarkable device performs half the task of converting horizontally polarized light into vertically polarized light or vice versa. Horizontally and vertically polarized lights are said to be **plane polarized** because their electric fields always oscillate back and forth in one plane as they move through space. But the quarter-wave plate turns plane polarized light into **circularly polarized** light. We encountered circular polarization before in the radio transmissions from FM stations. In circularly polarized light, the electric field actually rotates about the direction in which the light is traveling.

Now the light passes through an objective lens that focuses it onto the reflective layer of the disc. On its way to the reflective layer, the light enters the plastic surface of the disc. At its entry point, the beam is still more than 0.5 mm in diameter, which explains why dust or fingerprints on the disc's surface don't cause much trouble. While contamination may block some of the laser light, most of it continues onward to the reflective layer.

The light comes to a tight focus just as it arrives at the reflective layer. Although it might seem that all of the light should converge together to a single point on that surface, it actually forms a spot roughly one wavelength in diameter. This spot size

Fig. 13.2.3 In the optical system of a CD or DVD player, light from a laser diode passes through a polarization beam splitter, a collimating lens, a quarter-wave plate, and an objective lens before focusing on the reflective layer inside the disc. Reflected light turns 90° at the polarization beam splitter and focuses on an array of photodiodes.

Fig. 13.2.4 When a converging lens focuses a laser beam, the light doesn't meet at a single point in space. Instead, it reaches a narrow waist with a diameter roughly equal to the wavelength of the light.

is limited by the wave nature of light. No matter how perfectly you try to focus light, you can't make a spot that's much smaller than the light's wavelength. Instead, the beam forms a narrow waist and then spreads apart (Fig. 13.2.4). This *beam waist* is about one wavelength of the light in diameter and a few wavelengths long, depending on the f-number of the converging lens. Since that waist is less than 2 microns long, the player's autofocusing system must keep the objective lens just the right distance from the reflective layer.

This fundamental limitation on how tightly a beam of light can be focused is another example of diffraction; the focusing lens truncates the light wave and inevitably introduces spreading. But even reaching this ideal focusing limit requires careful design and fabrication of the optical elements. While most other optical systems fall short of their ideal limits, a CD or DVD player's optical system does as well as can be done within the constraints of diffraction itself. Its optics are essentially perfect and are said to be "diffraction limited."

The amount of light that reflects from the layer depends on whether or not the laser spot hits a pit. This reflected light follows the optical path in reverse. It's collimated by the objective lens and then passes through the quarter-wave plate again. The plate now finishes the job it started earlier; the light ends up plane polarized, but with the opposite polarization from what it had when it left the laser. Horizontally polarized light is now vertically polarized or vice versa.

The reflected light then passes through the collimating lens, which makes the light converge, and then strikes the polarization beam splitter. Because the light's polarization has changed, the beam splitter no longer allows the beam to pass directly through. Instead, it turns this reflected beam 90° and directs it toward the detector array. This clever redirection scheme is important for two reasons. First, it conserves laser light by allowing most of it to travel from the laser diode to the detector. Second, it prevents reflected light from returning to the laser diode, where that light would be amplified and cause the laser diode to misbehave.

The light comes to a focus on an array of **photodiodes**—special diodes that conduct current when they're exposed to light. In the dark, a photodiode is like any other diode, carrying current only in one direction and preventing charge from traveling backward across the depletion region of its p-n junction. However, when it's exposed to light, a photodiode can carry a backward current that's proportional to the light intensity. The photons provide the energy needed to send charges backward across the depletion region.

The array of photodiodes allows the player to detect pits via the reflected light intensity and also to determine whether the objective lens is properly positioned relative to those pits. For generality, Fig. 13.2.3 omits optical elements involved in autofocusing and autotracking. However, because of those elements, the pattern of light hitting the detector array indicates which way the objective lens should move, if necessary. That lens is attached to coils of wire that are suspended near permanent magnets. By varying the currents flowing through those coils, the player uses magnetic forces to move its objective lens about rapidly and keep it in the right place over the disc.

► Check Your Understanding #3: That's Why It's Called a *Laser* Disc

Why must the CD or DVD player use a laser diode rather than a more conventional light source such as an incandescent bulb?

Optical Fibers

Optical playback of prerecorded discs is fine for music and movies, but it isn't of much use when you want up-to-the-minute information. The Internet and World Wide Web require communication links that operate at lightning speed. Yet even here, optics and light have an important role to play. The fastest way to send enormous amounts of information is to use optical fibers.

An optical fiber is a glass conduit that guides light from one place to another. Nearly every photon that enters the fiber at one end emerges from the other end moments later. In its simplest form, the fiber is made from two different glasses: a solid core of one glass surrounded by a cladding of the other glass. Both glasses are so incredibly transparent that light can travel through them for kilometers with little loss. For comparison, look through the edge of an ordinary piece of window glass and you'll see how dark the glass looks. It absorbs far too much light to be suitable for optical fibers. They're made of the purest glasses known.

If both glasses are almost perfectly transparent, what keeps the light from leaking out of the sides of the fiber? The answer is a phenomenon known as total internal reflection. As light tries to move from the inner glass core to the outer glass cladding, it's reflected perfectly and thus can't escape.

Total internal reflection is an extreme case of refraction. When light encounters the boundary between two materials with different indices of refraction, refraction causes that light to bend (Fig. 13.2.5). If the material it enters has a smaller index of refraction than the one it leaves, the light bends away from a line perpendicular to the boundary. The amount of this bend depends on the two indices of refraction and on the angle at which the light approaches the boundary. As long as the approach angle is steep enough, light will succeed in entering the second material. But if the approach angle is too shallow, the light won't enter the second material at all. Instead, it will reflect perfectly from the boundary. In fact, total internal reflection is far more efficient at reflecting light than a conventional metal mirror.

To keep the light inside the fiber's core, the core glass must have a higher index of refraction than the cladding glass. As light in the high-index core encounters the boundary with the low-index cladding, it experiences total internal reflection and bounces back into the core (Fig. 13.2.6a, p. 430). As long as the fiber doesn't bend too sharply, the light bounces back and forth inside the core and can't escape. As a result, light entering the fiber core through one of its cut ends follows the fiber all the way to its other cut end.

But a large-diameter fiber (typically 50 μm or more) has a problem. Light rays bouncing through it at slightly different angles travel different distances during their passage through the fiber. Light heading almost straight down the center of the fiber rarely bounces and takes less time to complete its trip than light that bounces many times. Because this wide fiber has many bouncing paths or "modes" in which light can travel through it, a short pulse of light going through the fiber gets stretched out in time (Fig. 13.2.6a). This pulse broadening severely limits the rate at which information can be sent through a multimode fiber.

To reduce stretching problems, the core of a better performance optical fiber has a graded refractive index. The core glass is specially treated so that its index of refraction decreases smoothly away from its center toward the cladding. Instead of bouncing abruptly when it reaches the boundary between core and cladding, light in this graded index environment turns smoothly back toward the core (Fig. 13.2.6b). The path differences between the modes in a graded index multimode fiber aren't so different and a short pulse of light isn't stretched very much in a fiber of moderate length.

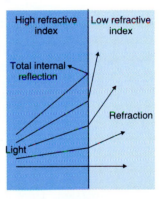

Fig. 13.2.5 When light traveling through one material enters a second material with a lower refractive index, it bends toward the boundary between those materials. If its approach angle is too shallow, the light will bend so much that it will simply reflect from the boundary. This effect is called total internal reflection.

Fig. 13.2.6 Light trying to leave the high-index core of an optical fiber at a shallow angle undergoes total internal reflection at the boundary with the low-index cladding. (*a*) In an ordinary multimode fiber, a pulse of light can follow many paths through the core and becomes spread out in time. (*b*) A core with a graded refractive index exhibits less temporal spreading because the reflection process is more gradual. (*c*) However, the least spreading occurs in a single-mode fiber. The small core diameter of this fiber provides light with only one mode of travel so that the only spreading occurs because of ordinary dispersion.

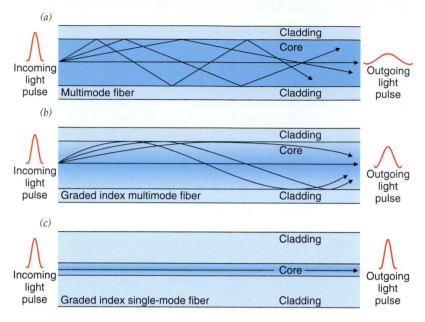

But in very long fibers, even small path differences add up and short pulses become blurred in multimode fibers. Therefore, the highest performance optical conduits are single-mode fibers. These fibers have very narrow graded index cores that permit light to travel only in one mode—effectively right down the center of the fiber (Fig. 13.2.6*c*). The core is typically only 9 μm in diameter. A pulse of light entering this narrow core broadens very little in time during its passage.

What little broadening occurs in a single-mode fiber isn't caused by the light taking different paths; it's caused by ordinary dispersion in the glass. To carry information, the light wave must change with time and thus must have a range of frequencies and wavelengths. As usual, the shorter wavelengths travel slower than the longer wavelengths and the pulses get stretched out in time. To minimize these dispersion effects, the highest-speed optical fibers operate at wavelengths that minimize dispersion. They also operate at wavelengths that minimize loss of light through absorption in the glass. These two wavelengths coincide at 1550 nm in so-called "dispersion shifted" fiber, so this infrared wavelength is commonly used in long-haul optical communication.

> **Check Your Understanding #4: If It Looks Like a Mirror . . .**
>
> When you look into an aquarium from the front, the sides of the aquarium often look like mirrors. Those sides are actually clear glass, so why do they reflect light so well?

Optical Communication

A typical optical communication transmitter uses a 1550-nm laser diode to generate short pulses of light. These pulses carry information from the transmitter to a receiver somewhere far away. The transmitter produces its pulses by varying the current passing through the laser diode. Light emerging from the laser diode is focused into the exposed core of a single-mode optical fiber, and it follows the core all the way to the other end of the fiber. When the light emerges, it's gathered by a lens and focused onto the receiver's photodiode. Each pulse of light causes a pulse of current to flow through the photodiode, allowing the receiver to begin processing the information.

A laser diode and a single-mode optical fiber can send billions of bits of data per second for 50 km or 100 km without significant errors. At longer distances, the gradual absorption of light in the glass makes it difficult to receive the information reliably. The easiest solution to this problem is to receive the data before the light becomes too weak and then to retransmit it with another laser diode.

But instead of interrupting the optical transmission with a receiver and retransmitter, some long-haul communication systems employ erbium-doped fiber amplifiers (EDFAs). An EDFA is a piece of optical fiber that has about 0.01% erbium ions in its glass core. When the EDFA is exposed to 980-nm or 1480-nm light, it becomes a laser amplifier for 1550-nm light. As the weakened pulses of light from a long fiber pass through the fiber amplifier, the amplifier duplicates photons and brightens the pulses. These amplified pulses then continue through ordinary fiber before being amplified again. Undersea optical cables often splice fiber amplifiers into the fibers every 50 km or so. These amplifiers allow light to travel thousands of kilometers through a continuous path, from one side of an ocean to the other.

To get the most out of a single optical fiber, many communication systems use several laser diodes operating at somewhat different wavelength ranges around 1550 nm. Light from these diodes is merged together and focused into the fiber. When the light emerges at the far end of the fiber, its different wavelength ranges are split apart and directed onto individual receivers. The different wavelength ranges are like different channels, so that this wavelength-division multiplexing allows one fiber to carry far more information than it could with light from a single laser. Remarkably enough, an EDFA can amplify all of these different channels at once because erbium ions can copy a wide range of wavelengths.

Check Your Understanding #5: Making Something Out of Nothing

If light weakens gradually as it passes through a long optical fiber, why can't you simply wait until the very end of the fiber to amplify what little light remains and then send the amplified light into a receiver?

birefringence, phonographs

Epilogue for Chapter 13

In this chapter we looked at several common objects and studied the ways in which they use optics. In *cameras*, we saw how a converging lens can form a real image of an object and how that real image can be used to record a scene on a piece of film or an image sensor chip. We learned about the roles of focal length and f-number in determining the size and brightness of the real image and explored some of the complications that must be considered when designing lenses.

In *optical recording and communication*, we looked at how laser light is affected by various optical devices and at what happens to it when it's focused to a small spot. We learned about total internal reflection and at how this effect makes it possible to send light hundreds or thousands of kilometers through a fiber of ultra-clear glass.

Explanation: Focusing Sunlight

The magnifying glass forms a real image of the sun on the sheet of paper. Because the sun is so far away, the image distance between the lens and the paper is equal to the focal length of the lens. The real image is relatively small, yet it contains all of the sunlight that strikes the entire surface of the lens. With so much light focused onto a small region of the paper, the radiative heat transfer from the sun to the paper becomes quite large and the paper's temperature rises until it begins to smoke or burn. The lens's image also contains light from other parts of the sky, and this light can be seen surrounding the sun's disk. Because of the way light is bent during the formation of the real image, the clouds in the sky appear upside down and backward on the sheet of paper.

Chapter Summary

How Cameras Work: A camera lens projects a real image of the scene in front of the camera onto the film inside the camera. This real image is formed when the lens bends all of the light reaching it from one part of the scene onto one part of the film. For this imaging process to work well, the distance between the lens and the film must be adjusted so that the light converges together (focuses) just as it reaches the film. If the film is too close to or too far from the lens, the image is blurry. The depth of focus depends on the effective diameter of the lens—its aperture. The smaller the lens's aperture, the less critical the focus but the less light the lens gathers. A long focal length lens brings the light to a focus far behind it, forming a relatively large but dim real image on the film. To brighten this image, the long focal length lens must have a large aperture so that it gathers lots of light. The f-number, the ratio of a lens's focal length to its aperture, characterizes the brightness of the lens's image.

How Optical Recording and Communication Work: A CD or DVD player uses a beam of laser light to read the pits on a thin reflective layer inside the disc. As the disc rotates, the pits pass through the focused laser beam and the amount of reflected light fluctuates up and down. The player monitors the reflected light and uses it both to recreate the music and to keep its optical system properly aligned. This continual realignment allows the player to follow the spiral track of pits as the disc turns and to keep the laser beam tightly focused on the reflective layer. The optical system, which includes a laser diode, several photodiodes, and a variety of lenses and other optical elements, is so well designed and executed that the spot it creates on the reflective layer is limited in smallness only by diffraction effects due to the wave nature of light.

Laser light can also carry information long distances through an optical fiber. Made from a core of extremely transparent glass, clad in a second glass, this fiber confines light via total internal reflection. As light attempts to leave the core at a shallow angle, it is perfectly reflected from the boundary with the cladding and continues through the fiber for great distances.

Important Laws and Equations

1. The Lens Equation: One divided by the focal length of a lens is equal to the sum of one divided by the object distance and one divided by the image distance, or

$$\frac{1}{\text{focal length}} = \frac{1}{\text{object distance}} + \frac{1}{\text{image distance}}, \tag{13.1.1}$$

Check Your Understanding—Answers

Section 13.1 CAMERAS

1. It is a real image of the room lights, created by the converging magnifying glass.

Why: A magnifying glass is a converging lens and can form a real image. It brings light spreading outward from the room lights together onto the sheet of paper as a real image.

2. Light from the more distant background focuses closer to the lens than light from your friend. Since the camera is focusing on your friend, the background appears out of focus.

Why: When the aperture of your camera lens is wide open, focusing becomes critical. Objects in front of or behind your friend are out of focus on the film and appear blurry.

3. On a bright day, your camera needs only a small aperture to gather enough light for an exposure, so the depth of focus is large. On a dark day, it uses the largest aperture available to gather light and has a small depth of focus.

Why: Light gathering and depth of focus go hand in hand. If you must open the aperture of your camera's lens to gather enough light for an exposure, focusing will become critical and your photographs will have small depths of focus.

4. The magnifying glass has only one glass element and suffers from chromatic aberration (and many other image imperfections).

Why: A single converging lens can't form a high-quality image on a flat surface because it has chromatic aberration, spherical aberration, coma, and astigmatism. If the lens is small, these failings are often invisible. But in a large magnifying glass, they are all readily apparent and you can't form a sharp image of the entire light fixture.

5. The lens is initially creating a real image near your eye. This real image moves past you as the lens approaches the photograph. Finally, the lens is close enough to create an enlarged virtual image of the photograph.

Why: A magnifying glass forms either a real or virtual image, depending on object distance and the lens's focal length. If the lens and photograph are separated by more than the focal length, the image is real and you see it near your eye. You can touch this inverted image or project it on a piece of paper. If the lens and picture are separated by less than the focal length, the upright image is virtual and you see it on the far side of the lens.

Section 13.2 OPTICAL RECORDING AND COMMUNICATION

1. Your cell phone's microphone converts the sound of your voice into a fluctuating current. This current is measured periodically and represented as a sequence of numbers. These numbers are then represented by radio waves and transmitted through space. The numbers eventually work their way to a receiving cell phone, where the radio waves are converted back into numbers and the numbers are converted back into a fluctuating current. Finally, the current is amplified and sent through a speaker to create sound.

Why: Many modern communication devices use a digital representation of sound. In radio wave communication, digital representations are particularly useful because they are immune to the noise that plagues analog radio transmissions and also permit more efficient use of the radio bandwidth.

2. The pits should become about half as deep as they are now.

Why: The pits in the reflective layer should be about a quarter of a wavelength deep so that light reflected from the bottom of a pit interferes destructively with light reflected from an adjacent flat region. If the light wavelength changes to 400 nm, about half the present value, then the pits should change to about half their present depth.

3. The light in a CD or DVD player must be coherent so that it can be focused to a single diffraction-limited spot on the reflective layer and experience destructive interference when it encounters a pit.

Why: You can't focus light from an incandescent bulb to a diffraction-limited spot. Each photon leaving the bulb is independent and will focus at a somewhat different point in space. The bulb's incoherent light also doesn't experience the strong interference effects of laser light.

4. The light you see is experiencing total internal reflection as it tries to leave the glass at a shallow angle. You see perfect reflections of the fish inside the aquarium.

Why: Light has trouble passing from water or glass into the air at a shallow angle. When it tries, it experiences total internal reflection and bounces off the surface without losing any intensity at all.

5. As the light is absorbed, the number of photons in a pulse gradually diminishes. When that number gets to zero, there's nothing left to amplify.

Why: Each pulse of light in the fiber contains a limited number of photons. The glass absorbs many of these photons during their long passage through a fiber, so they must be amplified before their numbers become so small that there's a chance that none of them makes it at all.

Check Your Figures—Answers

Section 13.1 CAMERAS

1. The image will form at a distance twice the focal length behind the lens.

Why: Since the object distance is twice the focal length, we can use Eq. 13.1.1 to find the image distance:

$$\frac{1}{\text{image distance}} = \frac{1}{\text{focal length}} - \frac{1}{\text{object distance}}$$
$$= \frac{1}{\text{focal length}} - \frac{1}{2 \cdot \text{focal length}}$$
$$= \frac{1}{2 \cdot \text{focal length}}.$$

The image distance is twice the focal length of the lens.

Exercises

1. Your eye contains a converging lens and a light-sensitive surface, the retina. Explain how your eye resembles a camera.

2. On a bright, sunny day you can use a magnifying glass to burn wood by focusing sunlight onto it. The focused sunlight forms a small circular spot of light that heats the wood until it burns. Why is the spot of light circular?

3. One part of a fiber optic communication system uses a lens to focus light from a semiconductor laser onto the end of an optical fiber. The tiny light source and fiber are on opposite sides of the lens, and each is 1.0 cm away from the lens. Why must the lens have a focal length of 0.5 cm?

4. Why is a laser beam's focus delayed by its entry into the plastic of a CD or DVD? (Draw a picture.)

5. Light from a distant object approaches the objective lens of a telescope as a group of parallel light rays. Draw a picture of the light rays from three stars passing through a converging lens to show that they focus at three separate locations.

6. If you hold your camera up to a small hole in a fence, you will be able to take a picture of the scene on the other side. However, the exposure time will have to be quite long, and the depth of focus will be surprisingly large. Explain.

7. Your new 35-mm camera comes with two lenses, a 50-mm focal length "normal" lens and a 200-mm focal length "telephoto" lens. The minimum f-number for the 50-mm lens is 1.8. Although the 200-mm lens has much larger glass elements in it, its lowest f-number is 4. Why is the telephoto lens's f-number so much larger?

8. If you're taking a photographic portrait of a friend and want objects in the foreground and background to appear blurry, how should you adjust the camera's aperture and shutter speed?

9. Your eye is similar to a camera. Why does squinting your eye increase your depth of focus and make it easier to focus?

10. Why is it easiest for the lens of your eye to form sharp images on your retina when the scene in front of you is bright and the iris of your eye is very small?

11. People who need glasses find it hardest to see clearly without them in dimly lit situations when the irises of their eyes are wide open. Why?

12. As you zoom the lens of your video camera inward to take a close-up of one person, what characteristic of the lens changes and does it increase or decrease?

13. As you zoom a video camera in for a close-up of a person (see Exercise 12), how does the distance between the effective center of the lens and the real image of that person change?

14. Sports photographers often use very long lenses on their cameras. How do the lengths of these lenses affect the photographs they produce?

15. Two similar looking magnifying glasses have different magnifications. One is labeled 2× (two times) and the other 4× (four times). Which lens has the longer focal length?

16. Which magnifying glass from Exercise 15 must you hold closer to the object you're looking at in order to see a virtual image of that object far in the distance?

17. If you look into a large glass marble, you'll find that decorations in the glass on its far side will appear greatly enlarged. Why?

18. Like a CD or DVD, a digital audiotape represents sound in digital form. The air pressures associated with sound are recorded as a series of numbers on the magnetic tape. Why is a digital audiotape essentially free of noise?

19. A CD or DVD could store much more music if it recorded fewer air pressure measurements each second. Why must so many measurements be made each second?

20. Scientists measure the distance to the moon by bouncing laser light from reflectors left on the moon by the Apollo astronauts. This light is sent backward through a telescope so that it begins its trip to the moon from an enormous opening. This procedure reduces the size of the laser beam when it reaches the moon. Explain.

21. As it passes through the lens that focuses it onto a CD's aluminum layer, the player's laser beam is more than 1 mm in diameter. Why does its large size as it leaves the lens allow the beam to focus to a smaller spot?

22. Why can light from a blue laser form a narrower beam waist than light from an infrared laser?

23. The objective lens in a DVD player must move quickly to keep the laser beam focused on the disc's reflective layer. Why must this lens have a very small mass?

24. Why don't portable CD or DVD players play properly if you shake them back and forth too quickly?

25. Sometimes the surfaces of a glass of water look mirrored when you observe them through the water. Explain.

26. When you look into the front of a square glass vase filled with water, its sides appear to be mirrored. Why do the sides appear so shiny?

27. To prepare a sample for an electron microscope, the operator embeds the sample in clear plastic and then shaves off ultrathin sheets with a diamond knife. The operator can tell how thick those sheets are by observing their colors through a microscope. Why do the sheets appear colored?

28. When using color to determine the thickness of an electron microscope sample (see Exercise 27), the operator must consider the plastic's index of refraction. Why?

29. Some of the laser light striking a DVD's reflective layer hits the flat region around a pit and reflects back toward the photodiode. How does this reflected wave actually *reduce* the amount of light detected by the photodiode?

Problems

1. Your camera has a 35-mm focal length lens. When you take a photograph of a distant mountain, how far from that lens is the real image of the mountain?

2. If you use a 35-mm focal length lens to take a photograph of flowers 2 m from the lens, how far from that lens does the real image of the flowers form?

3. You're trying to take a photograph of two small statues with a 200-mm telephoto lens. One statue is 4 m from the lens, and the other statue is 5 m from the lens. The real image of which statue forms closer to the lens? How much closer?

4. When you place a saltshaker 30 cm from your magnifying glass, a real image forms 30 cm from the lens on the opposite side. You can see this image dimly on a sheet of paper. What is the focal length of the magnifying glass?

5. When light from your desk lamp passes through a 50.0-mm focal length lens, it forms a sharp real image on a sheet of paper located 50.5 mm from the lens. How far is the desk lamp from the lens?

Cases

1. The clear plastic safety reflector on a bicycle has a flat front surface, but its back interior surface is covered with the corners of cubes. These cube corners or "corner cubes" stick up from the surface into the air like hundreds of small triangular pyramids. What makes the reflector so useful is that light striking it is sent directly back toward its source.

   a. When light hits the reflector at a right angle to its front surface, most of that light propagates to the back of the reflector where it encounters one of the corner cubes. It hits the surface of the cube at a shallow angle and instead of leaving the plastic, the light reflects perfectly. In fact, it reflects perfectly from all three faces of the corner cube and ends up returning directly toward the source of the light. But the back surface of the reflector is clear plastic followed by air. Why does light reflect perfectly off each surface of the corner cube?

b. While the reflector works best for light that hits it at a right angle to its surface, it also works for light that arrives at other angles. But when that angle becomes too shallow, the reflector stops working and light passes through its back surface. Explain why the light is able to get through in this case.

c. How does refraction at the front surface of the reflector help to widen the range of angles over which the reflector is able to send light directly back toward its source?

d. Why won't a flat mirror work well as a safety reflector?

2. The lens of your eye resembles that of a camera—light from the scene in front of you focuses to a real image on your retina. But unlike a camera lens, the lens of your eye can actually change its focal length by changing its curvature. The more curved the lens's surfaces are, the more strongly it bends light together and the shorter its focal length. The lens's variable focal length allows you to focus the real image of a particular object onto your retina without having to change the distance between the lens and the retina.

a. Explain why focusing on a distant object requires less lens curvature than focusing on a nearby object.

b. The lens of a farsighted person has trouble becoming curved enough to form a real image of a nearby object on the retina. Explain why eyeglasses containing converging lenses help a farsighted person see nearby objects.

c. The lens of a nearsighted person has trouble becoming flat enough to form a real image of a distant object on the retina. Explain why glasses containing diverging lenses, lenses that bend light rays apart, help a nearsighted person see distant objects.

d. The iris of your eye changes diameter to control how much light reaches your retina. The brighter the light, the smaller the iris's opening. Why is your eye's depth of focus greater in bright light than in dim light?

3. The inexpensive disposable cameras that are sold in grocery stores produce remarkably good pictures considering the financial constraints on their design. Despite their apparent simplicity, these cameras are sophisticated devices.

a. A disposable camera has no focus adjustment yet it produces reasonably sharp images. How does its small lens contribute to this result?

b. The camera lens is normally made from a single piece of clear plastic with a refractive index of 1.55. If they accidentally used a plastic with a refractive index of 1.45 and left the rest of the camera unchanged, the photographs would be blurry. Why?

c. The plastic used in the camera lens exhibits very little dispersion. How would dispersion affect the photographs?

d. The shutter in a disposable camera is a thin metal plate with a hole in it. When you take a picture, a spring pulls this plate past the lens and the hole allows light to

reach the film. How do the spring's tension and the plate's mass affect the exposure time?

4. A video camera uses a converging lens to form a real image of a scene on an electronic imaging device. This device forms a video signal out of the pattern of light on its surface.

a. Why must the lens of the video camera be a converging lens, rather than a diverging lens—a lens that bends light rays apart?

b. The camera focuses automatically by analyzing the contrast in the video image. As you shift the camera from a distant object to one that's closer, the lens moves. Does it move toward the imaging device or away from it? Explain.

c. The camera has an automatic iris—a diaphragm that controls the lens's aperture. When you turn the camera toward a bright scene, the iris reduces the lens's aperture. What effect does this change have on the camera's depth of focus?

d. The camera has a zoom lens that changes its focal length at the touch of a button. When you zoom in on a person for a closeup, does the focal length increase or decrease?

e. As the lens changes its focal length, it also moves toward or away from the electronic image device. As you zoom out to view more of the scene, does the lens move toward or away from the imaging device?

5. A slide projector is essentially the reverse of a camera. Light from an illuminated slide passes through a converging lens and forms a real image on the screen.

a. Since the slide is the object, the object distance is the separation between the slide and the lens. Compare this object distance to the lens's focal length when the projector casts a real image on a screen far at the other side of a room.

b. You focus the projector by moving the lens toward or away from the slide. If you want to focus the real image on a closer screen, which way should you move the lens?

c. The projector has a zoom lens that changes its focal length when you turn a dial. This lens makes it possible to change the size of the image on the screen. Zooming the lens also moves it toward or away from the slide. As the lens's focal length increases, should it move toward or away from the slide to keep the real image focused on the screen?

d. As you change the focal length of the zoom lens and move it toward the slide, the real image on the screen grows larger. Draw pictures of the light rays to show why moving the lens toward the slide causes the real image to grow.

e. Why must the slide be upside down in the projector in order to produce an upright real image on the screen?

6. Doctors use endoscopes to look inside patients through narrow openings. An endoscope contains a 1- or 2-meter long bundle of optical fibers that are carefully arranged and bonded together at their ends but that are independent and flexible in between. Each optical fiber is a pipe for light— light entering one end emerges from the other. When one end of the endoscope's fiber bundle is illuminated with a particular pattern of light, that same pattern of light appears at the bundle's other end.

a. The core of each optical fiber is made from a glass with a large refractive index. This core is wrapped in a glass with a lower refractive index. Each time light tries to leave the core at a shallow angle, it is perfectly reflected. Why?

b. The endoscope uses an objective lens to form a real image of nearby tissue on one end of the fiber bundle.

Why must the objective lens have a very short focal length in order to form a real image of nearby tissue?

c. The doctor uses an eyepiece to view light emerging from the other end of the bundle. How does the focal length of that eyepiece affect the magnification of the endoscope?

d. To obtain a large depth of focus, should the aperture of the endoscope's objective lens be large or small?

*e. To illuminate the patient's tissue, the endoscope has a second bundle of fibers that carries light from a bulb to the tissue. It uses a converging lens to form a real image of the bulb's filament on the entry surface of the fibers. If the bulb's filament and the fibers are 4 cm apart and the lens is halfway between them, what should the lens's focal length be?

CHAPTER 14

Modern Physics

In recent years, scientists have been looking deeper into the atom to see how it's made, farther into space to see how the universe works, and more carefully into objects to see how complicated things can be understood in terms of simple laws. Among the most important tools that these scientists have to work with are quantum theory and the theory of relativity. This chapter will look at some of the ways in which modern physics contributes to our lives.

EXPERIMENT: Radiation Damaged Paper

One path of modern physics involves the control and use of high-energy radiation. While our access to most forms of high-energy radiation is restricted, there's one source that anyone can use: the sun. Because the sun's ultraviolet light is energetic enough to damage chemical bonds and rearrange molecules, it can provide us with a glimpse of the effects that occur with X rays and beyond.

To see sun damage for yourself, expose some sheets of colored construction paper to direct sunlight for a few days. Cover the paper with some opaque objects such as coins and place it outdoors. Don't cover the paper with glass because glass absorbs enough ultraviolet light to slow the damage process. After a day or two, you should find that the exposed portions of the paper have lightened; the sun's ultraviolet radiation has destroyed some of the dye molecules in the paper. If you find that nothing happens, the dye is evidently robust enough to tolerate ultraviolet light for a while.

Try the experiment again with different papers and different colors. Can you *predict* which papers will fade fastest? *Observe* the results and see if they *verify* your predictions. Try to *measure* the rates at which dye molecules are damaged. Do they seem to have a half-life? How could you tell?

This same optical bleaching appears on items that are displayed in shop win-

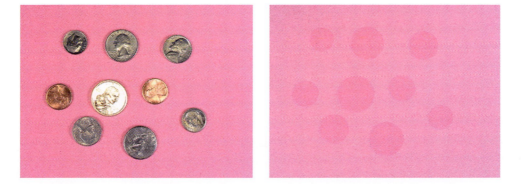

dows and on outdoor furniture. It was once the only method people had for whitening fabrics. The sun's ultraviolet light also damages your skin when you sit in the sun: a sunburn isn't thermal damage, it's radiation damage.

Chapter Itinerary

Fortunately ultraviolet light can't penetrate far into your body. In *nuclear weapons,* we'll look at more penetrating forms of radiation. We'll also explore the structures of atomic nuclei and see how taking them apart or joining them together can release enormous amounts of energy. In *medical imaging and radiation,* we'll examine high-energy radiation and see how it's used to help rather than hurt. We'll study the ways in which X rays and gamma rays are produced and how they interact with the atoms and molecules in a patient. We'll also look at the particle accelerators that produce high-energy particles for radiation therapy. Finally, we'll discuss the bases for CT imaging and MRI, which make it possible to prepare detailed maps of patients' insides without ever touching their bodies.

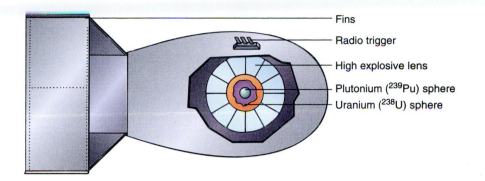

Fins
Radio trigger
High explosive lens
Plutonium (^{239}Pu) sphere
Uranium (^{238}U) sphere

SECTION 14.1   Nuclear Weapons

The atomic bomb is one of the most remarkable and infamous inventions of the 20th century. It followed close on the heels of various developments in the understanding of nature, developments that in many ways made the invention of nuclear weapons inevitable. By the late 1930s, scientists had discovered most of the principles behind nuclear energy and were well aware of how those principles might be applied. The onset of World War II prompted concern that Germany would choose to follow the military path of nuclear energy. Propelled by fear, curiosity, and temptation, the scientists, engineers, and politicians of that time brought nuclear weapons into existence.

Questions to Think About: Where is nuclear energy stored in the atoms? From where did this nuclear energy come? How do nuclear weapons release nuclear energy? Why are nuclear weapons so difficult to build? Why do we associate uranium and plutonium with nuclear weapons? How much uranium or plutonium does it take to build a bomb?

Experiments to Do: Since uranium and plutonium aren't sold in hardware stores, you won't be able to build your own bomb. However, you can get a feel for the way a chain reaction works by playing with a box of dominoes. If you stand the dominoes on end on a flat table, each of them will have extra gravitational potential energy that it can release by tipping over. If you spread the dominoes widely about the table and then give the table a gentle shake, they'll tip over one by one.

However, if you pack the dominoes tightly together, so that one falling domino can knock over others, they'll no longer be independent. As you jiggle the table, nothing will happen until the first domino falls, but then many or even all of the dominoes will tip over in quick succession. You will have created a chain reaction, where a single event triggers an ever-increasing number of subsequent events. What characteristics of the dominoes and their arrangement determine whether or not such a chain reaction occurs? Can you envision a scenario in which a single tipping domino could trigger the release of an enormous amount of stored energy? Another chain reaction, this time in the decay of atomic nuclei, is what makes nuclear weapons possible.

Background

At the end of the nineteenth century, "classical physics" reigned supreme. Here classical physics means the rules of motion and gravitation identified by such people as Galileo, Newton, and Kepler, and the rules of electricity and magnetism developed by others including Ampère, Coulomb, Faraday, and Maxwell. It was generally felt that most of physics was well understood: physicists knew all of the laws governing the behavior of objects in our universe, and what was left was to apply those laws to more and more complex examples. It was a time when physicists didn't know what they didn't know.

However, a few nagging problems remained—specific difficulties that couldn't be explained by the rules of classical physics. Among these were the spectrum of light emitted by a black body, the photoelectric effect in which electrons are ejected from metals by light, and the apparent absence of an ether or medium in which light traveled. At the beginning of the twentieth century, the whole of classical physics collapsed under the weight of these seemingly trivial difficulties, and a largely new understanding of the universe emerged. The major advances took 25 years, from 1901 to 1926, and the time since has largely been spent applying those new laws to more and more complicated examples.

The two main developments, both essential to the making of the atomic bomb, were the discoveries of quantum physics and relativity. Often these are called *quantum theory* and the *theory of relativity*. But while the word *theory* might imply that they're somehow on shaky ground, they're not theories in the sense of hypotheses waiting to be tested. In fact, they've been confirmed countless times since they were developed and have been shown to have enormous predictive power. Rather, they're theories in the sense of being carefully constructed and codified rules that model the behavior of the physical universe in which we live. Between them, these two theories made the discovery of nuclear forces and nuclear energy inevitable. Finally, given peoples' love for gadgets and power, they also made the development of nuclear weapons inevitable.

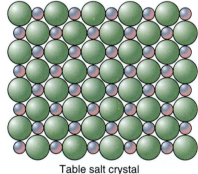

Table salt crystal
(Sodium chloride)

Chlorine negative ion
Diameter: 1.81×10^{-10} meter

Sodium positive ion
Diameter: 0.97×10^{-10} meter

Check Your Understanding #1: In Theory, It Means That . . .
If something is referred to as a "theory," how likely is it to be true?

The Nucleus

Though the name "atomic" bomb has stuck for more than half a century, the more correct name would be "nuclear" bomb. The items that are responsible for the energy released by nuclear weapons are not atoms, but tiny pieces of atoms—their nuclei (plural for nucleus). But before we can discuss nuclei, let's put them into context. Let's start by looking at atoms.

To get an idea of just how tiny atoms are, imagine magnifying a grain of table salt, 1 mm on a side, until it was the size of the State of Colorado. That grain would then appear as an orderly arrangement of spherical particles, each about the size of a grapefruit (Fig. 14.1.1). These spherical particles would be single atoms, and there would be about 7.2 million of them along each edge of the grain.

Like most solids, table salt is a crystal and its atoms are bound to one another by their outermost components: their electrons. Electrons dominate the chemistry

Fig. 14.1.1 A salt crystal is an orderly array of positively charged sodium ions and negatively charged chlorine ions. These ions are held together by the attractive forces between oppositely charged ions. The ions are so small that there are about 7.2 million ions on the edge of a 1-mm-wide salt crystal.

of atoms and molecules. Sodium is a reactive metal because of its electrons, and chlorine is a reactive gas because of its electrons. When mixed, these two chemicals react violently to form table salt and release a considerable amount of light and heat. This, then, is a true "atomic bomb."

Obviously, something is missing here. If a crazy person could buy a kilogram or two of sodium and a tank of chlorine from the chemistry stockroom and destroy an entire city, rural living would be a whole lot more popular. Fortunately the energy released by chemical reactions is fairly limited. A kilogram of chemical explosives just can't do that much damage. But atomic bombs tap an entirely different store of energy deep within the atoms.

While all of nuclear weaponry is often attributed to Einstein's famous equation, $E = mc^2$, that notion is vastly oversimplified. Nonetheless, this equation is quite significant. One of Einstein's discoveries at the beginning of the 20th century was that matter and energy are in some respects equivalent. In certain circumstances mass can become energy or energy can become mass. This equivalence is part of the theory of relativity and has some interesting consequences. It implies that an object can reduce its mass by transferring energy to its surroundings. Thus, if you weigh an object before and after it undergoes some internal transformation, you can use any weight loss to determine how much energy was released from the object by that transformation.

Because of this equivalence, mass and changes in mass can be used to locate energy that's hidden within normal matter. This technique is important in nuclear physics, but it also applies to chemistry. When sodium and chlorine react to form table salt, their combined mass decreases by a tiny amount. What's missing is some chemical potential energy, which becomes light and heat and escapes from the mixture. In leaving, this chemical potential energy reduces the mass of the sodium and chlorine mixture by about 1 part in 10 billion. That tiny change in mass is too small to detect with present measuring devices, although scientists are working on techniques that will soon make it possible to measure mass changes due to chemical bonds.

However, electrons are by far the lightest part of an atom and thus have relatively little mass to release as energy. Most of an atom's mass is located in its **nucleus.** The nucleus is fantastically small—only a little more than 10^{-15} m in diameter. If you were to peer into one of the grapefruit-sized sodium ions of our giant salt crystal, you would see a tiny particle at its center. There, just at the threshold of visibility, would be the ion's nucleus, only 1 micron in diameter. The remaining 99.9999999999999% of the ion is occupied only by its 10 electrons, moving about in their orbitals.

The sodium nucleus contains 11 protons and 12 neutrons (Fig. 14.1.2). Each of these nuclear particles or **nucleons** has about 2000 times as much mass as an electron, so that 99.975% of the sodium ion's mass is in this nucleus. Thus, while the electrons are certainly important to chemistry and matter as we know it, their contribution to the ion's mass is insignificant. The ion is mostly empty space, lightly filled with fluffy electrons and having a tiny nuclear lump at its center.

The nucleons that make up this nucleus experience two competing forces. The first of these forces is the familiar electrostatic repulsion between like electric charges. Because each proton in the nucleus has a single positive charge, they're constantly trying to push one another out of the nucleus. However, the second of these forces is attractive and holds the nucleus together. This new force is called the **nuclear force,** and at short distances it dominates the weaker electrostatic repulsion. However, the nuclear force only attracts the nucleons toward one another when they're touching. As soon as they're separated, they're on their own.

Sodium nucleus
(11 protons, 12 neutrons)

Neutron
Diameter: 1×10^{-15} meter

Proton
Diameter: 1×10^{-15} meter

Fig. 14.1.2 At the center of a sodium ion is a tiny nucleus containing about 99.975% of the ion's mass. It consists of 11 positively charged protons and 12 uncharged neutrons. The protons repel one another at any distance, but the protons and neutrons are bound together by the highly attractive nuclear force as long as they touch one another.

Fig. 14.1.3 This hopping toy stores energy as its spring is compressed and retains that energy while its suction cup grips its base. When the suction cup lets go, the toy leaps into the air.

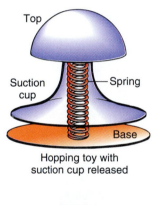

Hopping toy with suction cup released

Hopping toy with suction cup holding

Fig. 14.1.4 A spring and a suction cup combine to form a toy that hops suddenly after a long wait. The spring tries to separate the top from the base, while the suction cup tries to hold the two parts together. Energy you store in the spring is released when the leaking suction cup eventually allows the spring to expand.

The competition between these two forces, the repulsive electrostatic force between like charges and the attractive nuclear force between nucleons, is analogous to what happens in a familiar toy (Figs. 14.1.3 and 14.1.4). This hopping toy has a suction cup attached to a spring, so that the spring tries to separate the toy's top from its base while the suction cup tries to keep the two parts together. When the two parts are well separated, only the spring exerts a force. But when the two parts touch, the suction cup begins to act and holds the two parts together.

What makes the hopping toy exciting is that its suction cup leaks. Eventually, the suction cup lets go and allows the spring to toss the toy into the air. But suppose that the suction cup didn't leak. Once pushed together, the pieces would never separate and the spring would retain its stored energy indefinitely. To get the suction cup to let go, you would have to pull it away from the base. Only then could the spring release its stored energy.

In effect, an energy barrier would be preventing the leak-free toy from hopping. Until you did a little work on it by pulling the suction cup off the base, it wouldn't be able to release its stored energy. Another example of a system that needs energy to release energy is a bottle of champagne, where you must push on the cork to help it out of the neck. After that initial investment of energy, a great deal of energy is released as gas blasts the cork across the room.

The nucleus is in a similar situation. The attractive nuclear force prevents the nucleus from coming apart, despite the enormous amount of electrostatic potential energy it contains. The nuclear force creates an energy barrier that prevents the nucleons from separating. Unless something adds energy to the nucleus to help the nucleons break free of the nuclear force, the nucleus will remain together forever. At least that's the prediction of classical physics.

However, quantum physics has an important influence on the behavior of the nucleus. One of the many peculiar effects of quantum physics is that you can never really tell exactly where an object is located, or at least not for long. That fuzziness is a manifestation of the **Heisenberg uncertainty principle,** which observes that some pairs of physical quantities, such as position and momentum or energy and time, are not entirely independent and cannot be determined simultaneously beyond a certain accuracy. This principle is a result of the partly wave and partly particle nature of objects in our universe. Since waves are normally broad things that occupy a region of space rather than a single point, objects in our universe normally don't have exact locations.

The smaller an object's mass, the fuzzier it is and the more uncertain its location. While the fuzzy nucleons in a nucleus will normally stay in contact with one another for an extremely long time, there's always a tiny chance that they'll find themselves temporarily separated by a distance that's beyond the reach of the nuclear force. The nucleons will then suddenly be free of one another, and electrostatic repulsion will push them apart in a process called **radioactive decay.** The quantum process that allows the nucleons to escape from the nuclear force without first obtaining the energy needed to surmount the energy barrier is called **tunneling** because the nucleons effectively tunnel through the barrier.

The more protons there are in a nucleus, the more they repel one another and the more likely they are to cause radioactive decay. Adding additional neutrons to the nucleus reduces this proton–proton repulsion by increasing the size of the nucleus without adding to its positive charge. However, adding too many neutrons also destabilizes the nucleus for reasons that we'll discuss in the next section. So constructing a stable nucleus is a delicate balancing act.

In nuclei with only a few protons, the attractive nuclear force wins big over the repulsive electrostatic force and the nucleons stick like crazy. These nuclei resemble hopping toys with weak springs and big suction cups: once brought together, the pieces never come apart. In fact, the average binding energy of the nucleons (the energy required to separate them from one another divided by the number of nucleons) would increase if these nuclei had even more protons and neutrons.

In nuclei with many protons, the electrostatic repulsion is so severe that the nuclear force can't hold the nucleons together for long. These nuclei decay rapidly. They resemble hopping toys with strong springs and small suction cups. The average binding energy of the nucleons would increase if these nuclei had fewer protons and neutrons.

In nuclei with roughly 26 protons, in between the two extremes we've just considered, the attractive nuclear force and repulsive electrostatic force are nicely balanced. These nuclei are extremely stable, and you can't increase the average binding energy of their nucleons by adding or subtracting nucleons. Smaller nuclei can release potential energy by growing to reach this intermediate size, while larger nuclei can release potential energy by shrinking toward the same goal.

For a small nucleus to grow, something must push more nucleons toward it. Electrostatic repulsion will initially oppose this growth, but once everything touches, the nuclear force will bind the particles together and release a large amount of potential energy. This coalescence process is called **nuclear fusion.**

For a large nucleus to shrink, something must separate its pieces beyond the reach of the nuclear force. Electrostatic repulsion will then push the fragments apart and release a large amount of potential energy. This fragmentation process is called **nuclear fission.**

The energies released when small nuclei undergo fusion or when large nuclei undergo fission are enormous compared to chemical energies. Uranium, a large nucleus, converts about 0.1% of its mass into energy when it breaks apart. Hydrogen, a tiny nucleus, converts about 0.3% of its mass into energy when it fuses with other hydrogen nuclei. Kilogram for kilogram, nuclear reactions release about 10 million times more energy than chemical reactions. Fortunately, they're much harder to start.

➤ Check Your Understanding #2: A Sticky Nuclear Problem

If you take two intermediate-sized nuclei and combine them to make a single uranium nucleus, will the process release or consume energy?

Radioactive Decay and the Fission Bomb

Natural radioactive decay was discovered accidentally by French physicist Antoine-Henri Becquerel (1852–1906) in 1896. Intrigued by the recent discovery of X rays, he began looking for materials that might emit X rays after exposure to light. To his surprise, he found that uranium fogged photographic plates, even through an opaque shield and even without exposure to light. His discovery was soon confirmed and elaborated on by Polish-born French physicist Marie Curie (1867–1934) and French chemist Pierre Curie (1859–1906). This wife and husband team discovered several new radioactive elements, including polonium (named after Marie's homeland) and radium.

In 1911, British physicist Ernest Rutherford (1871–1937) discovered that atoms have nuclei. He subsequently found that nuclei sometimes shatter when struck by energetic helium nuclei. And in 1932, British physicist James Chadwick (1891–1974) discovered a fragment of the nucleus, the neutron, which has no electric charge and can thus approach a nucleus without any electrostatic repulsion. It was soon discovered that neutrons stick to the nuclei of many atoms.

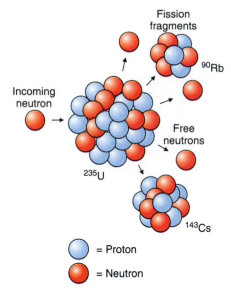

Fig. 14.1.5 When a neutron strikes a uranium nucleus, there's a good chance that the nucleus will fall apart into fragments. This process is called induced fission. Among the fragments of induced fission are other neutrons.

But the crucial discovery that made the atomic bomb possible was neutron-induced fission of nuclei. In 1934, Italian physicist Enrico Fermi (1901–1954) and his colleagues were trying to solve a particular riddle about the nucleus, a radioactive decay process called beta decay. They were adding neutrons to the nuclei of every atom they could get hold of. When they added neutrons to uranium, with its huge nucleus, they observed the production of some very short-lived radioactive systems. They thought that they had formed ultraheavy nuclei and even went so far as to give these new elements tentative names.

However, four years later, Austrian physicists Lise Meitner (1878–1968) and Otto Frisch (1904–1979) and German chemists Otto Hahn (1879–1968) and Fritz Strassmann (1902–1980) collectively showed ❏ that what Fermi's group had actually done was to fragment uranium into lighter nuclei (Fig. 14.1.5). Many of the fragments created by this **induced fission** were neutrons, which could themselves cause the destruction of other uranium nuclei.

It had suddenly become possible to create a **chain reaction** in which the fission of one uranium nucleus would induce fission in two nearby uranium nuclei, which would in turn induce fission in four other uranium nuclei, and so on (Fig. 14.1.6). The result would be a catastrophic nuclear process in which many or even most of the nuclei in a piece of uranium would shatter and release fantastic amounts of energy.

In a sense, the remaining work toward both the atomic bomb and the hydrogen bomb was a matter of technical details. Only four conditions had to be satisfied in order for an atomic or *fission bomb* to be possible:

1. A source of neutrons had to exist in the bomb to trigger the explosion.
2. The nuclei making up the bomb had to be **fissionable**—that is, they had to fission when hit by a neutron.

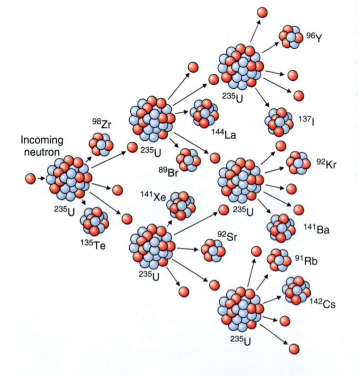

Fig. 14.1.6 A chain reaction occurs when the fragments of one fissioning nucleus induce fission in at least one additional nucleus. Such a chain reaction is particularly easy in ^{235}U, the light isotope of uranium, where each fissioning nucleus releases an average of 2.5 neutrons.

3. Each induced fission had to produce more neutrons than it consumed.
4. The bomb had to use the released neutrons efficiently so that each fission induced an average of more than one subsequent fission.

Meeting the first condition was easy. Many radioactive elements emit neutrons. But meeting the second and third conditions was more difficult. Here is where uranium fit into the picture. It was known to be fissionable, and it was known to release more neutrons than it consumed.

But not all uranium nuclei are the same. While a uranium nucleus must contain 92 protons, so that it forms a neutral atom with 92 electrons and has all of the chemical characteristics of uranium, the number of neutrons in that nucleus is somewhat flexible. Nuclei that differ only in the numbers of neutrons they contain are called **isotopes,** and natural uranium nuclei come in two isotopes: ^{235}U and ^{238}U, where the number specifies how many nucleons are in each nucleus. The ^{235}U nucleus contains 92 protons and 143 neutrons, for a total of 235 nucleons. In contrast, the ^{238}U nucleus contains 238 nucleons—92 protons and 146 neutrons.

It turns out that only ^{235}U is suitable for a bomb. It's marginally stable, with too many protons for the nuclear force to keep together, even with the diluting effects of 143 neutrons. ^{235}U has a radioactive **half-life** of 710 million years, meaning that in a large sample of these nuclei, approximately half will spontaneously fall apart over a period of 710 million years. Of those nuclei remaining after this period, another half will decay in the next 710 million years, and so on. But when struck by a neutron, ^{235}U fissions immediately and releases about 2.5 neutrons in the process.

^{238}U is slightly more stable than the lighter isotope, and its half-life is 4.51 billion years. But this nucleus absorbs most neutrons without undergoing fission. Instead, it undergoes a series of complicated nuclear changes that eventually convert it into plutonium, an element not found in nature. It becomes ^{239}Pu, a nucleus with 94 protons and 145 neutrons. As we'll see later on, plutonium itself is useful for making nuclear weapons.

So ^{238}U actually slows a chain reaction rather than encouraging it. Since only ^{235}U can support a chain reaction, natural uranium had to be separated before it could be used in a bomb. But ^{235}U is quite rare. The earth's store of uranium nuclei was created long ago, in the explosion of a dying star. That *supernova* heated the nuclei of smaller atoms so hot that they collided together and stuck. Uranium nuclei were formed, with the supernova's energy trapped inside them. They were incorporated in the earth during its formation about 4 or 5 billion years ago and have been decaying ever since. The only uranium isotopes that remain in any quantity are ^{235}U and ^{238}U. Since ^{235}U is less stable, it has dwindled to only 0.72% of the uranium nuclei. The remaining 99+% of the uranium is ^{238}U.

Separating ^{235}U from ^{238}U is extremely difficult. Since atoms containing these two nuclei differ only in mass, not in chemistry, they can only be separated by methods that compare their masses. Because the mass difference is relatively small, heroic measures are needed to extract ^{235}U from natural uranium. During World War II and the Cold War era, the United States government developed enormous facilities for separating the two uranium isotopes. The need for such installations is one of the major obstacles to the proliferation of nuclear weapons.

The last condition for sustaining a chain reaction is that the bomb must use neutrons efficiently, so that each fission induces an average of more than one subsequent fission. That means that the bomb's contents can't absorb neutrons wastefully and that it can't let too many of them escape without causing fission. A lump of relatively pure ^{235}U wouldn't absorb neutrons wastefully, but it might allow too many of them to escape through its surface. For a chain reaction to occur, the lump must

❑ **Austrian-born physicist Lise Meitner** moved to Berlin in 1907 and soon began a 30-year collaboration with chemist Otto Hahn. In 1934, she convinced Hahn to join her in studying nuclear processes and they made great progress. Unfortunately, Meitner's Jewish ancestry made her a target of Nazi academics restriction and she fled to Sweden in 1938. Meitner continued to guide their collaboration through letters. Only months after she left, Hahn and his assistant Fritz Strassmann found that neutron irradiation of heavy elements was creating smaller rather than larger nuclei. Meitner and her nephew Otto Frisch soon developed a model of nuclear fission based on these measurements. However, Hahn published the results without Meitner's name on the paper, ostensibly to avoid Nazi interference. As a result of this omission, the 1944 Nobel Prize in chemistry was awarded to Hahn alone. Hahn went on to claim that Meitner was simply his assistant, rather than the leader of their joint effort. In recognition of this gross injustice, element 109 was named meitnerium in 1994.

half-life

carbon dating

be large enough that each neutron has a good chance of hitting another nucleus before it leaves the lump. The lump also should have a minimal amount of surface. It should be a sphere.

But how large must that sphere be? Since atoms are mostly empty space, a neutron can travel several centimeters through a lump of uranium without hitting a nucleus. Thus a golf ball-sized sphere of uranium would allow too many neutrons to escape. For a bare sphere of ^{235}U, the **critical mass** required to initiate a chain reaction is about 52 kg, a ball about 17 cm in diameter. At that point, each fission will induce an average of one subsequent fission. But for an **explosive chain reaction,** in which each fission induces an average of much more than one fission, additional ^{235}U is needed: a **supercritical mass.** About 60 kg will do it.

By 1945, the scientists and engineers of the Manhattan Project had found ways to meet these four conditions and were prepared to initiate an explosive chain reaction. They had accumulated enough ^{235}U to construct a supercritical mass. Carefully machined pieces of ^{235}U would be put into the bomb so that they would join together at the moment the bomb was to explode. When the critical mass was reached, a few initial neutrons would start the chain reaction. When the uranium became supercritical, catastrophic fission would quickly turn it into a tremendous fireball.

However, assembly was tricky. The supercritical mass had to be completely assembled before the chain reaction was too far along; otherwise the bomb would begin to overheat and explode before enough of its nuclei had time to fission. In pure ^{235}U, the time it takes for one fission to induce the next fission is only about 10 ns (10 nanoseconds or 10 billionths of a second). In a supercritical mass, each generation of fissions is much larger than the previous generation, so it takes only a few dozen generations to shatter a significant fraction of the uranium nuclei. The whole explosive chain reaction is over in less than a millionth of a second, with most of the energy released in the last few generations (about 30 ns).

To make sure that the assembly was complete before the bomb exploded, it had to be done extraordinarily quickly. In the ^{235}U bomb called "Little Boy" (Fig. 14.1.7), exploded over Hiroshima on August 6, 1945 at 8:15 a.m., the supercritical mass was assembled when a cannon fired a cylinder of ^{235}U through a hole in a sphere of ^{235}U (Fig. 14.1.8). When the cylinder was centered in the hole, it completed a 60-kg sphere of uranium, housed in a tungsten carbide and steel container. This container confined the uranium, holding it together with its inertia as the explosion began. An explosive chain reaction started immediately, and by the time the uranium blew itself apart, 1.3% of the ^{235}U nuclei had fissioned. The energy released in that event was equal to the explosion of about 15,000 tons of TNT.

But "Little Boy" was actually the second nuclear explosion. Its concept was so foolproof and its ^{235}U so precious that "Little Boy" was dropped without ever being tested. However, the Manhattan Project had also developed a plutonium-based

Fig. 14.1.7 The Little Boy (top) used a cannon to fire a cylinder of uranium into an incomplete sphere of uranium. The Fat Man (bottom) used high explosives to crush a sphere of plutonium to extreme density. Little Boy destroyed Hiroshima, while Fat Man destroyed Nagasaki.

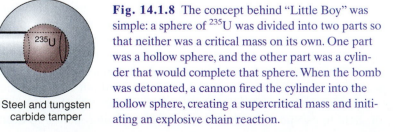

Fig. 14.1.8 The concept behind "Little Boy" was simple: a sphere of ^{235}U was divided into two parts so that neither was a critical mass on its own. One part was a hollow sphere, and the other part was a cylinder that would complete that sphere. When the bomb was detonated, a cannon fired the cylinder into the hollow sphere, creating a supercritical mass and initiating an explosive chain reaction.

bomb that involved a much more sophisticated concept. This bomb was much less certain to work, so it was tested once before it was used.

"The Gadget," as the first atomic bomb was called, didn't use ^{235}U. Instead, it used plutonium that had been synthesized from ^{238}U in nuclear reactors. A nuclear reactor carries out a controlled chain reaction in uranium, and neutrons from this chain reaction can convert ^{238}U into ^{239}Pu.

Like ^{235}U, ^{239}Pu meets the conditions for a bomb. The ^{239}Pu nucleus is relatively unstable, with a half-life of only 24,400 years. It fissions easily when struck by a neutron and releases an average of 3 neutrons when it does. Thus ^{239}Pu can be used in a chain reaction. For a bare sphere of ^{239}Pu, the critical mass is about 10 kg—a ball about 10 cm in diameter.

But ^{239}Pu has a problem. It's so radioactive and releases so many neutrons when it fissions that a chain reaction develops almost instantly. There is much less time to assemble a supercritical mass of plutonium than there is with uranium. The cannon assembly method won't work because the plutonium will overheat and blow itself apart before the cylinder can fully enter the sphere.

Thus a much more sophisticated assembly scheme was employed. When "The Gadget" (Fig. 14.1.9) was detonated at Alamogordo on June 16, 1945 at 5:29 a.m., over 2000 kg of carefully designed high explosives crushed or *imploded* a sphere of plutonium (Fig. 14.1.10). By itself, the 6.1-kg sphere wasn't large enough to be a critical mass; it was a **subcritical mass.** But it was surrounded by a *tamper* of ^{238}U whose massive nuclei reflected many neutrons back into the plutonium like marbles bouncing off bowling balls. And the implosion process compressed the plutonium well beyond its normal density. With the plutonium nuclei packed more tightly together, they were more likely to be struck by neutrons and undergo fission.

The scheme worked. The chain reaction that followed caused 17% of the plutonium nuclei to fission and released energy equivalent to the explosion of about 22,000 tons of TNT. The tower and equipment at the Trinity test site disappeared into vapor, and the desert sands turned to glass for hundreds of meters in all directions. A nearly identical device named "Fat Man" (Fig. 14.1.7) was dropped over Nagasaki on August 9, 1945 at 11:02 a.m.

In the years following the first fission bombs, development focused on how best to bring the fissionable material together. The longer a supercritical mass could be held together before it overheated and exploded, the larger the fraction of its nuclei that would fission and the greater the explosive yield. The crushing technique of "The Gadget" and "Fat Man" became the standard, and bombs grew smaller and more efficient at using their nuclear fuel. The implosion process reduced the amount of plutonium needed to reach a supercritical mass, so that very small fission bombs were possible. The smallest atomic bomb, the "Davy Crockett," weighed only about 220 N (50 pounds).

Fig. 14.1.9 The first atomic bomb, nicknamed "The Gadget," was detonated on a tower at a remote desert site near Alamogordo, New Mexico on June 16, 1945. Here employees of the top secret Manhattan Project are seen hoisting parts of the plutonium bomb onto the tower before the explosion.

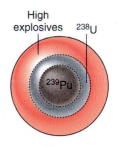

Fig. 14.1.10 The concept used in "The Gadget" and "Fat Man" was relatively sophisticated. Carefully shaped high explosives crushed a baseball-sized sphere of ^{239}Pu inside a neutron-reflecting shell of ^{238}U. The plutonium was compressed to high density and quickly reached supercritical mass, initiating an explosive chain reaction.

Check Your Understanding #3: Tickling the Dragon's Tail
What would happen if you slowly moved two 30-kg hemispheres of ^{235}U together to form a single sphere?

The Fusion or Hydrogen Bomb

Since fissionable material begins to explode as soon as it exceeds critical mass, this critical mass limits the size and potential explosive yield of a fission bomb. In search

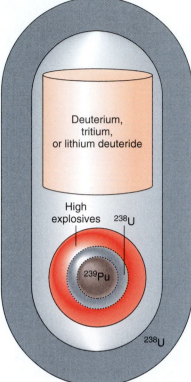

Fig. 14.1.11 A fusion bomb releases energy by fusing deuterium (^{2}H) and tritium (^{3}H) nuclei together to form helium (^{4}He) and neutrons. This fusion is initiated by heating the hydrogen to more than 100 million degrees Celsius with a fission bomb. High-energy neutrons released by the fusion process then induce fission in the ^{238}U tamper, releasing still more energy. Lithium (^{6}Li) produces tritium when exposed to neutrons.

of a way around this limit, bomb scientists took a look back at small nuclei and figured out how to extract energy by sticking them together.

The fission bomb brought to the earth, for the first time, temperatures that had previously only been observed in stars. Stars obtain most of their energy by fusing hydrogen nuclei together to form helium nuclei, a process that releases a great deal of energy. Because hydrogen nuclei are protons and repel one another with tremendous forces, hydrogen doesn't normally undergo fusion here on earth. To cause fusion, something must bring those protons close enough for the nuclear force to stick them together. The only practical way we know of to bring the nuclei together is to heat them so hot that they crash into each other. That is what happens in a *fusion bomb,* also called a *thermonuclear* or *hydrogen bomb.*

In a fusion bomb, an exploding fission bomb heats a quantity of hydrogen to about 100 million degrees Celsius (Fig. 14.1.11). At that temperature, hydrogen nuclei begin to collide with one another. To ease the nuclear fusion processes, heavy isotopes of hydrogen are used: *deuterium* (^{2}H) and *tritium* (^{3}H). While the normal hydrogen nucleus (^{1}H) contains only a proton, the deuterium nucleus contains a proton and a neutron. The tritium nucleus contains a proton and two neutrons. When a deuterium nucleus collides with a tritium nucleus, they stick to form a helium nucleus (^{4}He) and a free neutron. Because this process converts about 0.3% of the original mass into energy, the helium nucleus and the neutron fly away from one another at enormous speeds.

Since hydrogen won't explode spontaneously, even in huge quantities, a hydrogen bomb can be extremely large. A fission bomb is used to set it off, but after that, the sky's the limit. Some enormous fusion bombs were constructed and tested during the early days of the Cold War. These bombs usually consisted of a fission starter and a hydrogen follower, all wrapped up in a tamper of ^{238}U. The tamper confined the hydrogen as the fusion began. Once fusion was underway, converting deuterium and tritium into helium and neutrons, the neutrons collided with ^{238}U nuclei in the tamper. These fusion neutrons were so energetic that they were able to induce fissions even in ^{238}U nuclei and release still more energy. Overall, this structure is sometimes called a fission–fusion–fission bomb.

A variation on this bomb is the so-called "neutron" or "enhanced radiation" bomb. That bomb has no ^{238}U tamper so that the energetic neutrons from the fusion process travel out of the explosion and irradiate everything in the vicinity. This bomb is lethal to humans but not particularly destructive to property.

Tritium is a radioactive isotope created in nuclear reactors. It has too many neutrons to be a stable nucleus and slowly decays into a light isotope of helium (^{3}He) through an exotic nuclear process in which a neutron transforms into a proton, an electron, and an antineutrino. Because tritium has a half-life of 12.3 years, fusion bombs containing tritium require periodic maintenance to replenish their tritium.

Many fusion bombs use solid lithium deuteride instead of deuterium and tritium gases. Lithium deuteride is a salt containing lithium (^{6}Li) and deuterium (^{2}H). When a neutron from the fission starter collides with a ^{6}Li nucleus, the two fragment into a helium nucleus (^{4}He) and a tritium nucleus (^{3}H). In the bomb, lithium deuteride is quickly converted into a mixture of deuterium, tritium, and helium, which then undergoes fusion.

Check Your Understanding #4: It's Hard to Get Together

Why must hydrogen be heated to extremely high temperature to initiate fusion?

Heat, Radiation, and Fallout

Once a nuclear weapon has exploded—after its fissionable material has fissioned and its fusible material has fused—what then? First, a vast number of nuclei and subatomic particles emerge from the explosion at enormous speeds, many at nearly the speed of light. These particles crash into nearby atoms and molecules, heating them to fantastic temperatures and producing a local fireball around the bomb itself. They also cause extensive radiation damage in the surrounding area.

Second, a flash of light emerges from the explosion, caused partly by the fission and fusion processes themselves and partly by the ultrahot fireball that follows. This light is not only visible light, but also every portion of the electromagnetic spectrum from infrared to visible to ultraviolet to X rays to gamma rays. It burns things nearby, inside and out.

Third, the explosion creates a huge pressure surge in the air around the fireball. A shock wave propagates outward from the fireball at the speed of sound, knocking over everything in its path for a considerable distance. Fourth, the rarefied and superheated air rushes upward, lifted by buoyant forces, to create a towering mushroom cloud.

But the most insidious aftereffect of a nuclear explosion is *fallout,* the creation and release of radioactive nuclei. Fission converts uranium and plutonium nuclei into smaller nuclei. Each new nucleus has several dozen protons and its share of neutrons from the nucleus that fissioned. These new nuclei attract electrons and become seemingly normal atoms like iodine or cobalt. But while large nuclei such as uranium need extra neutrons to dilute their protons and reduce their electrostatic repulsions, intermediate and small nuclei like those of iodine and cobalt don't need as many neutrons. The new nuclei wind up with too many neutrons and are radioactive. They have half-lives that are anywhere from thousandths of a second to thousands of years.

Until they decay, the atoms that contain these nuclei are almost indistinguishable from normal atoms. They are radioactive isotopes of common atoms, and our bodies naively incorporate them into our tissues. There they sit, performing whatever chemical tasks our bodies require of them. But eventually these radioactive atoms fall apart and release nuclear energy. Because each radioactive decay that occurs near us or inside us releases perhaps a million times more energy than is present in a chemical bond, these decays cause chemical changes in our cells. They can kill cells or damage the cells' genetic information, potentially causing cancer.

▶ Check Your Understanding #5: Not So Good to Eat

Normal iodine ^{127}I is stable forever. But the fission product ^{131}I is radioactive and has a half-life of about 8 days. What are the consequences of eating ^{131}I?

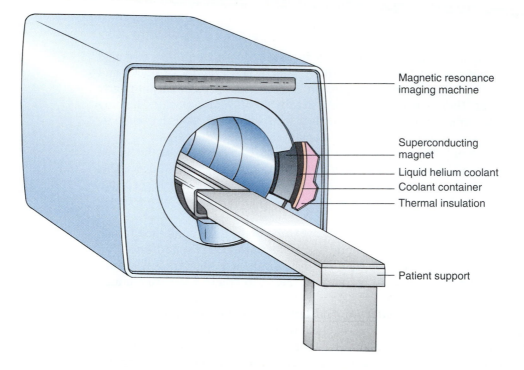

Magnetic resonance
imaging machine

Superconducting
magnet

Liquid helium coolant

Coolant container

Thermal insulation

Patient support

Medical Imaging and Radiation

Some of the most important recent advances in health care have occurred at the border between medicine and physics. As scientists have refined their understanding of atomic and molecular structure and learned to control various forms of radiation, they have invented tools that are enormously valuable for diagnosing and treating illness and injury. The developments continue, with new applications of physics appearing in clinical settings almost every time you turn around. In this section, we'll examine two of the most significant examples of medical physics: the imaging techniques that are used to detect problems and the radiation therapies that are used to treat them.

Questions to Think About: *What's different about bones and tissue that bones appear light in an X-ray image while tissue appears dark? How can a CT scan or an MRI image show a cross-sectional view of a living person without touching that person? If X rays are a form of electromagnetic radiation, why don't opaque materials absorb them? Why does a CT scan show primarily bone while an MRI image shows primarily tissue? The subatomic particles used in radiation therapy often have enormous energies. From where do these energies come?*

Experiments to Do: *One triumph of medical imaging is its ability to locate objects inside a person without actually entering the body. This is often done using X rays to look through the person from several different angles. From each vantage point, the imaging machine determines which objects are to the left or right of one another, but it can't tell how far away those objects are. Nonetheless, by piecing together information from many different observations, the imaging machine can locate each object exactly.*

You can experiment with this process by sprinkling a number of coins onto the surface of a small table, without looking to see where they come to rest. Close your eyes and bring your head to the height of the table's surface. Open only one eye and take a brief look at the coins. If the lighting is bright and uniform, and you don't move your head, you'll have trouble telling how far the coins are from your eye. While you'll know something about where each coin is, you won't know enough to locate it exactly on the table's surface. Now move your head to a new position around the table and take a second brief look. Again, you won't be able to tell how far away the coins are, but you'll learn more about their relative positions. How many such views will it take for you to learn exactly where the coins are? How does the presence of many opaque coins complicate the problem? Why does opening two eyes at once make it much easier for you to locate the coins?

X Rays

Since their discovery in 1895, X rays have played an important role in medical treatment. Their usefulness was obvious from the very evening they were discovered. It was November 8 and German physicist Wilhelm Conrad Roentgen (1845–1923) was experimenting with an electric discharge in a vacuum tube. He had covered the entire tube in black cardboard and was working in a darkened room. Some distance from the tube a phosphored screen began to glow. Some kind of radiation was being released by the tube, passing through the cardboard and the air, and causing the screen to fluoresce. Roentgen put various objects in the way of the radiation, but they didn't block the flow. Finally, he put his hand in front of the screen and saw a shadowed image of his bones. He had discovered X rays and their most famous application at the same time.

The first clinical use of X rays was on January 13, 1896, when two British doctors used them to find a needle in a woman's hand. In no time, X-ray systems became common in hospitals as a marvelous new technique for diagnosis. But this imaging capability was not without its side effects. Although the exposure itself was painless, overexposure to X rays caused deep burns and wounds that took some time to appear. Evidently the X rays were doing something much more subtle to the tissue than simply heating it.

X rays are a form of electromagnetic radiation, as are radio waves, microwaves, and light. These different forms of electromagnetic radiation are distinguished from one another by their frequencies and wavelengths—while radio waves have low frequencies and long wavelengths, X rays have extremely high frequencies and short wavelengths. But they're also distinguished by their photon energies. Because of its low frequency, a radio wave photon carries little energy. A medium-frequency photon of blue or ultraviolet light carries enough energy to rearrange one bond in a molecule. But a high-frequency X-ray photon carries so much energy that it can break many bonds and rip molecules apart.

In a microwave oven, the microwave photons work together to heat and cook food. The amount of energy in each microwave photon is unimportant because they don't act alone. But in radiation therapy, the X-ray photons are independent. Each one carries enough energy to damage any molecule that absorbs it. That's why X-ray burns involve little heat and appear long after the exposure—the molecular damage caused by X rays takes time to kill cells.

Check Your Understanding #1: Forms of Radiation

Which is more closely related to X rays: the beam of electrons traveling through a television picture tube or the infrared light from the hot filament of a toaster?

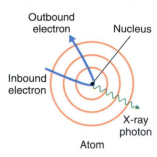

Fig. 14.2.1 When a fast-moving electron arcs around a massive nucleus, it accelerates rapidly. This sudden acceleration creates a bremsstrahlung X-ray photon, which carries off some of the electron's energy.

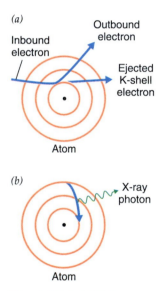

Fig. 14.2.2 (*a*) When a fast-moving electron collides with an electron in one of the inner orbitals of a heavy atom, it can knock that electron out of the atom. (*b*) An electron from one of the atom's outer orbitals soon drops into the empty orbital in a radiative transition that creates a characteristic X ray.

Making X Rays

Medical X-ray sources work by crashing fast-moving electrons into heavy atoms. These collisions create X rays via two different physical mechanisms: bremsstrahlung and X-ray fluorescence.

Bremsstrahlung occurs whenever a charged particle accelerates. This process is nothing really new, since we know that radio waves are emitted when a charged particle accelerates on an antenna. But in a radio antenna, the electrons accelerate slowly and emit low-energy photons. Bremsstrahlung usually refers to cases in which a charged particle accelerates extremely rapidly and emits a very-high-energy photon. In X-ray tube bremsstrahlung, a fast-moving electron arcs around a massive nucleus and accelerates so abruptly that it emits an X-ray photon (Fig. 14.2.1). This photon carries away a substantial fraction of the electron's kinetic energy. The closer the electron comes to the nucleus, the more it accelerates and the more energy it gives to the X-ray photon. However, the electron is more likely to miss the nucleus by a large distance than to almost hit it, so bremsstrahlung is more likely to produce a lower energy X-ray photon than a higher energy one.

In **X-ray fluorescence,** the fast-moving electron collides with an inner electron in a heavy atom and knocks that electron completely out of the atom (Fig. 14.2.2). This collision leaves the atom as a positive ion, with a vacant orbital close to its nucleus. An electron in that ion then undergoes a radiative transition, shifting from an outer orbital to this empty inner one and releasing an enormous amount of energy in the process. This energy emerges from the atom as an X-ray photon. Because this photon has an energy that's determined by the ion's orbital structure, it's called a **characteristic X ray.**

To discuss the energies carried by X-ray photons, we need an appropriate energy unit. While the joule is a useful unit of energy for describing collisions between bats and balls, it's too big to be practical for X rays and collisions of subatomic particles. Instead, we'll use **electron-volts** (abbreviated eV). One electron-volt (1 eV) is about 1.602×10^{-19} J and is the amount of kinetic energy an electron acquires as it accelerates through a voltage difference of 1 V.

Photons of visible light carry energies of between 1.6 eV (red light) and 3.0 eV (violet light). Because the ultraviolet photons in sunlight have energies of up to 7 eV, they are able to break chemical bonds and cause sunburns. But X-ray photons have much larger energies than even ultraviolet photons.

In a typical medical X-ray tube, electrons are emitted by a hot cathode and accelerate through vacuum toward a positively charged metal anode (Fig. 14.2.3). The anode is a tungsten or molybdenum disk, spinning rapidly to keep it from melting. The energy of the electrons as they hit the anode is determined by the voltage difference across the tube. In a medical X-ray machine, that voltage difference is typically about 87,000 V, so each electron has about 87,000 eV of energy. Since an electron gives a good fraction of its energy to the X-ray photon it produces, the photons leaving the tube can carry up to 87,000 eV of energy. No wonder X rays can damage tissue!

When the electrons collide with a target of heavy atoms, they emit both bremsstrahlung and characteristic X rays (Fig. 14.2.4). The characteristic X rays have specific energies so they appear as peaks in the overall X-ray spectrum. The bremsstrahlung X rays have different energies but are most intense at lower energies. Because lower energy X-ray photons injure skin and aren't useful for imaging or radiation therapy, medical X-ray machines use absorbing materials, such as aluminum, to filter them out.

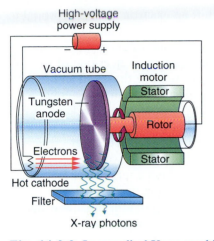

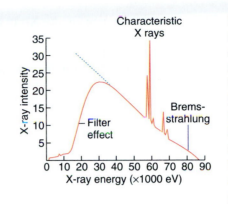

Fig. 14.2.4 When electrons with 87,000 eV of energy collide with tungsten metal, they emit X rays via bremsstrahlung and X-ray fluorescence. While the bremsstrahlung X rays have a broad range of energies, an absorbing filter blocks the low-energy ones. X-ray fluorescence produces characteristic X rays with specific energies.

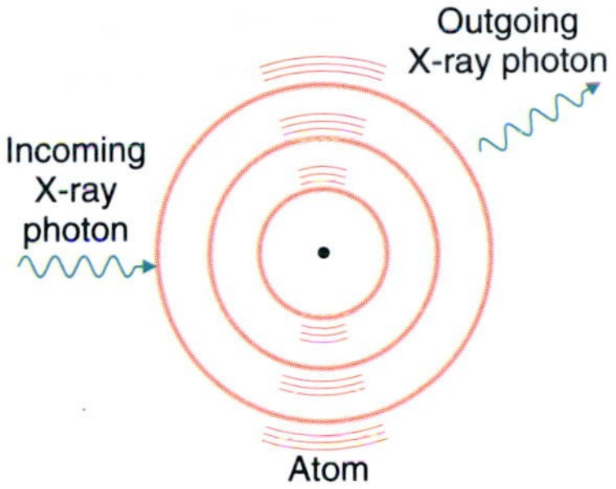

Fig. 14.2.3 In a medical X-ray machine, electrons from a hot filament accelerate toward a positively charged metal disk. They emit X rays when they collide with the disk's atoms. A motor spins the disk to keep it from melting. The filter absorbs useless low-energy X rays.

Check Your Understanding #2: The Origins of Synchrotron Radiation

The giant particle accelerators used in high-energy physics are often built as rings so that they can use the same electrically charged particles over and over again. As these particles travel in circles around the rings, they emit X rays. Explain.

Using X Rays for Imaging

X rays have two important uses in medicine: imaging and radiation therapy. In X-ray imaging, X rays are sent through a patient's body to a sheet of film or an X-ray detector. While some of the X rays manage to pass through tissue, most of them are blocked by bone. The patient's bones form shadow images on the film behind them. In X-ray radiation therapy, the X rays are again sent through a patient's body, but now their lethal interaction with diseased tissue is what's important.

X-ray photons interact with tissue and bone through four major processes: *elastic scattering*, the *photoelectric effect*, *Compton scattering*, and *electron–positron pair production*. **Elastic scattering** is already familiar to us as the cause of the blue sky: an atom acts as an antenna for the passing electromagnetic wave, absorbing and reemitting it without keeping any of its energy (Fig. 14.2.5). Because this process has almost no effect on the atom, elastic scattering isn't important in radiation therapy. However, it's a nuisance in X-ray imaging because it produces a hazy background: some of the X rays passing through a patient bounce around like pinballs and arrive at the film from odd angles. To eliminate these bouncing X-ray photons, X-ray machines use filters to block X rays that don't approach the film from the direction of the X-ray source.

The **photoelectric effect** is what makes X-ray imaging possible. In this effect, a passing photon induces a radiative transition in an atom: one of the atom's electrons absorbs the photon and is tossed completely out of the atom (Fig. 14.2.6). If the

Fig. 14.2.5 When an X-ray photon scatters elastically from an atom, the whole atom acts as an antenna. The passing photon jiggles all of the charges in the atom, and these charges absorb the photon and reemit it in a new direction.

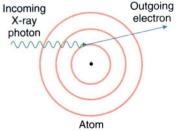

Fig. 14.2.6 In the photoelectric effect, an absorbed photon ejects an electron from an atom. Part of the photon's energy is used to remove the electron from the atom, and the rest becomes kinetic energy in the electron.

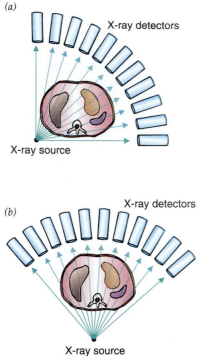

(a)

X-ray detectors

X-ray source

(b)

X-ray detectors

X-ray source

Fig. 14.2.7 A computed tomography or CT scan image is formed by analyzing X-ray shadow images taken from many different angles and positions. An X-ray source and an array of electronic X-ray detectors form a ring that rotates around the patient as the patient slowly moves through the ring.

atom were using the X-ray photon to shift an electron from one orbital to another, that photon would have to have just the right amount of energy. But because a free electron can have any amount of energy, the atom can absorb any X-ray photon that has enough energy to eject one of its electrons. Part of the photon's energy is used to remove the electron from the atom, and the rest is given to the emitted electron as kinetic energy.

However, the likelihood of such a *photoemission* event decreases as the ejected electron's energy increases. This decreasing likelihood makes it difficult for a small atom to absorb an X-ray photon. All of its electrons are relatively weakly bound and the X-ray photon would give the ejected electron a large kinetic energy. Rather than emit a high-energy electron, a small atom usually just ignores the passing X-ray photon.

In contrast, some of the electrons in a large atom are quite tightly bound and require most of the X-ray photon's energy to remove them. These electrons would depart with relatively little kinetic energy. Because the photoemission process is most likely when low-energy electrons are produced, a large atom is likely to absorb a passing X ray. Thus, while the small atoms found in tissue (carbon, hydrogen, oxygen, and nitrogen) rarely absorb medical X rays, the large atoms found in bone (calcium and phosphorus) absorb X rays frequently. That's why bones cast clear shadows onto X-ray film. Tissue shadows are also visible, but they're less obvious.

Although one shadow image of a patient's insides may help to diagnose a broken bone, more subtle problems may not be visible in a single X-ray image. For a better picture of what's going on inside the patient, the radiologist needs to see shadows from several different angles. Better yet, the radiologist can turn to a *computed tomography (CT) scanner*. This computerized device automatically forms X-ray shadow images from hundreds of different angles and positions and produces a detailed 3D X-ray map of the patient's body.

The CT scanner works one "slice" of the patient's body at a time. It sends X rays through this narrow slice from every possible angle, including the two shown in Fig. 14.2.7, and determines where the bones and tissues are in that slice (Fig. 14.2.8). The scanner then shifts the patient's body to work on the next slice.

Check Your Understanding #3: Aluminum X-Ray Windows

Aluminum atoms are much smaller than calcium atoms. While aluminum metal blocks visible light, it's relatively transparent to high-energy X rays. Explain.

Using X Rays for Therapy

Radiation therapy also uses X rays, but not the ones used for medical imaging. Even though tissue absorbs fewer imaging photons than bone, most imaging photons are absorbed before they can pass through thick tissue. For example, only about 10% of the imaging photons make it through a patient's leg even when they miss the bone. That percentage is good enough for making an image, but it won't do for radiation therapy because most imaging X rays would be absorbed long before they reached a deep-seated tumor. Instead of killing the tumor, intense exposure to these X rays would kill tissue near the patient's skin.

To attack malignant tissue deep beneath the skin, radiation therapy uses ex-

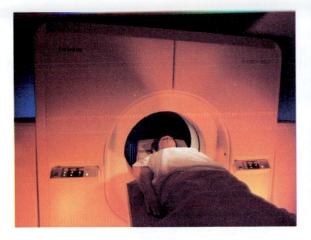

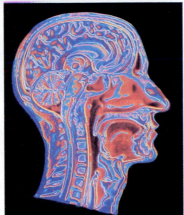

Fig. 14.2.8 The computed tomography (CT) scanner on the left uses X rays to image layer after layer of the patient's body. With the help of a computer, it produces a three-dimensional (3D) map of heavy elements in the patient. Part of that map, showing the patient's head, appears in the image on the right.

tremely high-energy photons. At photon energies near 1,000,000 eV, the photoelectric effect becomes rare in tissue and bone, and the photons are much more likely to reach the tumor. Photons still deposit lethal energy in the tissue and tumor, but they do this through a new effect: Compton scattering.

Compton scattering occurs when an X-ray photon collides with a single electron so that the two particles bounce off one another (Fig. 14.2.9). The X-ray photon knocks the electron right out of the atom. This process is different from the photoelectric effect because Compton scattering doesn't involve the atom as a whole and the photon is scattered (bounced) rather than absorbed. The physics behind this effect resembles that of two billiard balls colliding, although it's complicated by the theory of relativity. The fact that it occurs at all is proof that a photon carries both energy and momentum and that these quantities are conserved when a particle of light collides with a particle of matter.

Compton scattering is crucial to radiation therapy. When a patient is exposed to 1,000,000-eV photons, most of the photons pass right through them, but a small fraction undergo Compton scattering and leave some of their energy behind. This energy kills cells and can be used to destroy a tumor. By approaching a tumor from many different angles through the patient's body, the treatment can minimize the injury to healthy tissue around the tumor while giving the tumor itself a fatal dose of radiation.

But Compton scattering isn't the only effect that occurs when high-energy photons encounter matter. X rays with slightly more than 1,022,000 eV can do something remarkable when they pass through an atom: they can cause **electron–positron pair production.** A **positron** is the **antimatter** equivalent of an electron. Our universe is symmetrical in many ways, and one of its nearly perfect symmetries is the existence of antimatter. Almost every particle in nature has an antiparticle with the same mass but opposite characteristics. A positron or antielectron has the same mass as an electron, but it's positively charged. There are also antiprotons and antineutrons.

Antimatter doesn't occur naturally on earth, but it can be created in high-energy collisions. When an energetic photon collides with the electric field of an atom, the photon can become an electron and positron. In the previous section, we discussed matter becoming energy; pair production is an example of energy becoming matter. It takes about 511,000 eV of energy to form an electron or a positron, so the photon must have at least 1,022,000 eV to create one of each. Any extra energy goes into kinetic energy in the two particles.

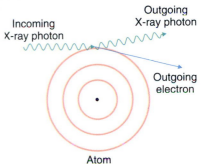

Fig. 14.2.9 In Compton scattering, an X-ray photon collides with a single electron and the two bounce off one another. The electron is knocked out of the atom.

The positron doesn't last long in a patient. It soon collides with an electron and the two annihilate one another—the electron and positron disappear and their mass becomes energy. They turn into photons with a total of at least 1,022,000 eV. So energy became matter briefly and then turned back into energy. This exotic process is present in high-energy radiation therapy and becomes quite significant at photon energies above about 10,000,000 eV. Not surprisingly, it also helps to kill tumors.

Check Your Understanding #4: More of a Good Thing

About how much energy would a photon need to create a proton–antiproton pair?

Gamma Rays

Producing very-high-energy photons isn't quite as easy as producing those used in X-ray imaging. In principle, a power supply could create a huge voltage difference through an X-ray tube so that very-high-energy electrons would crash into metal atoms and produce very-high-energy photons. But million-volt power supplies are complicated and dangerous, so other schemes are used instead.

One of the easiest ways to obtain very-high-energy photons is through the decay of radioactive isotopes. The isotope most commonly used in radiation therapy is cobalt 60 (^{60}Co). The nucleus of ^{60}Co has too many neutrons and that makes it unstable. It eventually decays, undergoing a series of transformations that produce two high-energy photons: one with 1,170,000 eV and one with 1,330,000 eV. These photons penetrate tissue well and are quite effective at killing tumors. ^{60}Co sources, carefully shrouded in lead containers, are often used in radiation therapy.

Although the process by which ^{60}Co produces those two high-energy photons is complicated, it shows that protons, electrons, and neutrons are not immutable and that there are other subatomic particles in our universe. The decay begins when one of the neutrons in a ^{60}Co nucleus abruptly turns into three particles: a proton, an electron, and a *neutrino* (or more precisely, an *antineutrino*). This process is called *beta decay* and can occur because neutrons that are by themselves or in nuclei with too many neutrons are unstable and radioactive. When one of the neutrons in a neutron-rich ^{60}Co nucleus decays, it leaves behind a positively charged proton and the nucleus becomes nickel 60 (^{60}Ni). The negatively charged electron and the neutral neutrino escape from the nucleus and travel outward.

The **neutrino** is a subatomic particle with no charge and little or no mass. Neutrinos aren't found in normal atoms. Though important in nuclear and particle physics, neutrinos are difficult to observe directly because they travel at or near the speed of light and hardly ever collide with anything. Without charge, they don't participate in electromagnetic forces and, unlike the electrically neutral neutron, they don't experience the nuclear force. They experience only gravity and the **weak force,** the last of the four **fundamental forces** known to exist in our universe. (The other three fundamental forces are the gravitational force, the electromagnetic force, and the **strong force**—that's a more complete version of the nuclear force that we discussed in the previous section.) Because it's weak and occurs only between particles that are very close together, the weak force rarely makes itself apparent. One of the few occasions where it plays an important role is in beta decay.

With almost no way to push or pull on another particle, a neutrino can easily pass right through the entire earth. Neutrinos are detected occasionally, but only with the help of enormous detectors. That's why physicists first showed that neutrinos are emitted from decaying neutrons by measuring energy and momentum before and after the decay. The proton and electron produced by the decay don't have the same total energy and momentum as the neutron had before the decay. Something must have carried away the missing energy and momentum, and that something is the neutrino.

Once ^{60}Co has turned into ^{60}Ni, the decay isn't quite over. The ^{60}Ni nucleus that forms still has extra energy in it. Nuclei are complicated quantum physical systems just as atoms are, and they have excited states, too. The ^{60}Ni nucleus is in an excited state, and it must undergo two radiative transitions before it reaches the ground state. These radiative transitions produce very-high-energy photons or **gamma rays** that are characteristic of the ^{60}Ni nucleus—one with 1,170,000 eV of energy and the other with 1,330,000 eV. These gamma rays are what make ^{60}Co radiation therapy possible.

> **Check Your Understanding #5: A Visit from the Snake Oil Salesman**
> If someone offered to sell you a bottle of neutrinos, you'd be foolish to buy it. What's wrong with the idea of a bottle of neutrinos?

Particle Accelerators

Electromagnetic radiation isn't the only form of radiation used to treat patients. Energetic particles such as electrons and protons are also used. Like tiny billiard balls, these fast-moving particles collide with the atoms inside tumors and knock them apart. As usual, this atomic and molecular damage tends to kill cells and destroy tumors.

However, obtaining extremely energetic subatomic particles isn't easy. High-voltage power supplies can be used to accelerate an electron or proton to about 500,000 eV, but that isn't enough. When a charged particle enters tissue, it experiences strong electric forces and is easily deflected from its path. To make sure that it travels straight and true, all the way to a tumor, the particle must have an enormous energy. Giving each charged particle the millions or even billions of electron-volts it needs for radiation therapy takes a particle accelerator.

Particle accelerators use metal cavities that behave like the tank circuits and antennas we discussed in Section 11.1. Almost any metal structure can act simultaneously as a capacitor and an inductor and thus have a natural resonance for sloshing charge. In the resonant cavities of a particle accelerator, this sloshing charge creates huge electric fields that change with time. Those electric fields push charged particles through space until they reach incredible energies.

One important type of particle accelerator is the *linear accelerator*. In this device, the electric fields in a series of resonant cavities push charged particles forward in a straight line (Fig. 14.2.10). Each of these cavities has charge sloshing back and forth rhythmically on its wall. When a small packet of charged particles enters the first cavity through a hole, it's suddenly pushed forward by the strong electric field inside that cavity (Fig. 14.2.10a). The packet accelerates forward and leaves the first resonant cavity with more kinetic energy than it had when it arrived: the electric field in that cavity has done work on the packet.

(a)
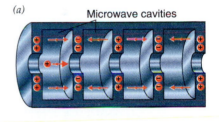
Microwave cavities

(b)
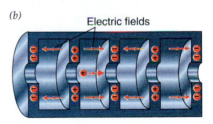
Electric fields

Fig. 14.2.10 In a linear accelerator, moving charged particles are pushed forward by electric fields that change with time. (*a*) While the moving positive charge is passing through the first of a series of microwave cavities, the field there pushes it forward. (*b*) By the time the moving charge has entered the second cavity, the fields have reversed and the field there pushes it forward again.

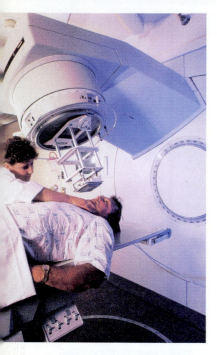

Fig. 14.2.11 This radiation therapy unit uses a linear accelerator to produce extremely high-energy subatomic particles. These particles penetrate deep inside the patient to destroy a cancerous tumor. The linear accelerator itself is hidden from view in the room behind this one. Its beam is steered by magnets in the rotatable arm toward a particular spot in the patient.

relativity theory

If the fields in the cavities were constant, the electric field in the second cavity would slow the packet down. In Fig. 14.2.10*a,* you can see that the electric field in the second cavity points in the wrong direction. But by the time the packet reaches the second cavity, the charge sloshing in its walls has reversed and so has the electric field (Fig. 14.2.10*b*). The packet is again pushed forward, and it emerges from the second cavity with still more kinetic energy.

Each resonant cavity in this series adds energy to the packet, so that a long string of cavities can give each of the packet's charged particles millions or even billions of electron-volts. This energy comes from microwave generators that cause charge to slosh in the accelerator's resonant cavities. The linear accelerator then only has to inject charged particles into the first cavity, using equipment resembling the insides of a television picture tube, and those charged particles will come flying out of the last cavity with incredible energies (Fig. 14.2.11).

However, this acceleration technique has a few complications. Most importantly, each cavity must reverse its electric field at just the right moment to keep the packet accelerating forward. For simplicity of operation, all the cavities have the same resonant frequency and reverse their electric fields simultaneously. Since the packet spends the same amount of time in each cavity and it speeds up as it goes from one cavity to the next, each cavity must be longer than the previous one.

But as the packet approaches the speed of light, something strange happens. The packet's energy continues to increase as it goes through the cavities, but its speed stops increasing very much. What this means is that the simple relationship between kinetic energy and speed given in Eq. 2.3.7 isn't valid for objects moving at almost the speed of light! This breakdown is part of Einstein's **relativity theory,** the rules governing motion at speeds comparable to the speed of light. As a further consequence of relativity, the packet can approach the speed of light but can't actually reach it. Though each charged particle's kinetic energy can become extraordinarily large, its speed is limited by the speed of light.

Because the packet's speed stops increasing significantly after it has gone through the first few cavities of the linear accelerator, the lengths of the remaining cavities can be constant. Only the first few cavities have to be specially designed to account for the packet's increasing speed inside them. The charged particles emerge from the accelerator traveling at almost the speed of light. They pass through a thin metal window that keeps air out of the accelerator and enter the patient's body. They have so much energy that they can penetrate deep into tissue before coming to a stop.

Check Your Understanding #6: Particle Recycling

Many research accelerators send each packet of electrons through the same series of resonant cavities several times. After the packet leaves the last cavity, magnets steer it around in a circle and send it back through the cavities again. With each pass through the cavities, the packet acquires more energy, so how can it possibly stay synchronized with the reversing electric fields in the cavities?

Magnetic Resonance Imaging

While X rays do an excellent job of imaging bones, they aren't as good for imaging tissue. A better technique for studying tissue is *magnetic resonance imaging* or *MRI*.

This technique locates hydrogen atoms by interacting with their magnetic nuclei. Since hydrogen atoms are common in both water and organic molecules, finding hydrogen atoms is a good way to study biological tissue.

The nucleus of a hydrogen atom is a proton. Like the electrons we studied in Section 9.3, a proton is effectively a spinning ball of charge and this moving charge gives it a magnetic field. The proton behaves like a tiny bar magnet, with a north pole at one end and a south pole at the other. If you put the proton in a magnetic field, it will tend to align itself with that field.

However, this classical view of the protons as little spinning magnetic balls isn't quite right. Because of the Heisenberg uncertainty principle, a proton's axis of rotation isn't well defined and a proton in a magnetic field exhibits only two distinct orientations: aligned or anti-aligned. For simplicity, we'll call the aligned protons "spin-up" and the anti-aligned ones "spin-down." And while protons would align perfectly with the field at absolute zero, thermal energy flips the protons about. Even in a strong magnetic field, spin-up protons only slightly outnumber spin-down protons.

When a patient enters the strong magnetic field of an MRI machine (Figs. 14.2.12, 14.2.13), the protons in the patient's body respond to the field and an excess of spin-up protons appears. Because aligning with a magnetic field reduces a bar magnet's potential energy, each of these spin-up protons has less energy than it would have if it were spin-down. While it's possible to convert one of the extra spin-up protons into a spin-down proton, that process takes energy.

An MRI machine uses radio wave photons to flip the extra spin-up protons. When a radio wave photon with just the right amount of energy passes a spin-up proton, the proton may absorb it in a radiative transition and become a spin-down proton. How much energy the radio wave photon needs depends on how much energy it takes to flip the spin. That energy in turn depends on how strong the magnetic field is around the proton. Since the energy needed to flip the proton's spin increases as the magnetic field gets stronger, the energy of the radio wave photon must also increase with the field.

If the protons in the patient's body were all experiencing exactly the same magnetic field, they would all require the same radio wave photon energy to flip. But the protons don't all experience the same field. The MRI machine introduces a slight spatial variation to its magnetic field. Because the magnetic field is different for different protons, only some of them can absorb radio waves of a particular energy. This selective absorption is how the MRI imager locates protons within a patient.

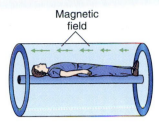

Fig. 14.2.12 An MRI machine places the patient in a strong magnetic field. This field varies spatially, so that protons at different places in the patient's body experience different fields and absorb different radio wave photons.

spins and quantum mechanics

sophisticated MRI

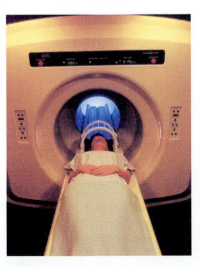
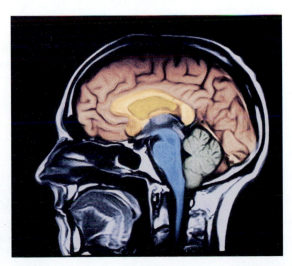

Fig. 14.2.13 The patient on the left is entering the intense magnetic field of a magnetic resonance imaging (MRI) system. Using the interactions between electromagnetic waves and protons that take place in magnetic fields, it produces a three-dimensional map of hydrogen atoms in the patient's body. A portion of this map, showing the patient's head, is displayed in the image on the right.

In its simplest form, an MRI machine applies a spatially varying magnetic field to the patient's body. It then sends various radio waves through the patient and looks for those radio waves to be absorbed by protons. Since only a proton that is experiencing the right magnetic field can absorb a particular radio wave photon, the MRI machine can determine where each proton is by which radio waves it absorbs. By changing the spatial variations in the magnetic field and adjusting the energies of the radio waves, the MRI machine gradually locates the protons in the patient's body. It builds a detailed three-dimensional map of the hydrogen atoms. A computer manages this map and can display cross-sectional images of the patient from any angle or position.

Check Your Understanding #7: Major Magnets

Some of the most modern MRI machines use extremely strong magnetic fields. As a machine's magnetic field gets stronger, what must happen to the radio waves that are used to flip the spins of protons in a patient?

Epilogue for Chapter 14

In this chapter we examined some of the applications of modern physics. In *nuclear weapons,* we examined nuclear fission to see how nuclear chain reactions can be used to release electrostatic potential energy stored in the large nuclei of uranium and plutonium atoms. We also studied nuclear fusion and found that when small nuclei of hydrogen bind together, they release a potential energy associated with the nuclear force. We learned about radioactive isotopes and fallout.

In *medical imaging and radiation,* we learned how X rays are produced and why those X rays pass more easily through tissue than through bone. We saw how X rays can be used to make images and explored the uses of gamma rays for radiation therapy. We looked at particle accelerators and concluded by examining magnetic resonance imaging.

Explanation: Radiation Damaged Paper

The paper fades because the ultraviolet light photons in sunlight have enough energy to shift electrons out of the orbitals that bond the dye molecules together. These dye molecules then fall apart and leave the paper without color. In some cases, the photons of light completely remove the electrons from the dye molecules. That process is the same photoelectric effect that makes it possible to distinguish tissue from bone in X-ray imaging.

Chapter Summary

How Nuclear Weapons Work: A fission bomb releases nuclear energy through a chain reaction in the fissionable isotopes of uranium or plutonium. In a chain reaction, each fission induces, on average, at least one subsequent fission. Assembling a supercritical mass must be done rapidly so that the bomb can't blow itself apart before most of the nuclei have undergone fission.

A fusion bomb uses heat from a fission bomb to initiate fusion in the heavy isotopes of hydrogen—deuterium and tritium. Because tritium has a short half-life, it must be replaced frequently. In some fusion bombs, the tritium is created during the explosion by neutron collisions with lithium.

How Medical Imaging and Radiation Work: Because X rays pass more easily through tissue than they do through bone, X rays form shadow images of a patient's bones. These X rays are produced when energetic electrons accelerate near metal nuclei and when they knock other electrons out of the metal atoms.

Radiation therapy is done with very-high-energy X rays and gamma rays because they pass more easily through tissue. These electromagnetic waves kill tumor cells by depositing energy in the atoms and molecules of those cells. Gamma rays are usually obtained from radioactive nuclei. Some radiation therapy is done with energetic particles that are given enormous energies by particle accelerators.

Magnetic resonance imaging uses the magnetic nature of hydrogen nuclei (protons) to locate hydrogen atoms in a patient's body. The patient is put in a strong magnetic field, and the protons tend to align with this field. The imaging machine then uses radio waves to flip the alignment of those protons. By making the magnetic field vary slightly from place to place, the machine is able to find the protons it's flipping. A computer records and analyzes the results so that it can present cross-sectional images of the hydrogen atoms in a patient's body.

Check Your Understanding—Answers

Section 14.1 NUCLEAR WEAPONS

1. That depends completely on the theory and the extent to which it has been compared to the real world.

Why: A great many theories have been formulated to explain the world around us. Each of these theories is an attempt at describing some particular behavior of our universe in terms of various rules or mechanisms. But until a theory has been tested by comparing it carefully to the system it's trying to explain, you can't tell if it's true or not. Some theories are eventually proved true, others false, and many remain uncertain. The theories of relativity and quantum physics have long since been proved true, although there is always the possibility that they may only be parts of a more complete theory.

2. It will consume energy.

Why: To merge two smaller nuclei and form a uranium nucleus, you will have to push the two together with considerable force because they contain many protons. You will have to do considerable work to bring these nuclei close enough for the nuclear force to bind them. The energy you invested in this new nucleus is the same energy that is released when it undergoes fission.

3. Before they actually touched, a chain reaction would occur.

Why: As soon as the hemispheres became close enough that each fission began to induce an average of one subsequent fission, a chain reaction would occur. You would have assembled a critical mass. The hemispheres would begin to get hotter and hotter and would eventually melt or explode. Because they would blow apart before most of the nuclei could fission, any explosion would be rather small. But you would suffer severe radiation injury. Just such an accident killed Louis Slotin during the Manhattan Project.

4. The protons in hydrogen nuclei repel one another quite strongly at short distances. They must be moving very quickly with lots of thermal energy for them to touch.

Why: In a gas, thermal energy takes the form of kinetic energy. The hotter the gas, the faster the particles are moving. At 100 million degrees Celsius, the hydrogen nuclei are moving so fast that they can overcome their electrostatic repulsion and touch. When they do, the nuclear force pulls them together and they fuse.

5. The ^{131}I is incorporated into your body and exposes you to radiation, particularly during the first few weeks before most of it has decayed away.

Why: Your body can't distinguish ^{127}I from ^{131}I because they're chemically identical. Since your body uses iodine in its functions, any ^{131}I that you ingest is likely to become part of your body's iodine supply. Over the next 8 days, about half of this iodine will decay and expose you to radiation. The remaining ^{131}I is also radioactive, and half of this amount will decay in the next 8 days. Thus after 16 days, only one-quarter of the original amount will remain. After 24 days, only one-eighth will remain. And so on.

Section 14.2 MEDICAL IMAGING AND RADIATION

1. The infrared light from the hot filament.

Why: Infrared light and X rays are both forms of electromagnetic radiation. All that distinguishes them is their frequencies and wavelengths, and the energy of their photons. The beam of electrons in a television tube is also a form of radiation, but it involves particles of matter, not electromagnetic waves.

2. Because a particle traveling in a circle is accelerating, it emits electromagnetic waves. In this case, those waves are X rays.

Why: A rapidly accelerating charged particle will emit an X ray, whether it's accelerating around the nucleus of a heavy atom or around the ring of a particle accelerator. In an accelerator, these X rays are called synchrotron radiation. Synchrotron radiation is useful in research and industry and is often deliberately enhanced by adding special magnets to the ring of the accelerator.

3. The electrons in an aluminum atom are so weakly bound that they are unlikely to absorb high-energy X-ray photons through the photoelectric effect.

Why: Like the small atoms in biological tissue, aluminum atoms rarely use the photoelectric effect to absorb high-energy X rays. This result makes it possible to use thin films of aluminum as windows and filters for X-ray sources.

4. About 2,000,000,000 eV.

Why: A proton has about 2000 times the mass of an electron, so producing a proton–antiproton pair should take about 2000 times the 1,000,000-eV energy required to produce an electron–positron pair.

5. The bottle couldn't confine neutrinos because it barely interacts with them.

Why: Because neutrinos experience only gravity and the weak force, the bottle couldn't trap them for long. Actually, you're being bathed in neutrinos from the sun all the time without even noticing it. They're as common as dirt and you can't do anything with them anyway.

6. The packet is traveling at almost the speed of light, so its speed barely changes as its energy increases.

Why: If the packet were speeding up with each trip through the cavities, it wouldn't stay synchronized with the reversing electric fields inside them. But the packet's speed is so nearly constant once it nears the speed of light that it can travel through the cavities over and over again without any problem.

7. Each radio wave photon must carry more energy (the radio waves must have higher frequencies).

Why: The stronger the magnetic field it's in, the more energy a proton needs to flip its spin. The MRI machine must use higher frequency, more energetic radio waves to cause such spin flips. There are many advantages to using extremely strong magnetic fields, including lower noise and better spatial resolution in the image. However, the magnetic fields of these advanced MRI machines are so strong that they can erase credit cards from across the room and rip steel objects right out of your pockets.

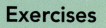

Exercises

1. Alchemists tried for centuries to turn base metals into precious metals through chemical reactions. Why were they doomed to fail?

2. The uranium sold by chemical supply companies is "spent uranium"—natural uranium from which the ^{235}U has been removed. There's no concern that this spent uranium will be used to make an atomic bomb. Why not?

3. Naturally occurring copper has two isotopes, ^{63}Cu and ^{65}Cu. What is different between atoms of these two isotopes?

4. Why is it extremely difficult to separate the two isotopes of copper, ^{63}Cu and ^{65}Cu?

5. If a nuclear reaction adds an extra neutron to the nucleus of ^{57}Fe (a stable isotope of iron), it produces ^{58}Fe (another stable isotope of iron). Will this change in the nucleus affect the number and arrangement of the electrons in the atom that's built around this nucleus? Why or why not?

6. If a nuclear reaction adds an extra proton to the nucleus of ^{58}Fe (a stable isotope of iron), it produces ^{59}Co (a stable isotope of cobalt). Will this change in the nucleus affect the number and arrangement of the electrons in the atom that's built around this nucleus? Why or why not?

7. When two medium-sized nuclei are stuck together during an experiment at a nuclear physics lab, the result is usually a large nucleus with too few neutrons to be stable. The nucleus soon falls apart. Why could more neutrons make it stable?

8. To stick the two nuclei together in Exercise 7, the nuclear physics laboratory had to accelerate them to enormous speeds and let them collide head on. Why was it so hard to get these nuclei together?

9. When a large nucleus is split in half during an experiment at a nuclear physics lab, the result is usually two medium-sized nuclei with too many neutrons to be stable. These nuclei eventually fall apart. Why don't these smaller nuclei need as many neutrons as they received from the original nucleus?

10. Why is it difficult or impossible to make very small atomic bombs?

11. When high explosives crush a sphere of plutonium so that it undergoes a chain reaction, the mass of the plutonium doesn't change, yet it exceeds "critical mass." Critical mass is clearly a misnomer. What *does* change about the plutonium as it's crushed that allows the chain reaction to proceed?

12. Suppose that there were a nonradioactive isotope that was fissionable and released several neutrons as it underwent fission. Building a nuclear weapon from this isotope would be extremely easy because it wouldn't require any explosives to detonate. Why wouldn't explosives be needed to detonate this weapon and how could it be detonated?

13. Explain the strategy of putting highly radioactive materials in a storage place for many years as a way to make them less hazardous. That wouldn't work for chemical poisons, so why does it work for radioactive materials?

14. Once fallout from a nuclear blast distributes radioactive isotopes over a region of land, why is it virtually impossible to separate many of those radioactive isotopes from the soil?

15. Burning chemical poisons in a gas flame often renders those poisons harmless. Why won't that strategy make radioactive materials less dangerous?

16. Sunscreen absorbs ultraviolet light while permitting visible light to pass. Why does this coating reduce the risk of chemical and genetic damage to the cells of your skin?

17. Why is it important to keep foods and drugs out of direct sunlight, even when they're not in danger of overheating?

18. Why are many drugs packaged in amber-colored containers that block ultraviolet light?

19. Museums often display priceless antique manuscripts under dim yellow light. Why not use white light?

20. The anode of an X-ray tube is made from atoms with relatively large nuclei and many electrons. Why would the X-ray tube have trouble creating X rays through the bremsstrahlung process if its anode were made of beryllium, a metallic element with an extremely small and light nucleus?

21. An X-ray tube with a beryllium anode (see Exercise 20) would also be unable to produce characteristic X rays via X-ray fluorescence. Since a beryllium atom has only four protons in its nucleus, it would emit only a visible or ultraviolet light photon when one of its outer electrons underwent a radiative transition to an empty inner orbital. Why must an X-ray machine's anode be made of atoms with many protons in their nuclei before the anode can emit X-ray photons through radiative transitions in those atoms?

22. An X-ray technician can adjust the energy of the X-ray photons produced by a machine by changing the voltage drop between the X-ray tube's cathode and anode. Explain.

23. Lead, with 82 electrons per atom, is an excellent absorber of X rays. Why?

24. Helium gas, which has atoms with only 2 electrons each, is almost completely transparent to X rays. In contrast, xenon gas, which consists of atoms with 54 electrons each, absorbs X rays reasonably well. Explain this difference.

25. The idea that electromagnetic waves behaved as particles during collisions was one of the surprising realizations of early quantum physics. It was strange enough that light appeared as photons or quanta of energy; it was even stranger that one of these photons had zero mass and yet carried momentum with it. However, that's what the 1922 experiments of American physicist Arthur Holly Compton (1892–1967) indicated. Why does the Compton effect, in which an X-ray photon scatters off a free electron and deflects the path of that electron, prove that the X-ray photon has momentum?

26. Electric charge is strictly conserved in our universe, meaning that the net charge of an isolated system can't change. Why doesn't the production of an electron–positron pair in a patient cause a change in the patient's net charge?

27. Which has more mass: a positron or an antiproton?

28. Scientists have always been curious about the nuclear reactions that occur deep inside the core of the sun. Unfortunately, the sun is so thick and opaque that no electromagnetic waves reach us directly from the core. However, we can observe neutrinos produced in the sun's core. Why are those neutrinos able to reach us when light can't?

29. Which subatomic particle isn't contained in an iron atom: a proton, a positron, an electron, or a neutron?

30. What is the electric charge of an antiproton?

31. Modern high-energy physics facilities study extremely high-energy subatomic particles. In designing the equipment for handling these particles, it's frequently safe to assume that the particles are traveling at almost exactly the speed of light. Given that the kinetic energies of these particles may change significantly during different experiments, why is this assumption about speed still reasonable?

32. Magnetic resonance imaging (MRI) differs from computed tomography imaging in that it involves no "ionizing radiation." What electromagnetic radiation is used in MRI and why aren't the photons of this radiation able to remove electrons from atoms and convert those atoms into ions?

33. Magnetic resonance imaging isn't good at detecting bone. Why not?

34. No magnetic metals such as iron or steel are permitted near a magnetic resonance imaging machine. In part, this rule is a safety precaution since those magnetic metals would be attracted toward the machine. But the magnetic fields from these magnetic metals would also spoil the imaging process. Why would having additional magnetic fields inside the imaging machine spoil its ability to locate specific protons inside a patient's body?

Cases

1. Carbon dating is a technique used to determine the ages of carbon-containing objects that obtain their carbon from contemporary plants. It's based on a radioactive isotope of carbon, ^{14}C (carbon–14), that's present in our atmosphere in small quantities.

a. The ^{14}C in our atmosphere is produced indirectly by cosmic rays—high-energy particles that enter our atmosphere from elsewhere in the universe. When these cosmic rays collide with nuclei in the atmosphere, they often produce neutrons, and some of these neutrons interact with the nuclei of nitrogen atoms. Why is it easy for a neutron to approach a ^{14}N nucleus in order to stick to it?

b. When it sticks to a ^{14}N nucleus, a neutron causes that nucleus to rearrange and release a proton. The result is a ^{14}C nucleus. Adding a neutron and subtracting a proton leaves the number of nucleons in the nucleus unchanged, but it converts one chemical element into another. Why is ^{14}C chemically different and distinguishable from ^{14}N?

*c. ^{14}C is radioactive with a half-life of 5730 years. That means there's a 50% chance of each ^{14}C nucleus spontaneously falling apart during a period of 5730 years. After 28,650 years, what fraction of the original ^{14}C nuclei in an object will remain?

d. From the 5730-year half-life of ^{14}C, it's possible to calculate the fraction of ^{14}C nuclei that will survive 1 year without decaying: 99.9879%. The fraction of nuclei that survive 2 years without decaying is 99.9879% of 99.9879%, or 99.9758%. And so on. If you know how many ^{14}C nuclei an object contained when it was new, how could you determine how old the object is?

e. Most carbon nuclei are not radioactive. The carbon in rocks and petroleum is about 99% ^{12}C and about 1% ^{13}C. But because ^{14}C is introduced into the atmosphere at a steady rate and plants obtain their carbon from the atmosphere, ^{14}C nuclei make up a small but important fraction of the carbon nuclei in a living plant. This fraction can be measured in a modern plant to determine its value in an ancient plant while that plant was alive. After the plant died, the fraction of carbon nuclei that were ^{14}C decreased as the ^{14}C nuclei decayed. How does measuring the fraction of ^{14}C nuclei in a plant's carbon allow scientists to determine how long ago it lived?

f. Scientists can also determine how long ago an animal lived if that animal ate plants to obtain carbon. But if the animal obtained a substantial fraction of its carbon from minerals, that dating process wouldn't work. Why not?

2. Radon is a colorless, odorless gas that's extremely radioactive. It has three naturally occurring isotopes—^{219}Rn, ^{220}Rn, and ^{222}Rn—that are made when radioactive elements in the earth's crust decay. While the nuclei of these isotopes each contain 86 protons, they contain different numbers of neutrons. Radon is thought to be the single most important cause of lung cancer in nonsmokers. It's particularly abundant in indoor air above geological formations that include radon-producing minerals.

a. ^{219}Rn, ^{220}Rn, and ^{222}Rn nuclei are found in the atoms of chemically inert gases. In fact, radon is one of the noble gas elements, which include helium, neon, argon, krypton, and xenon. It's not surprising that the atoms of all three isotopes are gases. It would be much more surprising if one isotope were normally a solid, one a liquid, and the third a gas. Why?

*b. The first isotope, ^{219}Rn, is produced when the radioactive element actinium (^{227}Ac) decays. The half-life of ^{219}Rn is extremely short, only 3.92 s. If there were 800 of these ^{219}Rn nuclei in a box, how long would it be before only about 25 of them were left?

c. The second isotope of radon, ^{220}Rn, is produced when the radioactive element thorium (^{232}Th) decays. This isotope has a half-life of 54.5 s and decays into yet another radioactive nucleus. The ensuing series of radioactive decays usually produces a stable isotope of lead (^{208}Pb). In the long series of unstable nuclei that appear briefly as thorium turns into lead, the only nucleus that forms a gaseous atom is ^{220}Rn. From that information, why can you safely assume that very little ^{220}Rn is found in building air, even if it's built above rocks that are rich in thorium?

d. The third radon isotope, ^{222}Rn, is produced when uranium (^{238}U) decays. This isotope has a half-life of 3.8 days. Why is this isotope much more likely than the others to be found in building air, particularly if that building is built above uranium-containing rocks?

3. There are two types of smoke detectors in common use: photoelectric and ionization. A photoelectric smoke detector uses a beam of light to "see" the smoke particles. Unfortunately, it isn't sensitive to the smallest smoke particles and requires considerable electric power to operate. An ionization smoke detector is more sensitive and consumes less power. It uses a radioactive material to ionize the air so that a small amount of electric current can pass through that air. When smoke is present, the ionized air molecules stick to the smoke particles, less current flows through the air, and the detector's alarm goes off. The radioactive material in an ionization smoke detector is an isotope of americium, a radioactive element that's not found in nature. The specific isotope of americium is ^{241}Am. This isotope is made by adding neutrons to ^{238}U in a nuclear reactor. In fact, ^{241}Am is made in exactly the same way that ^{239}Pu is made, except that ^{241}Am has been exposed to a few more neutrons.

a. ^{241}Am has a half-life of 458 years. What would happen if an ionization smoke detector used a radioactive isotope with a half-life of 458 days instead?

b. Many ^{241}Am nuclei were made during the supernova explosion that produced the earth's uranium. Why aren't there any naturally occurring ^{241}Am nuclei around today?

c. When uranium nuclei undergo spontaneous fission in the earth's crust, their fragments are often radioactive nuclei. Why aren't any of these fragments ^{241}Am nuclei?

d. An ^{241}Am nucleus decays into a ^{239}Np nucleus, an isotope of neptunium with a half-life of 2 million years. Both americium and neptunium are radioactive solids. Why is it important that the radioactive isotope used in the smoke detector not decay into a radioactive gas such as radon?

4. Physicists use accelerators to study the smallest objects in the universe: subatomic particles. They need these accelerators because enormous energies are required to resolve extremely fine details. Since a particle's wavelength decreases as its energy increases, higher energy particles are able to resolve tinier features.

a. In an electron accelerator, electrons pass through a long chain of resonant cavities, powered by strong microwave generators. The electric fields in these cavities push the electrons forward. Why must the electrons be bunched together in packets rather than sent into the cavities continuously?

b. A packet of electrons takes the same amount of time to pass through each cavity of the chain. The first few cavities of the chain are shorter than the rest. Why?

c. The remaining cavities of the chain are essentially equal in length. Why don't the cavities continue to get longer?

d. Some accelerators also work with positrons. They inject packets of positrons into the chain of cavities so that each positron packet is exactly one cavity behind an electron packet. Explain why this arrangement leads to the acceleration of both the electrons and the positrons.

e. When the packets of electrons and positrons emerge from the chain of cavities, they pass through a magnetic field. The electrons turn one way while the positrons turn the other way. Explain why these two particles turn in different directions.

f. The electrons and positrons are steered with magnets until they collide with one another head on. When they hit, an electron and positron annihilate each other in a burst of energy that makes new particles from empty space. Why is it that the more energy the electron and positron have, the more mass the particles they create can have?

5. Police forensic laboratories often use the neutron activation technique to look for trace elements on objects left at a crime scene or belonging to a suspect. In this technique, the objects are made temporarily radioactive by exposing them to neutrons. Instruments then study the radioactive decays of the activated atoms to determine what atoms they were.

a. Neutron activation is done most easily at a nuclear reactor facility. Why can't a lamp or an X-ray tube be used to add neutrons to an object's nuclei?

b. A neutron can attach itself to almost any nucleus in an object, even if the neutron is only moving with thermal energy. Why doesn't the nucleus push the neutron away before the neutron makes contact with it?

c. Once the neutron has attached itself to a nucleus, the nucleus may become radioactive. Detectors studying an object that has been exposed to neutrons detect electrons and gamma rays emerging from that object. Describe one likely mechanism by which neutron-activated nuclei may produce these electrons and gamma rays.

d. By measuring the energies of the gamma rays emitted by an object following neutron activation, the detectors can determine what elements are present in that object, even if those elements are there in extremely small quantities. For example, they can determine whether trace amounts of barium, residue of the primer in a bullet cartridge, are present on a suspect's glove. Why are the gamma rays emitted by a neutron-activated barium nucleus different from those emitted by a neutron-activated iron nucleus?

e. While neutron activation studies can tell what elements are present in an object, they can't tell what chemicals those elements are part of. Why can't neutron activation tell whether the barium atoms it detects are present as barium metal or as a barium compound such as barium chloride?

f. How can neutron activation techniques be used to detect an art forgery made with pigments that were not known to the painting's alleged artist?

g. An object is radioactive following neutron activation, but that radioactivity fades quickly. The object is normally safe in a few weeks or months; you just have to wait. Why won't heating or chemically treating the object speed the decrease in radioactivity?

6. Magnetic resonance imaging (MRI) is an outgrowth of a technique used by scientists to identify the chemical and magnetic environments of nuclei—usually hydrogen nuclei, which are individual protons. The original technique is called nuclear magnetic resonance, or NMR. When nuclear magnetic resonance was used in imaging, the word "nuclear" was dropped from its title because that word made the imaging technique sound far more dangerous than it actually is.

a. When a sample of a chemical containing hydrogen atoms is placed in a magnetic field, it will absorb only specific electromagnetic waves. How are these electromagnetic waves affecting the protons in the sample when the sample is absorbing them?

b. If the magnetic field in which the sample resides is increased, the frequency of the electromagnetic waves must also be increased or the sample will stop absorbing them. Why?

c. The electrons in the sample itself affect the magnetic field that the protons experience. In and around each proton are a number of electrons, which can move somewhat in response to electric fields. When the sample is placed in a magnetic field, the field induces a response that is analogous to a current in the sample, even if the sample isn't actually metallic. This current partially cancels the magnetic field that the proton experiences so that it absorbs electromagnetic waves at a lower frequency than it would absorb if it had no nearby electrons. Why does the "induced" current tend to cancel the magnetic field being applied to the sample rather than strengthen it?

d. The frequency of electromagnetic waves that a particular proton will absorb also depends on any nearby protons and whether they are spin-up or spin-down. How are these nearby protons able to change this frequency?

Appendix A

Vectors

Many of the quantities of physics are vectors, meaning that they have both magnitudes (amounts) and directions. Among these vector quantities are position, velocity, acceleration, force, torque, momentum, angular momentum, and electric and magnetic fields. Of these vector quantities, position is probably the easiest to visualize: you specify an object's position by giving its position vector—its distance and direction from a reference point. For example, you can specify the library's position with respect to your home by giving both its distance from your home (say, 3.162 km or 1.965 miles) and its direction from your home (18.43° east of due north). That position vector is all the information someone would need to travel from your home to the library.

In illustrations, such as Fig. A.1, vector quantities are drawn as arrows. The length of each arrow indicates the vector's magnitude, while the direction of the arrow indicates which way the vector points. Suppose that you live in a city with major east–west and north–south streets spaced 1 km apart. Figure A.1 shows four aerial views of your city. The vector **A** in Fig. A.1a shows the position of the library with respect to your home. It begins at your home and ends at the library, thus indicating both the magnitude and direction of the library's position.

Let's look at another position vector. The vector **B** in Fig. A.1b begins at the library and ends at your friend's home. This position vector shows the position of your friend's home with respect to the library and is 2.828 km long and points 45° east of south. If you happen to be at the library, you could use this vector to find your friend's home.

But how can you go from your home to your friend's home? To make this trip, you must add two vectors: you first follow vector **A** from your home to

the library and then follow vector **B** from the library to your friend's home. This combined trip is shown as the upper path in Fig. A.1c. But you could also go

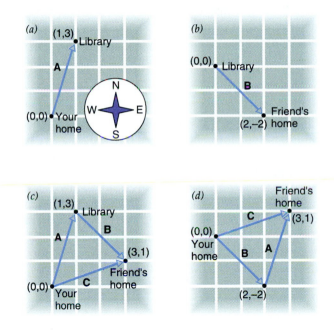

Fig. A.1 Four aerial views of your city, showing the major north–south and east–west streets that are spaced 1 km apart. (*a*) To go from your home to the library, you must travel 3.162 km at a direction 18.43° east of north—vector **A**. (*b*) To go from the library to your friend's home, you must travel 2.828 km at a direction 45° east of south—vector **B**. (*c*) You can go from your home to your friend's home either by going first to the library along vector **A** and then going to your friend's home along vector **B**, or you can travel directly along vector **C**, which is the sum of vectors **A** and **B**. (*d*) You can also reach your friend's home by going first along vector **B** and then along vector **A**. The sum of these two vectors is still vector **C**. However, you will not visit the library on that trip.

directly from your home to your friend's home by following a new vector in Fig. A.1c—vector **C**. This vector from your home to your friend's home is the sum of vectors **A** and **B** and is 3.162 km long and points 18.43° north of east. Using bold letters to indicate that **A, B,** and **C** are vectors, we can write **A** + **B** = **C,** meaning that vector **C** is the sum of vectors **A** and **B**.

Another interesting path from your home to your friend's home is shown in Fig. A.1d: you first travel along vector **B** and then along vector **A**. On this path, you would arrive at your friend's home without visiting the library. The first leg of this journey would take you into new territory, but the second leg would leave you at your friend's home. The sum of vectors **B** and **A** is still vector **C**, or **B** + **A** = **C**. Thus vectors added in any order yield the same sum.

While you can estimate the sum of two vectors by drawing arrows on a sheet of paper, obtaining an accurate sum requires some thought. Adding their magnitudes is unlikely to give you the magnitude of the sum vector, and adding their directions doesn't even make sense. To add two vectors, it helps to specify them in another form: as their *components* along two or three directions that are at right angles with respect to one another. In the present example of travel in a city, two right-angle directions are all we need. If height were also important, we'd need three right-angle directions.

For the two right-angle directions we need in the city, let's choose east and north. We can then specify vector **A** as its component toward the east and its component toward the north. Its east component is 1 km and its north component is 3 km, so vector **A,** the position of the library with respect to your home, is 1 km to the east and 3 km to the north.

This new form for vector **A,** a pair of distances, is often more convenient than the old form, a distance and a direction. If you go to the library by walking 3.162 km in the direction 18.43° east of north, you'll have to pass through a lot of other buildings and backyards. Walking to the library by heading 1 km east and 3 km north allows you to travel on the sidewalks.

If we designate the position of your home as 0 km east and 0 km north, then the position of the library is 1 km east and 3 km north. These positions are labeled in Fig. A.1a as (0, 0) and (1, 3), respectively. The first number in the parentheses is the distance east, measured in kilometers, and the second number is the distance north, also in kilometers.

To go from the library to your friend's home, you must go 2 km east and 2 km south. That's the new form of vector **B**. Because a southward position has a negative component along the northward direction, the position of your friend's home with respect to the library is 2 km east and −2 km north. In Fig. A.1b, the library is at (0, 0) and your friend's home is at (2, −2).

Now adding these vectors **A** and **B** is relatively easy. To go from your home to your friend's home by way of the library, you must move east first 1 km and then 2 km for a total of 3 km, and you must move north 3 km and then −2 km for a total of 1 km. Thus the position of your friend's home with respect to your home, vector **C**, is 3 km east and 1 km north. In Fig. A.1c, your home is at (0, 0) and your friend's home is at (3, 1).

Similarly, you could go from your home to your friend's home by following first vector **B** and then vector **A,** as shown in Fig. A.1d. This trip would take you through unknown regions, but you'd still arrive at the right place. You would move east first 2 km and then 1 km, and you'd move north first −2 km and then 3 km. In the end, you'd be 3 km east and 1 km north of your home, the position of your friend's home. Thus the sum of vectors **B** and **A** is still vector **C**.

As you can see, vectors are useful for specifying physical quantities in our three-dimensional world. When you encounter vectors, remember that their directions are just as important as their magnitudes and that these directions must be taken into account when you add two vectors together.

Appendix B

Units, Conversion of Units

When you return from a camping trip and begin to describe it to your friends, there are a number of physical quantities that you may find useful in your conversation. Distance, weight, temperature, and time are as important in everyday life as they are in a laboratory. And when you explain to your friends how far you hiked, how much your backpack weighed, how cold the weather was, and how long the trip took, you'll have to relate those quantities to standard units or your friends won't appreciate just how difficult the trip was.

Most physical quantities aren't simple numbers like 7 or 2.9. Instead, they have units like length or time and are specified in multiples of widely accepted standard units such as meters or seconds. When you say a door is 3.0 meters tall, you're comparing the door's height to the meter, a widely accepted standard unit of length. With this comparison, most people can determine just how tall the door is, even though they've never seen it.

But the meter isn't familiar to everyone; many people prefer to measure length in multiples of another standard unit—the foot. These people might be more comfortable hearing that the door is 9.8 feet tall. These quantities, 3.0 meters and 9.8 feet, are the same length.

Determining the door's height in feet doesn't require a new measurement because we can convert the one in meters to one that's in feet. To perform this conversion, we need to know how to express one particular length in both units. Any length will do. For example, Table B.1 states that 1 foot is the same length as 0.30480 meter. Because of that equality, we know that the following equation is true:

$$\frac{1 \text{ foot}}{0.30480 \text{ meter}} = 1.$$

We can multiply 3.0 meters, the height of the door, by this version of 1 and obtain the door's height in feet:

$$3.0 \text{ meters} \cdot \frac{1 \text{ foot}}{0.30480 \text{ meter}} = 9.8425 \text{ feet}.$$

Notice that the original units, meters, cancel and are replaced by the new units, feet. Since we only knew the door's height to two digits of precision in meters, we can't report the door's height to any more than two digits of precision in feet. So we round the result to 9.8 feet.

You can change the units of almost any physical quantity by multiplying that quantity by a version of 1. You should form this 1 by dividing new units by old units, where the number of new units in the numerator is equivalent to the number of old units in the denominator. You can obtain these pairs of equivalent quantities from Table B.1. When you do this multiplication, the old units will cancel and you'll be left with the physical quantity expressed in the new units.

One last note about units: when you use physical quantities in a calculation, make sure that you keep the units throughout the process. They're as important in the calculation as they are anywhere else. Some of the units may cancel, but in all likelihood the result of the calculation will have some units left and these units must be appropriate to the type of result you expect. If you're expecting a length, the units of your result should be meters or feet or another standard unit of length. If the units you obtain are seconds or kilograms, you've made a mistake in the calculation.

| Table B.1 | **Conversion of Units**[a]

1. Acceleration (SI unit: 1 meter/second^2 or 1 m/s^2)

$$1 \text{ foot/second}^2 = 0.30480 \text{ m/s}^2$$

2. Angle (SI unit: 1 radian)

$$1 \text{ degree } (1°) = 0.017453 \text{ radian}$$

3. Area (SI unit: 1 meter^2 or 1 m^2)

$$1 \text{ foot}^2 = 0.092903 \text{ m}^2$$
$$1 \text{ inch}^2 = 6.4516 \times 10^{-4} \text{ m}^2$$

4. Density (SI unit: 1 kilogram/meter^3 or 1 kg/m^3)

$$1 \text{ pound/foot}^3 = 16.018 \text{ kg/m}^3$$

5. Energy (SI unit: 1 joule or 1 J)

$$1 \text{ Btu} = 1054.7 \text{ J}$$
$$1 \text{ calorie, thermochemical} = 4.1840 \text{ J}$$
$$1 \text{ electron-volt } (1 \text{ eV}) = 1.6022 \times 10^{-19} \text{ J}$$
$$1 \text{ foot-pound} = 1.3558 \text{ J}$$
$$1 \text{ kilocalorie (food Calorie)} = 4184.0 \text{ J}$$
$$1 \text{ kilowatt-hour} = 3,600,000 \text{ J}$$

6. Force (SI unit: 1 newton or 1 N)

$$1 \text{ pound} = 4.4482 \text{ N}$$

7. Length (SI unit: 1 meter or 1 m)

$$1 \text{ angstrom } (1 \text{ Å}) = 10^{-10} \text{ m}$$
$$1 \text{ centimeter } (1 \text{ cm}) = 0.01 \text{ m}$$
$$1 \text{ fermi } (1 \text{ fm}) = 10^{-15} \text{ m}$$
$$1 \text{ foot} = 0.30480 \text{ m}$$
$$1 \text{ inch} = 0.02540 \text{ m}$$
$$1 \text{ kilometer } (1 \text{ km}) = 1000 \text{ m}$$
$$1 \text{ light year} = 9.4606 \times 10^{15} \text{ m}$$
$$1 \text{ micron } (1 \text{ } \mu\text{m}) = 10^{-6} \text{ m}$$
$$1 \text{ mil} = 2.5400 \times 10^{-5} \text{ m}$$
$$1 \text{ mile} = 1609.3 \text{ m}$$
$$1 \text{ millimeter } (1 \text{ mm}) = 0.001 \text{ m}$$
$$1 \text{ nanometer } (1 \text{ nm}) = 10^{-9} \text{ m}$$
$$1 \text{ picometer } (1 \text{ pm}) = 10^{-12} \text{ m}$$

8. Mass (SI unit: 1 kilogram or 1 kg)

$$1 \text{ gram } (1 \text{ g}) = 0.001 \text{ kg}$$
$$1 \text{ metric ton} = 1000 \text{ kg}$$
$$1 \text{ pound (at the earth's surface)} = 0.45359 \text{ kg}$$
$$1 \text{ slug} = 14.594 \text{ kg}$$

9. Power (SI unit: 1 watt or 1 W)

$$1 \text{ Btu/hour} = 0.29307 \text{ W}$$
$$1 \text{ horsepower} = 745.70 \text{ W}$$

10. Pressure (SI unit: 1 pascal or 1 Pa)

$$1 \text{ atmosphere} = 101,325 \text{ Pa}$$
$$1 \text{ millimeter of mercury (1 torr)} = 133.32 \text{ Pa}$$
$$1 \text{ pound/inch}^2 \text{ (1 psi)} = 6894.8 \text{ Pa}$$

11. Temperature (SI units: degrees Celsius or °C; kelvin units or K)

Because temperature in the three common units, °C, K, and °F, aren't multiples of one another, you must use special formulas to convert between them:

$$\text{Temperature in °C} = \tfrac{5}{9} \cdot (\text{Temperature in °F} - 32)$$
$$\text{Temperature in °C} = \text{Temperature in K} - 273.15$$
$$\text{Temperature in K} = \text{Temperature in °C} + 273.15$$

12. Time (SI unit: 1 second or 1 s)

$$1 \text{ day} = 86,400 \text{ s}$$
$$1 \text{ femtosecond (1 fs)} = 10^{-15} \text{ s}$$
$$1 \text{ hour} = 3600 \text{ s}$$
$$1 \text{ microsecond } (1 \text{ } \mu\text{s}) = 10^{-6} \text{ s}$$
$$1 \text{ millisecond (1 ms)} = 0.001 \text{ s}$$
$$1 \text{ minute} = 60 \text{ s}$$
$$1 \text{ nanosecond (1 ns)} = 10^{-9} \text{ s}$$
$$1 \text{ picosecond (1 ps)} = 10^{-12} \text{ s}$$

13. Torque (SI unit: 1 newton-meter or 1 N·m)

$$1 \text{ foot-inch} = 0.11298 \text{ N·m}$$
$$1 \text{ foot-pound} = 1.3558 \text{ N·m}$$

14. Velocity (SI unit: 1 meter/second or 1 m/s)

$$1 \text{ foot/second} = 0.30480 \text{ m/s}$$
$$1 \text{ kilometer/hour (1 km/h)} = 0.27778 \text{ m/s}$$
$$1 \text{ knot} = 0.51444 \text{ m/s}$$
$$1 \text{ mile/hour (1 mph)} = 0.44704 \text{ m/s}$$

15. Volume (SI unit: 1 meter^3 or 1 m^3)

$$1 \text{ cup} = 2.3659 \times 10^{-4} \text{ m}^3$$
$$1 \text{ fluid ounce} = 2.9574 \times 10^{-5} \text{ m}^3$$
$$1 \text{ foot}^3 = 0.028317 \text{ m}^3$$
$$1 \text{ gallon} = 0.0037854 \text{ m}^3$$
$$1 \text{ liter (1 L)} = 0.001 \text{ m}^3$$
$$1 \text{ milliliter (1 mL)} = 10^{-6} \text{ m}^3$$
$$1 \text{ quart} = 0.00094635 \text{ m}^3$$

[a] This table lists pairs of equivalent quantities, one in SI units and one in other units. Each pair can be used to convert measurements expressed as multiples of one unit into measurements expressed as multiples of the other unit. These pairs are grouped according to physical quantity and are given to a precision of five digits.

Appendix C

Formulas of Physics

1. Newton's Second Law of Motion (p. 8): An object's acceleration is equal to the force exerted on that object divided by the object's mass, or

$$\text{force} = \text{mass} \cdot \text{acceleration}. \quad \textbf{(1.1.1)}$$

2. Relationship Between Mass and Weight (p. 12): An object's weight is equal to its mass times the acceleration due to gravity, or

$$\text{weight} = \text{mass} \cdot \text{acceleration due to gravity}. \quad \textbf{(1.2.1)}$$

3. The Velocity of an Object Experiencing Constant Acceleration (p. 13): The object's present velocity differs from its initial velocity by the product of its acceleration times the time since it was at that initial velocity, or

$$\text{present velocity} = \text{initial velocity} + \text{acceleration} \cdot \text{time}. \quad \textbf{(1.2.2)}$$

4. The Position of an Object Experiencing Constant Acceleration (p. 15): The object's present position differs from its initial position by the product of its average velocity since it was at that initial position times the time since it was at that initial position, or

$$\text{present position} = \text{initial position} + \text{initial velocity} \cdot \text{time} + \tfrac{1}{2} \cdot \text{acceleration} \cdot \text{time}^2. \quad \textbf{(1.2.3)}$$

5. The Definition of Work (p. 25): The work done on an object is equal to the product of the force exerted on that object times the distance that object travels in the direction of the force, or

$$\text{work} = \text{force} \cdot \text{distance}. \quad \textbf{(1.3.1)}$$

6. Gravitational Potential Energy (p. 26): An object's gravitational potential energy is its mass times the acceleration due to gravity times its height above a zero level, or

$$\text{gravitational potential energy} = \text{mass} \cdot \text{acceleration due to gravity} \cdot \text{height}. \quad \textbf{(1.3.2)}$$

7. Newton's Second Law of Rotational Motion (p. 47): The torque exerted on an object is equal to the product of that object's moment of inertia times its angular acceleration, or

$$\text{torque} = \text{moment of inertia} \cdot \text{angular acceleration}. \quad \textbf{(2.1.1)}$$

The angular acceleration points in the same direction as the torque. This law doesn't apply to objects that are wobbling.

8. Relationship Between Force and Torque (p. 49): The torque produced by a force is equal to the product of the lever arm times that force, or

$$\text{torque} = \text{lever arm} \cdot \text{force}, \quad \textbf{(2.1.2)}$$

where we include only the component of the force that's at right angles to the lever arm.

9. Linear Momentum (p. 60): An object's linear momentum is its mass times its velocity, or

$$\text{linear momentum} = \text{mass} \cdot \text{velocity}. \quad \textbf{(2.3.1)}$$

10. The Definition of Impulse (p. 61): The impulse given to an object is equal to the product of the force exerted on that object times the length of time that force is exerted, or

$$\text{impulse} = \text{force} \cdot \text{time}. \quad \textbf{(2.3.2)}$$

11. Angular Momentum (p. 63): An object's angular momentum is its moment of inertia times its angular velocity, or

$$\text{angular momentum} = \text{moment of inertia} \cdot \text{angular velocity}. \quad \textbf{(2.3.4)}$$

12. The Definition of Angular Impulse (p. 64): The angular impulse given to an object is equal to the product of the torque exerted on that object times the length of time that torque is exerted, or

$$\text{angular impulse} = \text{torque} \cdot \text{time}. \quad \textbf{(2.3.5)}$$

13. Kinetic Energy (p. 66): An object's translational kinetic energy is one-half of its mass times the square of its speed, or

$$\text{kinetic energy} = \tfrac{1}{2} \cdot \text{mass} \cdot \text{speed}^2. \quad \textbf{(2.3.7)}$$

An object's rotational kinetic energy is one-half of its moment of inertia times the square of its angular speed, or

$$\text{kinetic energy} =$$
$$\tfrac{1}{2} \cdot \text{moment of inertia} \cdot \text{angular speed}^2. \quad \textbf{(2.3.8)}$$

14. Hooke's Law (p. 81): The restoring force exerted by an elastic object is proportional to how far it is from its equilibrium shape, or

$$\text{restoring force} = -\text{spring constant} \cdot \text{distortion}. \quad \textbf{(3.1.1)}$$

15. Acceleration of an Object in Uniform Circular Motion (p. 95): An object in uniform circular motion has an acceleration equal to the square of its speed divided by the radius of its circular trajectory, or

$$\text{acceleration} = \frac{\text{speed}^2}{\text{radius of circular trajectory}}. \quad \textbf{(3.3.1)}$$

16. The Ideal Gas Law (p. 128): The pressure of a gas is equal to the product of the Boltzmann constant times the particle density times the absolute temperature, or

$$\text{pressure} = \text{Boltzmann constant} \cdot \text{particle density} \cdot \text{absolute temperature}. \quad \textbf{(4.1.3)}$$

17. Bernoulli's Equation (p. 136): For an incompressible fluid in steady-state flow, the sum of its pressure potential energy, its kinetic energy, and its gravitational potential energy is constant along a streamline, or

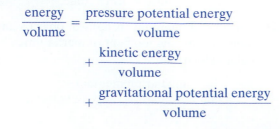

$$\frac{\text{energy}}{\text{volume}} = \frac{\text{pressure potential energy}}{\text{volume}}$$
$$+ \frac{\text{kinetic energy}}{\text{volume}}$$
$$+ \frac{\text{gravitational potential energy}}{\text{volume}}$$

$$= \text{pressure} + \tfrac{1}{2} \cdot \text{density} \cdot \text{speed}^2$$
$$+ \text{density} \cdot \text{acceleration due to}$$
$$\text{gravity} \cdot \text{height}$$
$$= \text{constant} \ (\textit{along a streamline}). \quad \textbf{(4.2.4)}$$

18. Poiseuille's law (p. 150): The volume of fluid flowing through a pipe each second is equal to $(\pi/128)$ times the pressure difference across that pipe times the pipe's diameter to the fourth power, divided by the pipe's length times the fluid's viscosity, or

$$\text{volume flow rate} =$$
$$\frac{\pi \cdot \text{pressure difference} \cdot \text{pipe diameter}^4}{128 \cdot \text{pipe length} \cdot \text{fluid viscosity}}. \quad \textbf{(5.1.1)}$$

19. The Law of Universal Gravitation (p. 176): Every object in the universe attracts every other object in the universe with a force equal to the gravitational constant times the product of the two masses, divided by the square of the distance separating the two objects, or

$$\text{force} = \frac{\text{gravitational constant} \cdot \text{mass}_1 \cdot \text{mass}_2}{(\text{distance between masses})^2}. \quad \textbf{(5.3.1)}$$

20. The Stefan–Boltzmann Law (p. 201): The power an object radiates is proportional to the product of its emissivity times the fourth power of its temperature times its surface area, or

$$\text{radiated power} =$$
$$\text{emissivity} \cdot \text{Stefan-Boltzmann constant} \cdot \text{temperature}^4 \cdot \text{surface area}. \quad \textbf{(6.2.1)}$$

21. Coulomb's Law (p. 269): The magnitudes of the electrostatic forces between two objects are equal to the Coulomb constant times the product of their two electric charges divided by the square of the distance separating them, or

$$\text{force} = \frac{\text{Coulomb constant} \cdot \text{charge}_1 \cdot \text{charge}_2}{(\text{distance between charges})^2}. \quad \textbf{(8.1.1)}$$

If the charges are like, then the forces are repulsive. If the charges are opposite, then the forces are attractive.

22. Power Consumption (p. 302): The electric power consumed by a device is the voltage drop across it times the current flowing through it, or

$$\text{power consumed} = \text{voltage drop} \cdot \text{current}. \quad \textbf{(9.1.1)}$$

23. Power Production (p. 303): The electric power provided by a device is the voltage rise across it times the current flowing through it, or

power provided = voltage rise · current.   **(9.1.2)**

24. Ohm's Law (p. 306): The voltage drop across a wire is equal to the current flowing through that wire times the wire's electric resistance, or

voltage drop = current · electric resistance.   **(9.2.1)**

25. Relationship Between Wavelength and Frequency (p. 371): The product of the frequency of an electromagnetic wave times its wavelength is the speed of light, or

speed of light = wavelength · frequency. **(11.3.1)**

26. Relationship Between Energy and Frequency (p. 397): The energy in a photon of light is equal to the product of the Planck constant times the frequency of the light wave, or

energy = Planck constant · frequency. **(12.2.1)**

27. The Lens Equation (p. 419): One divided by the focal length of a lens is equal to the sum of one divided by the object distance and one divided by the image distance, or

$$\frac{1}{\text{focal length}} = \frac{1}{\text{object distance}} + \frac{1}{\text{image distance}}.$$

(13.1.1)

Glossary

absolute temperature scale A scale for measuring temperature in which 0 K corresponds to absolute zero.

absolute zero The temperature at which all thermal energy has been removed from an object or system of objects. Because it's impossible to find and remove all the thermal energy from an object, absolute zero can be approached but is not actually attainable.

acceleration A vector quantity that measures how quickly an object's velocity is changing: the greater the acceleration, the more the object's velocity changes each second. It consists of both the amount of acceleration and the direction in which the object is accelerating. This direction is identical to the direction of the force causing the acceleration. The SI unit of acceleration is the meter-per-second^2.

acceleration due to gravity A physical constant that specifies how quickly a freely falling object accelerates and also relates an object's weight to its mass. At the earth's surface, the acceleration due to gravity is 9.8 m/s^2 (or 9.8 N/kg).

activation energy The energy required to initiate a chemical reaction. This energy serves to break or weaken the bonds in the starting chemicals so that the reaction can proceed to form the reaction products.

adverse pressure gradient A region of fluid flow in which the fluid must flow toward higher pressure. The fluid's momentum and kinetic energy carry it through this situation, although the fluid does slow down.

aerodynamic forces The forces exerted on an object by the motion of the air surrounding it. The two types of aerodynamic forces are lift and drag.

aerodynamics The study of the dynamic (moving) interactions of air with objects.

airfoil An aerodynamically engineered surface, designed to obtain particular lift and drag forces from the air flowing around it.

alternating current An electric current that periodically reverses its direction of flow.

ampere (A) The SI unit of electric current (identical to the coulomb-per-second). One ampere is defined as the passage of 6.25×10^{18} charged particles per second and is roughly the current flowing through a 100-W light bulb operating on household electric power.

amplifier A device that replicates an input signal as a larger output signal.

amplitude The maximal displacement of a particular oscillator away from its equilibrium position.

amplitude modulation A technique for representing sound or data by changing the amplitude (strength) of a radio transmission.

analog The representation of numbers as values of physical quantities such as voltage, charge, or pressure.

angular acceleration A vector quantity that measures how quickly an object's angular velocity is changing: the greater the angular acceleration, the more the object's angular velocity changes each second. It consists of both the amount of angular acceleration and the direction about which the angular acceleration occurs. This direction is identical to the direction of the torque causing the angular acceleration.

angular impulse The mechanical means for transferring angular momentum. One object gives an angular impulse to a second object by exerting a certain torque on the second object for a certain amount of time. In return, the second object gives an equal but oppositely directed angular impulse to the first object.

angular momentum A conserved vector quantity that measures an object's rotational motion. It is the product of that object's moment of inertia times its angular velocity.

angular position A quantity that describes an object's orientation relative to some reference orientation.

angular speed A measure of the amount an object rotates in a certain amount of time.

angular velocity A vector quantity that measures how rapidly an object's angular position is changing: the greater the angular velocity, the farther the object turns each second. It consists of both the object's angular speed and the direction about which the object is rotating. This direction points along the axis of rotation in the direction established by the right-hand rule.

anharmonic oscillator An oscillator in which the restoring force on an object is not proportional to its displacement from a stable equilibrium. The period of an anharmonic oscillator depends on the amplitude of its motion.

antiferromagnetic Composed of magnetic atoms that alternate back and forth in magnetic orientation so that their magnetic fields cancel.

antimatter Matter resembling normal matter, but with many of its characteristics such as electric charge reversed.

aperture The diameter or effective diameter of a lens or opening.

apparent weight The sum of the gravitational and fictitious forces on an object. All three quantities are vectors, so that apparent weight can be quite large if the two forces point in the same direction or quite small if they point in opposite directions.

Archimedes' principle The observation that an object partially or wholly immersed in a fluid is acted upon by an upward buoyant force equal to the weight of the fluid it's displacing.

atmospheric pressure The pressure of air in the earth's atmosphere. Atmospheric pressure reaches a maximum of about 100,000 Pa near sea level and diminishes with increasing altitude.

atom The smallest portion of a chemical element that retains the chemical properties of that element.

audio signal An electric representation of sound in which a current is proportional to the shift in air pressure.

axis of rotation The straight line in space about which an object or group of objects rotates. More specifically, the axis of rotation points in a particular direction along that line to reflect the sense of rotation according to the right-hand rule.

band A group of levels that involve similar paths through a solid and thus involve similar energies.

bandwidth The range of frequencies involved in a group of electromagnetic waves.

base of support A surface outlined on the ground by the points at which an object is supported.

Bernoulli's equation An equation relating the total energy of an incompressible fluid in steady-state flow to the sum of its pressure potential energy, kinetic energy, and gravitational potential energy.

binary The digital representation of numbers in terms of the powers of two. The number 6 is represented in binary as 110, meaning 1 four (2^2), 1 two (2^1), and 0 ones (2^0).

bit A single binary value, either a 0 or a 1.

black body spectrum The distribution of thermal electromagnetic radiation emitted by a black object. This distribution is the amount of radiation emitted at each wavelength and depends only on the temperature of the black object.

Boltzmann constant The constant of proportionality relating a gas's pressure to its particle density and temperature. It has a measured value of 1.381×10^{-23} Pa·m³/(particle·K).

boundary layer A thin region of fluid near a surface that, because of viscous drag, is not moving at the full speed of the surrounding airflow.

bremsstrahlung The process in which a rapidly accelerating charge emits electromagnetic radiation, usually an X-ray or gamma-ray photon.

buoyant force The upward force exerted by a fluid on an object immersed in that fluid. The buoyant force is actually caused by pressure from the fluid. That pressure is highest below the object so the force exerted upward on the object's bottom is greater than the force exerted downward on the object's top.

byte Eight binary bits that collectively can represent a number from 0 to 255. Bytes are often used to represent letters and other characters, where a convention associates each character with a specific number.

calibration The process of comparing a local reference object to a generally accepted standard.

capacitance The amount of separated charge on the plates of a capacitor divided by the voltage difference across those plates. Capacitance is measured in farads.

capacitor An electronic component that stores separated electric charge on a pair of plates that are separated by an insulating layer.

Celsius A temperature scale in which 0 °C is defined as the melting point of water and 100 °C is defined as the boiling point of water at sea level. Absolute zero is −273.15 °C.

center of gravity The unique point about which all of an object's weight is evenly distributed and therefore balanced. Because weight is proportional to mass, the center of mass is identical to the center of gravity for objects that are much smaller than the earth. For larger objects, the centers of mass and gravity differ slightly. An object suspended from its center of gravity will balance and will experience no net torque due to gravity. In many situations, you can accurately predict an object's behavior by assuming that all of the object's weight acts at its center of gravity.

center of mass The unique point about which all of the object's mass is balanced. The center of mass is the natural pivot point for a free object. In the absence of outside forces or torques, an object's center of mass travels at constant velocity while the object rotates at constant angular velocity about this center of mass.

center of percussion The special spot on a bat or racket where a collision with a second object will not cause any acceleration of the bat's handle.

centrifugal force A fictitious force outward from the center of a circular trajectory, due to an object's centripetal acceleration.

centripetal acceleration An acceleration that is always directed toward the center of a circular trajectory.

centripetal force A centrally directed force on an object. A centripetal force is not an independent force but rather the sum of other forces, such as gravity, acting on the object.

chain reaction A process in which one event triggers an average of one or more similar events so that the process becomes self-sustaining.

chaos Unpredictable behavior in which minute changes in a system's initial arrangement lead to very different final arrangements. These differences grow more dramatic with each passing second.

chaotic system A dynamic system that is exquisitely sensitive to initial conditions. Minute changes in how you set up a chaotic system can lead to wildly different final configurations.

characteristic X ray An X ray emitted by X-ray fluorescence from an atom. The energy of the characteristic X ray is determined by the atom's orbital structure.

charges Objects, particularly small particles, that carry electric charge.

chemical bond The forces between two or more atoms that hold those atoms together to form a molecule. Chemical bonds are electromagnetic in origin, caused by the attractions and repulsions between charged particles in the atoms involved.

chemical potential energy Energy stored in the chemical forces between atoms.

chemical reaction An encounter between two or more atoms or molecules that results in a rearrangement of the atoms to form different atoms or molecules.

circularly polarized A light wave in which the electric and magnetic fields rotate about the direction in which that wave is heading.

closed circuit A complete electric circuit through which electric current can flow continuously.

coaxial cable A two-conductor electric cable in which an insulated central conductor is surrounded by a cylindrical outer conductor. Electromagnetic waves can travel inside it.

coefficient of restitution The measure of a ball's liveliness, determined by bouncing the ball from a rigid, immovable surface. It's the ratio of the ball's rebound speed to its collision speed.

coefficient of volume expansion The fractional change in an object's volume caused by a temperature increase of 1 °C.

coherent light Light consisting of identical photons that together form a single electromagnetic wave.

collision energy The amount of kinetic energy removed from two objects as they collide.

color temperature The temperature at which a black object will emit thermal electromagnetic radiation with a particular distribution of wavelengths.

components The portions of a vector quantity that lie along particular directions.

compressible   A substance that changes density significantly as its pressure changes. A gas is compressible since its density is proportional to its pressure.

Compton scattering   The process in which a photon bounces off a charged particle, usually an electron. The photon and charged particle exchange energy and momentum during the collision.

condense   Transform from a gas to a liquid.

conduction   The transmission of heat through a material by a transfer of energy from one atom or molecule to the next. The atoms themselves don't move with the heat.

conduction band   The group of quantum levels in an insulator that lies above the Fermi level.

conduction level   A quantum level in an insulator that involves more energy than the Fermi level and that is normally unoccupied by electrons.

conserved quantity   A physical quantity, such as energy, that is neither created nor destroyed within an isolated system when that system undergoes changes. A conserved quantity may pass among the objects within an isolated system, but its total amount remains constant.

constructive interference   Interference in which two or more waves arrive at a location in phase with one another and produce a particularly strong effect.

convection   The transmission of heat by the movement of a fluid. Convection normally entails the natural circulation of the fluid that accompanies differences in temperatures and densities.

converging lens   A lens that bends the individual light rays passing through it toward one another so that they either converge more rapidly than before or at least diverge less rapidly from one another. A converging lens has a positive focal length.

corona discharge   A faintly glowing discharge that surrounds a highly charged metal in the presence of a gas. In the discharge, electric charge is transferred from the metal to the gas molecules.

coulomb (C)   The SI unit of electric charge. One coulomb is about 1 million times the charge you acquire by rubbing your feet across the carpet in winter.

Coulomb constant   The fundamental constant of nature that determines the electrostatic forces two charges exert on one another. Its value is 8.988×10^9 N·m^2/C^2.

Coulomb's law   The magnitudes of the electrostatic forces between two objects are equal to the Coulomb constant times the product of their two electric charges divided by the square of the distance separating them. If the charges are like, then the forces are repulsive. If the charges are opposite, then the forces are attractive.

crest   The peak upward excursion of the system during the passage of a wave.

critical mass   A portion of a fissionable material that is able to sustain a fission chain reaction. The amount of material required depends on its mass, shape, and density.

current   The amount of electric charge flowing past a point or through a surface per unit of time.

cycles-per-second   (1/s) The SI unit of frequency (identical to the hertz).

cyclotron motion   The circular or spiral motion of a charged particle in a magnetic field. The charged particle tends to loop around the magnetic flux lines.

decimal   The digital representation of numbers in terms of the powers of ten. The number *124* is represented in decimal as 124, meaning 1 hundred (10^2), 2 tens (10^1), and 4 ones (10^0).

density   The mass of an object divided by its volume. The SI unit of density is the kilogram-per-meter^3.

depletion region   The nonconducting region near a p-n junction in which all of the valence levels are filled with electrons and there are no conduction level electrons.

destructive interference   Interference in which two or more waves arrive at a location out of phase with one another and produce a particularly weak effect.

diffraction   Wave effects that limit the focusability of light and alter the way in which it travels after passing through an opening.

digital   The representation of numbers by decomposition into digits that are then individually represented by specific values of physical quantities such as voltage, charge, or pressure.

diode   A semiconductor device that allows charge to flow through it only in one direction. Diodes are commonly formed by bringing n-type silicon into contact with p-type silicon.

direct current   An electric current that always flows in one direction.

direction   The line or course on which something is moving, is aimed to move, or along which something is pointing or facing.

discharge A flow of electric current through a gas.

dispersion The dependence of light's speed through a material on the frequency of that light.

distance The length between two positions in space.

doped Deliberately contaminated with an impurity that changes its physical properties.

drag forces The frictionlike forces exerted by a fluid and a solid on one another as the solid moves through the fluid. These forces act to reduce the relative velocity between the two.

dynamic In motion or associated with motion. A dynamic system is a system in motion. Dynamic pressure in a fluid is associated with that fluid's motion.

dynamic electricity A process involving moving electrically charged particles.

dynamic stability An object's stability when it's in motion.

dynamically stable An object that can be in a stable equilibrium when in motion.

elastic limit The most extreme distortion of an object from which it can return to its original size and shape without permanent deformation.

elastic potential energy The energy stored by the forces within a distorted elastic object.

elastic scattering The process in which two particles bounce off one another without losing any of their kinetic energies.

electric charge An intrinsic property of matter that gives rise to electrostatic forces between charged particles. Electric charge is a conserved physical quantity. There are two types of electric charge, called positive and negative.

electric charges Objects, particularly small particles, that carry electric charge.

electric circuit A complete loop of conductors, loads, and power sources through which an electric current can flow continuously.

electric conductor A material that permits electric charges to flow through it.

electric current The movement or flow of electric charge.

electric field An attribute of each point in space that exerts forces on electrically charged particles. An electric field has a magnitude and direction proportional to the force it would exert on a unit of positive charge at that location. While electric fields are often created by nearby charges, they can also be created by other electromagnetic phenomena.

electric insulator A material that prevents any net movement of electric charge through it.

electric resistance The measure of how much an object impedes the flow of electric current.

electromagnet A coil of wire, with or without an iron core, that becomes a magnet when an electric current flows around the coil.

electromagnetic induction The process whereby a changing magnetic field induces a current to flow in a conductor.

electromagnetic waves Waves consisting of electric and magnetic fields that travel through empty space at the speed of light. These waves carry energy and momentum and are emitted and absorbed as particles called photons. Radio waves, microwaves, infrared, visible, and ultraviolet light, X rays, and gamma rays are examples of electromagnetic waves.

electron–positron pair production The formation of an electron and a positron during an energetic collision.

electrons The tiny negatively charged particles that make up the outer portions of atoms and that are the main carriers of electricity and heat in metals.

electron-volt (eV) A unit of energy equal to the energy obtained by an electron as it moves through a voltage difference of 1 V. One electron-volt is equal to about 1.602×10^{-19} J.

electrostatic force The force experienced by a charged particle in the presence of other charged particles.

electrostatic potential energy Energy stored in the forces between electric charges.

elementary unit of electric charge The basic quantum of electric charge, equal to about 1.6×10^{-19} C.

emissivity A surface's capacity to emit or absorb thermal radiation, relative to that of a perfectly black object at the same temperature.

energy The capacity to do work. Each object has a precise quantity of energy, which determines exactly how much work that object could do in an ideal situation. The SI unit of energy is the joule.

English System of Units An assortment of antiquated units that were used throughout the English colonies and remain in common use in the United States today. Units in this system include feet, ounces, and miles-per-hour.

entropy The physical quantity measuring the amount of disorder in a system. The system's entropy would be zero at absolute zero.

equilibrium   The state of an object in which zero net force (or zero net torque) acts on it. An object that is stationary or in uniform motion is in equilibrium.

equilibrium position   The point at which an object experiences zero net force and doesn't accelerate.

escape velocity   The speed a spacecraft needs in order to follow a parabolic orbital path and escape forever from a particular celestial object.

evaporate   Transform from a liquid to a gas.

excited state   A configuration of a system in which the electrons (or other particles) are not all in the lowest energy orbitals or levels.

explosive chain reaction   A chain reaction in which each fission induces an average of much more than one subsequent fission and the fission rate skyrockets.

Fahrenheit   A temperature scale in which 32 °F is defined as the melting point of water and 212 °F is defined as the boiling point of water at sea level. Absolute zero is −459.67 °F.

farad (F)   The SI unit of electric capacitance. A 1-farad capacitor will have a voltage difference between its plates of 1 volt when storing 1 coulomb each of separated positive and negative charge.

feedback   The process of using information about a system's current situation to control changes you are making in that system.

Fermi level   The highest energy level normally occupied by electrons in a solid.

Fermi particles   A class of fundamental particles that includes electrons, protons, and neutrons and that obeys the Pauli exclusion principle.

ferrimagnetic   Composed of magnetic atoms that alternate back and forth in magnetic orientation but have different magnetic strengths so that their magnetic fields don't cancel completely.

ferromagnetic   Composed of magnetic atoms that all have the same magnetic orientation within a magnetic domain.

fictitious force   Not a real force at all, but the object's mass resisting acceleration. An object undergoing acceleration experiences what appears to be a force in the direction exactly opposite the direction of acceleration. The amount of fictitious force is proportional to the amount of acceleration.

first law of thermodynamics   The change in a stationary object's internal energy is equal to the heat transferred into that object minus the work that object does on its surroundings. This law restates the conservation of energy.

fissionable   Able to undergo induced fission.

fluid   A substance that has mass but no fixed shape. A fluid can flow to match its container. Gases and liquids are both fluids.

fluorescence   An emission of light that immediately follows an absorption of light.

f-number   The ratio of a lens's focal length to its effective aperture.

focal length   The distance after a converging lens at which the real image of a distant object forms.

force   An influence that if exerted on a free body results chiefly in an acceleration of the body and sometimes in deformation and other effects. A force is a vector quantity, consisting of both the amount of force and its direction. The SI unit of force is the newton.

frame of reference   A location from which motion is observed and measured. Such a frame of reference may be traveling at a constant velocity or it may be accelerating.

frequency   The number of cycles completed by an oscillating system in a certain amount of time.

frequency modulation   A technique for representing sound or data by changing the exact frequency of a radio transmission.

friction   The force that resists relative motion between two surfaces in contact. Frictional forces are exerted parallel to the surfaces in the directions opposing their relative motion.

fundamental forces   The four basic forces that act between objects in the universe—the gravitational force, the electromagnetic force, the strong force, and the weak force.

fundamental vibrational mode   The slowest and often broadest vibration that an object can support.

gamma rays   Extremely high-energy photons of electromagnetic radiation, often produced during radioactive decays.

gas   A form of matter consisting of tiny, individual particles (atoms or molecules) that travel around independently. A gas takes on the shape and volume of its container.

gravitational constant   The fundamental constant of nature that determines the gravitational forces two masses exert on one another. Its value is $6.6720 \times 10^{-11} \text{ N·m}^2/\text{kg}^2$.

gravitational potential energy Potential energy stored in the gravitational forces between objects.

gravity The gravitational attraction of the mass of the earth, the moon, or a planet for bodies at or relatively near its surface. All objects exert gravitational forces on all other objects.

ground state The lowest energy configuration of a system in which all of the electrons (or other particles) are in the lowest energy orbitals or levels.

half-life The time needed for half the nuclei of a particular radioactive isotope to undergo radioactive decay.

hard magnetic material A material that is relatively difficult to magnetize and that retains its magnetization once the magnetizing field is removed. Hard magnetic materials are suitable for permanent magnets.

harmonic An integer multiple of the fundamental frequency of oscillation for a system. The second harmonic is twice the frequency of the fundamental, and the third harmonic is three times the frequency of the fundamental. In principle, harmonics can continue forever.

harmonic oscillator An oscillator in which the restoring force on an object is proportional to its displacement from a stable equilibrium. The period of a harmonic oscillator doesn't depend on the amplitude of its motion.

heat The energy that flows from one object to another as a result of a difference in temperature between those two objects.

heat engine A device that converts thermal energy into work as heat flows from a hot object to a cold object.

heat exchanger A device that allows heat to flow naturally from a hotter material to a colder material without any actual exchange of those materials.

heat pump A device that pumps heat against its natural direction of flow, transferring it from a cold object to a hot object. To satisfy the second law of thermodynamics, a heat pump must convert some ordered energy into thermal energy.

Heisenberg uncertainty principle A quantum physical law that states that an object's position and momentum can't be sharply defined at the same time. This principle gives objects with small masses a fuzzy character.

henry (H) The SI unit of inductance. A 1-henry inductor will experience a 1-ampere change in the current flowing through it each second when subjected to a 1-volt voltage drop.

hertz (Hz) The SI unit of frequency (identical to cycles-per-second).

higher-order vibrational mode A vibrational mode that is more complicated than the fundamental mode and in which different parts of the system move in opposite directions.

Hooke's law The general law covering spring and elastic behavior. Hooke's law states that a spring exerts a restoring force that is proportional to the distance the spring is distorted from its equilibrium length.

horizontal polarization An electromagnetic wave in which the electric field always points left or right (horizontally). The magnetic field always points vertically.

ideal gas law The law relating the pressure, temperature, and particle density of an ideal gas. An ideal gas is one that is composed of perfectly independent particles. The particles don't stick and bounce perfectly from one another.

image distance The distance between the lens and the real or virtual image that the lens creates.

impedance A measure of a system's opposition to the passage of a current or a wave.

impedance mismatch An abrupt change in the opposition to a wave's passage, typically accompanied by reflections.

impulse The mechanical means for transferring momentum. One object gives an impulse to a second object by exerting a certain force on the second object for a certain amount of time. In return, the second object gives an equal but oppositely directed impulse to the first object.

incandescence The emission of thermal radiation from a hot object.

incoherent light Light consisting of individual photons, each its own independent electromagnetic wave.

incompressible A substance that doesn't change density significantly as its pressure changes. Liquids and solids are incompressible since their densities change very little as their pressures change dramatically.

index of refraction The factor by which the speed of light in a material is reduced from its speed in empty space, equal to the speed of light in empty space divided by light's speed in the material.

induced drag   The drag force that occurs when a wing deflects the stream of air passing across it to obtain lift.

induced fission   A fission event that's caused by a collision, usually with a neutron.

inductance   The voltage drop across an inductor divided by the rate at which current through that inductor is changing with time. Inductance is measured in henries.

inductor   An electronic component that stores magnetic energy in a coil of wire and opposes changes in current in that wire.

inert gas   A gas consisting of atoms that are chemically inactive and rarely bond permanently with other atoms or molecules. Inert gases include helium, neon, argon, krypton, and xenon.

inertia   A property of matter by which it remains at rest or in uniform motion in the same straight line unless acted upon by some outside force.

inertial frame of reference   A frame of reference that is not accelerating and is thus either stationary or traveling at constant velocity. The laws of motion accurately describe any situation that is observed from an inertial frame of reference.

infrared light   Invisible light having wavelengths longer than about 750 nanometers.

in phase   The relationship between two waves in which they complete the same portions of their oscillatory cycles at the same time and place.

integrated circuit   A thin wafer of semiconductor that has been carefully structured, modified, and processed so that its surface contains the interconnected transistors, resistors, and/or capacitors needed to perform some electronic task.

interference   A wave phenomenon in which waves passing through the same location from different directions reinforce or oppose one another.

internal kinetic energy   The portion of an object's kinetic energy that involves only the relative motion of particles within the object and that excludes the object's overall translation or rotation.

internal potential energy   The portion of an object's potential energy that involves only forces between particles within the object and that excludes the object's interactions with its surroundings.

ion   An atom or molecule with a net electric charge.

isotopes   Chemically indistinguishable atoms containing nuclei that differ only in their numbers of neutrons.

joule (J)   The SI unit of energy and work (synonymous with the newton-meter). Lifting 1 liter of water upward 10 centimeters near the earth's surface requires about 1 joule of work.

joule-per-second (J/s)   The SI unit of power (synonymous with the watt).

Kelvin   The SI scale of absolute temperature, in which 0 K is defined as absolute zero. The spacing between units is the same as that used in the Celsius scale.

kilogram (kg)   The SI unit of mass. (The standard kilogram is a platinum–iridium cylinder kept at the International Bureau of Weights and Measures near Paris.) A liter of water has a mass of about 1 kilogram.

kilogram-meter^2 (kg·m^2)   The SI unit of moment of inertia. One kilogram-meter^2 is roughly the moment of inertia of your forearm as it pivots about your elbow.

kilogram-per-meter^3 (kg/m^3)   The SI unit of density. One kilogram-per-meter^3 is about the density of air at 2000 m (about 1 mile) above sea level.

kinetic energy   The form of energy contained in an object's translational and rotational motion.

laminar flow   Smooth, predictable fluid flow in which the velocity at any specific point doesn't change with time in either magnitude or direction and nearby portions of the fluid remain nearby as they travel along.

laser amplifier   A device that amplifies weak incoming light to produce brighter outgoing light. The outgoing light is a brighter copy of the incoming light.

laser medium   An assembly of excited atoms or atomlike systems that is capable of amplifying passing light through stimulated emission.

laser oscillator   A laser amplifier that is surrounded by mirrors so that it can amplify one or more spontaneously emitted photons to form an intense beam of coherent light.

law of universal gravitation   Every object in the universe attracts every other object in the universe with a force equal to the gravitational constant times the product of the two masses, divided by the square of the distance separating the two objects.

laws of thermodynamics   The four laws that govern the movement of heat between objects.

lens   A transparent optical device that uses refraction to bend light, often to form images.

lens equation   The equation relating a lens's focal length to the object and image distances.

Lenz's law   When a changing magnetic field induces a current in a conductor, the magnetic field from that current always opposes the change that induced it.

level   One of the paths that an electron can take in a solid that is allowed by quantum physics.

lever arm   The directed distance from the pivot or axis of rotation to the point at which the force is exerted.

lift forces   Forces exerted by a fluid on a solid that are at right angles to the fluid flow around that solid.

light   See visible, infrared, and ultraviolet light.

linear momentum   A conserved vector quantity that measures an object's motion. It is the product of that object's mass times its velocity.

liquid   A form of matter consisting of particles (atoms or molecules) that are touching one another but that are free to move relative to one another. A liquid has a fixed volume but takes the shape of its container.

luminescence   The emission of light by any means other than thermal radiation.

magnetic dipole   A pair of equal but opposite poles separated by a distance.

magnetic domains   Regions of uniform alignment within a magnetic material.

magnetic field   An attribute of each point in space that exerts forces on magnetic poles. A magnetic field has a magnitude and direction proportional to the force it would exert on a unit of north magnetic pole at that location.

magnetic flux   The total amount of magnetic field passing through a surface or region of space.

magnetic monopole   An isolated magnetic pole, either north or south. None has ever been observed.

magnetic pole   A property of nature that gives rise to magnetostatic forces between magnetic poles. There are two types of poles, north and south.

magnetic poles   Objects that carry magnetic pole.

magnetostatic force   The force experienced by a magnetic pole in the presence of other magnetic poles.

magnitude   The amount of some physical quantity.

Magnus force   A lift force experienced by a spinning object as it moves through a fluid. The Magnus force points toward the side of the ball moving away from the onrushing airstream.

mass   The property of a body that is a measure of its inertia or resistance to acceleration, that is commonly taken as a measure of the amount of material it contains, and that causes it to have weight in a gravitational field. The SI unit of mass is the kilogram.

mechanical advantage   The process whereby a mechanical device redistributes the amounts of force and distance that go into performing a particular amount of mechanical work.

meter (m)   The SI unit of length. (One meter is formally defined as the distance light travels through empty space in 1/299,792,458th of a second.) One meter is about the length of a long stride or about 3.28 feet.

meter^2 (m^2)   The SI unit of area. One meter^2 is about twice the area of an opened newspaper.

meter^3 (m^3)   The SI unit of volume. One meter^3 is about the volume of a four-drawer file cabinet.

meter-per-second (m/s)   The SI unit of velocity or speed. One meter-per-second is a typical walking pace or about 2.2 mph.

meter-per-second^2 (m/s^2)   The SI unit of acceleration. One meter-per-second^2 is about the acceleration of an elevator as it first begins to move upward.

microwaves   Electromagnetic waves with wavelengths between about 1 meter and 1 millimeter.

mode   A basic pattern of distortion or oscillation.

molecule   A particle formed out of two or more atoms. A molecule is the smallest portion of a chemical compound that retains the chemical properties of that compound.

moment of inertia   The property of a body that is a measure of its rotational inertia. An object's moment of inertia is determined by its mass and how far that mass is from the axis of rotation. The SI unit of moment of inertia is the kilogram-meter^2.

momentum   Linear momentum.

natural resonance   A mechanical process in which an isolated object's energy causes it to perform a certain motion over and over again. The rate at which this motion occurs is determined by the physical characteristics of the object.

net electric charge   The sum of all charges on an object, both positive and negative. Positive charges increase the net charge while negative charges decrease it. Net charge can be negative.

net force   The sum of all forces acting on an object, considering both the magnitude of each individual force and its direction. The magnitude of the net force is often less than the sum of the magnitudes of the individual forces, since they often oppose one another in direction.

net magnetic pole   The sum of all poles on an object, both north and south. Since there are no isolated magnetic poles, an object's net magnetic pole is always zero.

net torque   The sum of all torques acting on an object, considering both the magnitude of each individual torque and its direction. The magnitude of the net torque is less than the sum of the magnitudes of the individual torques, since they often oppose one another in direction.

neutral   Having zero net electric charge.

neutrinos   Chargeless and nearly massless particles created during radioactive decays and other nuclear events. They rarely interact with matter.

neutrons   The electrically neutral subatomic particles that, together with protons, make up atomic nuclei.

newton (N)   The SI unit of force (synonymous with the kilogram-meter-per-second^2). Ten United States quarters have a weight equal to about 1 newton. The common English unit of force, the pound, is about 4.45 newtons.

newton-meter (N·m)   The SI unit of energy and work (synonymous with the joule). Also the SI unit of torque, exerted by a 1-newton force located 1 meter from the axis of rotation. One newton-meter is about the torque exerted on your shoulder by the weight of a baseball held in your outstretched arm.

newton-per-meter^2   The SI unit of pressure (identical to the pascal).

Newton's first law of motion   An object that is free from all outside forces travels at a constant velocity, covering equal distances in equal times along a straight-line path.

Newton's first law of rotational motion   An object that is not wobbling and is free from all outside torques rotates with constant angular velocity, spinning steadily about a fixed axis.

Newton's second law of motion   An object's acceleration is equal to the force exerted on that object divided by the object's mass. This equality can be manipulated algebraically to state that the force on the object is equal to the product of the object's mass times its acceleration (Eq. 1.1.1).

Newton's second law of rotational motion   An object's angular acceleration is equal to the torque exerted on that object divided by the object's moment of inertia. This equality can be manipulated algebraically to state that the torque on the object is equal to the product of the object's moment of inertia times its angular acceleration (Eq. 2.1.1). The law doesn't apply to wobbling objects.

Newton's third law of motion   For every force that one object exerts on a second object, there is an equal but oppositely directed force that the second object exerts on the first object.

Newton's third law of rotational motion   For every torque that one object exerts on a second object, there is an equal but oppositely directed torque that the second object exerts on the first object.

normal   Directed exactly away from (perpendicular to) a surface. A line that is normal to a surface meets that surface at a right angle.

normal force   Support force.

n-type semiconductor   A semiconductor such as silicon that contains impurity atoms such as phosphorus, arsenic, antimony, or bismuth that create conduction level electrons in the semiconductor.

nuclear fission   The shattering of a heavy nucleus into smaller fragments. During fission, the positively charged fragments repel one another and release energy.

nuclear force   An attractive force that binds nucleons together once they touch one another.

nuclear fusion   The merging of two small nuclei to form a larger nucleus. During fusion, the nuclear force binds the nucleons together and releases energy.

nucleon   A general name given to the particles that make up atomic nuclei: protons and neutrons.

nucleus   The positively charged central component of an atom, containing most of the atom's mass and about which the electrons orbit. Plural is nuclei.

object distance   The distance between the lens and the object that it is imaging.

ohm (Ω)   The SI unit of electric resistance. A 1-ohm resistor exhibits a voltage drop of 1 volt when 1 ampere of current flows through it.

Ohm's law   The law relating the voltage drop across an electric conductor to the electric current passing through it and to its electric resistance.

open circuit   An incomplete electric circuit where a gap in the electric conductors stops electric current from flowing.

orbit   The path an object takes as it moves in the presence of a centripetal force.

orbital period   The time required to complete one full orbit.

orbitals   One of the orbits that an electron can take in an atom that is allowed by quantum physics.

ordered energy   Energy that can easily be used to do work.

oscillation   A repetitive and rhythmic movement or process that usually takes place about an equilibrium situation.

oscillator   A system experiencing an oscillation.

out of phase   The relationship between two waves in which they complete opposite portions of their oscillatory cycles at the same time and place.

parallel   (wiring arrangement) An arrangement in which the current reaching two or more electric devices divides into separate parts to flow through those devices and then joins back together as it leaves them.

particle density   The number of particles in an object divided by its volume. The particle density of water is about 3.35×10^{28} molecules-per-meter^3. The particle density of air at sea level is about 2.687×10^{25} molecules-per-meter^3.

pascal (Pa)   The SI unit of pressure (identical to the newton-per-meter^2). Atmospheric pressure at sea level is about 100,000 pascals. A 1-millimeter-high water droplet exerts a pressure of about 10 pascals on your hand.

Pauli exclusion principle   An observed property of nature that no two indistinguishable Fermi particles ever travel in the same orbital or level.

period   The time required to complete one full cycle of a repetitive motion.

phosphor   A solid that luminesces (emits light) when energy is transferred to it by light or by a collision with a particle.

photoconductor   A solid that is an electric insulator in the dark but that becomes an electric conductor when exposed to light of the correct wavelength.

photodiode   A diode that permits current to flow backward across the p-n junction when exposed to light. Light provides the energy needed to move charges across the junction's depletion region in the wrong direction. The current flowing in the reverse direction through a photodiode is proportional to the light intensity.

photoelectric effect   The process in which an atom absorbs a photon in a radiative transition that ejects one of the electrons out of the atom.

photon   A particle of light, having energy and momentum but no mass.

Planck constant   The fundamental constant of quantum physics, equal to the energy of an object divided by the frequency of its wave. It is about 6.626×10^{-34} J·s.

plane polarized   A light wave in which the electric field (and the magnetic field) fluctuates back and forth in a plane as the light travels through space.

plasma   A gaslike phase of matter consisting of electrically charged particles, such as ions and electrons. The strong electrostatic interactions between its particles distinguish a plasma from a gas.

p-n junction   The interface between an n-type semiconductor and a p-type semiconductor that gives the diode its one-directional characteristic for electrons.

Poiseuille's law   The volume of fluid flowing through a pipe each second is equal to ($\pi/128$) times the pressure difference across that pipe times the pipe's diameter to the fourth power, divided by the pipe's length times the fluid's viscosity.

polarized   A distribution of electric charge that is nonuniform so that the object has a negatively charged region and a positively charged region.

poles   Objects that carry a magnetic pole.

position   A vector quantity that specifies the location of an object relative to some reference point. It consists of both the length and the direction from the reference point to the object.

positron   The antimatter counterpart of the electron. The positron is positively charged.

potential energy   The stored form of energy that can produce motion. Potential energy is stored in the forces between or within objects.

power   The measure of how quickly work is done on an object. The SI unit of power is the watt.

precession   The change in orientation of a spinning object's rotational axis that occurs when it's subject to an outside torque.

pressure   The average amount of force a fluid exerts on a certain region of surface area. Pressure is reported as the amount of force divided by the surface area over which that force is exerted.

pressure drag   The drag force that results from higher pressures at the front of an object than at its rear.

pressure gradient   A distribution of pressures that varies continuously with position.

pressure potential energy   The product of a fluid's volume times its pressure. However, this energy isn't really stored in the water. Instead, it's energy that's provided by a pump (or other source) when the fluid is delivered.

primary colors of light   The three colors of light— red, green, and blue—that are sensed by the three types of color-sensitive cone cells in our eyes. Mixtures of these three colors of light can make our eyes perceive any possible color.

protons   The positively charged subatomic particles found in atomic nuclei.

p-type semiconductor   A semiconductor such as silicon that contains impurity atoms such as boron, aluminum, gallium, indium, or tellurium that create empty valence levels in the semiconductor.

quantized   Existing only in discrete units or quanta. Quantized physical quantities are only observed in integer multiples of the elementary quanta.

radian   The natural unit in which angles are measured. There are 2π radians in a full circle, so 1 radian is $180/\pi$ degrees or approximately $57.3°$.

radian-per-second (1/s)   The SI unit of angular velocity or angular speed. An object turning at 1 radian-per-second completes a full revolution in just less than 6.3 seconds.

radian-per-second^2 (1/s^2)   The SI unit of angular acceleration.

radiation   The transmission of heat through the passage of electromagnetic radiation between objects.

radiation trapping   The phenomenon in which a particular wavelength of light has trouble propagating through a material that eagerly absorbs and emits it. The light passes from one atom to the next and makes little headway.

radiative transition   The shift of an atom or other system from one state to another through the emission or absorption of an electromagnetic wave.

radio waves   Electromagnetic waves, usually with wavelengths longer than about 1 m.

radioactive decay   The spontaneous decay of a nucleus into fragments.

ramp   An inclined plane that allows work to be done over a longer distance, thereby requiring less force.

Rayleigh scattering   The redirection of light due to its interaction with small particles of matter.

real image   A pattern of light, projected in space, that exactly reproduces the pattern of light at the surface of the original object. A real image forms after the lens that creates it and can be projected onto a surface.

rebound energy   The amount of kinetic energy returned to two objects as they push apart following a collision.

reflection   The redirection of all or part of a light wave so that it returns from a boundary between materials.

refraction   The bending of light's path that occurs when the light crosses a boundary between materials and experiences a change in speed.

relative motion   The movement of one object from the perspective of another object. Two objects that are moving relative to one another have different velocities.

relativity theory   The physical rules governing motion at speeds comparable to the speed of light. The general theory of relativity considers motion near massive objects.

resistor   An electronic component that impedes the flow of electric current, converting some of its energy into heat.

resonant energy transfer   The gradual transfer of energy to or from a natural resonance caused by small forces timed to coincide with a particular part of each oscillatory cycle.

restoring force   A force that acts to return an object to its equilibrium shape. A restoring force is directed toward the position the object occupies when it's in its equilibrium shape.

Reynolds number A dimensionless number that characterizes fluid flow through a system. At low Reynolds numbers, a fluid's viscosity dominates the flow while at high Reynolds numbers, a fluid's inertia dominates.

right-hand rule The convention whereby the specific direction of an object's angular velocity is established. According to this rule, if the fingers of your right hand are curled to point in the direction of the object's rotation, your thumb will point in the direction of the angular velocity.

rotational inertia A property of matter by which it remains at rest or in steady rotation about the same rotational axis unless acted upon by some outside torque.

rotational motion Motion in which an object rotates about an axis. The orientation of an object undergoing only rotational motion will change, but its position will remain unchanged.

schematic diagram A symbolic picture of the conceptual structure of an electronic device.

second (s or sec) The SI unit of time. (One second is formally defined as the duration of 9,192,631,770 periods of the radiation corresponding to the transition between two hyperfine levels of the ground state of the cesium-133 atom.)

second law of thermodynamics The entropy of a thermally isolated system of objects never decreases. This law recognizes that creating disorder is easy; restoring order is hard.

semiconductor An insulatorlike material in which there is a small energy gap above the Fermi level, so that only a modest amount of energy is needed to shift an electron from a valence level to a conduction level.

separated electric charge Isolated quantities of positive and negative charge. Separated electric charge has electrostatic potential energy, stored in the forces attracting those charges back together.

series (wiring arrangement) An arrangement in which the current reaching two or more electric devices flows sequentially through one device after the next before leaving them.

shock wave A narrow region of high pressure and temperature that forms when the speed of an object through a medium exceeds the speed at which sound, waves, or other vibrations travel in that medium.

short circuit A defect in a circuit that allows current to bypass the load it's supposed to operate.

SI Units A system of units (Systéme Internationale d'Unités) that carefully defines related units according to powers of 10. SI units are now used almost exclusively throughout most of the world, with the notable exception of the United States.

signal An electric representation of information.

simple harmonic motion The regular, repetitive motion of a harmonic oscillator. The period of simple harmonic motion doesn't depend on the amplitude of oscillation.

sliding friction The forces that resist relative motion as two touching surfaces slide across one another.

soft magnetic material A material that is relatively easy to magnetize and that loses its magnetization once the magnetizing field is removed. Soft magnetic materials are suitable for electromagnets.

solid A form of matter consisting of particles (atoms or molecules) that touch and that are not free to move relative to one another. A solid has a fixed volume and shape.

sound In air, sound consists of pressure waves, patterns of compressions and rarefactions that travel outward from their source at the speed of sound.

speed A measure of the distance an object travels in a certain amount of time. The SI unit of speed is the meter-per-second.

speed of light The speed with which an electromagnetic wave travels through space. In empty space, a vacuum, the speed of light is exactly 299,792,458 m/s.

speed of sound The speed at which sound's compressions and rarefactions travel in a medium such as air or water.

spontaneous emission of radiation Light emission that occurs when an excited atom or atomlike system releases stored energy randomly through a radiative transition. The photon that results is independent and unique.

spring constant As a measure of the stiffness of an elastic object, the spring constant relates the object's distortion to the restoring force it exerts. The larger the spring constant, the stiffer the spring.

sputtering   Ejection of atoms from a surface caused by the impact of energetic ions, atoms, or other tiny projectiles.

stable equilibrium   A state of equilibrium to which an object will return if it's disturbed. At equilibrium, the object is free of net force or torque. However, if an object is moved away from that equilibrium state, the net force or torque that will then act on it will tend to return it to equilibrium.

stalling   An aerodynamic problem that occurs when the airflow over a wing separates from the surface, leaving a pocket of turbulent air between it and the wing.

standard units   Agreed upon amounts of various physical quantities, which define a system in which those quantities are subsequently measured.

standing wave   A wave that oscillates in place, with energy going back and forth between potential and kinetic forms.

state   A possible arrangement of electrons (or other particles) in a system.

static   Stationary or independent of motion. A static system is a system in which nothing is moving. Static pressure in a fluid is caused by effects other than the fluid's motion.

static electricity   Any stationary arrangement of electric charges, though often associated with the separated electric charge produced by sliding friction.

static friction   The forces that resist relative motion as outside forces try to make two touching surfaces begin to slide across one another.

static stability   An object's stability when it's not in motion.

statically stable   An object that can be in a stable equilibrium when at rest.

steady-state flow   A situation in a fluid where the characteristics of the fluid at any fixed point in space don't change with time.

Stefan–Boltzmann constant   The constant of proportionality relating a surface's radiated power to its emissivity, temperature, and surface area. It has a measured value of 5.67×10^{-8} J/(s·m²·K⁴)

Stefan–Boltzmann law   The equation relating a surface's radiated power to its emissivity, temperature, and surface area.

stiffness   A measure of how rapidly a restoring force increases as the system exerting that force is distorted.

stimulated emission of radiation   Light emission that occurs when an excited atom or atomlike system releases stored energy through a radiative transition by duplicating a photon passing through that system.

streamline   The path followed by a particular portion of a flowing fluid.

strong force   The fundamental force that gives structure to nuclei and nucleons and is the basis for the nuclear force.

subatomic particles   The fundamental building blocks of the universe, from among which atoms and matter are constructed.

subcritical mass   A portion of fissionable material that is too small to sustain a chain reaction.

sublimation   The process by which atoms or molecules go directly from a solid to a gas.

superconductor   An electric conductor that permits electrons to flow without losing any of their kinetic energy to thermal energy. Electrons will continue to flow in a superconductor indefinitely. Materials only become superconducting at extremely low temperatures.

supercritical mass   A portion of fissionable material that is well in excess of a critical mass so that it undergoes an explosive chain reaction.

support force   A force that is exerted when two objects come into contact. Each object exerts a force on the other object to keep the two from passing through one another. Support forces are always normal, or perpendicular, to the surfaces of objects.

surface area   The extent of a two-dimensional surface bounded by a particular border. The SI unit of surface area is the meter².

surface waves   Disturbances in the stable equilibrium shape of a surface.

sympathetic vibration   The transfer of energy between two natural resonances that share a common frequency of oscillation.

tank circuit   A simple resonant circuit consisting of a capacitor and an inductor. Energy flows back and forth between these two devices repetitively.

temperature   A measure of the average internal kinetic energy per particle in a material. In a gas, temperature measures the average kinetic energy of each atom or molecule.

terminal velocity   The velocity at which an object moving through a fluid experiences enough drag force to balance the other forces on it and keep it from accelerating.

thermal conductivity   The measure of a material's capacity to transport heat by conduction from its hotter end to its colder end.

thermal energy   A disordered form of energy contained in the kinetic and potential energies of the individual atoms and molecules that make up a substance. Because of its random distribution, this disordered energy can't be converted directly into useful work. Other names for thermal energy include internal energy and heat.

thermal equilibrium   A situation in which no heat flows in a system because all of the objects in the system are at the same temperature.

thermal motion   The random motions of individual particles in a material due to the internal or thermal energy of that material.

third law of thermodynamics   As an object's temperature approaches absolute zero, its entropy approaches zero. This law points out that absolute zero is the unattainable state in which an object has no disorder.

thrust   A forward, propulsive force.

tidal forces   The differences between one celestial object's gravity at particular locations on the surface of a second object and the average of that gravity for the entire second object. Tidal forces tend to stretch the second object into an egg shape.

torque   An influence that if exerted on a free body results chiefly in an angular acceleration of the body. A torque is a vector quantity, consisting of both the amount of torque and its direction. The SI unit of torque is the newton-meter.

total internal reflection   Complete reflection of a light wave that occurs when that wave tries unsuccessfully to leave a material with a large refractive index for a material with a small refractive index at too shallow an angle.

trajectory   The path taken by an object as it moves.

transformer   A device that uses magnetic fields to transfer electric power from one circuit to another circuit. The two circuits are electrically isolated since no charges actually travel between the two circuits.

transistor   An electronic component that allows a tiny amount of electric charge, either moving or stationary, to control the flow of a large electric current.

translational motion   Motion in which an object moves as a whole along a straight or curved line.

traveling wave   A wave that moves through space, with energy always evenly distributed between potential and kinetic forms.

trough   The peak downward excursion of the system during the passage of a wave.

tunneling   Because of the Heisenberg uncertainty principle, small objects have somewhat ill-defined positions and occasionally move through energy barriers to places they can't reach classically. That process is tunneling.

turbulent flow   Irregular, fluctuating, nonpredictable fluid flow in which the velocity at any specific point changes with time in either magnitude or direction and nearby portions of the fluid become separated.

ultraviolet light   Invisible light having wavelengths shorter than about 400 nanometers.

uniform circular motion   Motion at a constant speed around a circular trajectory. An object undergoing uniform circular motion is accelerating toward the center of the circle and experiences an outward fictitious force, "centrifugal force."

unstable equilibrium   An equilibrium situation to which the object will not return if it's disturbed. At equilibrium, the object is free of net force or torque. However, if the object is moved away from that equilibrium situation, the net force or torque that will then act on it will tend to accelerate it further away from the equilibrium situation.

valence band   The group of quantum levels in an insulator that lies below the Fermi level.

valence level   A quantum level in an insulator that involves less energy than the Fermi level and that is normally occupied by electrons.

vector quantity   A quantity, characterizing some aspect of a physical system, that consists of both a magnitude and a direction in space.

velocity   A vector quantity that measures how quickly an object's position is changing: the greater the velocity, the farther the object travels each second. It consists of both the object's speed and the direction in which the object is traveling. The SI unit of velocity is the meter-per-second.

vertical polarization   An electromagnetic wave in which the electric field always points up or down (vertically). The magnetic field always points horizontally.

vibration   A spontaneous repetitive and rhythmic movement about an equilibrium position.

vibrational antinode   A region of a vibrating object that is experiencing maximal motion.

vibrational node   A region of a vibrating object that is not moving at all.

virtual image   A pattern of light that appears to come from a particular region of space and reproduces the pattern of light at the surface of the original object. A virtual image forms before the lens that creates it and can't be projected onto a surface.

viscosity   The measure of a fluid's resistance to relative motion within that fluid.

viscous drag   A drag force that results from viscous forces on a moving surface immersed in a fluid.

viscous forces   The forces exerted within a fluid that oppose relative motion. Layers of fluid that are moving across one another exert viscous forces on each other.

visible light   Light having wavelengths between about 400 nanometers (violet) and 750 nanometers (red). This small portion of the electromagnetic spectrum is all that we are able to detect with our eyes.

volt (V)   The SI unit of voltage (identical to the joule-per-coulomb). The voltage on the positive terminal of a common battery is about 1.5 volts above that on its negative terminal.

voltage   The electrostatic potential energy of each unit of positive electric charge at a particular location.

voltage drop   The amount of electrostatic potential energy that each coulomb of positive charge loses in passing through a device. It's equal to the voltage of the charges entering the device minus the voltage of the charges leaving that device.

voltage rise   The amount of electrostatic potential energy that each coulomb of positive charge receives in passing through a device. It's equal to the voltage of the charges leaving the device minus the voltage of the charges entering that device.

volume   The extent of a three-dimensional region of space bounded by a particular enclosure. The SI unit of volume is the meter^3.

vortex   A whirling region of fluid that is moving in a circle above a central cavity.

wake   The trail left behind by an object as it moves through a fluid.

wake deflection force   A lift force experienced by a spinning ball when it deflects its turbulent wake to one side. The wake deflection force points toward the side of the ball moving away from the onrushing airstream.

water hammer   The impact of a moving mass of water that is suddenly stopped.

watt (W)   The standard SI unit of power, equal to the transfer of 1 joule-per-second. One watt is the power used by the bulb of a typical flashlight.

wave velocity   The speed and direction of the moving crests of a wave. In a typical water wave in the open ocean, the entire pattern of crests and troughs moves as a whole with the wave velocity.

wavelength   A structural characteristic of a wave, corresponding to the distance separating adjacent peaks or troughs.

weak force   The fundamental force that allows electrons and neutrinos to interact and that's responsible for beta decay.

weight   (near the earth's surface) The downward force exerted on an object due to its gravitational interaction with the earth. An object's weight is equal to the product of that object's mass times the acceleration due to gravity. The direction of the weight is always toward the center of the earth.

work   The mechanical means of transferring energy. Work is defined as the force exerted on an object times the distance that object travels in the direction of the force. A large force exerted for a short distance or a small force exerted for a long distance can perform the same amount of work. The SI unit of work is the joule.

X rays   Very high-energy photons of electromagnetic radiation.

X-ray fluorescence   The process in which an electron in one of the outer orbitals of an atom undergoes a radiative transition to an empty inner orbital, emitting an X-ray photon.

zeroth law of thermodynamics   Two objects that are each in thermal equilibrium with a third object are also in thermal equilibrium with one another. This law is the basis for a meaningful system of temperatures.

Solutions to Selected Exercises and Problems

Chapter 1

E.1 The dolphin's inertia carries it upward, even though its weight makes it accelerate downward and gradually stop rising.

E.3 Your feet accelerate upward rapidly when they hit the ground and the snow continues downward, leaving your feet behind.

E.5 Any collision in which the car accelerates forward, such as when the car is hit from behind by a faster moving car.

E.7 As it turns left, the car accelerates left. The loose objects remain behind and end up on the right side of the dashboard.

E.9 Backward, in the direction opposite your forward velocity.

E.11 The anvil's large mass slows its acceleration, so the hot metal is squeezed between the moving hammer and the stationary anvil.

E.13 The pad's inertia tends to keep it in place. If you pull the paper away too quickly, the pad won't be able to accelerate with the paper.

E.15 Regardless of their horizontal components of velocity, all objects fall at the same rate. The ball and bullet descend together.

E.17 The forward component of his velocity remains constant as he falls, and he follows an arc that carries him forward over the rocks.

E.19 In the absence of air effects, a ball hit at 45° above horizontal will travel farthest. A ball hit higher or lower won't travel as far.

E.21 Its magnitude must equal the suitcase's weight.

E.23 Zero net force.

E.25 The astronaut exerts an upward force of 850 N on the earth.

E.27 Both forces have exactly the same magnitude.

E.29 The wall exerts a support force to accelerate you backward.

E.31 A person does more work. Movements that appear large compared to the ant's height still involve small distances and little work.

E.33 Yes, it pushed the wall inward and the wall dented inward.

E.35 As it rolls on the surface, it always accelerates downhill.

E.37 Its greatest acceleration is at the steepest point; its greatest speed is at the bottom of the slide.

E.39 The kinetic energy becomes gravitational potential energy.

P.1 3200 N.
P.3 11.13 m/s.

P.5 Your Mars weight would be about 38% of your earth weight.

P.7 About 0.64 s (0.32 s on the way up and 0.32 s on the way down).

P.9 48 N.
P.11 About 0.20 s.

P.13 4800 N.
P.15 9800 N.

P.17 14,700,000 J.
P.19 12.5 J.

P.21 8000 N.

Chapter 2

E.1 The angle by which the front, center seat would have to be rotated, as viewed from above, to have each seat's orientation.

E.3 It would rotate about its center of mass, not its geometric center. It would appear to wobble as it turned.

E.5 The wheel has angular inertia, as measured by its moment of inertia, making it hard to start and stop spinning.

E.7 A force exerted at the hinges produces no torque about them.

E.9 The farther the water is from the water wheel's pivot, the more torque its weight produces on the wheel.

E.11   Your force far from the hinges produces a large torque. To oppose this torque, the nut must exert a huge force near the hinges.

E.13   The weights of your chest and your feet exert torques in opposite directions about your knees. They partially balance one another.

E.15   It reduces the car's moment of inertia so that the car can undergo rapid angular accelerations and change directions quickly.

E.17   By pushing far from the pivot, you exert more torque on the lid.

E.19   Your small effort exerted on the crowbar far from its pivot produces a large force on the box, located near the crowbar's pivot.

E.21   Skidding sideways does work against sliding friction, converting some of the skier's kinetic energy into thermal energy.

E.23   The pin's surface turns with the crust and doesn't slide across it.

E.25   The nearer the frictional force is to the pivot, the less torque it produces to slow the yo-yo's rotation. Slipperiness reduces friction.

E.27   A static frictional force from the pavement pushes you forward.

E.29   Pressing the wheels more tightly against the pavement increases the maximum force that static friction can exert on the wheels.

E.31   The chalk experiences sliding friction as you write and leaves visible wear chips on the blackboard.

E.33   To avoid accelerating when pushed on with a force, the villain would have to have infinite mass. That's impossible.

E.35   Angular momentum.

E.37   The collapsing star's angular momentum can't change. Since its moment of inertia decreases, its angular velocity must increase.

E.39   As you descend, you land hard and your knees and legs must convert your kinetic energy into thermal energy. Injuries can occur.

E.41   Sliding friction converts some into heat, but a slippery pole converts a considerable fraction into kinetic energy.

P.1   About 122 N·m to the left.

P.3   Three times as much torque

P.5   12.5 N·m, slowing the blade.

P.7   Its energy would be only 0.2 times as large as before.

P.9   2400 kg·m/s forward.   **P.11**   3600 J.

P.13   450 kg·m/s to the right.

Chapter 3

E.1   As you draw the string away from its equilibrium shape, it experiences a restoring force proportional to its displacement.

E.3   The top of the curl supports more weight than the bottom.

E.5   15 N.

E.7   A scale must be calibrated regularly with standard weights to remain accurate. Typical scales are out of calibration and inaccurate.

E.9   Correct.

E.11   You support your clothes and the scale supports you.

E.13   It has a coefficient of restitution near 0 (lots of inertial friction).

E.15   A 30% loss of rebound height is a 30% loss of gravitational potential energy—equal to the energy that became thermal energy.

E.17   You do work on the sand as you step on it, but the sand doesn't return this energy to you as you lift your foot back up again.

E.19   An increase in the balls' coefficients of restitution.

E.21   Your relative velocity is small—in your frame of reference the car in front of you is barely moving, so the impact is very gentle.

E.23   Because the trains' relative velocity is zero, a person jumping between them views them both as essentially stationary.

E.25   During a bounce, the work done on a RIF ball—to store energy in it—involves a smaller force exerted for a longer distance.

E.27   At high pressure, the shoes are stiffer and exert larger forces when distorted. They accelerate more rapidly and bounce faster.

E.29   The force that the hammer exerts on the nail during their collision would diminish if the hammer's surface distorted easily.

E.31   While you can feel accelerations, you can't feel velocity.

E.33   When the rattle accelerates, the beads inside it continue on and hit the walls of the rattle. The rattle then makes noise.

E.35   The sharper the curve, the more centripetal force the train needs to accelerate around the curve. If the track can't supply it . . . disaster.

E.37   Static friction between the tires and the road accelerates the car inward around the curve. At excessive accelerations, the car skids.

E.39   The nail exerts an enormous force on the hammer to slow it down. The hammer pushes back, driving the nail into the wood.

E.41   At the bottom of each swing.

E.43   As the salad undergoes rapid centripetal acceleration, the water travels in straight lines and runs off the salad.

E.45   After going over a bump, the car begins to accelerate downward and your apparent weight is briefly less than your real weight.

E.47   As the car accelerates upward, you must pull upward on the briefcase extra hard to make it accelerate upward too.

E.49 As your hands accelerate, the water is left behind.

E.51 As a sprinter accelerates, the ground must push toward the sprinter's center of mass to avoid exerting a torque on the sprinter.

E.53 The leftward frictional force on your feet keeps you balanced.

E.55 As long as the inward force from the U-shaped surface points toward the skater's center of mass, the skater doesn't begin rotating.

E.57 The torque you exert on the crank is much larger than the torque the blades exert on the batter.

E.59 Each second, the blade pushes the dough a short distance. To do significant work, the force it exerts on the dough must be large.

P.1 15,000 N/m.

P.3 10 mm.

P.5 It must turn at about 31 m/s.

P.7 20 m/s^2.

Chapter 4

E.1 It will sink. Its average density is larger than that of helium.

E.3 As water enters the car, the car's average density will increase.

E.5 The upper liquid is less dense than the lower liquid. It floats.

E.7 One kg of gasoline takes up more space than 1 kg of water.

E.9 An upward force equal in magnitude to the fish's weight.

E.11 Air pressure decreases steadily with altitude.

E.13 Air pressure holds the dimple down when the jar has a vacuum inside, but the pressure difference vanishes when the jar is opened.

E.15 Cooling the air in the container reduced its pressure. The resulting pressure imbalance across the lid pushes the lid inward.

E.17 As it cooled, the air became more dense.

E.19 It won't travel as far upward because some of its kinetic energy becomes gravitational potential energy on the way up.

E.21 The pressure imbalance on the base of the dam is larger.

E.23 The animals are too short for gravity to produce a big pressure difference between top and bottom.

E.25 The pressure outside your lungs is above atmospheric pressure.

E.27 Gas only accelerates toward lower pressure.

E.29 As wind slows in the bag, its pressure rises and inflates the bag.

E.31 It's stored as elastic potential energy in the skin of the balloon.

P.1 101,400 Pa.

P.3 Three times as much as before.

P.5 0.94 times as much as before.

P.7 122.4 kg or 0.1224 m^3.

P.9 28.4 N.

P.11 About 250,000 Pa above atmospheric pressure.

P.13 141 m/s or 510 km/h. **P.15** About 3,100,000 Pa.

Chapter 5

E.1 The increased flow in the cold water pipe requires a larger pressure difference in it, leaving too little pressure to operate the shower.

E.3 Flow through an artery scales with the fourth power of radius.

E.5 In warm weather, the molecules in a viscous liquid have more thermal energy and are able to move past one another more easily.

E.7 Squeezing highly viscous frosting through a narrow "pipe" requires an enormous pressure difference across that pipe.

E.9 Air near the stationary bridge surface is slowed by viscous forces, creating a slower moving boundary layer.

E.11 The pressure at the can bottom rises suddenly due to water hammer. The pressure at the can top falls or remains the same.

E.13 In turbulent flow, adjacent regions of water become separated.

E.15 The bucket stops the water, so its pressure rises. The higher pressure above the bucket pushes it downward hard.

E.17 The water slows down just upstream and downstream of the posts and speeds up on the sides of the posts.

E.19 The dimpled ball will hit first.

E.21 The front runner drags the air forward so that you experience less drag while running through forward-moving air.

E.23 It balances the backward force due to air drag.

E.25 Both objects are slowed by similar pressure drags. But the spear's larger mass and momentum keep it from slowing as quickly.

E.27 It increases pressure drag and reduces your terminal velocity.

E.29 An airfoil, round in front and tapered behind, would be better.

E.31 Lift supports the skier and drag pulls the skier backward.

E.33 Without backspin, the ball can't obtain lift and won't go as far.

E.35 Small disturbances in the airflow around the sides of the nonspinning volleyball produce lift forces that push the ball to the side.

E.37 The water speeds up around the curve and its pressure drops. The resulting pressure imbalance pushes the spoon into the stream.

E.39 Air travels faster over your hand than under it, so the pressure above your hand is less than the pressure under your hand.

E.41 Turbulent air pockets form behind its moving blades, so swirling vortices and eddies flow out of the fan.

E.43 By pushing exhaust downward, they obtain upward thrusts.

E.45 The fan pushes air backward and air pushes the boat forward.

E.47 They are equivalent. **E.49** Yes.

E.51 The ball transfers momentum between you so that you acquire momentum in the opposite direction from your friend.

E.53 Directly toward the earth.

E.55 A rocket must do much less work against the moon's gravity.

P.1 For a fish with an obstacle length (width) of about 2 cm, turbulence will appear if it moves faster than about 0.1 m/s.

P.3 About 84 times as long.   **P.5** About 400,000 Pa.

P.7 About 18.3 m/s.   **P.9** About 0.76 m/s.

P.11 About 0.00028 times your earth weight.

P.13 About 4.7×10^{14} N.

Chapter 6

E.1 Your body temperature would decrease.

E.3 The work you do in pushing the bread in the direction it moves.

E.5 They are poor conductors of heat. The metal carries heat better.

E.7 The grate lets convection carry air upward past the wood.

E.9 The rising hot air of convection draws the flame upward.

E.11 Heat rises with convection and ignites the rest of the stick.

E.13 The glass permits thermal radiation to heat the pie directly.

E.15 The blackened pan absorbs thermal radiation from the oven while the shiny pan reflects that thermal radiation.

E.17 The naphthalene sublimes and the mothball becomes gaseous.

E.19 From the color of its light; whiter is hotter.

E.21 The cooler the object, the longer the wavelengths of its black body spectrum. The spectrum of a very cold object peaks in the microwave portion of the electromagnetic spectrum.

E.23 In a three-way bulb, any filament that's on is at full temperature. A dimmed bulb's filament operates below full temperature.

E.25 Hotter skin emits brighter infrared with shorter wavelengths.

E.27 The sidewalk has a different coefficient of volume expansion than the ground and would break when the temperature changed.

E.29 The metal lid has a larger coefficient of volume expansion than the glass jar, and it pulls away from the jar when you heat them both.

E.31 They are heat pumps and transfer heat to the surrounding air.

E.33 Without the fan, the air conditioner won't be able to transfer its waste heat to the surrounding air. It will stop pumping heat.

E.35 Its temperature increased. Gravity did work on the gas, and this work appeared in the gas as thermal energy. The gas became hot.

E.37 The gas still in the container does work pushing the other gas into the siphon and uses up some of its thermal energy. It cools.

E.39 Though not forbidden by the laws of motion, it's extraordinarily unlikely for the vase fragments to reassemble themselves.

E.41 This uneven distribution of thermal energies is extraordinarily unlikely. It would violate the second law of thermodynamics.

E.43 Though not forbidden by the laws of motion, such uneven distributions of snow are remarkably unlikely.

E.45 Heat flowing into or out of the pavement during weather changes allows the pavement to do work as it tears itself apart.

E.47 As heat flows from the hot spots to the cold spots, by way of huge convection cells, some heat is becoming ordered kinetic energy.

E.49 Without temperature differences, heat won't naturally flow. Such heat flow is what powers a heat engine such as the winds.

E.51 The hotter the burned gas, the larger the fraction of heat that can be converted to work as it flows to the outdoor air.

E.53 Converting burned fuel entirely into work would violate the second law of thermodynamics.

P.1 18,800 W.   **P.3** 73,500,000 W.

Chapter 7

E.1 It would run slow.

E.3 The period will decrease as the pendulum gets shorter.

E.5 Push on it as it flexes away from you (so you do work on it).

E.7 Above the ground, the ball experiences a *constant* restoring force, its weight, not one that's *proportional* to its displacement.

E.9 Dividing sound's roundtrip time by its speed gives distance.

E.11 To oscillate, the string's energy must shift back and forth between two forms. Without tension, it has no elastic potential energy.

E.13 Because the hammer rebounds from the string as the string moves away from it, the hammer bounces weakly and receives only a small fraction of the rebound energy.

E.15 Less mass, more tension, or less length.

E.17 Once the rubber band can't stretch any more, increasing its tension doesn't make it longer. It behaves like a violin string, with a pitch that rises as its tension increases.

E.19 The copper wrap adds mass to the string to lower its pitch.

E.21 The longer rods have more mass vibrating and are less stiff.

E.23 A piccolo's air columns are half the length of those in a flute, so they vibrate at twice the pitch or one octave higher.

E.25 A small change in the pipe's length alters the stiffness and mass of the air column and changes its vibrational frequency slightly.

E.27 Ocean water can't enter or leave the Mediterranean Sea quickly enough to allow its high tide to be very different from its low tide. The tide simply rearranges the water levels within the sea itself.

E.29 Wind pushes the waves in the direction of its velocity, so the waves acquire momentum in the direction of that velocity.

E.31 Resonant energy transfer occurs when your steps are synchronized with the rhythmic motion of the coffee sloshing in the cup.

E.33 The wave's energy is proportional to its circumference and its height. As its circumference grows, its height must diminish.

E.35 Crests and troughs don't travel along the string. Instead, the center of the string becomes alternately a crest and a trough.

E.37 With more tension, the restoring forces in the string are stiffer and the accelerations are more rapid. Everything happens faster.

E.39 There isn't enough water in front of the largest waves to complete their crests as they pass over the sand bar.

Chapter 8

E.1 Because they are identical, the socks acquire like charges as the result of sliding friction with clothes in the dryer. Like charges repel.

E.3 The charged paint particles electrically polarize the surface being coated, so that the paint particles are attracted to it and stick.

E.5 You'd have to compare the objects with a reference. The object that repelled the reference would have its charge.

E.7 The positive balloon has a higher voltage than the negative one.

E.9 Their voltages change. The positive balloon's voltage increases; the negative balloon's voltage decreases.

E.11 5000 levels.

E.13 All in lowest energy level.

E.15 Heating would diminish the contrast of the images.

E.17 The red light photons don't have enough energy to shift electrons from filled valence levels to empty conduction levels.

E.19 The electric field should point directly downward.

E.21 The forces between magnetic poles decrease rapidly with distance. The two closest poles experience the strongest forces.

E.23 The forces between a positive charge and the hairbrush decrease with their separation. Thus the electric field that acts on that positive charge also decreases with distance from the hairbrush.

E.25 The increased charge on the treetop exerts particularly strong forces on nearby charge. In other words, the charge on the treetop produces a strong local electric field.

E.27 The tray's pole is north.

E.29 The changing magnetic field induces currents in the aluminum, making the aluminum magnetic in such a way that it repels the coil.

E.31 Magnetic drag pulls the copper along with the moving magnet.

E.33 Repulsion upward and a magnetic drag force leftward.

P.1 3.3×10^{-9} C. **P.3** 8.3×10^{19} N.

P.5 94,800 m.

Chapter 9

E.1 When on, the switch should close the circuit by connecting the two wires electrically. When off, it should open the circuit. The broken switch is unable to disconnect the wires and open the circuit.

E.3 Current flows from the power source, through the fence wire, through the animal, through the earth, and back to the power source.

E.5 Current flows very briefly until you're positively charged, too.

E.7 Current arrives through one wire and leaves through the other.

E.9 Batteries, because something is supplying power to the current.

E.11 Current flows from the battery's positive terminal, through the hobbyist's tongue, to the battery's negative terminal. Energy is being transferred from the battery to the hobbyist's tongue by the current.

E.13 Each wire of the overloaded cord has a large voltage drop and consumes considerable electric power. It produces thermal energy.

E.15 The voltage drop in each wire is proportional to the current.

E.17 As you draw more current through the building's wires, the voltage drops in those wires increase. Less power reaches the lights.

E.19 It will rise by 10%.

E.21 The power deposited in a metal is proportional to its electric resistance, so high-resistance metals heat more.

E.23 9 A.

E.25 Aluminum isn't intrinsically magnetic and can't be magnetized.

E.27 The domains aligned with the new applied field grow while those that are anti-aligned with that field shrink.

E.29   An AC electromagnet would induce currents in the scrap iron and steel, leading to repulsive rather than attractive forces.

E.31   The distances between magnetization reversals are wider on the lower frequency tape than on the higher frequency tape.

E.33   The ring must be easily magnetized by small currents in its coil.

E.35   It's attracted to the poles of the magnetic recording particles.

P.1   1 A.   **P.3**   0.15 W.

P.5   Each battery supplies 3 W.

P.7   About 13,333 s, or 3.7 hours.

P.9   0.05 Ω.   **P.11**   0.6 V.

P.13   240,000,000 W.

Chapter 10

E.1   The microphone can't provide enough power.

E.3   The phonograph produces a much smaller voltage rise than a CD player would provide. The amplifier expects a larger voltage.

E.5   The resistor extracts power from the current passing through it, turning some of that current's energy into thermal energy.

E.7   When it carries no current, the resistor has no voltage drop.

E.9   The one with the thicker insulating layer.

E.11   As its plate area grows smaller, the capacitor has less room to store separated charge and thus stores less of it at a given voltage.

E.13   The gate's charge must be very near the channel so that it can attract opposite charge and draw that charge into the channel.

E.15   With positive charge on its gate, the MOSFET allows current to flow through the calculator. With no charge, the current flow stops.

E.17   When current experiences a voltage drop as it flows through a MOSFET, some of its energy is converted into thermal energy.

E.19   There are 6 hundreds. **E.21**   *219* and *85*.

E.23   Yes. There are 0 thousands in both decimal representations.

E.25   Instead of reporting 2 twos, the binary representation should report it as 1 four, the next higher power of two.

E.27   Form a circuit in which the MOSFET controls current flowing between the battery and the bulb. Touch the MOSFET's gate to the capacitor plate. If the bulb lights, the plate has positive charge on it.

E.29   It takes time for a MOSFET to change the charge on a wire in order to store or retrieve a bit.

E.31   Each charge change consumes a certain amount of energy. The more such changes occur each second, the more power is consumed.

E.33   Current must flow twice as long to charge the longer wire.

Chapter 11

E.1   As charges accelerate in the wires, they emit radio waves.

E.3   As charges accelerate in the computer, they emit radio waves.

E.5   The inductor's magnetic field contains energy, and both reach their largest values when the capacitor's energy is zero.

E.7   The inductor opposes a more rapid decrease in current.

E.9   Their sideband waves include a range of frequencies near the carrier wave frequency. These sideband waves mustn't overlap.

E.11   The AM station changes the power of its transmission in order to represent air pressure fluctuations with the radio wave.

E.13   Amplitude modulation introduces additional frequencies that extend as much as 5 kHz above and below the carrier wave.

E.15   The environment is full of 2.45-GHz microwaves leaking from people's microwave ovens. They would interfere with your radios.

E.17   The work done by the power supply becomes the light's energy.

E.19   Electrons in the picture tube complete one full circle of cyclotron motion, regardless of outward speed, and focus on the screen.

E.21   Electrons would steer in the magnetic field and miss the anode.

E.23   The magnetic field changes the electron beams' trajectories.

E.25   Water trapped in the ceramic would absorb microwaves, and the ceramic would become extremely hot. It might even shatter.

E.27   Releasing the microwaves into the room wouldn't be healthy.

E.29   Both involve field cancellation in electromagnetic waves.

P.1   0.333 m.   **P.3** 0.461×10^{-7} m (461 nm).

Chapter 12

E.1   About 160 nm.   **E.3**   About 4.7×10^{14} Hz.

E.5   The wavelength scale should shift to the left.

E.7   Rayleigh scattering deflects blue light in *all* directions.

E.9   The light refracts as it enters and leaves the glass.

E.11   The large particles scatter light better than air molecules.

E.13   Light bends and disperses on entry, reflects from the back surface, and bends and disperses more on exit from the diamond.

E.15 Each surface in the granulated sugar reflects some of the light passing through it. These random reflections make the sugar white.

E.17 Some light reflects from each surface of a lime particle.

E.19 Light reflects from the top *and* bottom surfaces of an oil film, and the two reflections interfere with one another. The type of interference depends on the film's thickness and the light's wavelength.

E.21 Squeezing the plates together moves the rings where constructive interference occurs for a particular color.

E.23 Horizontally polarized light reflects more strongly at shallow angles than at right angles.

E.25 Green and blue.

E.27 Mercury atoms themselves produce the tubes' ultraviolet light.

E.29 Fluorescence from brighteners replaces the missing blue light.

E.31 Excited mercury atoms emit primarily invisible ultraviolet light.

E.33 Each mercury atom in a narrow tube is near the phosphors.

E.35 The radio station's photons are identical—part of a single wave.

E.37 No, it produces all the photons at once with a single antenna.

E.39 They'll absorb the microwaves the maser is trying to produce.

E.41 Initial photons that were emitted at random are evidently passing through enough excited atoms to be duplicated many times.

P.1 3.37×10^{-19} J.   P.3 7.95×10^{-15} J.

Chapter 13

E.1 Both form real images of objects on light-sensitive surfaces.

E.3 The lens equation gives 0.5 cm as the focal length of the lens when the image and object distances are both 1.0 cm.

E.7 To have an f-number of 1.8, the elements in the 200-mm lens must be four times the diameter of the elements in the 50-mm lens.

E.9 Reducing the size of your eye's lens increases its depth of focus.

E.11 The depth of focus is smallest when the whole lens is used.

E.13 The distance increases.

E.15 The lower magnification 2× glass has the longer focal length.

E.17 The glass marble acts as a magnifying glass, producing enlarged virtual images of nearby objects.

E.19 The disc wouldn't be able to represent high-pitched sound well.

E.21 The lens's large opening reduces diffraction effects.

E.23 The low mass lens can be accelerated rapidly by modest forces.

E.25 Light is experiencing total internal reflection inside the glass.

E.27 Waves reflected from a sheet's front and back surfaces interfere.

E.29 Waves reflected by the pits and flats interfere destructively.

P.1 35 mm.

P.3 The real image of the more distant statue forms 2.2 mm closer.

P.5 5 m.

Chapter 14

E.1 Changing one element into another requires *nuclear* reactions.

E.3 Nuclei of ^{65}Cu atoms contain two more neutrons than those of ^{63}Cu.

E.5 No, the number of electrons is set by the number of protons.

E.7 Neutrons would experience the attractive nuclear force while diluting the repulsion between protons.

E.9 With fewer protons, the diluting effect of neutrons matters less.

E.11 Its nuclei move closer together and block neutrons better. The chance decreases that a neutron will escape from the sphere.

E.13 With time, radioactive nuclei decay spontaneously and the materials become less and less hazardous.

E.15 Chemical reactions have no effect on nuclei.

E.17 Photons in sunlight can cause chemical damage.

E.19 Photons in blue or ultraviolet light can cause chemical damage to the molecules in the manuscripts. Yellow light generally can't.

E.21 To emit an X-ray photon, an electron must shift into an orbital that's extremely tightly bound near a nucleus with a large positive charge. That can't occur in an atom with a small nucleus.

E.23 Almost any X-ray matches the energy of one of lead's many electrons and thus can cause efficient photoelectron emission.

E.25 Since the velocity of the electron changes during the collision, its momentum changes. Because momentum is conserved, the photon must have a momentum and that momentum must change, too.

E.27 An antiproton has more mass than a positron.

E.29 A positron isn't found in an iron atom.

E.31 Near the speed of light, changes in a particle's kinetic energy are accompanied by only modest changes in the particle's speed.

E.33 MRI detects hydrogen. Bone contains little hydrogen.

Photo Credits

Chapter 1

Page 1: Courtesy Lou Bloomfield. Page 4: Chris Trotman/Duomo Photography, Inc. Page 8 (top): Jeff Christensen/Gamma Liaison. Page 8 (bottom): Steven E. Sutton/Duomo Photography, Inc. Pages 9, 18: Courtesy Lou Bloomfield. Page 19: Richard Megna/Fundamental Photographs. Page 21: Courtesy Jerry Ohlinger's Movie Material Store. Page 29: Larry Mulvehill/Photo Researchers.

Chapter 2

Page 39: Courtesy Lou Bloomfield. Page 46: Dugald Bremmer/Stone. Page 57: © Bettman/Corbis Images. Page 65: Steven E. Sutton/Duomo Photography, Inc.

Chapter 3

Pages 77, 84, 87 (top), 88: Courtesy Lou Bloomfield. Page 87 (bottom): © C.E. Miller/Northpoint. Page 95: Stuart Dee/The Image Bank. Page 96 (top): Harry DiOrio/The Image Works. Page 96 (bottom): Bob Torrez/Stone. Page 98: Chad Slattery/Stone. Page 103: Courtesy Dr. David Jones. Page 104: Wesley Hitt/Gamma Liaison.

Chapter 4

Page 117: Courtesy Lou Bloomfield. Page 127: Coco McCoy/Rainbow. Page 128: Corbis Images. Page 136 (top): Courtesy Department of Water and Power of the City of Los Angeles. Page 136 (bottom): J. Szkodzinski/The Image Bank.

Chapter 5

Page 145: Courtesy Lou Bloomfield. Page 153 (left): Charles Krebs/The Stock Market. Page 153 (right): Kaz Mori/The Image Bank. Page 154: Courtesy Lou Bloomfield. Page 155: Courtesy Peter Bradshaw, Stanford University. Page 159: © Bettman/Corbis Images. Page 161: Courtesy Rebus, Inc. Page 163: Michael Dwyer/Stock, Boston. Page 166, 168: Courtesy Thomas Miller. Page 171: Courtesy Smithsonian Institution. Page 175 (top): Courtesy NASA. Page 175 (bottom): Courtesy New York Public Library.

Chapter 6

Page 187: Courtesy Lou Bloomfield. Page 190: Courtesy Travis Industries, Inc. Page 194: Courtesy Bryant Heating and Cooling. Pages 199, 203, 204: Courtesy Lou Bloomfield. Page 211: Courtesy Bryant Heating and Cooling. Page 212: Courtesy Lou Bloomfield. Page 219: Courtesy of BMW Corporation.

Chapter 7

Page 229: Courtesy Lou Bloomfield. Page 234: Robert Mathena/Fundamental Photographs. Page 236: Gregory G. Dimijian, M.D./Photo Researchers. Page 238 (left): Tibor Bognar/The Stock Market. Page 238 (right): Courtesy Lou Bloomfield. Page 243 (top): Steve Bronstein/The Image Bank. Page 243 (bottom): AP/Wide World Photos. Page 250: Everett Johnson/Leo de Wys, Inc. Page 253: Fritz Herle/Photo Researchers. Page 254: Don & Liysa King/The Image Bank.

Chapter 8

Pages 263, 269, 280, 284, 285, 288: Courtesy Lou Bloomfield. Page 286: Courtesy Dornier-System GmbH, Friedrichshafen. Page 287: © AFP/Corbis Images.

Chapter 9

Pages 295, 299, 301, 311, 316: Courtesy Lou Bloomfield. Page 317: Courtesy San Diego Magnetics.

Chapter 10

Pages 327, 331, 332, 335, 343, 344, 346: Courtesy Lou Bloomfield. Page 342: Courtesy Intel Corporation.

Chapter 11

Pages 353, 363, 364, 366, 374: Courtesy Lou Bloomfield.

Chapter 12

Page 383: Courtesy Lou Bloomfield. Page 387: Galen Rowell/Mountain Light Photography, Inc. Page 390: Steve Satushek/The Image Bank. Pages 391, 392, 400, 406: Courtesy Lou Bloomfield. Page 404: Courtesy SDL, Inc.

Chapter 13

Pages 413, 416, 418, 421 (top), 422: Courtesy Lou Bloomfield. Page 419, 421 (bottom): Courtesy Canon USA, Inc.

Chapter 14

Pages 439, 444: Courtesy Lou Bloomfield. Page 448 (top): Courtesy U.S. Dept. of Energy. Page 448 (bottom): Courtesy Defense Nuclear Agency. Page 449: Courtesy Los Alamos Scientific Laboratory, University of California. Page 457 (left): Deep Light Productions/Photo Researchers. Page 457 (right): Alfred Pasieka/Photo Researchers. Page 460: Doug Martin/Photo Researchers. Page 461 (left): Med. Illus. SBHA/Stone. Page 461 (right): Charles Thatcher/Stone.

Index